Precision Frequency Control

Volume 1
Acoustic Resonators and Filters

Precision Frequency Control

Volume 1
Acoustic Resonators and Filters

Edited by

EDUARD A. GERBER
ARTHUR BALLATO

U.S. Army Electronics Technology and Devices Laboratory
Fort Monmouth, New Jersey

1985

ACADEMIC PRESS, INC.
(Harcourt Brace Jovanovich, Publishers)
Orlando   San Diego   New York   London
Toronto   Montreal   Sydney   Tokyo

COPYRIGHT © 1985, BY ACADEMIC PRESS, INC.
ALL RIGHTS RESERVED.
NO PART OF THIS PUBLICATION MAY BE REPRODUCED OR
TRANSMITTED IN ANY FORM OR BY ANY MEANS, ELECTRONIC
OR MECHANICAL, INCLUDING PHOTOCOPY, RECORDING, OR
ANY INFORMATION STORAGE AND RETRIEVAL SYSTEM, WITHOUT
PERMISSION IN WRITING FROM THE PUBLISHER.

ACADEMIC PRESS, INC.
Orlando, Florida 32887

United Kingdom Edition published by
ACADEMIC PRESS INC. (LONDON) LTD.
24–28 Oval Road, London NW1 7DX

Library of Congress Cataloging in Publication Data

Main entry under title:

Precision frequency control.

    Includes bibliographies and indexes.
    Contents: v. 1. Acoustic resonators and filters --
v. 2. Oscillators and standards.
    1. Piezoelectric devices. 2. Electric resonators.
3. Oscillators, Crystal. 4. Acoustic surface wave
devices. I. Gerber, Eduard A. II. Ballato, Arthur.
TK7872.P54P74 1984   621.3815'363   84-14630
ISBN 0-12-280601-8 (v. 1 : alk. paper)

PRINTED IN THE UNITED STATES OF AMERICA

85 86 87 88   9 8 7 6 5 4 3 2 1

Contents

CONTRIBUTORS ix
PREFACE xi
CONTENTS OF VOLUME 2 xv
INTRODUCTION xxi

1 Properties of Piezoelectric Materials

	List of Symbols	2
1.1	An Overview	2
	by Larry E. Halliburton and Joel J. Martin	
	1.1.1 Structural Requirements	3
	1.1.2 Electroelastic Relations	4
	1.1.3 Materials of Current Interest	10
1.2	Physical Properties of Quartz	23
	by Larry E. Halliburton, Joel J. Martin, and Dale R. Koehler	
	1.2.1 Crystallography	23
	1.2.2 Crystal Growth and Extended Defects	25
	1.2.3 Point Defects	32
	1.2.4 Thermal Properties	39
	1.2.5 Material Evaluation Techniques	40

2 Theory and Properties of Piezoelectric Resonators and Waves

	List of Symbols for Sections 2.1 and 2.2	48
2.1	Bulk Acoustic Waves and Resonators	50
	by Thrygve R. Meeker	
	2.1.1 Introduction	50
	2.1.2 Basic Quasi-Static Theory of a Piezoelectric Elastic Material	52

		2.1.3	Linear Theory	53
		2.1.4	Nonlinear Theory	56
		2.1.5	The Christoffel Plane-Wave Solutions for the Linear Quasi-Static Piezoelectric Crystal	58
		2.1.6	Thickness Modes	63
		2.1.7	Contour Modes in Thin Plates and in Thin and Narrow Bars	77
		2.1.8	Theory for Combined Thickness and Contour Modes	90
		2.1.9	Electrical Effects in Piezoelectric Resonators	102
		2.1.10	Equivalent Electrical Circuits for Piezoelectric Resonators	103
		2.1.11	Properties of Modes in Crystal Resonators	107
		2.1.12	Piezoelectric Materials	107
		2.1.13	Conclusion	110
	2.2	Properties of Quartz Piezoelectric Resonators		110
		by Thrygve R. Meeker		
		2.2.1	Temperature Coefficient of Resonance Frequency	110
		2.2.2	Dependence of Crystal Inductance on Temperature	112
		2.2.3	Tabulation of Properties of Quartz Resonators	113
		2.2.4	Conclusion	113
		List of Symbols for Section 2.3		118
	2.3	Surface Acoustic Waves and Resonators		119
		by William R. Shreve and Peter S. Cross		
		2.3.1	Introduction	119
		2.3.2	Resonator Design	126
		2.3.3	Fabrication	137
		2.3.4	State-of-the-Art Performance	140
		2.3.5	Conclusion	144

3 Radiation Effects on Resonators
by James C. King and Dale R. Koehler

3.1	Introduction			147
3.2	Radiation Effects and Modeling			148
	3.2.1	Substitutional Al^{3+} Defect Center		148
	3.2.2	Frequency Changes		149
	3.2.3	Optical Effects		153
	3.2.4	Elastic Modulus Changes		153
3.3	Dynamics of Radiation Effects			154
	3.3.1	Hydrogen and Transient Effects		154
	3.3.2	ESR and IR Studies		155
	3.3.3	Trap Characterization		156
	3.3.4	Material Quality and Anelastic Losses		157
	3.3.5	Thermal Effects		158

4 Resonator and Device Technology
by John A. Kusters

4.1	Resonator Material Selection			161
4.2	Sawing			163
	4.2.1	Natural Quartz		165
	4.2.2	Cultured Quartz		166

CONTENTS vii

4.3	X-Ray Orientation	166
4.4	Mechanical Operations	168
4.5	Cleaning	170
4.6	Vacuum Deposition	170
4.7	Mounting and Sealing	174
4.8	Special Fabrication Considerations for SAW Devices	178
4.9	Novel Resonator Techniques	180
4.10	Environmental Effects	182

5 Piezoelectric and Electromechanical Filters

	List of Symbols for Sections 5.1 and 5.2		186
5.1	General		187
	by Robert C. Smythe		
5.2	Bulk-Acoustic-Wave Filters		188
	by Robert C. Smythe		
	5.2.1	Introduction	188
	5.2.2	Crystal Filters	189
	5.2.3	Electromechanical Filters	221
	List of Symbols for Sections 5.3 and 5.4		228
5.3	Surface-Acoustic-Wave Filters		230
	by Robert S. Wagers		
	5.3.1	Introduction	230
	5.3.2	Interdigital Transducer Admittance	233
	5.3.3	Relation of Normal-Mode Theory Admittance to the Impulse Model	239
	5.3.4	Limitations on the Use of Electrostatic Fields	240
	5.3.5	Electromechanical Coupling Constant k	241
	5.3.6	Electrical Q and Insertion Loss	242
	5.3.7	Bulk-Wave Modeling of Interdigital Transducers	244
	5.3.8	Advanced Bulk-Wave Models	249
5.4	SAW Bandpass and Bandstop Filters		257
	by Robert S. Wagers		
	5.4.1	Introduction	257
	5.4.2	Impulse-Response Realizations	260
	5.4.3	SAW Bandpass Filter Capabilities	263
	5.4.4	SAW Bandstop Filters	266

6 Long-Term Stability and Aging of Resonators
by Eduard A. Gerber

6.1	Low-Frequency Bulk-Wave Devices		271
6.2	High-Frequency Bulk-Wave Devices		273
	6.2.1	Causes of Aging	273
	6.2.2	Progress through Holder Design	274
	6.2.3	Progress through Mounting and Crystal Plate Design	275
	6.2.4	Isolation of Aging Causes	277
	6.2.5	Influence of Temperature	279
	6.2.6	Influence of Radiation	279

6.3	Surface-Wave Devices		279
	6.3.1	SAW Resonators	280
	6.3.2	SAW Delay Lines	283

Bibliography
by Eduard A. Gerber and Arthur Ballato

Introduction	285
General Bibliography	286
Chapter Bibliographies	293

INDEX TO VOLUMES 1 AND 2   417

Contributors

Numbers in parentheses indicate the pages on which the authors' contributions begin.

ARTHUR BALLATO (285), U.S. Army Electronics Technology and Devices Laboratory, Fort Monmouth, New Jersey 07703

PETER S. CROSS[1] (119), Hewlett-Packard Laboratories, Palo Alto, California 94304

EDUARD A. GERBER[2] (271, 285), U.S. Army Electronics Technology and Devices Laboratory, Fort Monmouth, New Jersey 07703 (Retired)

LARRY E. HALLIBURTON (1), Physics Department, Oklahoma State University, Stillwater, Oklahoma 74078

JAMES C. KING (147), Sandia National Laboratories, Albuquerque, New Mexico 87185

DALE R. KOEHLER (23, 147), Sandia National Laboratories, Albuquerque, New Mexico 87185

JOHN A. KUSTERS (161), Hewlett-Packard Laboratories, Santa Clara, California 95050

JOEL J. MARTIN (1), Physics Department, Oklahoma State University, Stillwater, Oklahoma 74078

THRYGVE R. MEEKER (47), AT&T Bell Laboratories, Allentown, Pennsylvania 18103

WILLIAM R. SHREVE (119), Hewlett-Packard Laboratories, Palo Alto, California 94304

ROBERT C. SMYTHE (185), Piezo Technology Incorporated, Orlando, Florida 32854

ROBERT S. WAGERS (230), Central Research Laboratories, Texas Instruments Incorporated, Dallas, Texas 75265

---

[1]Present address: Spectra Diode Laboratories, San Jose, California 95134.
[2]Present address: 11 Community Drive, West Long Branch, New Jersey 07764.

Preface

The editors take pleasure in presenting this two-volume work on precision frequency control. The title encompasses the spectrum of frequency-determining and frequency-selective devices, subject to the constraint imposed by the adjective. A simple circuit consisting of an inductance and a capacitance can function as a frequency-controlling element. Its precision, however, is completely insufficient for modern electronic equipment. Different physical phenomena must be utilized to meet today's requirements. The discussion and explication of these phenomena and their applications are the main purposes of these books.

The aims are twofold: first, to offer a concise compendium of the state of the art to researchers and specialists engaged in a rapidly expanding and complex field of technology. It will enable them to work efficiently in their fields and to develop devices that meet the requirements of the equipment and systems engineer.

A second purpose of the books is to furnish information concerning properties and capabilities of frequency-control devices to users of these devices, such as equipment and systems designers. The volumes will also be very useful for technical managers who will be able to find, in a single publication, a description of the world of precision frequency control, written by experts, and an entree to the full literature of the field.

The idea of these books originated several years ago when the editors recognized that the literature in the field of frequency control was increasing at an explosive rate and that it would be extremely difficult, particularly for a novice in this field, to attain without guidance an essential level of knowledge in a reasonable time. Another incentive for compiling this text is the fact that there is no single book available on the world market that treats all precision frequency-control devices and allows the reader to weigh the advantages or disadvantages of the various technical approaches against one another.

The number of experimental observations and theoretical investigations in the field of precision frequency control has increased steadily over the past 60 years and has led, particularly during the past few years, to a deluge of original publications that is becoming more and more difficult to absorb in its totality, even for the trained specialist. In view of this, our aim is not to attempt to offer a textbook on the subject, but rather to provide a tutorial and coherent treatment of the more recent developments in the field, supported by an extensive literature reference list covering approximately the past fifteen years. The individual chapters are written by experts in their respective specialities. The editors feel that the fundamentals of this field, starting with the seminal works of the Curies, Voigt, Cady, Townes, Ramsey, and others, are very well represented in older textbooks and in many voluminous review papers and handbook articles whose titles the reader will find in the bibliography.

The material of the work is presented in two volumes, "Acoustic Resonators and Filters" (Volume 1) and "Oscillators and Standards" (Volume 2). The reader will find in the introduction to the bibliography, included in both volumes, some suggestions on how to use the chapter bibliographies to best advantage. The 16 chapters of the text can be read independently of one another. Their topics have been chosen to maximize the readability of the book, with lengths governed jointly by the number of publications pertinent to each chapter and by the importance the editors attach to each topic, although obviously it is impossible to discuss in the text all of the more than 5000 publications referenced. The selection of specific areas discussed is to a certain extent subjective, but we feel that they give a good indication of the overall progress in our field.

The reader will find glossaries of letter symbols—whenever necessary—at the beginning of each chapter and, in certain instances, introducing a section. These characters, as well as graphic symbols used in the book, correspond as much as possible to those specified in the following IEEE Standards:

IEEE 260	1978	Letter Symbols for Units of Measurement
IEEE 280	1968	Letter Symbols for Quantities Used in Electrical Science and Electrical Engineering
IEEE 315	1975	Graphic Symbols for Electrical and Electronics Diagrams
IEEE 176	1978	Piezoelectricity
IEEE 177	1966	Definitions and Methods of Measurement for Piezoelectric Vibrators

Copies of these standards may be obtained from The Institute of Electrical and Electronics Engineers, 345 East 47th Street, New York, New York 10017.

The editors wish to express their sincere thanks to the authors of the various chapters for their cooperation and enjoyable collaboration, the editorial and production staffs of Academic Press for their patience and support, Mrs. Carolyn

Clever for her typing, and the personnel of the U.S. Army Electronics Technology and Devices Laboratory, Fort Monmouth, for much encouragement and assistance. They wish to thank, in particular, Mrs. Gloria Gatling for doing a careful and patient job of typing the final version of the bibliography, Miss Betsy Hatch for her computer work, and Mr. Ted Lukaszek for his steady and manifold help prior to and during the preparation of the text. Finally, the editors gratefully acknowledge stimulating discussions over many years with R. D. Mindlin (Columbia University) and R. A. Sykes (AT&T Bell Laboratories), who, by the synthesis of their theoretical and experimental work and by their education of students and colleagues, helped substantially to make a science out of an art in the field of crystal frequency control.

E. A. Gerber
West Long Branch, New Jersey

A. Ballato
Long Branch, New Jersey

Contents of Volume 2

7 Resonator and Device Measurements
by Erich Hafner

      List of Symbols
7.1   Introduction
7.2   The Crystal Unit and Its Equivalent Circuit
      7.2.1   The Resonator-Equivalent Electrical Circuit
      7.2.2   Device Properties of the Crystal Unit
      7.2.3   The Characteristic Parameters of a Crystal Unit
7.3   Crystal-Resonator Measurements
      7.3.1   General
      7.3.2   Resonator Measurement Instruments
      7.3.3   Measurement Methods
7.4   Summary and Conclusions
      Appendix I: The Generalized Equivalent Circuit
      Appendix II: Two-Port Relations for the IEC-444 π Network
      Appendix III: Two-Port Relations for a Transmission Bridge

8 Precision Oscillators

      List of Symbols for Section 8.1
8.1   Bulk-Acoustic-Wave Oscillators
      by Warren L. Smith
      8.1.1   General Characteristics of Crystal-Controlled Oscillators
      8.1.2   Circuit Configurations for Crystal-Controlled Oscillators
      8.1.3   Temperature-Control Techniques for Precision Oscillators
      8.1.4   Temperature-Compensation Methods for Semiprecision Oscillators
      8.1.5   Miniature Integrated-Circuit Oscillators
      List of Symbols for Section 8.2
8.2   Surface-Acoustic-Wave Oscillators
      by Thomas E. Parker
      8.2.1   Introduction
      8.2.2   The Basic SAW Oscillator—A Physical Point of View
      8.2.3   Frequency Stability
      8.2.4   Multifrequency Oscillators
      8.2.5   Electronic Amplifiers and Other External Components

        8.2.6   Advantages and Disadvantages of SAW Oscillators
        8.2.7   Conclusion
        List of Symbols for Section 8.3
  8.3 Quartz Frequency Standards and Clocks—Frequency Standards in General
      by Warren L. Smith
        8.3.1   Design Considerations
        8.3.2   Measurement and Specification of Frequency Stability

9 Temperature Control and Compensation
by Marvin E. Frerking

        List of Symbols
  9.1 Temperature Control
        9.1.1   Thermal Loss
        9.1.2   Warm-Up Considerations
  9.2 Temperature Compensation
        9.2.1   Analog Temperature Compensation
        9.2.2   Digital Temperature Compensation
        9.2.3   Microprocessor Temperature Compensation

10 Microwave Frequency and Time Standards
by Helmut Hellwig

        List of Symbols
  10.1 Concepts, Design, and Performance
        10.1.1   Historical Perspective
        10.1.2   Concept of an Atomic Resonator
        10.1.3   Design Principles
        10.1.4   Performance Principles
        10.1.5   Active and Passive Electronic Systems
        10.1.6   Phase-Lock Servos
        10.1.7   Frequency-Lock Servos
        10.1.8   Electronic Systems
  10.2 Passive Beam Standards
        10.2.1   Beam Generation
        10.2.2   Spatial State Selection
        10.2.3   Microwave Interrogation
        10.2.4   Detection of Atoms
        10.2.5   The Cesium-Beam Standard
        10.2.6   Other Passive Beam Standards
        10.2.7   New Horizons
  10.3 Gas-Cell Standards
        10.3.1   Gas-Cell Principles
        10.3.2   Optical State Selection
        10.3.3   The Rubidium Gas-Cell Standard
        10.3.4   Other Gas-Cell Standards
        10.3.5   New Horizons

- 10.4 Hydrogen Masers
  - 10.4.1 Hydrogen-Maser Principles
  - 10.4.2 Active and Passive Masers
  - 10.4.3 Frequency Stability and Accuracy
  - 10.4.4 Other Masers
  - 10.4.5 New Horizons
- 10.5 Other Microwave Frequency Standards
  - 10.5.1 The Ammonia Maser
  - 10.5.2 Trapped Ions
- 10.6 Comparison of Frequency Standards
- 10.7 Applications
  - 10.7.1 Metrology and Science
  - 10.7.2 Technology

11 Laser Frequency Standards
by Rudolf Buser and Walter Koechner

- 11.1 Introduction
- 11.2 The Potential Role of Lasers
- 11.3 Basic Laser Configuration
- 11.4 Stabilization of Lasers
  - 11.4.1 He–Ne Laser Stabilized with a Ne Cell
  - 11.4.2 He–Ne Laser with an I_2 Cell
  - 11.4.3 He–Ne Laser and Methane Absorption Cell
  - 11.4.4 CO_2 Laser with a CO_2 Absorption Cell
- 11.5 Measurement of Optical Oscillation Frequencies
- 11.6 Future Prospects and Problems

12 Frequency and Time—Their Measurement and Characterization
by Samuel R. Stein

- List of Symbols
- 12.1 Concepts, Definitions, and Measures of Stability
  - 12.1.1 Relationship between the Power Spectrum and the Phase Spectrum
  - 12.1.2 The IEEE Recommended Measures of Frequency Stability
  - 12.1.3 The Concepts of the Frequency Domain and the Time Domain
  - 12.1.4 Translation between the Spectral Density of Frequency and the Allan Variance
  - 12.1.5 The Modified Allan Variance
  - 12.1.6 Determination of the Mean Frequency and Frequency Drift of an Oscillator
  - 12.1.7 Confidence of the Estimate and Overlapping Samples
  - 12.1.8 Efficient Use of the Data and Determination of the Degrees of Freedom
  - 12.1.9 Separating the Variances of the Oscillator and the Reference
- 12.2 Direct Digital Measurement
  - 12.2.1 Time-Interval Measurements
  - 12.2.2 Frequency Measurements
  - 12.2.3 Period Measurements

12.3 Sensitivity-Enhancement Methods
    12.3.1 Heterodyne Techniques
    12.3.2 Homodyne Techniques
    12.3.3 Multiple Conversion Methods
12.4 Conclusion

13 Frequency and Time Coordination, Comparison, and Dissemination
by David W. Allan

    List of Acronyms
13.1 Introduction
    13.1.1 Historical Perspectives and Methods of Comparison
    13.1.2 Time and Frequency Standards
13.2 Terrestrial Time and Frequency Comparison or Dissemination Methods
    13.2.1 High and Medium Frequency
    13.2.2 Low- and Very-Low-Frequency Transmissions
    13.2.3 Other Methods
13.3 Extraterrestrial Time and Frequency Comparison or Dissemination Methods
    13.3.1 Operational-Satellite Techniques
    13.3.2 Experimental-Satellite Techniques
    13.3.3 Deep-Space Radio-Source Techniques
13.4 Coordinate Time for the Earth
13.5 Levels of Sophistication and Accuracies for the Users
    13.5.1 Typical User Applications
    13.5.2 Sophisticated and High-Accuracy Techniques
13.6 Summary

14 Other Means for Precision Frequency Control
by Fred L. Walls

14.1 Introduction
14.2 Low-Frequency Devices
    14.2.1 Quartz Tuning Forks
    14.2.2 Other Low-Frequency Devices
14.3 Microwave Devices
    14.3.1 Superconducting Cavities
    14.3.2 Dielectrically Loaded Cavities

15 Special Applications
by Fred L. Walls and Jean-Jacques Gagnepain

15.1 Microbalances, Thin-Film Measurement, and Other Mass-Loading Phenomena
15.2 Measurements of Force, Pressure, and Acceleration
    15.2.1 Force
    15.2.2 Pressure
    15.2.3 Acceleration
15.3 Temperature Measurements

16 Specifications and Standards
by Erich Hafner

    16.1   Specifications
    16.2   Standards
    16.3   Practices in the United States
    16.4   International Standardization

Bibliography
by Eduard A. Gerber and Arthur Ballato

Introduction
General Bibliography
Chapter Bibliographies

Introduction

The history of precision frequency control provides a good example of how technological maturity follows upon the prior accomplishment of scientific groundwork. The foundations of modern frequency control began with discovery of the piezoelectric effect by the brothers Curie in 1880, which found theoretical treatment in Voigt's classic book (1910). Founded on these accomplishments, the development of devices using the piezoelectric effect started during World War I and has proceeded since at an accelerating rate. Quartz crystals used for frequency control developed from rather simple, unevacuated, pressure-mounted units of the 1920s and 1930s to the present highly sophisticated plated units operating in ultrahigh vacua with temperature-compensating or temperature-controlling arrangements. Influences of the environment, such as mounting structure, pressure, and acceleration, have been greatly reduced by using doubly rotated crystal plates. Similarly, great progress has been made in the development of frequency-control devices based on atomic or molecular processes since Essen built and described the first cesium-beam frequency standard in 1957. They have progressed from the original 8-ft giant to the currently commercially available equipment of modest size and weight.

It is no accident that the flowering of our field has coincided with the advent of the space age. No stretch of the imagination is required to see the demands placed on oscillator stability by rocket and satellite environments; and in few applications is the need for precision so severe. Concurrently, similar requirements were imposed in the fields of communication and guidance systems, both commercial and military. For instance, systems for frequency- and time-division multiplex communication, satellite-assisted positioning, as well as remote surveillance and collision avoidance, would be impossible without precision frequency-control and timing devices.

In the dozen years from the launching of the first artificial satellite about the earth to the first manned lunar landing, the frequency-control field and its correlate areas of selection, signal processing, timing, and time distribution experienced an enormous period of development and growth. The advances made during this time turned out, in retrospect, to be only a prelude to the developments of the next decade. The interval following the first Apollo landing initiated what might justly be called the golden age of frequency control. The editors made no predictions as to the extent and duration of this exciting period— certainly it is continuing; but one may well question if we shall soon see a decade in which the development of both accuracy and precision will experience such favorable conditions as have been met within the area of frequency control.

The attribute "precision" in the title restricts the contents of this work to those devices whose Q value and frequency stability far exceed that of an ordinary LC circuit. Consequently, the reader will not find ceramic resonators and filters discussed. Material on polycrystalline and similar devices is included only if it bears on the behavior of high-Q devices (e.g., the theory of vibration of anisotropic bodies). On the other hand, superconductive LC devices with their high Q are properly included. One other remark regarding selection of material is pertinent: The main application of bulk-wave monocrystalline devices is to frequency control, whereas surface-acoustic-wave devices are being used in many other fields. We therefore discuss the fundamental properties of bulk-wave devices and their materials to a fuller extent. As far as surface-wave monocrystalline devices are concerned, only those aspects of material and resonator properties considered pertinent to precision frequency control are covered.

1 Properties of Piezoelectric Materials

Larry E. Halliburton
Joel J. Martin

Oklahoma State University
Stillwater, Oklahoma

Dale R. Koehler

Sandia National Laboratories
Albuquerque, New Mexico

	List of Symbols			2
1.1	An Overview			2
	by Larry E. Halliburton and Joel J. Martin			
	1.1.1	Structural Requirements		3
	1.1.2	Electroelastic Relations		4
		1.1.2.1	Dielectric Constants	5
		1.1.2.2	Elastic Constants	6
		1.1.2.3	Piezoelectric Constants	7
	1.1.3	Materials of Current Interest		10
		1.1.3.1	Quartz	11
		1.1.3.2	Lithium Niobate and Lithium Tantalate	11
		1.1.3.3	Bismuth Germanium Oxide	19
		1.1.3.4	Aluminum Phosphate	20
1.2	Physical Properties of Quartz			23
	by Larry E. Halliburton, Joel J. Martin, and Dale R. Koehler			
	1.2.1	Crystallography		23
		1.2.1.1	Structure	23
		1.2.1.2	Coordinate Systems	24
		1.2.1.3	Twinning and Structural Phase Transitions	25
	1.2.2	Crystal Growth and Extended Defects		25
		1.2.2.1	Improvements in Quartz Growth	27
		1.2.2.2	Nature of Extended Defects	29
	1.2.3	Point Defects		32
		1.2.3.1	Aluminum-Related Centers	33
		1.2.3.2	Oxygen Vacancy Centers	35
		1.2.3.3	Electrodiffusion	35
		1.2.3.4	Acoustic and Dielectric Loss	36
		1.2.3.5	Fundamental Radiation Response Mechanisms	38

1.2.4	Thermal Properties		39
1.2.5	Material Evaluation Techniques		40
	1.2.5.1	Determination of Q Value	41
	1.2.5.2	Determination of Sweeping Effectiveness	42
	1.2.5.3	Prediction of Radiation Hardness	43

LIST OF SYMBOLS

c_p	heat capacity	T_{ij}	stress tensor
c_{ijkl}	elastic stiffness constant	u_i	strain-induced distortion
d_{ijk}	piezoelectric strain coefficient	α	thermal diffusivity, or infrared extinction coefficient
$\mathbf{D}$	electric displacement		
e_{ijk}	piezoelectric stress coefficient	λ	thermal conductivity
$\mathbf{E}$	electric field	δ	phase angle
K_{ij}	dielectric constant tensor	δ_{ij}	Kronecker delta
$\mathbf{P}$	electric polarization	ε_0	permittivity of free space
Q	quality factor	ε_{ij}	permittivity tensor
s_{ijkl}	elastic compliance constant	τ	relaxation time
S_{ij}	strain tensor	χ_{ij}	electric susceptibility tensor
t	sample thickness	ω	angular frequency

1.1 AN OVERVIEW[§]

Certain crystals exhibit the phenomenon known as piezoelectricity, wherein electrical polarization is produced by mechanical stress. This induced polarization is proportional to the stress and changes sign with it. Piezoelectricity is a reversible phenomenon; the direct piezoelectric effect occurs when an applied stress produces an electric polarization, and the converse piezoelectric effect occurs when an applied electric field produces a strain.

The brothers Pierre and Jacques Curie, in 1880, were the first to experimentally demonstrate piezoelectric behavior in a series of crystals, including quartz and Rochelle salt. Considerable experimental and theoretical activity immediately followed this announcement, and the publication of Voigt's "Lehrbuch der Kristallphysik" in 1910 marked the establishment of piezoelectricity as a phenomenologically well-understood segment of fundamental physics. More recently, Cady (1964) prepared a monumental and easily-read treatise on the theory and applications of piezoelectricity.

§ Section 1.1 was written by Larry E. Halliburton and Joel J. Martin.

1 PROPERTIES OF PIEZOELECTRIC MATERIALS

In this chapter, the fundamental properties of piezoelectric materials are described, with particular attention being given to those materials having application in precision frequency control. We begin by establishing the structural requirements for the occurrence of piezoelectricity and then review the relationships between the various electric and elastic quantities used in characterizing piezoelectric materials. This is followed by summaries of the known properties of specific materials, including an in-depth survey of α-quartz.

1.1.1 Structural Requirements

For a crystal to be piezoelectric, it must be noncentrosymmetric (i.e., have no center of symmetry). This is easily seen from the definition of piezoelectricity given above. Suppose that a centrosymmetric crystal becomes polarized upon the application of a stress. Then let the whole system, including the crystal and the stress, be inverted through the center of symmetry. Both the stress and the crystal remain unchanged, but the induced electric polarization will be reversed. For this to happen, the only possible electric polarization is zero, and the requirement that piezoelectric crystals be noncentrosymmetric is proven. Before proceeding, however, it must be emphasized that although knowledge of a crystal's structure allows one to predict whether piezoelectric behavior is possible, the magnitudes of any piezoelectric constants depend on the detailed electronic bonding of the crystal.

A systematic approach to determining which crystals are noncentrosymmetric, and thus piezoelectric candidates, is provided by the science of crystallography. The basic description of the structure of crystals is discussed in a large number of texts; one of the most useful being the book by Megaw (1973). The reader needing more structural information is referred to the "International Tables for X-ray Crystallography" (Henry and Lonsdale, 1952) and "Crystal Structures" (Wyckoff, 1966). Crystals are usually grouped into seven systems: triclinic, monoclinic, orthorhombic, tetragonal, trigonal, hexagonal, and cubic. These range from the lowest (triclinic) to the highest (cubic) symmetry, and piezoelectric behavior can be found in each system.

The seven crystal systems can be divided into 32 classes (or point groups) depending upon point symmetry. Of these, 11 are centrosymmetric classes and 20 are piezoelectric classes. An exceptional case is class 432 from the cubic crystal system; it is noncentrosymmetric but is not piezoelectric. In addition, of the 21 noncentrosymmetric classes, there are 11 classes having no plane of symmetry. This allows both right and left forms to exist for these latter cases. Such enantiomorphous forms are the mirror image of each other, and neither type can be made to look exactly like the other by a simple rotation.

TABLE 1-1

Classification of Crystal Structures with Regard to Centrosymmetric, Piezoelectric, and Enantiomorphic Behavior[a]

Crystal system	Centrosymmetric classes	Piezoelectric classes	Classes with enantiomorphism
Triclinic	$\bar{1}$	1	1
Monoclinic	$2/m$	$2, m$	2
Orthorhombic	mmm	$222, mm2$	222
Tetragonal	$4/m, 4/mmm$	$4, \bar{4}, 422,$ $4mm, \bar{4}2m$	4, 422
Trigonal	$\bar{3}, \bar{3}m$	$3, 32, 3m$	3, 32
Hexagonal	$6/m, 6/mmm$	$6, \bar{6}, 622,$ $6mm, \bar{6}m2$	6, 622
Cubic	$m3, m3m$	$23, \bar{4}3m$	23, 432

[a] The international symbols (short form) are used to denote the various point-group classes. (Adopted from the IEEE Standard of Piezoelectricity, 1978.)

Table 1-1 uses the short form of the international symbols to summarize the assignments of the various classes. For example, from this table one sees that $LiNbO_3$, belonging to class $3m$, is piezoelectric but not enantiomorphic; whereas α-quartz, belonging to class 32, is both piezoelectric and enantiomorphic.

1.1.2 Electroelastic Relations

Applications of piezoelectric materials involve the relationships between electric field, electric displacement, electric polarization, stress, and strain. The first three of these quantities are vectors while the last two are second-rank tensors. Zwikker (1954) and especially Cady (1964) and Nye (1957) provide good introductions to the tensor formulation of the physical properties of crystals. We give in the following a very brief introduction to the mathematical representation of these electric and elastic fields and define the various coupling coefficients. A more elaborate and useful mathematical treatment of piezoelectricity is found in the book by Tiersten (1969).

Before proceeding to specific definitions, it is important to briefly discuss coordinate systems. Crystallographic axes, which are usually defined as parallel to the edges of the unit cell, form a natural coordinate system for each material. In many crystals these crystallographic axes are not orthogonal and thus are not convenient for computation. It is customary to always introduce a right-handed Cartesian coordinate system, even in the case of

1 PROPERTIES OF PIEZOELECTRIC MATERIALS

enantiomorphic forms such as left quartz. The IEEE has published standard conventions for establishing the crystallographic axes and the right-handed Cartesian coordinate system for each of the 32 crystal classes (IEEE Standard on Piezoelectricity, 1978). The positive sense of each of the orthogonal X, Y, Z axes is also defined in the IEEE Standard by reference to static piezoelectric measurements (i.e., determining the sign of the potential difference created during a "squeeze" test). A detailed description of this process is presented in Section 1.2.1.2. for α-quartz.

1.1.2.1 DIELECTRIC CONSTANTS

In an anisotropic (crystalline) material the electric displacement **D** is related to the electric field **E** and the electric polarization **P** by the equation

$$\mathbf{D} = \varepsilon_0 \mathbf{E} + \mathbf{P}, \tag{1-1}$$

where ε_0 is the universal constant representing the permittivity of free space. Here, as in the remainder of this chapter, we base all equations on the International System of Units (SI). In linear dielectrics, we can express **P** in terms of the field **E** within the crystal by

$$P_i = \varepsilon_0 \chi_{ij} E_j, \tag{1-2}$$

where χ_{ij} is the electric susceptibility tensor of the crystal. We are using in this chapter the Einstein convention where repeated indices imply summation. If a material has a spontaneous polarization, such as ferroelectrics, or an induced polarization due to applied strain, then Eq. (1-2) is interpreted as the increase in polarization. Combining Eqs. (1-1) and (1-2) gives

$$D_i = \varepsilon_{ij} E_j, \tag{1-3}$$

where ε_{ij}, the permittivity tensor, is defined as

$$\varepsilon_{ij} = \varepsilon_0 (\delta_{ij} + \chi_{ij}). \tag{1-4}$$

The Kronecker delta δ_{ij} is 1 if $i = j$ and zero otherwise. Finally, the dielectric constant tensor K_{ij} is given by

$$K_{ij} = \varepsilon_{ij}/\varepsilon_0 = \delta_{ij} + \chi_{ij}. \tag{1-5}$$

The χ_{ij}, ε_{ij}, and K_{ij} are second-rank tensors, and the requirement that they be symmetric restricts the possible number of independent tensor elements to six. In the case of the triclinic crystal system, all six independent tensor elements are nonzero and unique, whereas three nonzero elements (all unique) are needed for orthorhombic crystals. Only three nonzero elements are needed for the remaining systems; tetragonal, trigonal, and hexagonal systems require two unique values, and cubic systems have only one unique value. A superscript T or S attached to the appropriate tensor symbol informs the reader as to which quantity, stress or strain, is held constant during measurement.

1.1.2.2 ELASTIC CONSTANTS

Stress is represented by the symbol T and in the case of elasticity is defined as a pair of opposing forces acting per unit area on opposite sides of an infinitesimal surface element within the material. If the body is in equilibrium, these forces are an equal and opposite pair. Each force can be resolved relative to the surface element into a normal component known as *compressive* or *tensile stress* and a tangential component known as *shear*.

For example, consider a cube of side l located inside a solid, as shown in Fig. 1-1. The origin of the orthogonal x_1, x_2, x_3 coordinate system is at the center of this volume element, and in the limit as l approaches zero, the stress system can be defined at this point in the material. Each of the three components of the force on each of the three sides 1, 2, and 3 of the volume element are illustrated by arrows. The force component acting in the x_1 direction on the "1" face (and equivalently, a similar force acting outward on the opposite face) produces a tensile stress T_{11}. Similarly, the force component acting on the "1" face in the x_2 direction when coupled with its opposing force on the opposite face produces a shear stress T_{21}. There are

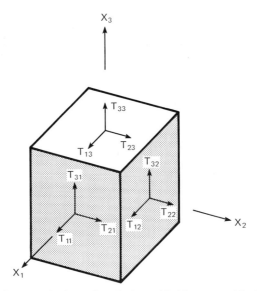

FIG. 1-1 The forces on the faces of a cube located inside a stressed body. Stress labels are placed next to the force vectors.

1 PROPERTIES OF PIEZOELECTRIC MATERIALS

a total of nine stress components, and their labels are placed next to the force vectors in Fig. 1-1. Because the net torque must be zero, the second-rank stress tensor is symmetric, and there are only six independent components. Since the stress tensor does not represent a property of the crystal structure, it does not necessarily conform to the crystal symmetry. The units of stress are N/m^2 or, equivalently, pascals (Pa).

Strain is represented by the symbol S and is a measure of the distortion, or deformation, of a solid body. It is specified at every point within the object. Consider a point initially at coordinates x_1, x_2, x_3 in a crystal, and suppose that after application of a complex stress, this point in the crystal is displaced to the position $x_1 + u_1, x_2 + u_2, x_3 + u_3$. The quantities u_1, u_2, u_3 represent the distortion, and the variation of distortion with position gives, in general, the nine components $\partial u_i/\partial x_j$ that form the basis of the second-rank strain tensor. To eliminate rotation effects, the components of the strain tensor are defined as

$$S_{ij} = \tfrac{1}{2}(\partial u_i/\partial x_j + \partial u_j/\partial x_i). \qquad (1\text{-}6)$$

This results in a symmetric strain tensor having six independent components.

Assuming Hooke's Law is valid, the relationships between stress and strain in a nonpiezoelectric material are given by

$$T_{ij} = c_{ijkl} S_{kl}, \qquad (1\text{-}7)$$

$$S_{ij} = s_{ijkl} T_{kl}, \qquad (1\text{-}8)$$

where the c_{ijkl} are the elastic stiffness constants and the s_{ijkl} are the elastic compliance constants.

Both elastic stiffness and compliance are fourth-rank tensors. There are 81 components in general for fourth-rank tensors; but because the stress and strain tensors are symmetric, the stiffness and compliance tensors must be symmetric, and each contains only 21 independent components for the lowest-symmetry (triclinic) case. Higher-symmetry crystal systems have considerably fewer independent nonzero components: six for trigonal crystals and three for cubic crystals. The superscript E or D for an elastic constant denotes whether the electric field or the electric displacement has been held fixed during measurement.

1.1.2.3 PIEZOELECTRIC CONSTANTS

In piezoelectric crystals, an applied stress produces an electric polarization. This is known as the direct piezoelectric effect, and the contribution to the polarization is described by the equation

$$P_i = d_{ijk} T_{jk} \qquad (1\text{-}9)$$

or

$$P_i = e_{ijk}S_{jk}. \tag{1-10}$$

Similarly, a piezoelectric crystal becomes deformed (i.e., develops strain) when placed in an electric field. This latter phenomenon is called the converse piezoelectric effect. The strain contribution is written as

$$S_{jk} = d_{ijk}E_i, \tag{1-11}$$

or the corresponding stress contribution is written as

$$T_{jk} = e_{ijk}E_i. \tag{1-12}$$

The d_{ijk} and e_{ijk} are known as the piezoelectric strain coefficients and piezoelectric stress coefficients, respectively. The first subscript represents the direction of electric polarization or electric field, whereas the last two subscripts describe the type of stress or strain.

In general, the total strain experienced by a crystal is the sum of two contributions, one due to the applied stress [Eq. (1-8)] and the other due to the applied electric field via the piezoelectric effect [Eq. (1-11)]. Thus, we have

$$S_{ij} = S^E_{ijkl}T_{kl} + d_{kij}E_k. \tag{1-13}$$

Also, the total electric polarization is the sum of two contributions, one due to the applied stress acting through the piezoelectric effect [Eq. (1-9)] and the other due to the applied electric field [Eq. (1-2)]. This gives

$$P_i = d_{ikl}T_{kl} + \varepsilon_0 \chi^T_{ik}E_k. \tag{1-14}$$

The electric displacement is preferred as the electric variable in place of polarization since it is more useful from an engineering and experimental point of view (IRE Standards on Piezoelectric Crystals, 1949). By adding $\varepsilon_0 E_i$ to the left side of Eq. (1-14) and $\varepsilon_0 \delta_{ik}E_k$ to the right side, we have

$$D_i = d_{ikl}T_{kl} + \varepsilon^E_{ik}E_k. \tag{1-15}$$

Equations (1-13) and (1-15) are one form of the piezoelectric equations of state, also referred to as the piezoelectric constitutive equations. Among the alternative forms of these equations are the following pair:

$$T_{ij} = c^E_{ijkl}S_{kl} - e_{kij}E_k, \tag{1-16}$$

$$D_i = e_{ikl}S_{kl} + \varepsilon^S_{ik}E_k. \tag{1-17}$$

Since the stress and strain tensors are symmetric, we must have $d_{ijk} = d_{ikj}$ and $e_{ijk} = e_{ikj}$. Because of this symmetry, it is convenient to convert the piezoelectric coefficients from tensor to matrix notation. For the d_{ijk}

1 PROPERTIES OF PIEZOELECTRIC MATERIALS

and e_{ijk}, we retain the first index (subscript) but replace the last two according to the following rules.

tensor notation, jk: 11   22   33   23,32   31,13   12,21

matrix notation, p;   1   2   3   4   5   6

This gives, for example, $e_{122} = e_{12}$ and $e_{231} = e_{25}$. In the case of the piezoelectric strain coefficients, we have $d_{122} = d_{12}$ and $2d_{231} = d_{25}$.

In addition to the piezoelectric coefficients, the shorthand matrix notation is widely used for the stress and strain tensors and the stiffness and compliance tensors. The following briefly summarizes this notation, including the necessary factors of two and four:

$$\left.\begin{aligned} d_{ijk} &= d_{ip}, \\ e_{ijk} &= e_{ip}, \\ S_{jk} &= S_p, \\ T_{jk} &= T_p, \end{aligned}\right\} \quad \text{when} \quad j = k, \quad p = 1, 2, 3, \qquad \begin{aligned} &(1\text{-}18)\\ &(1\text{-}19)\\ &(1\text{-}20)\\ &(1\text{-}21) \end{aligned}$$

$$\left.\begin{aligned} s_{jkmn} &= s_{pq}, \\ c_{jkmn} &= c_{pq}, \end{aligned}\right\} \quad \text{when} \quad j = k, \quad m = n, \quad p, q = 1, 2, 3, \qquad \begin{aligned} &(1\text{-}22)\\ &(1\text{-}23) \end{aligned}$$

$$\left.\begin{aligned} 2s_{jkmn} &= s_{pq}, \\ c_{jkmn} &= c_{pq}, \end{aligned}\right\} \quad \text{when} \quad j = k, \quad m \neq n, \quad p = 1, 2, 3, \quad q = 4, 5, 6, \qquad \begin{aligned} &(1\text{-}24)\\ &(1\text{-}25) \end{aligned}$$

$$\left.\begin{aligned} 2d_{ijk} &= d_{ip}, \\ e_{ijk} &= e_{ip}, \\ 2s_{jk} &= s_p, \\ T_{jk} &= T_p, \end{aligned}\right\} \quad \text{when} \quad j \neq k, \quad p = 4, 5, 6, \qquad \begin{aligned} &(1\text{-}26)\\ &(1\text{-}27)\\ &(1\text{-}28)\\ &(1\text{-}29) \end{aligned}$$

$$\left.\begin{aligned} 4s_{jkmn} &= s_{pq}, \\ c_{jkmn} &= c_{pq}, \end{aligned}\right\} \quad \text{when} \quad j \neq k, \quad m \neq n, \quad p, q = 4, 5, 6. \qquad \begin{aligned} &(1\text{-}30)\\ &(1\text{-}31) \end{aligned}$$

The stress and strain tensors become 1×6 matrices, the piezoelectric tensors become 3×6 matrices, and the stiffness and compliance tensors become 6×6 matrices. The reader should be warned that while the matrix notation is compact, it does not have the transformation properties of tensors. Consequently, the full notation must be used to carry out any transformation of coordinates.

In the case of triclinic (lowest) symmetry, there are 18 independent piezoelectric coefficients. Trigonal systems require fewer independent constants, two for class 32 and four for class $3m$. Only one independent constant is needed for the cubic class 23. In all cases, the number of nonzero piezoelectric coefficients is greater than the number of independent constants. Table 1-2 gives the complete piezoelectric matrices for α-quartz (class 32), lithium

TABLE 1-2

Piezoelectric Matrices for 3 of the 20 Possible Piezoelectric Crystal Classes

Classification	Matrix
Trigonal System Class 32 Example: α-quartz	$\begin{pmatrix} d_{11} & -d_{11} & 0 & d_{14} & 0 & 0 \\ 0 & 0 & 0 & 0 & -d_{14} & -2d_{11} \\ 0 & 0 & 0 & 0 & 0 & 0 \end{pmatrix}$
Trigonal System Class 3m Example: lithium niobate	$\begin{pmatrix} 0 & 0 & 0 & 0 & d_{15} & -2d_{22} \\ -d_{22} & d_{22} & 0 & d_{15} & 0 & 0 \\ d_{31} & d_{31} & d_{33} & 0 & 0 & 0 \end{pmatrix}$
Cubic System Class 23 Example: bismuth germanium oxide	$\begin{pmatrix} 0 & 0 & 0 & d_{14} & 0 & 0 \\ 0 & 0 & 0 & 0 & d_{14} & 0 \\ 0 & 0 & 0 & 0 & 0 & d_{14} \end{pmatrix}$

niobate (class 3m), and bismuth germanium oxide (class 23). An additional source of information is the IEEE Standard on Piezoelectricity (1978), where all the nonzero matrix elements of the elastic, piezoelectric, and dielectric matrices for the 32 crystal classes are illustrated.

The reader is referred to Cady (1964), Nye (1957), and Tiersten (1969) for more information concerning the general nature of the relationships between the thermal, electrical, and mechanical properties of anisotropic crystals. These books discuss secondary effects and provide expressions relating the electroelastic constants measured under different conditions, (i.e., isothermal versus adiabatic).

1.1.3 Materials of Current Interest

Although a great many piezoelectric materials have been characterized over the last 50 years, only a few have found wide application in precision frequency control devices. Alpha-quartz is the only material normally used for precision bulk piezoelectric resonators, whereas $LiNbO_3$, $LiTaO_3$, $Bi_{12}GeO_{20}$, and α-quartz are the crystals most commonly used for surface-acoustic-wave (SAW) devices. Recently, berlinite ($AlPO_4$) has also been suggested as a useful SAW material. Whatmore (1980) reviewed the current status of many of the materials considered for use in SAW devices and illustrated how the physical properties of a material limit device performance.

In the remainder of this section, we present information about the crystal structure, growth conditions, defects, and thermal behavior of the more

1 PROPERTIES OF PIEZOELECTRIC MATERIALS 11

prominent piezoelectric materials. Values of the dielectric, stiffness, compliance, and piezoelectric constants for these high-interest materials (along with their temperature coefficients when available) are given in Tables 1-3 through 1-7. Volume III/11 of the Landolt–Börnstein New Series (1979) contains a much larger tabulation of data on piezoelectric materials and is an excellent resource for the interested reader.

1.1.3.1 QUARTZ

Because of its importance in precision frequency control, we have devoted the entirety of Section 1.2 to a presentation of the physical properties of quartz. Crystallographic considerations, crystal growth and the nature of extended defects, characteristics of point defects, thermal properties, and quartz evaluation techniques are the topics receiving attention.

The values of the various physical constants for α-quartz are listed in Table 1-3. Kahan (1982) analyzed the presently accepted sets of elastic constants and their temperature coefficients, and he made a strong case for the need to experimentally redetermine these constants in α-quartz.

1.1.3.2 LITHIUM NIOBATE AND LITHIUM TANTALATE

Lithium niobate and lithium tantalate were first reported to be ferroelectric by Matthias and Remeika (1949), but it was not until mid-1960s that their unique electrooptic and electroacoustic properties were generally explored and their potential as device materials was developed. Ballman (1965) and Fedulov et al. (1965) were the first to successfully grow $LiNbO_3$ and $LiTaO_3$ by the Czochralski technique, and an early study by Peterson et al. (1964) suggested that $LiNbO_3$ possessed interesting electrooptic properties. These initial results prompted researchers at Bell Telephone Laboratories to undertake a systematic investigation of the growth and crystal structure of $LiNbO_3$. In a series of five well-known back-to-back papers, research groups at Bell described growth techniques and large-scale crystal imperfections (Nassau et al., 1966a), preparation of single-domain crystals (Nassau et al., 1966b), single crystal x-ray and neutron diffraction studies at room temperature (Abrahams et al., 1966a,b), and a polycrystal x-ray diffraction study between 24 and 1200°C (Abrahams et al., 1966c).

During the past 15 years, lithium niobate and, to a lesser extent, lithium tantalate have been widely studied from both fundamental and device viewpoints. An in-depth and extremely useful review of much of the physics and chemistry of lithium niobate was prepared by Rauber (1978). Although directed toward electrooptic applications, the book by Kaminow (1974) also provides an excellent introduction to the usefulness of $LiNbO_3$ and

TABLE 1-3

Physical Properties of α-Quartz

Density ($\times 10^3$ kg/m^3)

$\rho = 2.65$

Thermal expansion ($\times 10^{-6}/°C$)

α_{11}	α_{33}	β_{11}	β_{33}	
9.8	5.6	0.0204	0.0094	Kim and Smith (1969)
13.71	7.48	0.0065	0.0029	Bechmann et al. (1962)

Permittivity constants ($\times 10^{-10}$ F/m)

ε^S_{11}	ε^S_{33}	ε^T_{11}	ε^T_{33}	
0.3997	0.4103	0.4073	0.4103	Bechmann (1958)
		0.40025	0.41054	Fontanella et al. (1974)

Temperature coefficients of permittivity constants ^{a}($\times 10^{-4}/°C$)

$T\varepsilon_{11}$	$T\varepsilon_{33}$	
0.28	0.39	Westphal (1961)

Elastic stiffness constants ($\times 10^9$ N/m^2)

c^E_{11}	c^E_{33}	c^E_{44}	c^E_{12}	c^E_{13}	c^E_{14}	c^E_{66}	
86.74	107.2	57.94	6.99	11.91	−17.9	39.88	Bechmann (1958)
86.80	105.75	58.20	7.04	11.91	−18.04	39.88	McSkimin (1962)
83.63	77.60	57.32	4.47	−0.88	−18.88	39.58	Kahan (1982)

Temperature coefficients of elastic stiffness constants^a ($\times 10^{-4}/°C$)

Tc_{11}	Tc_{33}	Tc_{44}	Tc_{12}	Tc_{13}	Tc_{14}	Tc_{66}	
−0.485	−1.60	−1.77	−30.00	−5.50	1.01	1.78	Bechmann et al. (1962)
−0.496	−1.92	−1.72		−6.51	0.89	1.67	Adams et al. (1970)
−0.443	−1.600	−1.754	−26.90	−5.50	1.17	1.876	Zelenka and Lee (1971)

Elastic compliance constants ($\times 10^{-12}$ m^2/N)

s^E_{11}	s^E_{33}	s^E_{44}	s^E_{12}	s^E_{13}	s^E_{14}	s^E_{66}	
12.77	9.60	20.04	−1.79	−1.22	4.50	29.12	Bechmann (1958)

Temperature coefficients of elastic compliance constants^a ($\times 10^{-4}/°C$)

Ts_{11}	Ts_{33}	Ts_{44}	Ts_{12}	Ts_{13}	Ts_{14}	Ts_{66}	
0.155	1.40	2.10	−13.70	−1.66	1.34	−1.45	Bechmann et al. (1962)
0.085	1.397	2.111	−12.965	−1.688	1.406	−1.519	Zelenka and Lee (1971)

1 PROPERTIES OF PIEZOELECTRIC MATERIALS   13

TABLE 1-3 (Continued)

Piezoelectric strain constants ($\times 10^{-12}$ C/N)

d_{11}	d_{14}	
2.3	−0.67	Cady (1964)
2.31	−0.727	Bechmann (1958)
2.27		Bottom (1970)
2.31	−0.670	Zubov and Firsova (1974)

Temperature coefficients of piezoelectric strain constants^a ($\times 10^{-4}/°C$)

Td_{11}	Td_{14}	
−2.0	17.7	Cook and Weissler (1950)
−2.15	12.9	Bechmann (1951)

Piezoelectric stress constants (C/m^2)

e_{11}	e_{14}	
0.173	0.04	Cady (1964)
0.171	0.0403	Bechmann (1958)
0.1711		Graham (1972a)

Temperature coefficients of piezoelectric stress constants^a ($\times 10^{-4}/°C$)

Te_{11}	Te_{14}	
−1.6	−14.4	Bechmann (1966)

^{a} The temperature coefficient of quantity x is defined as $Tx = 1/x \, dx/dT$. Higher-order coefficients of the elastic constants are given by Bechmann et al. (1962).

LiTaO$_3$ and gives the reader a broader perspective as to the range of properties exhibited by these materials. Values for many of the physical constants characterizing LiNbO$_3$ and LiTaO$_3$ are tabulated in Tables 1-4 and 1-5, respectively.

Lithium niobate and its isomorph lithium tantalate crystallize in a trigonal structure. The melting points of LiNbO$_3$ and LiTaO$_3$ are 1260 and 1560°C, respectively, and their Curie temperatures are 1197 and approximately 620°C. At room temperature, the structure belongs to point group 3m and space group R3c. Above the Curie temperature, it loses the polar axis and changes to point group $\bar{3}$m and space group R$\bar{3}$c. One can view this lower-symmetry ABO$_3$ structure as a collection of distorted oxygen octahedra with the centers of the octahedra being occupied by the repeating sequence of A, B, and vacancy along the crystal's c axis. Abrahams et al. (1966a) determined the unit cell dimensions and the position parameters for LiNbO$_3$ at 24°C. Their lattice constants, when referred to an equivalent hexagonal

TABLE 1-4

Physical Properties of Lithium Niobate

Density ($\times 10^3$ kg/m^3)

$\rho = 4.64$

Thermal expansion ($\times 10^{-6}/°C$)

α_{11}	α_{33}	β_{11}	β_{33}	
15.4	7.5	0.0053	−0.0077	Kim and Smith (1969)

Permittivity constants ($\times 10^{-10}$ F/m)

ε_{11}^S	ε_{33}^S	ε_{11}^T	ε_{33}^T	
3.90	2.57	7.44	2.66	Warner et al. (1967)
3.92	2.47	7.54	2.54	Smith and Welsh (1971)
3.89	2.10			Teague et al. (1975)

Temperature coefficients of permittivity constants[a] ($\times 10^{-4}/°C$)

$T\varepsilon_{11}^S$	$T\varepsilon_{33}^S$	$T\varepsilon_{11}^T$	$T\varepsilon_{33}^T$	
3.23	6.27	3.82	6.71	Smith and Welsh (1971)

Elastic stiffness constants ($\times 10^9$ N/m^2)

c_{11}^E	c_{33}^E	c_{44}^E	c_{12}^E	c_{13}^E	c_{14}^E	c_{66}^E	
203	245	60	53	75	9	75	Warner et al. (1967)
203.0	242.4	59.5	57.3	75.2	8.5	72.8	Smith and Welsh (1971)

Temperature coefficients of elastic stiffness constants[a] ($\times 10^{-4}/°C$)

Tc_{11}	Tc_{33}	Tc_{44}	Tc_{12}	Tc_{13}	Tc_{14}	Tc_{66}	
−1.74	−1.53	−2.04	−2.52	−1.59	−2.14	−1.43	Smith and Welsh (1971)

Elastic compliance constants ($\times 10^{-12}$ m^2/N)

s_{11}^E	s_{33}^E	s_{44}^E	s_{12}^E	s_{13}^E	s_{14}^E	s_{66}^E	
5.78	5.02	17.0	−1.01	−1.47	−1.02	13.6	Warner et al. (1967)
5.831	5.026	17.10	−1.150	−1.452	−1.000	13.96	Smith and Welsh (1971)

Temperature coefficients of elastic compliance constants[a] ($\times 10^{-4}/°C$)

Ts_{11}	Ts_{33}	Ts_{44}	Ts_{12}	Ts_{13}	Ts_{14}	Ts_{66}	
1.66	1.60	2.05	0.28	1.94	1.33	1.43	Smith and Welsh (1971)

TABLE 1-4 (Continued)

Piezoelectric strain constants ($\times 10^{-12}$ C/N)

d_{22}	d_{31}	d_{33}	d_{15}	
21	−1	6	68	Warner et al. (1967)
22.4	−1.2	18.8	78.0	
20.8	−0.85	6.0	69.2	Smith and Welsh (1971)

Temperature coefficients of piezoelectric strain constants[a] ($\times 10^{-4}$/°C)

Td_{22}	Td_{31}	Td_{33}	Td_{15}	
2.34	19.1	11.3	3.45	Smith and Welsh (1971)

Piezoelectric stress constants (C/m²)

e_{22}	e_{31}	e_{33}	e_{15}	
2.5	0.2	1.3	3.7	Warner et al. (1967)
2.43	0.23	1.33	3.76	Smith and Welsh (1971)

Temperature coefficients of piezoelectric stress constants[a] ($\times 10^{-4}$/°C)

Te_{22}	Te_{31}	Te_{33}	Te_{15}	
0.79	2.21	8.87	1.47	Smith and Welsh (1971)

[a] The temperature coefficient of quantity x is defined as $Tx = 1/x \, dx/dT$.

unit cell, are $a_H = 5.1483$ Å and $c_H = 13.8631$ Å. A similar study by Abrahams and Bernstein (1967) gives lattice constants of $a_H = 5.1453$ Å and $c_H = 13.7835$ Å for LiTaO₃ at 25°C.

The Czochralski crystal growing technique as applied by Nassau et al. (1966a) to LiNbO₃ consists of first preparing starting material from a mix of Li₂CO₃ and Nb₂O₅. An rf induction heater then is used to produce a melt in a platinum crucible from which the crystal can be pulled. If temperatures and temperature gradients are carefully controlled, crystals over 2-cm diameter and at least 10-cm long can be pulled at rates approaching 2 cm/hr. Use of an after heater, or efficient heat shields, and a lengthy slow-cooling period help prevent cracking of the crystals during the initial return to room temperature.

Two techniques are normally used to obtain single-domain crystals, (i.e., to pole the crystals) (Nassau et al., 1966b). One method consists of applying a current to the crystal during growth. This requires electrical insulation of the seed support shaft relative to the crucible, but it is very convenient for commercial production facilities. The other method can be

TABLE 1-5

Physical Properties of Lithium Tantalate

Density ($\times 10^3$ kg/m^3)

$\rho = 7.45$

Thermal expansion ($\times 10^{-6}$/°C)

α_{11}	α_{33}	β_{11}	β_{33}	
16.2	4.1	0.0059	−0.0100	Kim and Smith (1969)

Permittivity constants ($\times 10^{-10}$ F/m)

ε^S_{11}	ε^S_{33}	ε^T_{11}	ε^T_{33}	
3.63	3.81	4.52	3.98	Warner et al. (1967)
		4.69	3.90	Yamada et al. (1969)
3.77	3.79	4.74	3.84	Smith and Welsh (1971)

Temperature coefficients of permittivity constants[a] ($\times 10^{-4}$/°C)

$T\varepsilon^S_{11}$	$T\varepsilon^S_{33}$	$T\varepsilon^T_{11}$	$T\varepsilon^T_{33}$	
3.29	11.6	2.11	11.47	Smith and Welsh (1971)

Elastic stiffness constants ($\times 10^9$ N/m^2)

c^E_{11}	c^E_{33}	c^E_{44}	c^E_{12}	c^E_{13}	c^E_{14}	c^E_{66}	
233	275	94	47	80	−11	93	Warner et al. (1967)
228	271	96	31	74	−12	98	Yamada et al. (1969)
229.8	279.8	96.8	44.0	81.2	−10.4	92.9	Smith and Welsh (1971)

Temperature coefficients of elastic stiffness constants[a] ($\times 10^{-4}$/°C)

Tc_{11}	Tc_{33}	Tc_{44}	Tc_{12}	Tc_{13}	Tc_{14}	Tc_{66}	
−1.03	−0.96	−0.43	−3.41	−0.50	6.67	−0.47	Smith and Welsh (1971)

Elastic compliance constants ($\times 10^{-12}$ m^2/N)

s^E_{11}	s^E_{33}	s^E_{44}	s^E_{12}	s^E_{13}	s^E_{14}	s^E_{66}	
4.87	4.36	10.8	−0.58	−1.25	0.64	10.9	Warner et al. (1967)
4.86	4.36	10.5	−0.29	−1.24	0.63	10.3	Yamada et al. (1969)
4.93	4.317	10.46	−0.519	−1.280	0.588	10.90	Smith and Welsh (1971)

Temperature coefficients of elastic compliance constants[a] ($\times 10^{-4}$/°C)

Ts_{11}	Ts_{33}	Ts_{44}	Ts_{12}	Ts_{13}	Ts_{14}	Ts_{66}	
1.11	1.24	0.60	−3.83	2.14	7.74	0.64	Smith and Welsh (1971)

1 PROPERTIES OF PIEZOELECTRIC MATERIALS

TABLE 1-5 *(Continued)*

Piezoelectric strain constants ($\times 10^{-12}$ C/N)

d_{22}	d_{31}	d_{33}	d_{15}	
7	-2	8	26	Warner et al. (1967)
8.5	-3.0	9.2	26	Yamada et al. (1969)
7.5	-3.0	5.7	26.4	Smith and Welsh (1971)

Temperature coefficients of piezoelectric strain constants[a] ($\times 10^{-4}$/°C)

Td_{22}	Td_{31}	Td_{33}	Td_{15}	
-1.32	3.27	2.74	-1.31	Smith and Welsh (1971)

Piezoelectric stress constants (C/m²)

e_{22}	e_{31}	e_{33}	e_{15}	
1.6	0.0	1.9	2.6	Warner et al. (1967)
2.0	-0.1	2.0	2.7	Yamada et al. (1969)
1.67	-0.38	1.09	2.72	Smith and Welsh (1971)

Temperature coefficients of piezoelectric stress constants[a] ($\times 10^{-4}$/°C)

Te_{22}	Te_{31}	Te_{33}	Te_{15}	
-0.60	0.87	1.54	-1.32	Smith and Welsh (1971)

[a] The temperature coefficient of quantity x is defined as $Tx = 1/x\, dx/dT$.

used after growth. It consists of applying platinum electrodes and then heating the crystal to slightly above the Curie temperature. An electric potential is maintained between the electrodes, and the poling current flows through the crystal as the temperature is slowly decreased below the Curie temperature. Poling is an important process for $LiNbO_3$ and $LiTaO_3$ since it is crucial to have single-domain crystals for all precision frequency control applications.

Following the initial development of the general growth techniques, investigators found that $LiNbO_3$ crystals grown from a congruent composition melt were of significantly higher quality than crystals grown from a stoichiometric melt (Byer et al., 1970; Carruthers et al., 1971). Phase diagrams of the system $Li_2O-Nb_2O_5$ provided by Lerner et al. (1968) suggested that the congruent melts should be lithium deficient, and Chow et al. (1974) precisely determined the congruently melting composition of $LiNbO_3$ to be between 48.5 and 48.6 mole % Li_2O.

Growth of $LiTaO_3$ crystals is similar to that of $LiNbO_3$, and the same problems have been encountered. The congruent melting composition of

LiTaO$_3$ is 48.75 mole % (Miyazawa and Iwasaki, 1971). Brandle and Miller (1974) developed a diameter control for LiTaO$_3$ growth. Fukuda et al. (1979) further developed the techniques of growing LiTaO$_3$ for SAW applications and Matsumura (1981) grew large-diameter X-axis LiTaO$_3$ crystals. The crystal growth and fundamental properties of mixed LiNb$_{1-y}$Ta$_y$O$_3$ crystals was reported by Shimura and Fujino (1977). As an alternative to the Czochralski technique, Kolb and Laudise (1976) investigated the possibility of hydrothermal growth of LiNbO$_3$ and LiTaO$_3$ but concluded that much development work is needed before the process could be practical.

Crystals grown from the melt can be highly nonstoichiometric, and a number of crystal properties of LiNbO$_3$, including the Curie temperature and birefringence, have been found to depend strongly on the Li/Nb ratio. The deviations from the ratio Li/Nb = 1 only occur on the Li-deficient side of equilibrium, but very little is known about the defect structures in LiNbO$_3$ that allow incorporation of Li vacancies. Lerner et al. (1968) proposed a simple replacement of Li$^+$ ions by Nb^{5+} ions with charge compensation then supplying four Li vacancies. Nassau and Lines (1970) concluded this was unlikely and proposed instead a more extended stacking disorder of cations along the c axis; a simple example is a Li–Nb–Li sequence being replaced by a Nb–Li–Nb sequence and four extra Li vacancies nearby for charge comparison.

Besides the Li/Nb ratio, the metal/oxygen ratio also can vary greatly in LiNbO$_3$. During growth in air, the crystals lose oxygen and may appear tan or brown. Annealing to near 1000°C in a nitrogen atmosphere or in a vacuum also removes oxygen and will turn the crystals black. These colored crystals can be cleared by simply annealing in oxygen at 1000°C for several hours. Glass et al. (1972) suggested that the reducing treatments produce Nb^{4+} ions that in turn give rise to the dark coloration; however, more evidence must be obtained before accepting this explanation. Sherman and Lemanov (1971) investigated the absorption of elastic waves in reduced LiNbO$_3$, and they describe possible loss mechanisms.

The major structural defects in LiNbO$_3$ and LiTaO$_3$ are twin formation, grain boundaries, and dislocations. Very little is presently known about these extended defects except that they are nearly always present in every crystal. Even in the "best" crystals, dislocation densities are in the 10^3 to 10^4/cm^2 range. Nassau et al. (1966a) reported on the ferroelectric-domain structure and dislocations in LiNbO$_3$ that were revealed by etching. Levinstein et al. (1966) carried out a similar study in LiTaO$_3$. Further studies of etching of LiNbO$_3$ were reported by Nassau et al. (1965), Niizeki et al. (1967), and Ohnishi and Iizuka (1975). Sugii et al. (1973) and Sugii and Iwasaki (1973) used x-ray topography to study dislocations, subgrain boundaries, and ferroelectric-grain boundaries in LiNbO$_3$.

The thermal properties of $LiNbO_3$ and $LiTaO_3$ have not been extensively studied. Kim and Smith (1969) determined the thermal expansion coefficients for both materials over the 0 to 500°C temperature range, and their results are given in Tables 1-4 and 1-5. Sugii et al. (1976) measured the temperature variation of the lattice parameters of $LiNbO_3$ and $LiTaO_3$. Zhdanova et al. (1968) found heat capacity and thermal conductivity values at 300 K of $c_p = 95.7$ J mole^{-1} K^{-1} and $\lambda = 4.2$ W m^{-1} K^{-1}, respectively, for $LiNbO_3$. Since their thermal conductivity versus temperature results are decreasing more slowly than the T^{-1} dependence expected for lattice conductivity, the reported value of λ may be as much as 25% too high.

1.1.3.3 BISMUTH GERMANIUM OXIDE

Bismuth germanium oxide ($Bi_{12}GeO_{20}$) crystallizes in a body-centered cubic (bcc) structure belonging to point group 32 and space group $I23$ (Abrahams et al., 1967). The lattice constant at 25°C is 10.1455 Å (Bernstein, 1967). Bismuth germanium oxide is a member of the selenite family in which an MO_2 compound stabilizes Bi_2O_3 in the bcc structure. This material is both optically active and piezoelectric, with a primary application being long-time delay lines because of its low surface-wave velocity. Values for many of the physical constants needed to predict elastic-wave properties are given in Table 1-6.

The first large crystals of $Bi_{12}GeO_{20}$ were grown by Ballman (1967) using the Czochralski method of pulling from the melt. This method continues to be the technique most commonly used. A stoichiometric mixture of Bi_2O_3 and GeO_2 is placed in a platinum crucible, and the melting point of 930°C is reached by using an rf induction heater. One of the refinements in crystal growth has been the development of an automatic crystal puller with optical diameter control using a laser beam (Gross and Kersten, 1972).

Bismuth germanium oxide is transparent from 0.450 to 7.5 μm; this short-wavelength cutoff causes the crystals to appear pale yellow. Photoconductivity is induced by visible light, but the dark resistivity is approximately 10^{10} Ω cm. There appear to be no measurements of the thermal properties of $Bi_{12}GeO_{20}$. Because of the large molecular weight, the thermal conductivity is expected to be very low.

Little information is available about either point or extended defects in $Bi_{12}GeO_{20}$. Ballman (1967) reported that the slow attack of the melt on the platinum crucible results in fine metallic inclusions in the crystal. According to Gross and Kersten (1972), the concentration of these inclusions is reduced by a factor of 10 when diameter control is used during growth. Measurements of the surface-acoustic-wave propagation loss on $Bi_{12}GeO_{20}$ indicates that the attenuation is inherent to the crystal and is not due to imperfections (Slobodnik and Budreau, 1972).

TABLE 1-6

Physical Properties of Bismuth Germanium Oxide, $Bi_{12}GeO_{20}$

Density ($\times\ 10^3$ kg/m³)

$\rho = 9.23$

Permittivity constants ($\times\ 10^{-10}$ F/m)

ε^S	ε^T	
3.4		Onoe et al. (1967)
3.4	3.54	Ballman (1967)
3.36		Kraut et al. (1970)

Elastic stiffness constants ($\times\ 10^9$ N/m²)

c_{11}^E	c_{44}^E	c_{12}^E	
120	25	39	Onoe et al. (1967)
128.48	25.52	29.42	Kraut et al. (1970)
128	25.5	30.5	Slobodnik and Sethares (1972)

Piezoelectric stress constant (C/m²)

e_{14}	
0.71	Onoe et al. (1967)
0.983	Kraut et al. (1970)
0.99	Slobodnik and Sethares (1972)

1.1.3.4 ALUMINUM PHOSPHATE

AlPO$_4$, also known as berlinite, crystallizes in a trigonal structure with point group 32 and is structurally similar to quartz. To convert from quartz to berlinite, simply replace half the silicons with aluminum (Al^{3+}) and half with phosphorus (P^{5+}), such that each oxygen ion links an aluminum and a phosphorus ion, and then allow slight relaxations of the oxygens to accommodate the different ionic radii of the two cations. As further evidence of their similarities, the α–β phase transition occurs at 581°C for berlinite and 573°C for quartz. Schwarzenbach (1966) determined atomic positions for berlinite; his unit cell parameters are $a_0 = 4.9429$ Å and $c_0 = 10.9476$ Å at 20°C. Although not extensively studied, values for many of the physical constants of AlPO$_4$ have been measured and are tabulated in Table 1-7.

TABLE 1-7

Physical Properties of Berlinite, $AlPO_4$

Density ($\times 10^3$ kg/m^3)
$\rho = 2.62$

Thermal expansion ($\times 10^{-6}$/°C)

α_{11}	α_{33}	β_{11}	β_{33}	
15.9	9.7	0.015	0.015	Chang and Barsch (1976)

Permittivity constants ($\times 10^{-10}$ F/m)

ε_{11}	ε_{33}	
0.54 (at 5 MHz)		Mason (1950)
0.44		Stanley (1954)
0.52 (at 1 MHz)		Kolb and Laudise (1981)

Elastic stiffness constants ($\times 10^9$ N/m^2)

c^E_{11}	c^E_{33}	c^E_{44}	c^E_{12}	c^E_{13}	c^E_{14}	c^E_{66}	
105.0	133.5	23.1	29.3	69.3	−12.7	37.9	Mason (1950)
64.0	85.8	43.2	7.2	9.6	−12.4	28.4	Chang and Barsch (1976)
69.3	88.6	43.0	10.5	13.5	−13.0	29.4	Bailey et al. (1982)

Temperature coefficients of elastic stiffness constants ($\times 10^{-4}$/°C)

Tc_{11}	Tc_{33}	Tc_{44}	Tc_{12}	Tc_{13}	Tc_{14}	Tc_{66}	
−0.76	−2.18	−1.57	−15	−4.0	0.72	1.03	Chang and Barsch (1976)

Piezoelectric strain constants ($\times 10^{12}$ C/N)

d_{11}	d_{14}	
3.33	1.55	Mason (1950)
3.52		Kolb and Laudise (1981)
2.87	2.20	Bailey et al. (1982)

Piezoelectric stress constants (C/m^2)

e_{11}	e_{14}	
−0.27	0.12	Mason (1950)
−0.30	0.13	Chang and Barsch (1976)
0.14	0.02	Bailey et al. (1982)

Temperature coefficients of piezoelectric stress constants[a] ($\times 10^{-4}$/°C)

Te_{11}	Te_{14}	
−2.7	−5.6	Chang and Barsch (1976)

[a] The temperature coefficient of quantity x is defined as $Tx = 1/x \, dx/dT$.

The first systematic growth of berlinite crystals from solution was reported by Stanley (1954). He used two methods:

(1) Seeds were introduced into solutions of $NaAlO_2$ and H_3PO_4, and the crystals grew as the temperature was held constant at 165°C.

(2) Crystals were grown from similar solutions by slowly increasing the temperature from 133 to 155°C at a rate of 0.5°C/day.

Measurements of the elastic constants of berlinite between 80 and 298 K and the thermal expansions from 293 to 950 K were reported by Chang and Barsch (1976). Based on these results, Jhunjhunwala *et al.* (1977) and O'Connell and Carr (1977a) made theoretical predictions of the SAW properties of berlinite. This in turn led Morency *et al.* (1978) to experimentally measure the SAW properties of berlinite, finding that it is temperature compensated and has piezoelectric coupling much higher than that of quartz. More recently, Bailey *et al.* (1982) measured the elastic, dielectric, and piezoelectric constants of berlinite and compared calculated and experimental SAW velocities and piezoelectric couplings.

Although berlinite has the highly desirable elastic and piezoelectric characteristics that would make it invaluable in many signal-processing devices, the fact that large, high-quality crystals have not been readily grown has prevented widespread use of this material. A major complicating factor in the growth of berlinite is the lack of natural crystals for use as seeds and as nutrient. Also, berlinite's negative solubility tends to dissolve crystals as they are cooled from the growth temperature, or to introduce cracks if they are cooled too quickly, and presents problems not encountered in the growth of quartz.

Despite the many problems, the promise that berlinite would be as good or better a SAW material as quartz has stimulated significant efforts to grow larger and higher-quality crystals. Kolb and Laudise (1978), in an expansion of the initial work of Stanley, successfully grew berlinite from seeds by warming a saturated solution and also by transport from a cooler region of an autoclave to the warmer region. Kolb *et al.* (1979) systematically studied the solubility of $AlPO_4$ and then grew a series of crystals on different seed orientations. The perfection of these latter crystals was assessed by light scattering, etching, and x-ray topography. Independently, Ozimek and Chai (1979) grew a number of berlinite crystals under varying conditions and characterized them by etching, by fabrication of piezoelectric vibrators, and by infrared, ultraviolet, and Raman spectroscopy. Detaint *et al.* (1980) grew berlinite crystals from which Y-rotated resonators were cut and studied. Kolb *et al.* (1981b) investigated the possibility of using HCl as a solvent for berlinite growth, and Kolb and Laudise (1981) measured the pressure–volume–temperature behavior of hydrothermal solutions and applied the results to berlinite growth. Aucoin et al. (1980) described a more controlled growing technique for berlinite. Thus, in spite of the many difficulties, considerable improvement is being made in berlinite growth by a number of research laboratories.

A few additional basic research programs have been pursued in this material. Among these are a Raman scattering study by Shand and Chai (1980) to monitor the H_2O in berlinite and an investigation of radiation effects in berlinite by Halliburton et al. (1980b). Also, the α–β phase transition in berlinite has been studied by electron paramagnetic resonance (Lang et al., 1977), Raman spectroscopy (Nicola et al., 1978), and Brillouin scattering (Ecolivet and Poignant, 1981).

1.2 PHYSICAL PROPERTIES OF QUARTZ[§]

The use of α-quartz for precision frequency control applications is widespread. In this section we present a survey of the various physical properties of quartz that are relevant to these applications.

1.2.1 Crystallography

1.2.1.1 STRUCTURE

Silica (SiO_2) crystallizes into a number of different structures (Megaw, 1973). Quartz, tridymite, and cristobalite are the better known forms, while the crystalline polymorphs coesite, stishovite, and keatite are much rarer. The only form of SiO_2 having application in frequency control is low quartz, commonly known as α-quartz. It belongs to the trigonal crystal system with point group 32. Both right- and left-handed α-quartz exist, corresponding to space groups $P3_221$ and $P3_121$, respectively.

In general, the α-quartz structure consists of SiO_4 tetrahedra that share each of their corners with another tetrahedron. The four oxygen ions surrounding a silicon are divided into two types, those with long bonds and short bonds to the central silicon. The Si–O long bond is 1.612 Å, the Si–O short bond is 1.606 Å, and the Si–O–Si bond angle is 143.65° (Le Page et al., 1980). Cohen and Sumner (1958) measured the lattice constants of natural and cultured quartz samples. Their results give unit cell parameters at 25°C of $a_0 = 4.9134$ Å and $c_0 = 5.4050$ Å. The variation of these unit cell dimensions between 86 and 298 K was reported by Danielsson et al. (1976). Positional parameters (i.e., atomic coordinates) for α-quartz have been measured with increasing precision (Young and Post, 1962; Zachariasen and Plettinger,

§ Section 1.2 was written by Larry E. Halliburton, Joel J. Martin, and Dale R. Koehler.

1965; Le Page and Donnay, 1976), and the parameter variations between 94 and 298 K have been determined (Le Page et al., 1980).

1.2.1.2 COORDINATE SYSTEMS

Trigonal crystals such as quartz are characterized by an axis of threefold symmetry. This unique axis is always taken to be the c axis, or optic axis. In addition, three equivalent twofold axes (a_1, a_2, and a_3) lie 120° apart in a plane perpendicular to the c axis. These four axes form a natural coordinate system, although containing a redundant axis, and crystal planes can be described by the Miller–Bravais indices $hkil$, where the first three indices refer to the three twofold axes and the fourth index refers to the c axis (Megaw, 1973).

For many applications, it is convenient to introduce a Cartesian coordinate system. The IEEE Standard on Piezoelectricity (1978) defines the Cartesian coordinates for α-quartz as follows:

(1) The Z axis is parallel to the c (optic) axis. The positive direction is arbitrary.

(2) The X axis is chosen to lie along one of the three equivalent a axes. The positive direction is chosen so that the d_{31} piezoelectric constant is positive for right-handed quartz. In practice, the sign of the X axis is determined by the polarity of the voltage produced when the sample is released from compression along the X axis. The side with the positive potential is the $+X$ side. In the case of left-handed quartz, the $+X$ direction is chosen so that d_{11} is negative, and the $+X$ side has a negative potential when the crystal is released from compression. A simple piezoelectric squeeze tester used to determine the $+X$ side has been described by Bond (1976).

(3) The Y axis is chosen to form a right-handed coordinate system for both right- and left-handed quartz crystals.

This 1978 IEEE Standard has reversed the direction of the $+X$ axis for α-quartz and now brings quartz into the same general system as used for other piezoelectric crystals. It is important to note, however, that the current standard is not being universally followed. For example, at the present time many quartz-growing companies continue to designate the reference surface of their quartz bars by the pre-1978 method. The interested reader should also refer to the comprehensive review by Donnay and Le Page (1978) for discussions of the various conventions used to describe the enantiomorphs of α-quartz.

1.2.1.3 TWINNING AND STRUCTURAL PHASE TRANSITIONS

Above 573°C, α-quartz changes to the hexagonal β-quartz structure. Samples that have been heated above 573°C will return to the α-quartz structure upon cooling but most likely will be electrically twinned. This structural phase transition and the resulting twinning rule out the possibility of growing quartz from the melt for electronic applications. It also requires that electrodiffusion (see Section 1.2.3.3) must be carried out at temperatures below 573°C.

Both optical (Brazilian) and electrical (Dauphiné) twins occur in α-quartz. In an optically twinned sample both right-handed and left-handed regions are present. Since creation of an optical twin requires the breaking of the strong Si–O bonds, it is a growth defect. On the other hand, electrical twinning is the existence of regions of the same stone with the same handedness but reversed X axes. Only slight atomic displacements are required to go from one electrical twin to the other, and it is not necessary to break Si–O bonds. Electrical twinning can occur as a growth defect or can be produced when the sample is cycled through the α–β transition. Also, electrical twins have been induced by laser heating and by stress (Newnham et al., 1975; Anderson et al., 1976; Anderson et al., 1977). Polarized light (Cady, 1964) and etching techniques (Willard, 1946; Cady, 1964) can be used to identify optical and electrical twinning.

Because of projected shortages of high-quality natural stones during World War II, attempts were made to detwin quartz (Thomas and Wooster, 1951). With the current availability of large cultured quartz stones today, twinning does not seem to be a serious problem for the crystal manufacturers, although stress- or thermal-induced twinning may occur during the more advanced crystal fabrication processes.

1.2.2 Crystal Growth and Extended Defects

In Germany during World War II, R. Nacken investigated the possibility of hydrothermally growing quartz crystals suitable for electronic applications. His work was continued at several American laboratories after the war, and this led to the commercial production of cultured quartz in the United States beginning in 1958. By 1971, the use of cultured quartz had surpassed that of natural crystals, and today, cultured quartz has replaced natural quartz in nearly all electronic applications. Although the hydrothermal technology employed today is not unlike that of the original production plants (Laudise and Sullivan, 1959), continuing investigations and development in this field have resulted in greatly improved quartz material.

Publications discussing the evolution of the technology outside the United States are also available (Rabbetts, 1967; Regreny and Autmont, 1970; Regreny, 1973; Yoda, 1972) and indicate the international nature of the quartz-growing industry.

In the hydrothermal growth process (Laudise and Nielsen, 1961), crystallization proceeds via the creation within an autoclave of a lower zone at elevated temperature in which the quartz nutrient (natural crystals) dissolves and a lower-temperature zone at the top of the autoclave in which crystallization onto seed plates occurs. A small research-size autoclave is illustrated in Fig. 1-2, and two commonly used orientations for seed plates are shown in Fig. 1-3. The desired temperature profile throughout an autoclave is achieved by heater positioning and control and by the use of a baffle plate separating the two zones. Either a sodium hydroxide or a sodium carbonate solution is used to dissolve the quartz nutrient in the lower zone and to transport material to the cooler seeds in the upper zone. Growth temperatures are usually 340–350°C, the pressure is in the range 1.0–1.3×10^8 Pa (i.e., 15,000 to 20,000 psi), and the temperature gradient is 5–30°C.

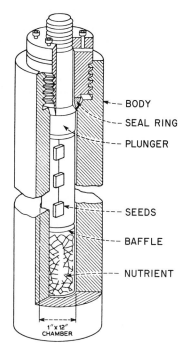

FIG. 1-2 A research-sized autoclave for use in hydrothermal crystal growth. (Courtesy of A. F. Armington.)

1 PROPERTIES OF PIEZOELECTRIC MATERIALS

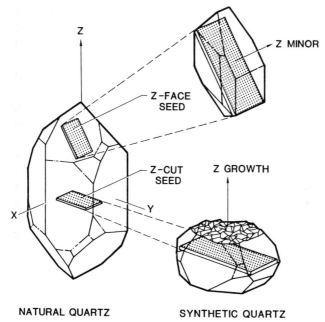

FIG. 1-3 Orientations of two seed plates commonly used in synthesizing quartz.

The initial growth-development programs in the 1950s and early 1960s provided solubility data and determined the general effects of varying growth temperature, temperature gradient, solution density, and seed orientation. It was quickly discovered that final product quality was strongly dependent on growth rate, and thus, early crystal-growing efforts attempted to optimize all of these parameters subject to the constraint of high-production output.

1.2.2.1 IMPROVEMENTS IN QUARTZ GROWTH

King et al. (1962) greatly improved the mechanical Q of cultured quartz by simply adding small amounts of lithium salts to the hydrothermal growth solution. In a subsequent study, Laudise et al. (1965) demonstrated that although the Q is improved markedly when lithium is added to the growth solution, no significant increase of Li^+ in the grown material occurs. The lack of correlation with Li^+ concentration suggested that Q at room temperature depends upon some other impurity, such as hydrogen. Infrared absorption data at 2.86 μm did indeed show that the addition of lithium to the solution keeps hydrogen, in the form of either OH^- or H_2O, from being included in the quartz (Ballman *et al.*, 1966). It is more generally believed that adding the lithium to the growth solution minimizes the incorporation of all other impurities in the crystal and gives a much more uniform stone.

Another significant step in improving cultured quartz was the introduction of "Premium Q Grade" material by Sawyer Research Products (Eastlake, Ohio) (Capone *et al.*, 1971). This material had Q values greater than 2.5×10^6 and was more resistant to radiation than any other untreated quartz of the time. The important features of the growth of the Premium Q Grade quartz were selection of good raw material, addition of lithium to the growth solution, a slow growth rate, and careful control of absolute temperature as well as the temperature gradient between the dissolving and growth zones. Strict adherence to these conditions resulted in cultured quartz well suited for use in high-precision devices.

A systematic search for conditions under which high-Q quartz ($Q > 10^6$) can be grown at high rates (more that 2.5 mm/day) was made by Lias *et al.* (1973). They found that such results could be achieved by increasing the growth temperature, the important factor being that the solid solubility of the protons that cause the acoustic loss in quartz decreases as the growth temperature increases. However, the conventional wisdom today remains that the highest-Q quartz ($Q > 2 \times 10^6$) is grown slowly (i.e., less than 0.7 mm/day). The growth rate for quartz is defined as the increase in thickness of the stone, including both sides, per unit time in the direction perpendicular to the seed plate.

To effect economy in the fabrication of AT-cut resonators, Barns *et al.* (1976) determined optimum growth conditions and procedures for commercial production of quartz from minor rhombohedral (01$\bar{1}$1), or z-face, seed plates. They found that good quality quartz ($Q > 10^6$) can be grown routinely from this seed orientation at rates near 0.9 mm/day. Cracking as a result of internal strain was a problem and accounted for as much as a 15–20% yield loss. However, these losses were minimized by careful selection of low-strain seeds using a novel polariscope developed specifically for quartz. They observed that defects in seed plates usually propagated into the grown quartz, and they found that strain increased with growth rate.

Attention has been devoted to the importance of long-term raw material supplies for use in the hydrothermal crystallization of quartz. The synthetic quartz industry in the past has used small pieces (2–3-cm size) of natural quartz obtained from vein deposits in Brazil, but the tenuous nature of this single-supply situation prompted a development program for alternatives. The resulting search and evaluation efforts (Kolb *et al.*, 1976) uncovered several quartz sources in the United States, including a number of North American pegmatitic quartz regions. Sand and silica glass were also used as nutrient, and although with the appropriate modification of the process variables some success was achieved, difficulties persist in growing production-size crystals in this manner.

Barnes et al. (1978) investigated the procedures necessary to grow low-dislocation and dislocation-free quartz (see also Section 1.2.2.2). They found that dislocations in new-growth regions propagate from preexisting dislocations in seeds and from particulate inclusions incorporated by the growing crystal. Thus, selection of dislocation-free seeds, careful seed preparation, and avoidance of particulate inclusions are necessary conditions for minimizing dislocation densities. The use of noble-metal-lined autoclaves greatly reduced inclusions of alkali iron silicates (e.g., tuhualite or acmite). Croxall et al. (1982) and Baughman (1982) reported successful growth of high-quality quartz beginning with fused silica instead of natural crystalline material. In their work, Croxall et al. (1982) converted high-purity fused silica to α-quartz by heating in an autoclave for 24 h at normal operating conditions. They then inserted carefully selected seeds into the autoclave and carried out a normal growth run. The resulting crystals had an aluminum content less than 0.1 ppm and a dislocation density ≤ 10 lines/cm^2.

A major technological development in many of the present quartz-production facilities has been the introduction of computerized control of the growing process. Rudd et al. (1969) reported on such an application. Special algorithms, based on operating conditions known to produce good crystals, permit the digital system to achieve a quasi-analog control over the autoclave during the growing process. Pressure and temperature conditions are, of course, monitored, but by additionally changing the temperature during the growth cycle, the growth rate is changed, and an optimization of Q throughout the growth region can be effected.

Armington et al. (1981) reported successful operation of a completely computerized hydrothermal growth facility established by the Air Force at Hanscom AFB, Massachussetts. This is essentially a research operation and is primarily being used to investigate growth conditions for the production of high-quality, low-drift, radiation-tolerant quartz. Initial emphasis is being placed on improving the purity of quartz by modification of growth conditions and nutrients, by examining the effects of seed quality, and by using noble-metal liners.

1.2.2.2 NATURE OF EXTENDED DEFECTS

Dislocations and fault surfaces are the dominant structural defects found in quartz. Both have been widely studied by a variety of experimental techniques, and a number of correlations between techniques have been made.

Especially important are the studies that relate the concentration of extended defects to resonator performance.

Arnold (1957) discovered that etching of cultured quartz in 48% hydrofluoric acid created tunnels that were very deep compared to their width. Nielsen and Foster (1960) also observed these tunnels resulting from etching and determined that they tend to lie in the general direction of growth. Dislocations were suggested as a possible origin of the tunnels. This was verified by Hanyu (1964) who compared etch tunnels observed optically with x-ray diffraction topographs and concluded that the etch tunnels occurred by preferential etching along dislocations.

The large-scale linear defects (i.e., dislocations that are prominent in the x-ray topographs and that form the long, narrow etch tunnels) were studied in more detail by Spencer and Haruta (1966) and Lang and Miuscov (1967). Using x-ray diffraction topography, they found that most of these dislocations originate at defect sites or inclusions located on the seed surface. Usually these dislocations are surrounded by large numbers of impurities, with the result that x-irradiation causes distinct visible coloration in those regions of quartz having large dislocation densities. None of the dislocations make more than a 30° angle with the local growth direction. In Z-growth material, however, exact alignment with the [0001] direction does not occur; the distribution of orientations makes a cone whose axis is [0001] and halfangle is 10°.

Iwasaki (1977) further correlated the etch tunnels with x-ray topographic images. Figure 1-4 shows this correspondence for $+X$-growth quartz. Spencer and Haruta (1966), in measurements on AT-cut, 5-MHz, fifth-overtone resonators, determined that high densities of dislocations (10^3–10^4 lines/cm^2) as determined by x-ray topography correlated with high acoustic loss ($Q \approx 10^5$). However, when the Q varied by a few percent, the difference in defect density was not discernible. Furthermore, this relation between Q and defect density does not hold when various lithium salts are added to the growth solution. In this latter case the Q can be increased by a factor of two or three with no perceptible decrease in defect density.

Balascio and Lias (1980) suggested that a relative measure of the anticipated mechanical strength of quartz is the etch-tunnel density. In a series of growth runs in production autoclaves, they correlated the etch-tunnel density with specific process variables and their regulation. Control of the initial seed-crystal interface and maintenance of a uniform growth rate throughout a run were found to greatly affect the formation of etch tunnels. Armington *et al.* (1980) used x-ray topography to observe linear defects (i.e., dislocations) and the resulting strain in crystals grown on different seed orientations. The density of dislocations was found to vary significantly with the orientation of the seed plate, being a minimum for Z seed plates.

1 PROPERTIES OF PIEZOELECTRIC MATERIALS 31

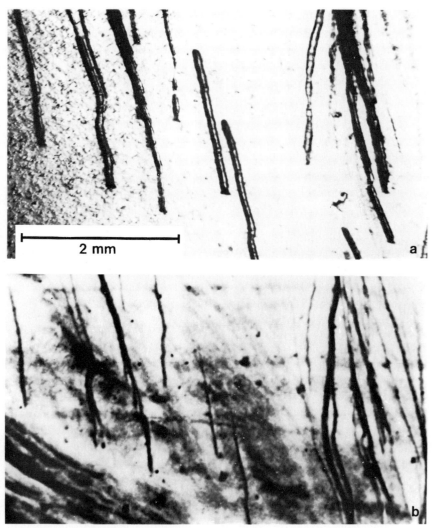

FIG. 1-4 Correspondence between etch tunnels and x-ray topographic image in +X region: (a) optical photograph and (b) x-ray topograph. [From Iwasaki (1977).]

Correlated aluminum impurity concentration measurements showed that increased strain enhances the incorporation of impurities in quartz.

Lang and Miuscov (1967) observed fault surfaces within cultured quartz, indicating cellular growth. The "cobbles" on the (0001) surfaces of quartz stones are the external manifestation of the cellular growth and the outcrops of the fault surfaces coincide with the grooves between cobbles and thus the cell boundaries. Cell diameters range up to 2 mm, and impurities tend to segregate preferentially in the cell walls. Additional studies of fault surfaces along with investigations of sub-boundaries within a growth region and growth sector boundaries have been reported (Homma and Iwata, 1973; Yoshimura and Kohra, 1976; Iwasaki and Kurashige, 1978; Iwasaki, 1980). Bye and Cosier (1979) used an x-ray double-crystal topographic technique to correlate the presence of growth striations and sub-boundaries with higher equivalent series resistance of 1.4-MHz resonators.

Moriya and Ogawa (1978, 1980) used light-scattering tomography to study growth defects in cultured quartz and compared the results with infrared absorption and x-ray topographic measurements. The optical technique probes inhomogeneities in the index of refraction caused by the irregular arrangement of atoms in areas of high impurity concentration or in the area surrounding dislocations. X-ray topography also probes these irregular arrangements of atoms by sensing the resulting strain fields but requires thin slices of crystals and more elaborate experimental apparatus. The light-scattering tomographs clearly show growth striations, growth sector boundaries and sub-boundaries, and edge dislocations.

The use of etching to produce chemically polished AT-cut resonators has been investigated by Vig *et al.* (1977). Chemical polishing is a more controlled process than mechanical polishing, and in many cases the etching reveals the presence of tunnels. These workers found that the number of etch-induced tunnels varied significantly from sample to sample in cultured material. It is noteworthy that vacuum-swept cultured quartz showed no etch tunnels. This fact suggests that the interstitial impurities, which can be removed by the electrolysis, are in large degree responsible for the formation of the etch tunnels. Also, Vig *et al.* (1977) compared the mechanical strength of chemically polished and mechanically polished AT-cut 20-MHz blanks, finding the chemically polished blanks to be superior in strength. Additional etching studies of singly and doubly rotated quartz plates have been reported by Vig *et al.* (1979).

As an important application of the etch tunnels, Kusters and Adams (1980) discovered significantly improved aging performance in crystals fabricated from blanks receiving heavy etching. The aging-rate slope in these etched crystals follows a normal pattern of aging but starts at a considerably lower initial level, and no frequency microjumps characteristic of new crystals are observed. These investigators suggest that the tunnels formed by the etch help relieve internal lattice stresses introduced earlier during crystal growth or manufacturing of the blank.

1.2.3 Point Defects

Point defects in quartz have been the subject of continuous study for over twenty-five years. The considerable progress that has been made during

1 PROPERTIES OF PIEZOELECTRIC MATERIALS 33

this time is summarized in a number of review papers (Fraser, 1968; Weil, 1975; Kahan, 1977; Griscom, 1979; Halliburton et al., 1980a).

1.2.3.1 ALUMINUM-RELATED CENTERS

Trivalent aluminum ions easily substitute for silicon and as a result require charge compensation (i.e., a 3+ aluminum ion needs an additional positive-charged entity in the lattice to compensate for the 4+ charge of the replaced silicon). Four of the aluminum charge compensators known to exist in quartz are H^+, Li^+, and Na^+ ions at interstitial sites and holes trapped at oxygen ions. One of these charge compensators is normally located adjacent to each of the substitutional aluminum ions, and this gives rise to either $Al-OH^-$, $Al-Li^+$, $Al-Na^+$, or Al-hole centers. Schematic representations of the aluminum-associated centers are given in Fig. 1-5. These defect centers can be observed by widely varying techniques: infrared absorption in the case of $Al-OH^-$ centers, acoustic loss and dielectric loss in the case of $Al-Na^+$ centers, and electron spin resonance and acoustic loss in the case of Al-hole centers.

The $Al-OH^-$ center, shown in Fig. 1-5a, is formed when an interstitial proton bonds to an oxygen ion, thus forming an OH^- molecule adjacent to a substitutional aluminum. Stretching vibrations of the OH^- molecule lead to infrared absorption, and two bands, at 3367 and 3306 cm^{-1}, have been attributed to the $Al-OH^-$ center. A number of investigations of the infrared absorption of the $Al-OH^-$ centers have been made (Dodd and Fraser, 1965; Brown and Kahan, 1975; Lipson et al., 1978, 1979; Sibley et al., 1979).

Figure 1-5b shows the $Al-M^+$ center, where M^+ represents either Li^+ or Na^+. This type of center consists of an interstitial alkali ion located adjacent to a substitutional aluminum and can give rise to one or more characteristic acoustic loss peaks because of the stress-induced motion of the alkali ion from one equilibrium position to another about the aluminum ion. An acoustic loss peak near 50 K in 5-MHz, fifth-overtone, AT-cut quartz resonators has been assigned to the $Al-Na^+$ center by King (1959) and Fraser (1964, 1968). The acoustic loss associated with quartz is discussed further in Section 1.2.3.4.

The Al-hole center (sometimes written as the $[AlO_4]^0$ center) consists of a hole (i.e., a missing electron) trapped in a nonbonding p orbital of an oxygen ion located adjacent to a substitutional aluminum, as shown in Fig. 1-5c. Nuttall and Weil (1981) showed that the ground state of the Al-hole center corresponds to trapping the hole on a long-bond oxygen, and Schnadt and Schneider (1970) showed that only 0.03 eV of energy is required to transfer the hole from one type of oxygen to the other. This means that at room temperature the hole is rapidly jumping among the four oxygens

(a)

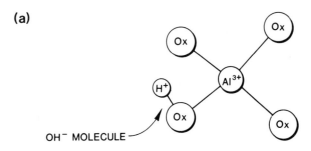

(b)

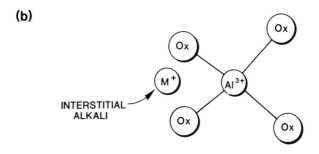

(c)

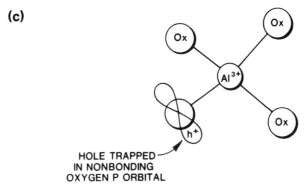

FIG. 1-5 Schematic representation of (a) the Al-OH$^-$ center, (b) the Al-M$^+$ center where M$^+$ is either Li$^+$ of Na$^+$, and (c) the Al–hole center. [From Halliburton et al. (1981).]

surrounding the aluminum ion. Markes and Halliburton (1979) investigated the production and stabilization conditions for Al–hole centers, and Koumvakalis (1980) correlated the Al–hole center ESR spectrum with a visible optical absorption (the smoky coloration). Acoustic loss peaks attributed to the Al–hole center are described in Section 1.2.3.4.

1.2.3.2 OXYGEN VACANCY CENTERS

Oxygen vacancy centers, better known as E' centers, are another class of point defects that has been extensively investigated in quartz. The model for the simplest of these defects, the E'_1 center, is an isolated oxygen vacancy having trapped a single unpaired electron. This electron is localized on only one of the two neighboring silicons and is in an sp^3 hybrid orbital extending into the vacancy. Weeks (1956) first observed the E'_1 center, and his work was later extended by Silsbee (1961). Nelson and Weeks (1960) and Arnold (1965) examined the ultraviolet optical absorption of samples containing E'_1 centers. The theoretical work of Feigl et al. (1974) and Yip and Fowler (1975) provided a clearer understanding of the electronic and ionic structure of the E'_1 center.

Two additional E'-type centers, the E'_2 and E'_4 centers, have been investigated. The E'_2 center, first discovered by Weeks (1963), and the E'_4 center, reported by Weeks and Nelson (1960), are both associated with hydrogen. Isoya et al. (1981), using a quantum chemistry computer program, showed that the model for the E'_4 center consists basically of an H^- ion trapped in the oxygen vacancy with an unpaired electron shared by the two adjoining silicons. A definitive model for the E'_2 center has not yet been established, although it most certainly must contain one or more oxygen vacancies and a proton.

Thus far, no acoustic-loss peaks in quartz have been associated with any of the E' centers.

1.2.3.3 ELECTRODIFFUSION

Quartz has large c-axis channels along which interstitial ions can migrate. King (1959), making use of this characteristic of quartz, was among the first to develop the electrodiffusion (sweeping) process as a method for changing the concentration of specific interstitial cations (i.e., H^+, Li^+, Na^+) within a given quartz crystal. Subsequent studies have shown that sweeping of the quartz prior to fabrication of resonators significantly enhances the radiation hardness of oscillators (Poll and Ridgeway, 1966; King and Sander, 1975; Pellegrini et al., 1978).

This sweeping technique consists of applying an electric field parallel to the c-axis of the crystal while maintaining the sample temperature in the 450–550°C range. Either a vacuum or an atmosphere of inert gas, air, or hydrogen surrounds the crystal. After the thermal energy frees the positive-charged species from their trapping sites, these cations are pulled along the large c-axis channels and out of the crystal by the electric field, and additional positive-charged species of a similar or different nature are taken into the crystal at the opposite electrode to maintain charge neutrality for the sample as a whole. For example, if either air or hydrogen gas surrounds the crystal, the sweeping process will remove interstitial alkali ions from the crystal and replace them with H^+ ions (Brown et al., 1980).

1.2.3.4 ACOUSTIC AND DIELECTRIC LOSS

In elastically oscillating systems, such as piezoelectric resonators, a portion of the energy in the system will be lost to internal damping forces. This anelasticity or internal friction was discussed in considerable detail by Berry and Nowick (1966) and by Wert (1966).

In a sinusoidally driven anelastic system, the stress and strain will differ in phase by an angle δ such that

$$\tan \delta = Q^{-1}, \qquad (1\text{-}32)$$

where Q^{-1} is the internal friction or loss. The quantity Q is the usual quality factor of an oscillating system. Loss can arise from intrinsic mechanisms, such as the thermal phonons, or from defects. Klemens (1965) and Mason (1965) discussed in detail the loss due to thermal phonons. In AT-cut quartz crystals, this intrinsic phonon loss is very low at cryogenic temperatures ($T < 10$ K), giving Q values as high as 10^8. As the temperature increases, the thermal phonon population grows, and the intrinsic phonon loss increases. The peak in the intrinsic phonon loss occurs near 20 K. This loss then decreases slightly with further increases in temperature and becomes nearly independent of temperature above 100 K. This behavior is illustrated by the solid curve in Fig. 1-6. Warner (1960) indicated that the maximum phonon-related Q of 15-mm diameter, 5-MHz, fifth-overtone, AT-cut blanks near room temperature is around 3×10^6.

Defects such as the Al–Na$^+$ center also cause anelastic loss. In this mechanism the defect undergoes a thermally activated reorientation that couples to an applied oscillating stress field. The induced loss (i.e., the increase in Q^{-1}) due to the reorientation is given by

$$\Delta Q^{-1} = D\omega\tau/(1 + \omega^2\tau^2), \qquad (1\text{-}33)$$

where D is the coupling factor, ω the angular frequency of the applied stress

1 PROPERTIES OF PIEZOELECTRIC MATERIALS

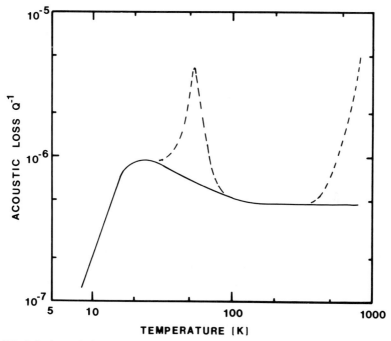

FIG. 1-6 Acoustic loss-versus-temperature spectrum of an AT-cut resonator in the as-grown state (dashed curve) and after hydrogen sweeping (solid curve).

field, and τ the relaxation time for the reorientation. Usually the temperature dependence of the relaxation time is given by

$$\tau = \tau_0 \exp(E/kT), \quad (1\text{-}34)$$

where τ_0 is the fundamental attempt time of the reorientation and E the activation energy or barrier height. The defect-associated loss ΔQ^{-1} then shows up as an absorption maximum in the Q^{-1}-versus-temperature spectrum at the temperature where the defect's hopping frequency is equal to the stress field's frequency (i.e., where $\omega\tau = 1$). This defect-associated loss mechanism also manifests itself in resonators as a frequency decrease at temperatures greater than the temperature of the absorption maximum. The magnitude of the frequency change is given by

$$\Delta f/f = Q_{\max}^{-1}/(1 + \omega^2\tau^2), \quad (1\text{-}35)$$

where $Q_{\max}^{-1}$ represents the value of Q^{-1} at the absorption maximum.

The Al–Na$^+$ center gives rise to an acoustic loss peak near 50 K in 5-MHz, fifth-overtone, AT-cut quartz blanks (King, 1959; Fraser, 1964, 1968). Euler et al. (1980) and Doherty et al. (1980) further characterized the effects of radiation and electrodiffusion on this 50-K Al–Na$^+$ acoustic-loss peak. The loss peak is observed in as-grown and in Na-swept crystals. Irradiation at room temperature removes the peak, but thermal annealing above 350°C restores it. The peak is not present in Li-swept or hydrogen-swept crystals.

Much less is known about the Al–Li$^+$ center, and no acoustic-loss peak that can be attributed to this center has been observed in high-quality quartz fabricated into AT-cut blanks. Similarly, no acoustic-loss peaks have been observed for Al–OH$^-$ or Al–OD$^-$ centers below 370 K in AT-cut blanks (Martin and Doherty, 1980). King (1959), in 5-MHz, fifth-overtone, AT-cut resonators, assigned a 100 K acoustic-loss peak and a broad acoustic loss from 125 to 160 K to the radiation-induced Al–hole center. Martin and Doherty (1980) verified these assignments and also showed that the Al–hole center has a third acoustic-loss peak at 25 K.

Above room temperature, interstitial alkali ions become thermally liberated from their trapping sites and diffuse along the c-axis channels. This diffusion causes an acoustic loss that exponentially increases with temperature (Fraser, 1964). Evidence that alkali ions are the responsible entities is obtained from hydrogen-swept quartz. Such material contains no alkali ions and does not show the high-temperature loss (Lipson *et al.*, 1981; Koehler, 1981). Figure 1-6 summarizes the acoustic-loss spectrum of an AT-cut resonator in the as-grown state and after hydrogen sweeping. The dashed curve represents the 50-K Al–Na$^+$ loss peak and the high temperature loss due to the alkali ion diffusion. These losses are removed by the sweeping, and the remaining solid curve can be described by the thermal-phonon-induced loss.

The dielectric loss associated with the Al–Na$^+$ center has been studied by Stevels and Volger (1962), Nowick and Stanley (1969), and Park and Nowick (1974). Two peaks are observed for Al–Na$^+$ centers, at 38 and 95 K for a frequency of 32 kHz, and the activation energies are 0.062 and 0.154 eV, respectively. The activation energy obtained from the 38 K dielectric-loss peak is in good agreement with the value of 0.059 eV obtained from the 50 K acoustic-loss peak (at 5 MHz) assigned to the Al–Na$^+$ center. Further study is needed before assigning a dielectric loss peak to the Al–Li$^+$ center (Nowick and Jain, 1980). Stevels and Volger (1962) and Taylor and Farnell (1964) reported possible Al–hole-center-related dielectric-loss peaks in irradiated samples. Snow and Gibbs (1964) measured the high-temperature dielectric loss due to the migration of alkali ions along the c-axis channels.

1.2.3.5 FUNDAMENTAL RADIATION RESPONSE MECHANISMS

The energy band gap of α-quartz is approximately 9 eV, and under normal circumstances the electrical transport is ionic. However, ionizing radiations

1 PROPERTIES OF PIEZOELECTRIC MATERIALS   39

(i.e., x-rays and γ rays as well as high-energy electrons and protons) create large numbers of uncorrelated electron–hole pairs that make a transient "electronic" contribution to the electrical conductivity. Hughes (1975) found that this electronic conductivity was independent of crystal direction and that it died away in 5 to 30 nsec following a pulse of radiation.

The creation of the uncorrelated electron–hole pairs by ionizing radiation is important in all insulator materials, but it is especially crucial in quartz because it leads to clearly observable effects in the behavior of interstitial cations and has major ramifications with regard to quartz device operation. In general, the radiation-induced electron–hole pairs lead to a freeing of the interstitials (i.e., H^+, Li^+, or Na^+) from their trapping sites. After being released, these interstitial cations contribute to the transient and steady-state Q^{-1} changes observed in quartz resonators (King and Sander, 1972).

The radiation-induced mobility of interstitials is the most important of the fundamental radiation response mechanisms in quartz. Not all interstitial ions behave in the same way, however. By monitoring OH^- infrared absorption bands at 77 K, Sibley et al. (1979) showed that protons can be induced to move within the lattice by ionizing radiation at temperatures as low as 10 K. Basically, there appears to be no lower-temperature limit for radiation-induced mobility of the hydrogen. In contrast, there is a critical temperature region below which interstitial alkali ions can not be induced to move by radiation. Markes and Halliburton (1979) and Halliburton et al. (1981) showed that the onset of radiation-induced mobility of the alkali interstitials occurs at approximately 200 K.

1.2.4 Thermal Properties

Expansion is the thermal property that most directly affects crystal-resonator performance. The frequency-versus-temperature characteristics of a resonator are determined almost entirely by the temperature dependence of the elastic constants and the thermal expansion of the crystal. As expected, the linear thermal expansion of α-quartz is highly anisotropic. White (1964) reported that at 283 K the coefficients of thermal expansion are 7.50×10^{-6} K^{-1} parallel to the c axis and 13.70×10^{-6} K^{-1} perpendicular to the c axis. He also measured the coefficients down to cryogenic temperatures. Touloukian et al. (1977), at the Thermophysical Properties Research Center at Purdue University, tabulated most of the available thermal expansion data. Corruccini and Gniewek (1961) also compiled and analyzed the thermal expansion data for low temperatures.

The early specific heat and thermal conductivity results for α-quartz were tabulated by Touloukian and Buyco (1970) and Touloukian et al. (1970), respectively. Specific heat, thermal conductivity, and thermal diffusivity

TABLE 1-8

Thermal Properties of α-Quartz[a]

	Temperature (K)			
	273	300	350	
Specific heat ($J\ kg^{-1}\ K^{-1}$)				
c_p	707	745	825	Touloukian and Buyco (1970)
Thermal conductivity ($W\ m^{-1}\ K^{-1}$)				
$\lambda_\parallel$	11.6	10.4	8.8	Touloukian et al. (1970)
$\lambda_\perp$	6.8	6.2	5.3	Touloukian et al. (1970)
Thermal diffusivity ($\times\ 10^{-7}\ m^2\ s^{-1}$)				
$\alpha_\parallel$	6.2	5.3	4.0	
$\alpha_\perp$	3.6	3.2	2.4	

[a] Values for the specific heat and thermal conductivity are given at three temperatures, and the thermal diffusivities are calculated from these values using the relationship $\lambda = c_p \alpha \rho$.

values taken from these tabulations are given in Table 1-8 for the temperatures 273, 300, and 350 K. Values of thermal conductivity λ and thermal diffusivity α are given for heat flow parallel and perpendicular to the c axis. The thermal diffusivities were calculated from the given specific heat and thermal conductivity values using a density of $2.65 \times 10^3\ kg/m^3$.

There have been several studies of radiation damage in α-quartz using low-temperature thermal conductivity techniques. Wasim and Nava (1979) observed a radiation-induced dip in the thermal conductivity between 5 and 7 K in natural quartz. Jalilian–Nosraty and Martin (1981) observed a similar dip in both irradiated unswept and unirradiated swept high-quality cultured quartz. They showed that the dip was a result of resonant phonon scattering by the Al–OH$^-$ centers. Laermans et al. (1980) found that very intense electron irradiations reduce the thermal conductivity of quartz and produce scattering similar to that observed in glasses. Radiation-induced increases in specific heat and decreases in thermal conductivity at low temperatures were reported by Hofacker and Loehneysen (1981).

1.2.5 Material Evaluation Techniques

The need for reliable procedures to evaluate the quality of quartz material prior to fabrication of devices is continually increasing as more and more applications are developed where sensitivity (high Q) and stability (minimum aging and radiation effects) are crucial operating criteria.

Accurate screening tests for use in material selection would greatly increase the uniformity and reliability of quartz-containing devices as well as save time and reduce costs associated with the fabrication processes. Specifically, the Q value, effectiveness of sweeping, radiation hardness (where appropriate), and mechanical strength are characteristics of the material that should be evaluated before processing is begun. A testing procedure has been implemented by the quartz-growing industry for classifying material according to Q value; however, none of the other parameters are routinely determined.

1.2.5.1 DETERMINATION OF Q VALUE

A room-temperature infrared test is presently used to determine the Q value of as-grown quartz bars. Dodd and Fraser (1965) found a correlation between bonded OH (i.e., the OH giving rise to the broad infrared absorption from 3700 to 3200 cm^{-1}) and the Q value of 5-MHz resonators operating near 100°C. They used the extinction coefficient at 3500 cm^{-1} as a measure of the bonded OH within the crystals. A further investigation of this correlation between infrared absorption and Q value was reported by Fraser et al. (1966).

Sawyer (1972) extended these earlier studies by concentrating on higher-Q material and emphasizing the precautions necessary for obtaining reliable and reproducible results. In order to minimize reflection and scattering effects at the sample surfaces, Sawyer took the extinction coefficient α at 3500 cm^{-1} to be the difference in absorptions at 3800 and 3500 cm^{-1}. His relationship between Q and α is the form

$$10^6/Q = 10^6/Q_0 + 7.47\alpha - 0.45\alpha^2, \qquad (1\text{-}36)$$

where the limiting value of Q (for $\alpha = 0$) is $Q_0 = 8.772 \times 10^6$. The extinction coefficient is measured at room temperature and is given by

$$\alpha = [\log_{10}(T_{3800}/T_{3500})]/t, \qquad (1\text{-}37)$$

where T_{3800} and T_{3500} are the transmitted light intensities at the two wavenumbers, respectively, and t is the thickness of the sample in centimeters. Although the general procedures described by Sawyer are widely used among quartz growers, the value of Q_0 and the coefficients of α and α^2 in Eq. (1-36) are often assigned slightly different values, and the specific wavenumbers at which the absorptions are measured may vary.

Brice and Cole (1978) suggested that the absorption at 3410 cm^{-1} should be used, instead of that at 3500 cm^{-1}, since it corresponds to a distinct peak and thus is self-locating. Additional factors in favor of using this line are its breadth, which minimizes problems with instrument resolution and calibration, and its lack of polarization effects. Brice and Cole (1979) suggested that the line peaking at 3585 cm^{-1} is not a good candidate for Q-value determinations because it is quite sensitive to instrument resolving power and polarization. However, the Toyo Communications Company (Kawasaki, Japan) has overcome such problems and routinely uses the 3585 cm^{-1} line for production-oriented measurements of Q (Asahara and Taki, 1972). In their procedure α is defined as

$$\alpha = [\log_{10}(T_{3900}/T_{3585})]/t, \tag{1-38}$$

and the Q value is calculated from

$$10^6/Q = 10^6/Q_0 + 7.44\alpha + 0.04\alpha^2, \tag{1-39}$$

where $Q_0 = 5.952 \times 10^6$.

It should be noted that as the Q value increases beyond 2.5×10^6, the infrared test becomes less precise. To increase the sensitivity of the test for special cases, such as Q values approaching 3×10^6, measurement of the 3410-cm^{-1} absorption peak could be made at 77 K. A relationship between this absorption and the Q value would have to be developed before routine implementation of such a test. The advantage in this latter procedure is the increase in peak absorption due to narrowing of the line at the lower temperature.

Another method for measuring the intrinsic Q value of quartz material was described by Fukuyo et al. (1977). They used a specially designed gap-type holder to measure the Q value of slim $-18.5°$-cut Y-bar resonators. All losses except those intrinsic to the material were eliminated in this procedure, and reliable Q values were obtained from a variety of quartz crystals. Sherman (1980) discussed problems in implementing this technique.

1.2.5.2 DETERMINATION OF SWEEPING EFFECTIVENESS

Sweeping quartz in an atmosphere of air or hydrogen results in removal of the interstitial alkali ions from the material, as discussed in Section 1.2.3.3. There are a number of physical characteristics of quartz that depend on the presence of these interstitial alkali ions, and measurement of these properties will provide information about the sweeping effectiveness. For example, Stevels and Volger (1962) and Park and Nowick (1974) showed that Na$^+$ in the form of Al–Na$^+$ centers can be monitored by making dielectric relaxation measurements. Similarly, the acoustic loss peak for the Al–Na$^+$ center can be monitored (King, 1959; Fraser 1968; Doherty et al., 1980). Unfortunately, neither of these two methods works for lithium.

1 PROPERTIES OF PIEZOELECTRIC MATERIALS   43

Another method of monitoring the interstitial alkali content is electrical conductivity. Hughes (1975) and Jain and Nowick (1982a,b) found significant radiation-induced conductivity that persists for long periods of time (up to several hours) following a radiation pulse. This delayed electrical conductivity is greatly reduced in swept samples and thus is attributed to interstitial alkali ions. Since the alkali concentration can vary by well over an order of magnitude from sample to sample even in the high-quality quartz, a single conductivity measurement is not sufficient to determine the percentage of alkalis removed by sweeping. Instead, one must determine the change in electrical conductivity by making measurements on the same sample before and after the sweeping. The conductivity, however, is a function of the absolute alkali-impurity content and Koehler (1981) reported on the use of post radiation-induced conductivity and high-temperature Q changes as measures of quartz radiation hardness. Since radiation-induced frequency and Q^{-1} changes are associated with the alkalis, these single conductivity measurements constitute a "measure" of sweeping effectiveness in a radiation-hardness sense.

As discussed in Section 1.2.3.5, the motion of interstitial hydrogen within the crystal is induced by radiation at all temperatures, whereas the motion of interstitial alkalis is induced by radiation only at temperatures above 200 K. Markes and Halliburton (1979) suggested a sweeping-evaluation test based on this temperature difference. Their procedure consists of using electron spin resonance (ESR) to monitor the intensity of the Al–hole center after the first and third steps of the following sequence of three irradiations: initial 77-K irradiation, room-temperature irradiation, and re-irradiation at 77 K. For a sample in which the sweeping process is complete (i.e., all the alkalis have been replaced by hydrogen), the Al–hole-center ESR spectrum will have the same intensity after the first 77-K irradiation as after the second 77-K irradiation. In the case of a partially swept sample, the ratio of the Al–hole-center ESR spectrum intensity after the first 77-K irradiation to that after the second 77-K irradiation is a sensitive indicator of the fraction of interstitial alkali ions replaced by hydrogen ions. In this procedure, the intermediate room-temperature irradiation is a crucial step, since it releases from the aluminum site any alkalis not removed during sweeping. The release of alkalis then allows additional Al–hole centers to be formed during the last irradiation. An advantage of this ESR sweeping test is that absolute results can be obtained after the sweeping is done (i.e., there is no need for a measurement before sweeping).

1.2.5.3 PREDICTION OF RADIATION HARDNESS

Both steady-state and transient frequency shifts and reductions in Q are observed following exposure of quartz resonators to ionizing radiation.

(These effects are described at length in Chapter 3 and will be discussed here only in regard to initial material selection criteria.) Aluminum ions and their charge compensators are the primary defects associated with the deleterious radiation effects in quartz. Steady-state frequency shifts can be related to the low-temperature acoustic-loss peaks of Al–Na$^+$ centers and Al–hole centers, while transient frequency shifts may be caused by the temporary dissociation of Al–OH$^-$ and Al–M$^+$ centers (King and Sander, 1972).

A prediction of the radiation hardness of a resonator must be based in considerable part on the aluminum content of the quartz. In the case of low radiation doses (<10 krad), germanium may also be an important impurity. Halliburton et al. (1981) showed that high-quality Z-growth cultured quartz can have aluminum concentrations ranging from 1 to 15 ppm (Si). Accurate measurement of aluminum concentration at this level presents considerable problems. Atomic absorption and mass spectrometry are standard analysis techniques, but they require very careful sample preparation and skillful operators to maintain the routine sensitivity needed for monitoring the low levels of aluminum in quartz.

The ESR method for determination of sweeping effectiveness, which was described in Section 1.2.5.2, also provides the aluminum concentration. In that procedure the irradiation at room temperature destroys all the Al–M$^+$ centers, which then allows the second 77-K irradiation to convert all the aluminum into Al–hole centers. By simultaneously measuring a standard reference sample containing a known number of spins, the concentration of Al–hole centers, and hence the concentration of aluminum, can be determined from the ESR spectra after this second 77-K irradiation.

Other methods for estimating the aluminum concentration in quartz are infrared absorption and dielectric loss measurements. After a room-temperature irradiation, the concentration of aluminum in the form of Al–OH$^-$ centers can be determined from the 3367- and 3306-cm^{-1} infrared bands taken at 77 K. Also, in as-grown crystals the concentration of aluminum in the form of Al–Na$^+$ centers can be determined from the corresponding dielectric-loss peak at 38 K (for a frequency of 32 kHz). However, neither of these last two methods measures all the aluminum in the crystal. The infrared measurement does not account for aluminum in the form of Al–hole centers, and the dielectric-loss measurement does not account for aluminum in the form of Al–Li$^+$ centers.

Two other techniques proposed to predict radiation hardness are measurement of radiation-induced conductivity and high-temperature resonator resistance (Q^{-1}), as mentioned in Section 1.2.5.2. Both effects, stemming from the presence of alkalis in the quartz and therefore correlating with the aluminum content, have been shown to be viable radiation hardness indices (Koehler, 1981). In the first technique, radiation frees the charge-

1 PROPERTIES OF PIEZOELECTRIC MATERIALS   45

compensating cations, and the ions are then responsible for the observed conductivity; whereas in the second technique, the thermal energy frees the cations, and the ions are then responsible for the observed acoustic-loss increases.

ACKNOWLEDGEMENTS

The authors would like to thank Drs. A. F. Armington, A. Kahan, J. C. King, and A. S. Nowick for helpful discussions and critical readings of the manuscript. The assistance of Ms. S. Bullock in final preparation of the chapter is greatly appreciated.

2 Theory and Properties of Piezoelectric Resonators and Waves

Thrygve R. Meeker

AT&T Bell Laboratories
Allentown, Pennsylvania

William R. Shreve
Peter S. Cross

Hewlett-Packard Laboratories
Palo Alto, California

	List of Symbols for Sections 2.1 and 2.2	48
2.1	Bulk Acoustic Waves and Resonators	50
	by Thrygve R. Meeker	
	2.1.1 Introduction	50
	2.1.2 Basic Quasi-Static Theory of a Piezoelectric Elastic Material	52
	2.1.3 Linear Theory	53
	2.1.4 Nonlinear Theory	56
	2.1.5 The Christoffel Plane-Wave Solutions for the Linear Quasi-Static Piezoelectric Crystal	58
	2.1.6 Thickness Modes	63
	2.1.6.1 Thickness Modes in a Plate with Infinite Length and Width and with Electric Field Parallel to the Thickness Direction (Thickness Excitation)	63
	2.1.6.2 Thickness Modes in a Plate with Infinite Length and Width and with Electric Field Perpendicular to the Thickness Direction (Lateral Excitation)	73
	2.1.6.3 Torsional Modes in Plates and Bars	75
	2.1.7 Contour Modes in Thin Plates and in Thin and Narrow Bars	77
	2.1.7.1 Extensional-Bar Mode in a Thin, Narrow Bar with Electric Field Parallel to the Bar Length	78
	2.1.7.2 Extensional-Bar Mode in a Thin, Narrow Bar with Electric Field Perpendicular to Bar Length	80
	2.1.7.3 Face-Shear Modes in Thin, Wide, Long Plates with Electric Field Perpendicular to Plate Surfaces	81
	2.1.7.4 Flexural Modes in Thin, Wide, Long Plates with Electric Field Perpendicular to the Plate Surface and Perpendicular to the Bending Axis (One Edge Clamped)	83

		2.1.7.5	Flexural Modes in Thin, Narrow Bars with Electric Field Perpendicular to the Bar Length and Parallel to the Bending Axis	87
		2.1.7.6	Modes Not Discussed in Detail	90
	2.1.8	Theory for Combined Thickness and Contour Modes		90
	2.1.9	Electrical Effects in Piezoelectric Resonators		102
	2.1.10	Equivalent Electrical Circuits for Piezoelectric Resonators		103
	2.1.11	Properties of Modes in Crystal Resonators		107
	2.1.12	Piezoelectric Materials		107
	2.1.13	Conclusion		110
2.2	Properties of Quartz Piezoelectric Resonators			110
	by Thrygve R. Meeker			
	2.2.1	Temperature Coefficient of Resonance Frequency		110
	2.2.2	Dependence of Crystal Inductance on Temperature		112
	2.2.3	Tabulations of Properties of Quartz Resonators		112
	2.2.4	Conclusion		113
	List of Symbols for Section 2.3			118
2.3	Surface Acoustic Waves and Resonators			119
	by William R. Shreve and Peter S. Cross			
	2.3.1	Introduction		119
		2.3.1.1	Background	120
		2.3.1.2	Comparison of SAWR and BAWR	123
	2.3.2	Resonator Design		126
		2.3.2.1	Grating Reflectors	126
		2.3.2.2	Transducers	130
		2.3.2.3	Cavity Design and Frequency Response	132
		2.3.2.4	Loss Mechanisms	135
	2.3.3	Fabrication		137
	2.3.4	State-of-the-Art Performance		140
		2.3.4.1	Frequency Response	140
		2.3.4.2	Stability	142
	2.3.5	Conclusion		144

LIST OF SYMBOLS FOR SECTIONS 2.1 AND 2.2

a_j	Direction cosines
A	Area
c, c_{ij}, c_{ijkl}	Elastic stiffness
$c^{c,E}, c_{ij}^{c,E}$	Elastic stiffness for thin plate
$\bar{c}, \bar{c}_{ij}, \bar{c}_{ijkl}$	Stiffened elastic constants
C_f	Free or low-frequency crystal capacitance
C_0	Clamped or high-frequency capacitance
C_1	Motional or mechanical capacitance
d, d_{ij}, d_{ijk}	Piezoelectric constant
D, D_j	Electric displacement
e, e_{ij}, e_{ijk}	Piezoelectric constant

2 PIEZOELECTRIC RESONATORS AND WAVES

$\bar{e}_{ij}, \bar{e}_{ijk}$	Normalized piezoelectric constant
e_{ij}, e_{ijk}	Normalized piezoelectric constant
$e_m^{(0)}, e_{ijk}^{(0)}$	Normal mode·piezoelectric constants
e^c, e_{ij}^c	Piezoelectric constant for contour modes of a thin plate
E, E_j	Electric field
f	Frequency
f_a	Antiresonance frequency
f_p	Parallel resonance frequency
f_r	Resonance frequency
f_s	Series resonance frequency
g, g_{ij}, g_{ijk}	Piezoelectric constant
h	Piezoelectric constant matrix or one-half the thickness, depending on the context
$h, 2h, t$	Thickness, depending on context
h_{ij}, h_{ijk}	Piezoelectric constant
I, I_j	Electric current
I_m	Moment of inertia
k_j	Wave number
$K^{(m)}, k^{(m)}, k_{ij}^{(m)}$	Piezoelectric coupling factor for the mth mode
$l, 2l$	Length, depending on context
L_1	Motional or mechanical inductance
M	Elastic moment
n	Overtone number
$P_i^{(m)}$	Power flow vector of the mth mode
Q_1	Motional or mechanical quality factor
r	Ratio of capacitances
R_1	Motional or mechanical resistance
s, s_{ij}, s_{ijkl}	Elastic compliance
S, S_j, S_{ij}	Elastic strain
t	Time or thickness, depending on context
T, T_j, T_{ij}	Elastic stress
$T_{ij}^{(0)}$	Normal mode elastic stress
u_j	Elastic displacement
$u_j^{(0)}$	Normal mode elastic displacement
$u_{n0}^{(m)}$	Elastic displacement magnitude for the mth mode
$U_i^{(0)}$	Normal mode elastic-displacement magnitude
$U_{kr}^{(j)}$	Christoffel wave amplitudes
v_n^b	Wave velocity in bars
$v^{(j)}, c^{(j)}, v_m^{(j)}$	Christoffel phase velocities
$V_n^{(m)}$	Wave velocity in plates
V	Electrical voltage
$w, 2w$	Width, depending on context
x_j	Coordinate axes
Y	Electrical admittance
Z	Electric impedance
Z_m	Mechanical impedance
β_{mn}	Dielectric impermittivity
$\beta_k^{(m)}$	Christoffel eigenvalues
Γ_{ik}	Christoffel stiffness
δ_{ij}	Kronecker delta
$\bar{\varepsilon}_{ij}$	Normalized dielectric constant

$\bar{\varepsilon}_{ij}$	Normalized dielectric constant
ε_{rs}	Dielectric constant
$\varepsilon^{c,s}, \varepsilon_{ij}^{c,s}$	Dielectric constant for contour modes of a thin plate
ρ	Density
φ	Electric potential
φ_0	One-half of applied voltage
ω	Angular frequency
Ω	Normalized frequency

2.1 BULK ACOUSTIC WAVES AND RESONATORS[§]

2.1.1 Introduction

In piezoelectric materials such as quartz, electrical current and voltage are coupled to elastic displacement and stress. This coupling makes it possible to electrically excite elastic wave motions in these materials. Confinement of this electrically excited elastic wave motion produces resonances of very high quality factor. The quality factor of any resonance is usually called its Q. The Q may be defined as the ratio of energy stored to energy dissipated per cycle at the resonance frequency. One of the distinct advantages of the piezoelectric resonator, as compared to a lumped-electrical-network resonator, for example, is that much higher Qs (10^4–10^7) are readily achievable. This uniquely high quality factor is one of the reasons for the application of piezoelectric technology in resonators for oscillators and filters.

Quartz has been a useful piezoelectric material for many years because it is relatively easy to fabricate into high-quality resonators. The quartz material is stable, and its inherent elastic anisotropy leads to resonators with very desirable dependences of resonance frequency on temperature, stress, acceleration, etc. The low piezoelectric coupling in quartz leads to restrictions on the bandwidth of crystal filters designed to operate without band-broadening termination inductors. The low dielectric constant of quartz leads to resonators with high impedances. However, the inherently small bandwidth and the high impedance have not prevented the widespread application of the quartz resonator.

Early workers (both experimental and theoretical) recognized the many possible resonance modes of the piezoelectric resonator (Cady, 1946; Heising, 1946; Mason, 1950). However, mathematical complexities made a systematic and comprehensive study of all (or even most) of the modes of a particular type of resonator very difficult. Consequently, most reports describe isolated classes of resonance modes. For any given resonator, one

---

[§] Sections 2.1 and 2.2 were written by Thrygve R. Meeker.

mode is usually desired and all the other modes are unwanted. In some cases two modes have been coupled deliberately in order to achieve a particular resonator property. In the quartz GT resonator (Mason, 1940; Heising, 1946), two plate extensional modes are coupled together by properly adjusting the length and width of the plate. The resulting resonator has a very small change in resonance frequency over a wide temperature range. The coupling of contour flexure with thickness shear in resonators is a very important practical consideration when a thickness-shear resonator is designed. This case of two coupled modes has been studied in considerable detail (Heising, 1946; Mindlin, 1974). A single, comprehensive, qualitative theory that includes many of the common modes, both wanted and unwanted, in plates and bars, was also reported (Meeker, 1977). (Some properties of the more useful of the many different kinds of quartz resonators will be discussed briefly in Section 2.2.)

In the past few years an increased interest in the nonlinear and stress- and drive-level dependences of the properties of the quartz resonator has developed. New resonator designs that allow reduced dependences of resonator frequency on applied stress or acceleration have emerged from the studies prompted by this interest. A later section of this chapter contains a discussion of some theoretical and practical aspects of nonlinear behavior of the piezoelectric resonator.

This chapter focuses attention on only one of the many ways to formulate an understanding of the properties and behavior of a dynamic physical system like a piezoelectric resonator. First, the independent and dependent parameters that describe the elastic and electric state of the piezoelectric material are defined. Second, differential equations that describe the acceleration of an infinitesimal region of the body caused by local forces and that describe the dynamic electric and magnetic field relations in the material are defined. Third, the solutions to the elastic and electric differential equations are written as space–time waves characterized by spatial wave numbers and by temporal frequencies. Fourth, dispersion relations that give the dependences of frequency on wave number making these waves be solutions to the differential equations are defined. For a particular frequency, the allowed wave numbers form a discrete infinite set of complex numbers. Fifth, combinations of the waves that satisfy the dispersion relations are combined to match elastic and electric conditions on the surfaces of the desired resonator system. The determinant of this set of boundary condition equations is called the *frequency equation*. Sixth, the roots of the frequency equation, which are the allowed resonance frequencies of the resonator, are given. These roots are usually a discrete infinite set of real numbers, if loss is not being considered in the theory. For the lossy case, the roots would be a discrete infinite set of complex numbers.

If bar-resonator theories are developed in terms of the solutions of an infinitely extended plate, then a very important class of unwanted modes can be easily overlooked—the torsions. The torsional-mode resonator also has useful properties for some applications. For this reason four types of simple modes should be considered in the design process of a resonator—two shears, the dilatations, and the torsions. All four of these types will be discussed in this chapter.

All parameters and equations in this chapter are expressed in MKS units.

2.1.2 Basic Quasi-Static Theory of a Piezoelectric Elastic Material

The basic quasi-static theory is based on definitions of four parameters that describe the elastic and electric state of the material. These basic state parameters are the elastic stress, elastic strain, electric displacement, and electric field. All parameters depend on position and time, as well as on temperature, pressure, acceleration, or other environmental conditions. The elastic strain represents the spatial variation of the elastic displacement; the electric field represents the spatial variation of the electric potential. Any two of the four parameters may be chosen to be the basic independent parameters of the theory. The remaining two parameters then become the dependent parameters. Two equations (called constitutive) give the relationships between the dependent and the independent parameters. The dependent parameters may be linear or nonlinear functions of the independent parameters. In the dynamic case an elastic differential equation describes the balance between the force on and the acceleration of an infinitesimal region of the body. An electric differential equation describes the balance between electric and magnetic fields. In the limit of zero frequency these two differential equations describe the static distribution of elastic and electric parameters. The static or equilibrium theory is not discussed in this chapter.

Specified values of the forces, elastic-particle velocities, electric potential, and electric displacement on the surface of the resonator are appropriate boundary conditions in this theory. The quasi-static description takes into account the very large difference between the velocity of the electric and elastic effects. A very accurate approximation results in which the magnetic part of the theory is totally decoupled and usually ignored. Most piezoelectric materials are nearly insulating dielectrics, and the absence of free charge in the bulk is also assumed.

The magnetic field effects may need to be retained in the theory to understand the properties of resonators operating in varying magnetic fields with very precise frequency requirements (Ballato and Lukaszek, 1980). Some

2 PIEZOELECTRIC RESONATORS AND WAVES

theoretical and experimental work on the properties of acoustic waves in semiconducting piezoelectric materials has been reported (Hutson, 1960; Hutson et al., 1961; Hutson and White, 1962; White, 1962; Kyame, 1951, 1954). The magnetic field effects and electrical conductivity effects are not discussed further in this chapter. The linearity or nonlinearity of the theory depends on the definitions of elastic strain and electric potential and on the assumed form of the constitutive relations. The linear and nonlinear theory are discussed in the next two sections of this chapter.

2.1.3 Linear Theory

The linear quasi-static differential equations, one set of particularly useful linear constitutive relations, and appropriate boundary conditions are summarized as follows. Here all quantities are tensors of the order indicated by the number of indices. Repeated indices imply summation, a comma implies differentiation with respect to the following space index, and a dot above the variable implies differentiation with respect to time. The variables of this theory are all tensors because of the way that they transform when the system coordinate frame is rotated (Love, 1944; Nye, 1960).

(1) Potentials
  (a) Elastic displacement: u_j
  (b) Electric potential: ϕ

(2) Fields
  (a) Elastic strain: $S_{ij} = \frac{1}{2}(u_{i,j} + u_{j,i})$
  (b) Electric field: $E_j = -\phi_{,j}$

(3) Constitutive relations
  (a) Elastic stress: $T_{ij} = c^E_{ijkl} S_{kl} - e_{kij} E_k$
  (b) Electric displacement: $D_j = e_{jkl} S_{kl} + \varepsilon^S_{jk} E_k$

(4) Differential equations
  (a) Newton's law for continuum: $T_{ij,i} = \rho \ddot{u}_j$
  (b) Maxwell's equation: $D_{i,i} = 0$

(5) Boundary conditions on plate surfaces
  (a) Electrical: ϕ and D_j
  (b) Mechanical: T_{ij} and u_j

Although the anisotropic descriptions of the elastic, dielectric, and piezoelectric constants in the constitutive relations have been known in detail for a long time, they are so important that is desirable to include them here for reference. The constitutive relations for the most general anisotropic material coordinate system are written in a compact matrix form as

$$\begin{bmatrix} T_1 \\ T_2 \\ T_3 \\ T_4 \\ T_5 \\ T_6 \\ D_1 \\ D_2 \\ D_3 \end{bmatrix} = \begin{bmatrix} c_{11} & c_{12} & c_{13} & c_{14} & c_{15} & c_{16} & -e_{11} & -e_{21} & -e_{31} \\ c_{21} & c_{22} & c_{23} & c_{24} & c_{25} & c_{26} & -e_{12} & -e_{22} & -e_{32} \\ c_{31} & c_{32} & c_{33} & c_{34} & c_{35} & c_{36} & -e_{13} & -e_{23} & -e_{33} \\ c_{41} & c_{42} & c_{43} & c_{44} & c_{45} & c_{46} & -e_{14} & -e_{24} & -e_{34} \\ c_{51} & c_{52} & c_{53} & c_{54} & c_{55} & c_{56} & -e_{15} & -e_{25} & -e_{35} \\ c_{61} & c_{62} & c_{63} & c_{64} & c_{65} & c_{66} & -e_{16} & -e_{26} & -e_{36} \\ e_{11} & e_{12} & e_{13} & e_{14} & e_{15} & e_{16} & \varepsilon_{11} & \varepsilon_{12} & \varepsilon_{13} \\ e_{21} & e_{22} & e_{23} & e_{24} & e_{25} & e_{26} & \varepsilon_{21} & \varepsilon_{22} & \varepsilon_{23} \\ e_{31} & e_{32} & e_{33} & e_{34} & e_{35} & e_{36} & \varepsilon_{31} & \varepsilon_{32} & \varepsilon_{33} \end{bmatrix} \begin{bmatrix} S_1 \\ S_2 \\ S_3 \\ S_4 \\ S_5 \\ S_6 \\ E_1 \\ E_2 \\ E_3 \end{bmatrix},$$

where

$$T_1 = T_{11}, \quad S_1 = S_{11}, \quad S_{ij} = \tfrac{1}{2}(u_{i,j} + u_{j,i}),$$

$$T_2 = T_{22}, \quad S_2 = S_{22}, \quad E_j = -\Phi, j,$$

$$T_3 = T_{33}, \quad S_3 = S_{33}, \quad c_{ij} = c_{ij}^E,$$

$$T_4 = T_{23}, \quad S_4 = 2S_{23}, \quad \varepsilon_{ij} = \varepsilon_{ij}^S,$$

$$T_5 = T_{13}, \quad S_5 = 2S_{13}, \quad T = c^E S - e^t E,$$

$$T_6 = T_{12}, \quad S_6 = 2S_{12}, \quad D = eS + \varepsilon^S E.$$

When numerical values are written in an equation such as this for a specific material in the most symmetrical coordinate system, some of the general elastic, electric, and piezoelectric constants are zero because of the crystal symmetry of the material being considered (Mason, 1950; Nye, 1960; Hearmon, 1961; IEEE, 1978). Equations such as this for all 32 crystal classes may be found in standards and textbooks (Cady, 1946; Mason, 1950; Berlincourt et al.; 1964; IEEE, 1978). The values of the constants all depend on the orientation of the material coordinate system with respect to the crystallographic axes of the material. All of these constants also depend on temperature, stress, etc., to some degree. Conventions for the signs of elastic and piezoelectric constants have been defined in standards (IEEE, 1978). It is very important that reports on the determination of material constants include a careful description of the conventions used to determine the constants from the measured properties of resonator and delay-line structures.

Values and temperature coefficients of material constants are usually reported for a coordinate system in which the material constants have maximum symmetry. This coordinate system might be called the *material*

2 PIEZOELECTRIC RESONATORS AND WAVES 55

coordinate system. Numerical values and temperature coefficients for the material constants of quartz were given in Chapter 1 of this volume. These temperature coefficients are derived by assuming that the material is in equilibrium at each temperature of interest. Since the crystal symmetry is independent of temperature (Nye, 1960) (when the temperature is the same everywhere in the crystal), these constants can be used in a simple linear theory to understand mode frequencies and elastic and electric mode shapes. Dimensional changes due to the changes in temperature must be included explicitly in this kind of theory, and all equations and parameter values of the theory exist in a sequence of homogeneous equilibrium thermal states. To understand the case in which the temperature is not the same in all regions of the material, a nonlinear theory has been developed in which the temperature coefficient of resonator or transducer parameters is related to the higher-order material constants and the strain induced by the temperature changes (Tiersten and Sinha, 1977, 1978a,b, 1979). This theory may also be used to understand the effects of thermal gradients that arise during the heating process itself (Holland, 1974a,b). This nonlinear theory will be discussed in more detail in Section 2.1.4. The material constants and temperature coefficients all change with coordinate system rotation like tensors of the appropriate order. For this reason the commonly used reduced index constants may only be used in the rotations by carefully observing the associated full indices. One way to use the reduced index constants is to develop algebraic equations for the rotations using the full index constants and then to abbreviate the indices (Cady, 1946; Mason, 1950). Although it is possible to describe an arbitrary coordinate system with a single three-dimensional transformation, it is usually easier to understand the generation of such a coordinate system by repeated single rotations about the appropriate coordinate axes. For practical experimental crystal orientation it is also convenient to use repeated rotations. Consequently, all recent standards describe various crystal orientations in terms of repeated single-axis rotations.

As the orientation of the coordinate system becomes more general (or the symmetry is reduced), the number of nonzero constants needed to describe the properties of plates and bars increases. These additional constants provide a greater opportunity for unwanted mode generation as well as for changes (desirable or undesirable) in the response of the resonator to external forces and other environmental changes.

The tensor properties of parameters of the simple theory account for conversion of the deceptively simple equations listed on p. 53 into a problem that has never been solved in closed form for a physically realizable three-dimensional resonator. Simple solutions for theoretical resonators with some dimensions infinite have been helpful in understanding some of the properties of real piezoelectric resonators. However, these solutions can be misleading because they obscure the great complexity of the mode spectrum of the real resonator. Furthermore, their elegant appearance sometimes makes it easy to forget the inherent approximations involved in their derivation. One way to fully appreciate the complexity of the equations summarized on p. 53 is to expand them fully by writing out all of the components.

Since no one has been able to solve the equations for the three-dimensional resonator in closed form, nearly all theoretical work on resonators involves various kinds of approximations. The discussions in later sections of this chapter are limited to the thin-plate approximation for elastic plates (Mindlin, 1955, 1961) and piezoelectric plates (Tiersten and Mindlin, 1961; Tiersten, 1969), to the simple thickness approximation in large-area piezoelectric plates (Lawson, 1941), to the length approximation in narrow, thin piezoelectric bars (Mason, 1948; Mason, 1950), to the multimode elastic-plate system (Mindlin and Spencer, 1967; Meeker, 1977), and to the multimode elastic-bar system (Lee, 1971a,b; Meeker, 1979a).

Recent applications of computer simulation of resonators using Green's function, variational, and finite element techniques are described by Holland (1968), Holland and EerNisse (1968, 1969), Allik and Hughes (1970), Cowdry and Willis), (1973), Kagawa et al. (1975), Kagawa and Yamabuchi (1976a,b, 1977), Matthaei (1978), Dworsky (1978), Tomikawa (1978), Vangheluwe (1978), and Milsom (1979). Although they are particularly useful in analyzing the modes for resonators with unusual shapes and geometries, these techniques will not be discussed further in this chapter.

2.1.4 Nonlinear Theory

Two different kinds of nonlinear resonator behavior have been investigated in considerable detail. A large dynamic excitation and an infinitesimal dynamic excitation superimposed on a large static excitation are both important practical situations. Recent work has been concentrated on developing an understanding of these two cases.

A review (Gagnepain and Besson, 1975) of theoretical and experimental work on the large finite excitation case also reported how to include the effects of loss in the theory. The very important case of a rotated Y-cut quartz resonator with high drive current has been studied in detail (Warner, 1960, 1963; Seed, 1962; Hammond et al., 1963; Gagnepain and Besson, 1975).

Much of the current work on nonlinear effects in resonators is based on earlier work on nonlinear elastic wave propagation (Thurston et al., 1966; Bateman et al., 1961; Thurston, 1964, 1965; Brugger, 1964, 1965a,b; Fowles, 1967; Graham, 1972). Fourteen third-order elastic constants for quartz were obtained from this early work and from experiments on wave propagation in samples being pressed hydrostatically or uniaxially (Thurston and Brugger, 1964; Thurston, 1965; Brugger, 1964, 1965a; McSkimin et al., 1965). Values for some third-order constants for quartz were also obtained from the dependence of resonance frequency on the value of an applied bias dc electric field (Hruska and Kazda, 1968). Some of the 23 fourth-order constants of quartz have been determined from measurements of the dependence of resonance frequency on drive level (Seed, 1962; Gagnepain and Besson, 1975). Some higher-order piezoelectric and dielectric constants of quartz have also been determined (Besson, 1974; Gagnepain and Besson, 1975).

In the finite-deformation nonlinear theory it is important to distinguish between the coordinates of a particle in the unbiased reference state and in the biased state. The basic nonlinear state parameters, constitutive equations, differential equations, and boundary conditions are as follows, when the coordinates of the unbiased reference state are chosen as the basic independent variables.

(1) Potentials
  (a) Elastic displacement: u_j
  (b) Electric displacement: ϕ

(2) Fields
  (a) Elastic strain: $2S_{ij} = (K + K^t) + (KK^t)$, where
  $K = J[(u_1, u_2, u_3)/(x_1, x_2, x_3)]$ (Jacobian)
  and K^t = transpose of K
  (b) Electric field: $E_j = -\phi_{,j}$

(3) Constitutive relations [higher-order and time-dependent (loss) terms are not shown] (comma in e subscripts separates electric and elastic indices)
  (a) Elastic stress

  $$T_j = c_{jk}S_k + \tfrac{1}{2}c_{jkl}S_kS_l + \tfrac{1}{6}c_{jklm}S_kS_lS_m - e_{mj}E_m - \tfrac{1}{2}e_{mn,j}E_mE_n$$
  $$- \tfrac{1}{6}e_{mnp,j}E_mE_nE_p - e_{m,jk}E_mS_k - \tfrac{1}{2}e_{m,jkl}E_mS_kS_l$$
  $$- \tfrac{1}{2}e_{mn,jl}E_mE_nS_l$$

  (b) Electric displacement

  $$D_h = e_{hj}S_j + \tfrac{1}{2}e_{h,jk}S_jS_k + \tfrac{1}{6}e_{h,jkl}S_jS_kS_l + e_{hm,j}E_mS_j$$
  $$+ \tfrac{1}{2}e_{hm,jk}E_mS_jS_k + \tfrac{1}{2}e_{hmn,j}E_mE_nS_j + \varepsilon_{hm}E_m$$
  $$+ \tfrac{1}{2}\varepsilon_{hmn}E_mE_n + \tfrac{1}{6}\varepsilon_{hmnp}E_mE_nE_p$$

  (c) Differential equations
  Newton's equation: $T_{ij,i} = \rho\ddot{u}_j$
  Maxwell's equation: $D_{i,i} = 0$

(4) Boundary conditions at resonator surfaces T_j and u_j; D_j and ϕ

In this case, the location of the material boundary is not known until the problem is solved, and the system variables at the boundaries must be defined in terms of the known deformation gradients and their values on the known boundary for the unbiased state. Consequently, the proper nonlinear problem involves changes from the linear problem in both the differential equations and the boundary conditions, as well as in the definitions of the independent variables and constitutive equations.

Theories for small infinitesimal dynamic excitations superimposed on a large finite elastic or electric bias have been described by Thurston (1964), Holland (1974a,b) Tiersten (1971), and Baumhauer and Tiersten (1973). The nonlinear effects often cause a small change in a corresponding linear effect, so that a perturbation on an associated linear eigenfunction expansion has been used to obtain a useful approximate theory for nonlinear behavior of quartz thickness-mode resonators.

These nonlinear theories have been applied to the development of an understanding of how the third- and fourth-order elastic constants of a material like quartz determine the response of a resonator to static or slowly varying forces applied to its boundaries, to acceleration of the mounting structure, and to thermal gradients. A new way of understanding the dependence of resonator frequency on temperature has been developed from the nonlinear equations (Tiersten and Sinha, 1978a; Sinha and Tiersten, 1978, 1979a,b).

Although many important details have not yet been studied, enough work on nonlinear effects in quartz has been reported to show how to proceed to develop an understanding of particularly useful cases.

2.1.5 The Christoffel Plane-Wave Solutions for the Linear Quasi-Static Piezoelectric Crystal

The linear differential equations listed on p. 53 have solutions corresponding to elastic plane waves travelling in various directions in the piezoelectric elastic medium. There is one plane wave for each sense of each direction. Consequently, all parameters are linear combinations of

$$\exp(j \text{ arg } +), \quad \exp(j \text{ arg } -), \quad (2.1\text{-}1)$$

or

$$\cos(\text{arg } +), \quad \sin(\text{arg } +), \quad \cos(\text{arg } -), \quad \sin(\text{arg } -), \quad (2.1\text{-}2)$$

where

$$\text{arg } + = \omega t - k_1 x_1 - k_2 x_2 - k_3 x_3,$$
$$\text{arg } - = \omega t + k_1 x_1 + k_2 x_2 + k_3 x_3,$$

2 PIEZOELECTRIC RESONATORS AND WAVES 59

ω being the angular frequency, t the time, and k_j the three wavenumbers in the x_j directions.

In the limit of no piezoelectric coefficients there are three elastic plane waves and one electric solution that is not a wave because of the quasi-static approximation being used. For the elastic system, Christoffel (1877) showed how the solution wave numbers and wave vectors depend on the direction of the propagating wave. The piezoelectric system was solved in a very similar way (Lawson, 1941).

For the piezoelectric case it is possible to separate the four characteristic equations (resulting from substituting wave solutions into the differential equations) into three equations in the elastic displacement components and one equation in the electric potential. The elastic constants in the three elastic equations are increased slightly by the piezoelectric and dielectric constants as a result of this separation process. Consequently, the only effects of the piezoelectricity are to increase the elastic constants slightly and to add a new equation involving the electric potential. This separation of variables and the resulting changes in constants are only possible when no quantities depend on the two directions perpendicular to the propagation direction of the wave. The thickness-stiffened constants only apply to this one-dimensional case. In the surface-wave formulation, for example, the system is not one-dimensional and the stiffened constants do not apply.

The Christoffel solutions for the piezoelectric case may be written compactly as

$$(\Gamma_{ik} - \rho v_r^{(j)2} \delta_{ik}) U_{kr}^{(j)} = 0. \quad (2.1\text{-}3)$$

In Eq. (2.1-3), $U_{kr}^{(j)}$ are the amplitudes of the three wave solutions for the propagation direction x_r, and $v_r^{(j)}$ are the corresponding wave phase velocities,

$$\Gamma_{ik} = a_j a_l \bar{c}_{ijkl}, \quad (2.1\text{-}4)$$

where a_j and a_l are the Cartesian components of the propagation direction and the stiffened elastic constants are given by

$$\bar{c}_{ijkl} = c_{ijkl}^E + \frac{e_{rrk}^2}{\varepsilon_{rr}^S}. \quad (2.1\text{-}5)$$

The electrical equation becomes

$$\varphi_{,rr}^{(n)} = \frac{\omega^2}{v_r^{(n)2}} \frac{e_{rrk}}{\varepsilon_{rr}^S} U_{kr}^{(n)}. \quad (2.1\text{-}6)$$

The zeros of the determinant of the coefficients of $U_{kr}^{(n)}$ in Eq. (2.1-3) are the conditions for a solution. This determinantal equation (often called the *dispersion relation*) gives the dependence of wave number on frequency so that space–time waves are solutions to the differential equations. The resulting three solutions are a quasidilatational wave with elastic displacement nearly along the propagation direction and two quasi-shear waves with elastic displacements nearly perpendicular to the propagation direction and to each other. For an isotropic material, the dilatational wave displacement is exactly along the propagation direction, and the two shear-wave displacements are exactly perpendicular to the propagation direction and to each other. The waves allowed by the Christoffel solution are combined to form solutions that satisfy appropriate boundary conditions on the surfaces of the resonator or transducer.

Although not considered very important in understanding bulk-wave resonator behavior, the direction of flow of energy or power (called the *Poynting vector* for purely electrical propagation) for each of the mode types has an important role in the performance of delay lines using anisotropic media and of surface acoustic wave (SAW) filters and resonators. In this chapter the power flow vector for elastic waves will be defined for completeness but will not be discussed further. The power flow vector for plane-wave propagation in the mth mode is given (Ballato, 1977) by

$$P_i^{(m)} = j\omega(T_{ik}^{(m)} u_k^{(m)*} - \phi^{(m)} D_i^{(m)*}), \qquad (2.1\text{-}7)$$

where the * indicates the complex conjugate. An expansion of Eq. (2.1-7) in terms of elastic, dielectric, and piezoelectric constants and the direction of wave propagation was described by Ballato (1977).

Although it is relatively straightforward to solve the three-dimensional equations for waves propagating in anisotropic piezoelectric plates, the mathematical difficulties of satisfying edge-type boundary conditions have stimulated the development of various approximate theories. The Mindlin approximation is discussed in Section 2.1.8.

In the next few sections the use of the Christoffel wave solutions in plate- and bar-resonator theories is discussed. Although the wave solutions in extensional and flexural thin and narrow bars are not plane waves, the theory for these cases is similar to that of the plate.

In considering various combinations of elastic mode type and electric field excitation direction, it is useful to use different forms of the constitutive relations listed on p. 53. Particular choices of constitutive relations often simplify the appropriate differential equations and boundary conditions. Four commonly used sets of constitutive relations (Mason, 1950; Berlincourt *et al.*, 1964) are as follows:

$$T_j = c_{jk}^E S_k - e_{mj} E_m, \qquad (2.1\text{-}8)$$

2 PIEZOELECTRIC RESONATORS AND WAVES

$$D_n = e_{nk} S_k + \varepsilon^S_{nm} E_m, \quad (2.1\text{-}9)$$

$$S_i = s^D_{ij} T_j + g_{ni} D_n, \quad (2.1\text{-}10)$$

$$E_m = -g_{mj} T_j + \beta^T_{mn} D_n, \quad (2.1\text{-}11)$$

$$T_j = c^D_{jk} S_k - h_{nj} D_n, \quad (2.1\text{-}12)$$

$$E_m = -h_{mk} S_k + \beta^S_{mn} D_n, \quad (2.1\text{-}13)$$

$$S_i = s^E_{ik} T_k + d_{mi} E_m, \quad (2.1\text{-}14)$$

$$D_m = d_{nk} T_k + \varepsilon^T_{nm} E_m. \quad (2.1\text{-}15)$$

Later sections of this chapter include discussions of the way in which different pairs of Eqs. (2.1-8)–(2.1-15) may be used in the development of an understanding of the electrical properties (impedance or admittance) of various types of crystal resonators. These discussions of resonator properties are usually simplified if a particular pair of the above constitutive equations is selected as a starting point. Since the superscripts on the elastic and dielectric constants also refer to this choice of starting equations, it is tempting to relate the superscripts to the resonator boundary conditions. However, it must be understood that Eqs. (2.1-8)–(2.1-15) are defined for the infinitely extended piezoelectric medium, and in fact all of the various constants are linearly related. All constants are relevant everywhere and at all times in the medium. The superscripts therefore refer to a choice of starting equations made for convenience and not to a basic state of the material system (Meeker, 1972).

The general form of these relations includes other effects, such as temperature gradients, heat flow, and magnetic effects. All of these other effects are considered small enough to be neglected in the discussions of this chapter. For the development of a very precise understanding of crystal-resonator behavior, it will be necessary to consider all of these neglected effects. As far as which set of constitutive relations is to be used in a particular theoretical formulation, convenience is the only guide. All of the different forms of the constitutive relations are related by simple linear transformations. Since all of the coefficients of the field quantities are linearly related and can be calculated directly from each other, there is no intrinsic reason other than convenience for using a particular set of the relations. The constitutive equations listed on p. 53 are especially useful because the differential equations are expressed in terms of elastic stresses and electric displacements. It is probably better to start with values of these coefficients and to transform them to the desired ones as needed. By doing this it is not necessary to store and retrieve values for all possible forms of the coefficients. If the tensor forms of the constitutive equations are written in matrix forms (Nye, 1960), then convenient matrix operations can be used to perform the desired transformations numerically. Thus, the matrix technique is a useful tool in computer calculations of resonator characteristics. In the following sections specific matrix transformations appropriate to several different resonator types are discussed. Various combinations of the three basic mode types (dilatation, slow shear, and fast shear) with different electrical and elastic boundary conditions lead to a very large number of different resonator types. In a given resonator design most of these modes contribute to the unwanted mode spectrum of the resonator, and a knowledge of their properties is needed to control the performance of the resonator. In the next two sections the derivation of equations for the admittance or impedance of thickness (Section 2.1.6) and contour (Section 2.1.7) modes in crystal resonators is discussed.

Two approaches to the development of an analytical understanding of a thickness-mode resonator are commonly used (Meeker, 1972). In the first approach, often used in the development of resonator theories, the boundary conditions are specifically defined at the beginning of the analysis. These boundary conditions lead to relationships among the basic material constants, and each new set of boundary conditions leads to a redefinition of the relevant material constants. In the second approach, often used in the development of transducer theories, the boundary conditions are left generally defined and appear in the solutions as forces and particle velocities on the surfaces of the resonator body. This second approach was used to develop the first exact equivalent circuit for a piezoelectric crystal resonator (Mason, 1948). If the appropriate boundary conditions are used in the solution derived by the second approach, then the results are of course the same as those found in the first approach. Although less general, the first approach is often used for the resonator case because it is simpler.

In this chapter the second approach will be described for the thickness-mode case, and the concept of a transducer impedance matrix will be developed. Only a brief discussion and the solution will be presented here. The first approach will be used to derive electrical impedance and admittance relations for the various resonator types considered.

In both of these approaches, orthogonal normal modes can be used to reduce the complexity of the derivations. The thickness modes in a plate with lateral-electric-field excitation will be analyzed using the normal mode technique as an illustration. The other modes considered in this chapter will be analyzed without using the normal modes. In Section 2.1.10 the associated equivalent electrical circuits are discussed. Specific equations for the properties of various modes in crystal resonators are reviewed in Section 2.1.11.

2 PIEZOELECTRIC RESONATORS AND WAVES　　　　　　　　63

2.1.6 Thickness Modes

2.1.6.1 THICKNESS MODES IN A PLATE WITH INFINITE LENGTH AND WIDTH AND WITH ELECTRIC FIELD PARALLEL TO THE THICKNESS DIRECTION (THICKNESS EXCITATION)

In this section the very important case of an infinite plate with electric field and electric displacement parallel to the thickness direction is discussed in detail.

For a plate with infinite length and width, two important cases can be considered. In the first case the plate surfaces are cut perpendicular to a propagation direction for which the three basic mode types (two shear and one dilatation) are uncoupled from each other. In this case three independent single-mode solutions can be formulated (Meeker, 1972). In the second more general case the plate surfaces are cut perpendicular to an arbitrary propagation direction in the anisotropic material, and the three basic modes are all coupled together by the differential equations and boundary conditions (Ballato, 1972a,b). Even in the anisotropic case, some plates for some materials can be oriented to have simple uncoupled modes. In piezoelectric plates the electrical coupling to each mode type also depends on the orientation of the plate. Consequently, some plates have basic modes that are not very strongly excited electrically (not at all in the ideal case). However, electrically inactive modes may be excited *elastically* by coupling to electrically active modes at the plate edges or *electrically* by fringing electrical fields with components in the exciting directions. Both kinds of coupling may be enhanced by nonlinear effects.

The following procedure (second approach) is useful in setting up a linear theory for the general case of an anisotropic plate of arbitrary orientation. First, a desired set of constitutive relations is selected. Then the relationship is determined between wave number and frequency (dispersion relation) so that waves satisfy the differential equations. Linear combinations of the three wave solutions for the same frequency are used to match the values of elastic stress, particle velocity, electric potential, and displacement at the two plate surfaces. The linear constants are then eliminated from these boundary condition equations. This elimination process allows the elastic stress (or force) and particle velocity components on one surface to be expressed in terms of the stress (or force) and particle velocity on the other surface and on the electric potential and electric displacement components. In this case the convenient constitutive relations are Eqs. (2.1-8) and (2.1-9).

This procedure leads to a result that can be expressed as

$$\begin{bmatrix} F_5^{0+} \\ F_4^{0+} \\ F_3^{0+} \\ F_5^{0-} \\ F_4^{0-} \\ F_3^{0-} \\ V^0 \end{bmatrix} = \begin{bmatrix} \dfrac{z_1}{j\tan\theta_1} & 0 & 0 & \dfrac{z_1}{j\sin\theta_1} & 0 & 0 & \dfrac{n_1}{j\omega C_0} \\ 0 & \dfrac{z_2}{j\tan\theta_2} & 0 & 0 & \dfrac{z_2}{j\sin\theta_2} & 0 & \dfrac{n_2}{j\omega C_0} \\ 0 & 0 & \dfrac{z_3}{j\tan\theta_3} & 0 & 0 & \dfrac{z_3}{j\sin\theta_3} & \dfrac{n_3}{j\omega C_0} \\ \dfrac{z_1}{j\sin\theta_1} & 0 & 0 & \dfrac{z_1}{j\tan\theta_1} & 0 & 0 & \dfrac{n_1}{j\omega C_0} \\ 0 & \dfrac{z_2}{j\sin\theta_2} & 0 & 0 & \dfrac{z_2}{j\tan\theta_2} & 0 & \dfrac{n_2}{j\omega C_0} \\ 0 & 0 & \dfrac{z_3}{j\sin\theta_3} & 0 & 0 & \dfrac{z_3}{j\tan\theta_3} & \dfrac{n_3}{j\omega C_0} \\ \dfrac{n_1}{j\omega C_0} & \dfrac{n_2}{j\omega C_0} & \dfrac{n_3}{j\omega C_0} & \dfrac{n_1}{j\omega C_0} & \dfrac{n_2}{j\omega C_0} & \dfrac{n_3}{j\omega C_0} & \dfrac{1}{j\omega C_0} \end{bmatrix} \begin{bmatrix} v_1^{0+} \\ v_2^{0+} \\ v_3^{0+} \\ v_1^{0-} \\ v_2^{0-} \\ v_3^{0-} \\ I^0 \end{bmatrix}, \quad (2.1\text{-}16)$$

where

$$F_j^{0\pm} = AT_j^0(x_3 = \pm h), \qquad I^0 = -A\dot{D}_3^0, \qquad n_m = Ae_m^0/2h,$$

$$v_k^{0\pm} = \pm \dot{u}_k^0(x_3 = \pm h), \qquad A = (2l)(2w), \qquad Z_m = A\sqrt{\varrho c^{(m)}},$$

$$V^0 = \int_{-h}^{h} E_3^0 \, dx_3, \qquad C_0 = \dfrac{A\varepsilon_{33}^s}{2h}, \qquad \theta_m = 2h\omega\sqrt{\varrho/c^{(m)}},$$

and $c^{(m)}$ are three eigenvalues of the Christoffel c_{3jk3}. In this equation the impedance matrix Z^0 expresses the linear relation between the generalized forces ($F_j^{i(0)}$, electrical voltage and elastic force) and the generalized flows ($v_j^{(0)}$, electrical current and elastic particle velocity) (Ballato, 1972a,b; Ballato et al., 1974). Here the impedance matrix refers to the voltage, force, current, and velocity of the normal modes of the transducer system. The concept of these normal modes is developed further in Section 2.1.6.2 for the thickness mode with lateral excitation. The impedance matrix Z^0 in the laboratory coordinate system (in which the values of these four parameters are observed) is found by a similarity transformation (Ballato, 1972a,b; Ballato et al., 1974; Hadley, 1961) developed from the eigenvectors of the

2 PIEZOELECTRIC RESONATORS AND WAVES

Christoffel solution:

$$[B] = \begin{bmatrix} \beta_1^{(1)} & \beta_1^{(2)} & \beta_1^{(3)} & 0 & 0 & 0 & 0 \\ \beta_2^{(1)} & \beta_2^{(2)} & \beta_2^{(3)} & 0 & 0 & 0 & 0 \\ \beta_3^{(1)} & \beta_3^{(2)} & \beta_3^{(3)} & 0 & 0 & 0 & 0 \\ 0 & 0 & 0 & \beta_1^{(1)} & \beta_1^{(2)} & \beta_1^{(3)} & 0 \\ 0 & 0 & 0 & \beta_2^{(1)} & \beta_2^{(2)} & \beta_2^{(3)} & 0 \\ 0 & 0 & 0 & \beta_3^{(1)} & \beta_3^{(2)} & \beta_3^{(3)} & 0 \\ 0 & 0 & 0 & 0 & 0 & 0 & 1 \end{bmatrix},$$

from which

$$[T] = [B][T^0], \qquad [Z] = [B][Z^0][B^t],$$
$$[v] = [B][v^0], \qquad [B^t][B] = [I].$$

For the single-mode system, the laboratory parameters are already normal modes, and the development of the impedance matrix is considerably simpler. The conditions on mode type for which the single-mode analysis is exact were reported by Meeker (1972). These conditions are the same as the conditions for pure mode propagation along the plate normal (i.e., y direction in rotated Y-cut quartz) and for which the $\beta_k^{(i)}$ matrix is diagonal.

Several important features of Eq. (2.1-16) will now be discussed briefly. First, the matrix elements depend only on the four quantities

$$\theta_m = (2h)\omega\sqrt{\rho/c^{(m)}}, \tag{2.1-17}$$

$$Z_m = A\sqrt{\rho c^{(m)}}, \tag{2.1-18}$$

$$C_0 = (A/2h)\,\varepsilon_{33}^S, \tag{2.1-19}$$

$$n_m = (A/2h)\,e_m^{(0)}, \tag{2.1-20}$$

where ρ is the material density, A the plate area, $c^{(m)}$ are the three wave velocities from the Christoffel solution of the differential equations

$$e_m^{(0)} = \beta_k^{(m)} e_{33k},$$

$\beta_k^{(m)}$ the eigenvectors of the Christoffel solution, and Z_m are elastic-wave impedances. Second, each elastic stress and associated particle velocity and the electric field and electric displacement can be considered a port of the transducer or resonator system. These equations then represent the effect of exciting one port on the output of the other ports. If the equations are constituted as in Eq. (2.1-16), then these equations are impedance equations for a general multiport network. This kind of network is very familiar to the electrical circuit designer, and many circuit ideas can be applied to the development of an understanding of the piezoelectric resonator or transducer. (This circuit concept will be developed further in a later section of this chapter on equivalent electrical circuits for the piezoelectric transducer or resonator.) Third, the conditions for zero electrical excitation of a particular port or mode can be defined as the zero element value at the appropriate place in the matrix in Eq. (2.1-16). Plate orientations that diagonalize the relations that follow Eq. (2.1-16) have simple uncoupled thickness modes. Each of the uncoupled solutions corresponds to the simple infinite-plate solution mentioned at the beginning of this section. Fourth, every parameter in the relations that follow Eq. (2.1-16) has a real and an imaginary part. The impedance matrix has two times the dimension indicated. This computational size is then 14×14—four parameters for each of the three elastic mode types and two parameters for one of the two electrical parameters.

In the next part of this section, an expression which may be solved for the resonance and antiresonance frequencies is derived using the first approach (Tiersten, 1963, 1969a, 1970). As mentioned at the end of Section 2.1.5. the boundary conditions are defined at the beginning of the analysis and are used to define specific combinations of material constants for the particular problem being considered.

The plate thickness is in the x_3 direction. The plate length and width are large enough so that the derivatives of all parameters P with respect to x_1 and x_2 are zero (i.e., $P_{,1} = P_{,2} = 0$). The electrodes are on the plate surfaces at $x_3 = \pm h$. Since $E_i = -\varphi_{,i}$, the electric fields E_1 and E_2 are zero. Variables S_1, S_2, and S_6 are zero. The constitutive relations reduce to

$$T_j = c_{j3}u_{3,3} + c_{j4}u_{2,3} + c_{j5}u_{1,3} - e_{3j}E_3, \qquad (2.1\text{-}21)$$

$$D_i = e_{i3}u_{3,3} + e_{i4}u_{2,3} + e_{i5}u_{1,3} + \varepsilon_{i3}E_3. \qquad (2.1\text{-}22)$$

where c_{ji} stands for c_{ji}^E, ε_{i3} stands for ε_{i3}^S, and i and j are 1, 2, and 3. For plane-wave solutions propagating in the x_3 direction.

$$u_3 = u_{n0}\exp[j(\omega t - k_3 x_3)], \qquad (2.1\text{-}23)$$

and the differential equations become the characteristic equations

$$\bar{c}_{55}u_{10} + \bar{c}_{54}u_{20} + \bar{c}_{53}u_{30} = \rho V_S^2 u_{10}. \qquad (2.1\text{-}24)$$

$$\bar{c}_{45}u_{10} + \bar{c}_{44}u_{20} + \bar{c}_{43}u_{30} = \rho V_S^2 u_{20}. \qquad (2.1\text{-}25)$$

$$\bar{c}_{35}u_{10} + \bar{c}_{34}u_{20} + \bar{c}_{33}u_{30} = \rho V_S^2 u_{30}. \qquad (2.1\text{-}26)$$

where $V_3^2 = \omega^2/k_3^2$ and the electrical differential equation

$$\varepsilon_{33}^S \varphi_{,33} = e_{33}u_{3,33} + e_{34}u_{2,33} + e_{35}u_{1,33} \qquad (2.1\text{-}27)$$

has been used to remove φ or E_3 from the three elastic equations. This exact substitution, only possible if the problem is one-dimensional, puts the total electrical effect into the definitions of the stiffnesses $\bar{c}_{ij}$, so that

$$\bar{c}_{ij} = c_{ij}^E + \frac{e_{3i}e_{3j}}{\varepsilon_{33}^S}. \tag{2.1-28}$$

Equations (2.1-24)–(2.1-26) are an eigenvalue problem, and three values of V_S^2, i.e.,

$$V_3^{(1)2}, \quad V_3^{(2)2}, \quad V_3^{(3)2}, \tag{2.1-29}$$

are the solutions. Corresponding to each $V_3^{(m)2}$ is a solution vector with components

$$u_{10}^{(m)}, \quad u_{20}^{(m)}, \quad u_{30}^{(m)}. \tag{2.1-30}$$

The three solution vectors are orthogonal. Equation (2.1-27) can be integrated directly to give the potential as

$$\varphi = \bar{e}_{33}u_{30} + \bar{e}_{34}u_{20} + \bar{e}_{35}u_{10} + K_1 x_3 + K_2. \tag{2.1-31}$$

where

$$\bar{e}_{ij} = e_{ij}/\varepsilon_{33}^S. \tag{2.1-32}$$

The boundary conditions are

$$T_3 = T_4 = T_5 = 0, \tag{2.1-33}$$

$$\varphi = \pm \varphi_0 \tag{2.1-34}$$

on the surfaces of the plate at $x_3 = \pm h$. A general solution to the above differential equations and boundary conditions is

$$u_{j0} = \sum_{n=1}^{3} P^{(n)}\beta_j^{(n)}S^{(n)} + \sum_{m=1}^{3} Q^{(m)}\beta_j^{(m)}C^{(m)}, \tag{2.1-35}$$

$$\varphi = \sum_{n=1}^{3} P^{(n)}\bar{e}_{3k}\beta_k^{(n)}S^{(n)} + \sum_{m=1}^{3} Q^{(m)}\bar{e}_{3k}\beta_k^{(m)}C^{(m)}. \tag{2.1-36}$$

where

$$S^{(n)} = \sin k_3^{(n)}x_3,$$

$$C^{(m)} = \cos k_3^{(m)}x_3,$$

$\beta_j^{(n)}$ are three components of each of the three eigensolutions of Eqs. (2.1-24)–(2.1-26),

$$K_2 = -\sum_{m=1}^{3} Q^{(m)}\bar{e}_{3k}\beta_k^{(m)}C^{(m)}, \tag{2.1-37}$$

and

$$K_1 h = \varphi_0 - \sum_{n=1}^{3} P^{(n)} \bar{e}_{3k} \beta_k^{(n)} S^{(n)}. \quad (2.1\text{-}38)$$

When Eqs. (2.1-35) and (2.1-36) are substituted into the six stress boundary conditions in Eq. (2.1-33) (three conditions for each of the two surfaces), two uncoupled sets of equations result; one set in $P^{(1)}$, $P^{(2)}$, and $P^{(3)}$ coupled to φ^0 and the other set in $Q^{(1)}$, $Q^{(2)}$, and $Q^{(3)}$ not coupled to φ^0. The electrically coupled equation corresponding to the antisymmetric solution is

$$\sum_{n=1}^{3} P^{(n)} \beta_k^{(n)} [\bar{c}_{j3k3} k_n^{(n)} h C_h^{(n)} - e_{3j3} \bar{e}_{3h} S_h^{(n)}] = 0. \quad (2.1\text{-}39)$$

where

$$C_h^{(n)} = \cos k_3^{(n)}, \quad S_h^{(n)} = \sin k_3^{(n)} h, \quad k_3^{(n)} = \frac{\omega}{V_3^{(n)}}. \quad (2.1\text{-}40)$$

Equation (2.1-39) has a solution for the values of ω for which the determinant of the $P^{(n)}$ is zero. This cubic determinantal frequency equation (Tiersten, 1963) is

$$\beta_k^{(n)} [\bar{c}_{j3k3} k_3^{(n)} h C_h^{(n)} - e_{3j3} \bar{e}_{3k} S_h^{(n)}] = 0. \quad (2.1\text{-}41)$$

There are three frequencies for which Eq. (2.1-41) is satisfied. These three frequencies correspond to the three modes (A, B, and C) mentioned in Section 2.1.5. In Eq. (2.1-41), h is one half of the plate thickness, and the three $k_3^{(n)}$ are defined in terms of the three velocities [Eq. (2.1-40)] associated with the Christoffel determinant in Eq. (2.1-4), so that all quantities except ω are known. Values of ω that make the determinant equal to zero are the thickness resonance frequencies. These three frequencies have been called the A, B, and C thickness modes of the plate (Koga, 1932; Lawson, 1941; Koga et al., 1958; Bechmann, 1961). The eigenvectors associated with each frequency identify the mode type. For the A mode the largest eigenvector component is along the propagation direction, and the mode is dilatational (the word "extensional" is often used but should be reserved to describe the length mode in a narrow thin bar). For the B and C modes the largest eigenvector component is perpendicular to the propagation direction, and these are shear modes. The higher-frequency shear mode is called the *fast shear mode* (largest associated wave velocity) and the lower-frequency shear mode is called the *slow shear mode*.

Another form of Eq. (2.1-41) (Coquin, 1964; Yamada and Niizeki, 1970a) is

$$k_3^{(1)} k_3^{(2)} k_3^{(3)} C_h^{(1)} C_h^{(2)} C_h^{(3)} \left[1 - \sum_{n=1}^{3} \frac{K^{(n)2} T_h^{(n)}}{k_3^{(n)} h} \right] = 0, \quad (2.1\text{-}42)$$

where

$$K^{(n)2} = \frac{(\beta_j^{(n)} e_{33j})^2}{v_3^{(n)} \varepsilon_{33}^S},$$

(2.1-43)

$$T_h^{(n)} = \tan k_3^{(n)} h.$$

The admittance of the thickness-excited thickness-mode resonator may be derived directly from Eqs. (2.1-21) and (2.1-22). This process was described in detail for the single-mode case (Meeker, 1972). For the three-mode resonator only the result will be given here (IEEE, 1978).

$$Y = j\omega C_0 \left[1 - \sum_{m=1}^{3} K^{(m)2} \frac{T_h^{(m)}}{k_3^{(m)} h} \right]^{-1},$$

(2.1-44)

where

$$C_0 = (2w)(2l)\varepsilon_{33}^S/(2h),$$
(2.1-45)

$$k_3^{(m)} = \omega \sqrt{\rho/\bar{c}^{(m)}},$$
(2.1-46)

$$K^{(m)2} = \frac{[\beta^{(n)} e_{lij} a_l a_i]^2}{\bar{c}^{(m)} \varepsilon_{rs}^S a_r a_s}.$$
(2.1-47)

The a_j are the direction cosines mentioned in Eq. (2.1-4).

Equation (2.1-42) can be solved graphically (Ballato, 1977) to obtain a good understanding of how the coupling between the three modes affects the mode resonance frequencies. Figure 2.1-1 shows a typical graphical solution of Eq. (2.1-42) to illustrate the technique. The three antiresonance frequencies and electromechanical couplings used in the plot are those for SC-cut thickness-excited quartz resonators. The three antiresonance frequencies (f_a as $Z \to \infty$) are 1.0, 1977/1797, and 3380/1797 MHz (Ballato, 1977). The corresponding electromechanical couplings are 0.0449, 0.0471, and 0.0333 (Ballato, 1977). For convenience the following functions are plotted:

$$G_c = 1 - k_c^2(\tan x_c)/x_c,$$
(2.1-48)

$$T_b = k_b^2(\tan x_b)/x_b,$$
(2.1-49)

$$T_a = k_a^2(\tan x_a)/x_a,$$
(2.1-50)

where

$$x_n = (\pi/2)(f/f_{0n}),$$
(2.1-51)

and k_n is the electromechanical coupling coefficient for the nth mode. Figure 2.1-1 shows that the resonance of the lowest frequency mode ($Z = 0$) is

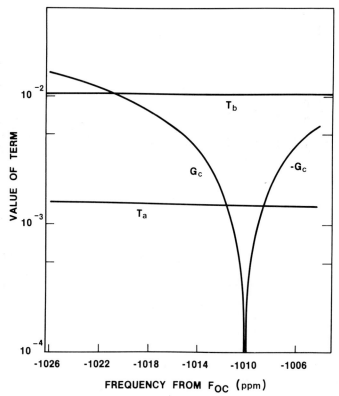

FIG. 2.1-1 Terms in electrical impedance of a thickness-excited, three-thickness-mode, SC-cut quartz resonator.

f_{0c} − 1010 Hz if the coupling to the other modes is ignored. The variable G_c is the normalized impedance of the uncoupled lowest-frequency mode. All values are plotted as logarithms to allow closely spaced values to be separated. The negative values of G_c are plotted as positive and labelled $-G_c$. If the coupling to mode B is considered, then the $Z = 0$ frequency drops to f_{0c} − 1021 Hz. Including the coupling to mode A lowers the resonance frequency to f_{0c} − 1022.5 Hz. Since f_{0c} was set at 1 MHz for the calculations and the plot, these frequency shifts are all in parts per million as well as in hertz.

Figure 2.1-1 also shows that a simple approximate equation should be quite accurate. This simple approximate equation provides a useful insight into the frequency shifts generated by the coupling to the other modes in the piezoelectric resonator. The approximate expression for Eq. (2.1-42) can be obtained if the arguments x_a and x_b are small enough at the frequency at which $k_c^2 \tan x_c/x_c$ has a value near one. In this case the important part of Eq. (2.1-42) can be written as

$$G_c - (k_a^2 + k_b^2) - (\pi^2/12)(k_a^2\alpha_a^2 + k_b^2\alpha_b^2) = 0, \qquad (2.1\text{-}52)$$

where $\alpha_a = f_{0c}/f_{0a}$, $\alpha_b = f_{0c}/f_{0b}$, and x_c is near $\pi/2$. The frequency in the B and A terms has been set equal to f_{0c}, and $\tan x_a/x_a$ and $\tan x_b/x_b$ have been replaced by the first two terms of their respective series expansions because of the small values of x_a and x_b at f_{0c}. The validity of this approximation is shown by the flatness of the T_a and T_b curves in Fig. 2.1-1. The C mode resonance frequency is shifted down as if it belonged to a single mode with an effective electromechanical coupling factor

$$k_{\text{eff}}^2 = \frac{k_c^2}{1 - (k_a^2 + k_b^2) - (\pi^2/12)(k_a^2\alpha_a^2 + k_b^2\alpha_b^2)}. \qquad (2.1\text{-}53)$$

Equation (2.1-52) shows that the lowest resonance frequency is shifted down as k_a and k_b increase and as f_{0a} and f_{0b} decrease. As α_a and α_b approach one, the simple approximation breaks down, and a more accurate solution is necessary. It should be noted that the temperature coefficient of the lowest resonance frequency depends in part on the higher-frequency-mode electromechanical coupling factors, which vary with temperature. Since each of the three modes has active odd harmonics, a frequency coincidence can occur for different harmonic orders for different modes. Any coupling between these isofrequency modes will produce frequency splitting and the resulting possibility of confusion of mode identification or of improperly assigned frequency values for material-constant determination. Designs of doubly rotated resonators operating on higher overtones must avoid these frequency coincidences at all temperatures of interest over the expected lifetime of the resonator. For fundamental-mode frequency separations as large as those in the quartz SC-cut resonator, this frequency coincidence only occurs for very high-order harmonics and can be ignored. Other cuts may not be free of this effect.

Figure 2.1-2 shows how the capacitance ratio ($r = C_0/C_1$) of the A, B, and C modes depends on the orientation angle φ of a doubly rotated quartz resonator. The value of r is related to the activity and excitation strength of a given mode, so that Figure 2.1-2 illustrates a basic problem in doubly rotated resonators, that is, the three modes are all excited and can have the same excitation strengths for some orientations. Since the frequencies of the three modes are not much different, this multiple excitation can cause serious problems in filter and oscillator applications.

The frequency separations and temperature coefficients of the A, B, and C modes depend on the crystallographic angle of the plate (Ballato, 1977).

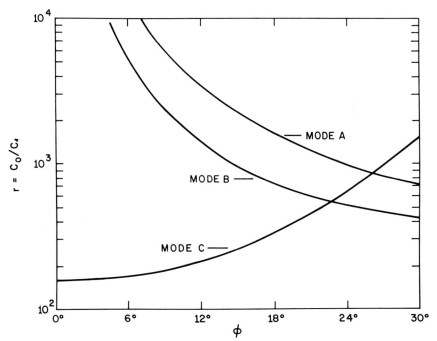

FIG. 2.1-2 Capacitance ratios for resonators using doubly rotated quartz cuts along the AT–SC locus.

For doubly rotated quartz resonators these relationships have been used to determine the orientation of the crystal plate (Miller, 1979; Warner 1981). Since the three modes have very different temperature coefficients of frequency, one mode can be used to sense the crystal temperature and provide a mechanism for temperature control while the other mode controls the frequency of a crystal oscillator (Kusters and Leach, 1978).

Although no real resonator can have a pure thickness mode because of coupling to contour modes at the plate edges, thin plates with large width and length have modes with frequencies nearly equal to the thickness-mode frequencies. The relative simplicity of the thickness-mode theory makes its use attractive for developing a basic understanding of this mode of the simple thickness resonator. The dependence of resonator frequency on electrode mass and stiffness, on plate material and thickness, and on temperature can be understood in terms of the simple thickness model. However, the existence of other modes and the mode shape itself (and therefore mode inductance) cannot be understood accurately in terms of the simple thickness model.

2 PIEZOELECTRIC RESONATORS AND WAVES

2.1.6.2 THICKNESS MODES IN A PLATE WITH INFINITE LENGTH AND WIDTH AND WITH ELECTRIC FIELD PERPENDICULAR TO THE THICKNESS DIRECTION (LATERAL EXCITATION)

If the plate thickness is along the x_3 direction of the crystal, then the infinite length and width imply that the derivative of all parameters with respect to x_1 and x_2 are zero. If the electric field is along x_1 and the x_3 surfaces of the plate are stress free and electrically open circuited (viz., unelectroded), then the three surface stresses, T_3, T_4, and T_5, and the electrical displacement D_3 are zero. Strictly considered, the existence of an electric field component E_1 and the infinite dimension along x_1 can only be consistent if the applied voltage is infinite. This troublesome singularity is removed by making the plate dimension large enough along x_1 to reduce the variation in elastic parameters along x_1 to zero but small enough along x_1 to allow the desired electric field E_1 with a finite applied voltage. Then the constitutive relations (Ballato, 1972b) become

$$T_{3j} = c^E_{3jk3} u_{k,3} + e_{33j}\varphi_{,3} - e_{13j}E_1, \tag{2.1-54}$$

$$D_3 = e_{3k3} u_{k,3} - \varepsilon^S_{33}\varphi_{,3} + \varepsilon^S_{31}E_1, \tag{2.1-55}$$

where $j = 1, 2, 3$ and $k = 1, 2, 3$.

For this one-dimensional case, integration of Eq. (2.1-6) leads to the equation for the potential as

$$\varphi = \bar{e}_{3k3} u_k + a_3 x_3 + a_1 x_1 + b_3, \tag{2.1-56}$$

where

$$\bar{e}_{3k3} = e_{3k3}/\varepsilon^S_{33}. \tag{2.1-57}$$

The definition of E_1 as $-\varphi_{,1}$ and the fact that u_k is independent of x_1 leads to a value for a_1 of

$$a_1 = -E_1. \tag{2.1-58}$$

Substituting Eq. (2.1-56) into Eq. (2.1-55), with the condition that $D_3 = 0$, leads to a value for a_3 of

$$a_3 = \bar{\varepsilon}^S_{31} E_1, \tag{2.1-59}$$

where

$$\bar{\varepsilon}^S_{31} = \varepsilon^S_{31}/\varepsilon^S_{33}. \tag{2.1-60}$$

These equations can be combined to give a set of reduced constitutive relations for this case as follows:

$$T_{3j} = \bar{c}_{3jk3} u_{k,3} - \tilde{e}_{13j} E_1, \tag{2.1-61}$$

$$D_1 = \tilde{e}_{1k3} u_{k,3} + \tilde{\varepsilon}_{11} E_1, \tag{2.1-62}$$

where

$$\bar{c}_{3jk3} = c^E_{3jk3} + (e_{33j} \bar{e}_{3k3}), \tag{2.1-63}$$

$$\tilde{e}_{1kj} = e_{1kj} - \bar{\varepsilon}^S_{31} e_{3kj}, \tag{2.1-64}$$

$$\tilde{\varepsilon}_{11} = \varepsilon^S_{11} - (\bar{\varepsilon}^S_{31} \varepsilon^S_{13}). \tag{2.1-65}$$

The expressions for T_{3j} and D_1 are substituted into the differential equations listed on p. 53 to give the Christoffel equations as

$$[\bar{c}_{3jk3} - c^{(i)} \delta_{ik}] \beta^{(i)}_h = 0. \tag{2.1-66}$$

Equation (2.1-66) can be solved for three $c^{(i)}$ and nine $\beta^{(i)}_k$ ($k = 1, 2, 3$ for each of the three $c^{(i)}$). The values $c^{(i)}$ are the eigenvalues of the $\bar{c}_{3jk3}$ matrix. Since $\beta^{(i)}_k$ are the components of the eigenvectors of the $\bar{c}_{3jk3}$ matrix, they are orthogonal in both i and k (Hadley, 1961). This double orthogonality can be expressed as

$$\beta^{(m)}_j \beta^{(n)}_k = \delta_{mn} \delta_{jk}. \tag{2.1-67}$$

If the $\beta^{(i)}_k$ are also normalized,

$$\beta^{(i)}_k \beta^{(i)}_k = 1, \tag{2.1-68}$$

then the elastic stresses and displacements and the piezoelectric constants may be transformed to normal coordinates (Ballato, 1972a,b; Ballato et al., 1974) as follows.

$$T^{(0)}_{3j} = \beta^{(j)}_i T_{3i}, \tag{2.1-69}$$

$$u^{(0)}_j = \beta^{(j)}_i u_i, \tag{2.1-70}$$

$$e^{(0)}_{33j} = \beta^{(j)}_i e_{33i}. \tag{2.1-71}$$

The inverses of these three equations can be used to express the constitutive relations [Eqs. (2.1-61) and (2.1-62)], the differential equations, and the boundary conditions in terms of the normal coordinates as

$$T^{(0)}_{3i} = c^{(i)} u^{(0)}_{i,3} - \tilde{e}^{(0)}_{13i} E_1, \tag{2.1-72}$$

$$D_1 = \tilde{e}^{(0)}_{1i3} u^{(0)}_{i,3} + \tilde{\varepsilon}_{11} E_1, \tag{2.1-73}$$

$$T^{(0)}_{3i'} = -\rho \omega^2 u^{(0)}_i, \tag{2.1-74}$$

$$D_3 = 0 \tag{2.1-75}$$

everywhere, and
$$T^{(0)}_{3i} = 0 \qquad (2.1\text{-}76)$$
at $x_3 = \pm h$. A solution to Eqs. (2.1-72)–(2.1-76) is
$$U_i^{(0)} = \frac{\tilde{e}^{(0)}_{13i} E_1}{c^{(i)} k_3^{(i)} c_h^{(i)}}. \qquad (2.1\text{-}77)$$

The electrical current is
$$I_1 = j\omega(2w) \int_{-h}^{h} D_1 \, dx_3. \qquad (2.1\text{-}78)$$

Using $V_1 = (2l)E_1$, the electrical admittance (I_1/V_1) becomes
$$Y_1 = j\omega \tilde{C}_0 \left[1 + \sum_{i=1}^{3} k^{(i)2} \frac{T_h^{(i)}}{k_3^{(i)} h} \right], \qquad (2.1\text{-}79)$$

where
$$\tilde{C}_0 = \tilde{\varepsilon}_{11}(2w)(2h)/(2l), \qquad (2.1\text{-}80)$$

$2h$ being the plate thickness, $2w$ the plate width, and $2l$ the plate length, and
$$k^{(i)2} = \frac{\tilde{e}^{(0)}_{13i} e^{(0)}_{1i3}}{\tilde{\varepsilon}_{11} c^{(i)}} \qquad (2.1\text{-}81)$$

with no sum over i. The use of the normal coordinates greatly simplifies the process of obtaining Eq. (2.1-79).

2.1.6.3 TORSIONAL MODES IN PLATES AND BARS

If the infinite-plate modes are used to develop a picture of the unwanted modes in bars and plates with finite widths or lengths, then a very important class of modes can be easily overlooked. Torsional modes do not appear naturally in infinite rectangular plates and therefore cannot arise when combinations of infinite-plate modes are used to represent modes in resonators with finite lateral dimensions. The use of modes in bars with circular cross-sections for the reference modes allows torsions to arise naturally in the theory, but this approach does not appear to have been discussed in the easily accessible literature.

Torsions are usually very weak in large-area plates but often become a significant problem in bars. It is therefore important to be able to estimate the frequency and activity of torsional modes in bars with various cross sections to understand the torsional unwanted modes in resonators. In some cases the torsional mode itself has interesting properties, and a useful

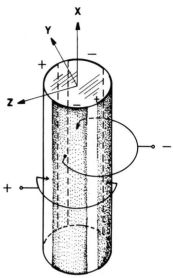

FIG. 2.1-3 Method for plating a quartz crystal to excite a torsional mode.

torsional resonator can be designed (Mason, 1950; Pozdnyakov, 1970, 1971; Hermann, 1977; Sedlacek, 1977; Paul and Sarma, 1978; Dinger, 1982). For example, the VP torsional-mode resonator is reported to have better properties than flexural- or extensional-mode resonators in the same frequency range (Pozdnyakov and Vasin, 1969).

In this section approximate expressions for the frequency of torsional-mode resonators with circular, elliptical, and rectangular cross-sections are discussed briefly. These frequency expressions (Cady, 1946) have been redefined in terms of the variable names used elsewhere in this chapter. In general, torsional-mode frequencies are similar to flexural-mode frequencies for the same structure and geometry (Cady, 1946). Since the torsions and flexures have similar asymmetries, they are often coupled.

Torsional modes can be excited by the same kind of asymmetrical electrode arrangements that excite flexural modes (Cady, 1946). Figure 2.1-3 shows a possible electrode configuration for a quartz cylinder with its length along the x axis (Mason, 1950). These electrodes excite the shear twist, which is necessary to excite the torsion. For the following list the length $(2l)$ of the resonator is along the x_3 direction of the rotated crystal axes.

(1) *Circular-cross-section torsional-mode resonators.* The torsional-mode frequency is

$$f_{c3}^{(n)} \approx \frac{n}{2(2l)} \sqrt{\frac{2}{\rho(s_{44} + s_{55})}}, \qquad (2.1\text{-}82)$$

where n is the overtone or harmonic number. In this case the frequency does not depend on the radius of the cylinder. Consequently, Eq. (2.1-82) should be a good approximation of the frequency of a bar with a square or nearly square cross section.

(2) *Elliptical-cross section torsional-mode resonators.* The torsional-mode frequency is

$$f^{(n)}_{E3} \approx \frac{n}{2(2l)} \sqrt{\frac{4}{\rho[(2w)^2 + (2h)^2][s_{44}/(2w)^2 + s_{55}/(2h)^2]}}. \quad (2.1\text{-}83)$$

The dimension is $2w$ along x_1 and $2h$ along x_2. Equation (2.1-83) can be used to estimate the torsional frequency of a rectangular-cross-section resonator.

(3) *Rectangular-cross-section torsional-mode resonators.* The frequency for the general case has not been written in closed form. For $s_{45} = 0$,

$$f^{(n)}_{R3} \approx \frac{nF(2h)}{2(2l)\sqrt{(2w)^2 + (2h)^2}} \sqrt{\frac{2}{\rho(s_{44} + s_{55})}}. \quad (2.1\text{-}84)$$

where

$$F = \sqrt{1 - 0.630\, h/w} \quad \text{for} \quad h/w < \tfrac{1}{3}$$

and $s_{44} = s_{55}$.

2.1.7 Contour Modes in Thin Plates and in Thin and Narrow Bars

Plate contour resonators have resonance frequencies determined primarily by a plate width or length. The resonance frequency is nearly independent of plate thickness. Although plate extensional and flexural contour resonators can be designed, most practical plate contour resonators use the face-shear mode and are called CT and DT resonators. A special case of contour resonator is the bar resonator. In this case the mode type is usually extensional (E mode) or flexural (F mode). The flexure-mode bars require special electrode configurations for electrical excitation because the motion and electric fields are antisymmetric through the thickness of the bar. The resonance frequency of extensional and flexural bars depends only slightly on the bar thickness. In the usual plate and bar contour resonator, the plate crystallographic angle, plate dimensions, and mounting structure are designed to control the unwanted mode spectrum of the completed resonator. Thickness modes are not usually a problem because of their high frequency, but other extensional, flexural, and face-shear modes can be excited in all three types. Consequently, design procedures for high-quality resonators require considerable experience. In the next few subsections of this chapter the derivation of expressions for electrical admittance or impedance for several types of contour-mode resonators are discussed briefly.

2.1.7.1 EXTENSIONAL-BAR MODE IN A THIN, NARROW BAR WITH ELECTRIC FIELD PARALLEL TO THE BAR LENGTH

In the case of the extensional-bar mode in a thin, narrow bar with the electric field paralell to the bar length [discussed by Berlincourt *et al.*, (1964 and IEEE (1978)], all elastic stresses on the lateral surfaces of a thin narrow bar are assumed equal to zero. The two electric-displacement components through the lateral surfaces are also assumed equal to zero. For this case, it is convenient to transform the constitutive equations listed on p. 53 into a different set, and it is convenient to use the matrix formulation of the constitutive equations (Bond, 1943; Nye 1960) as

$$T = c^E S - e^t E, \qquad (2.1\text{-}85)$$

$$D = eS + \varepsilon^S E. \qquad (2.1\text{-}86)$$

In these equations all quantities are represented by matrices of the appropriate sizes. Equations (2.1-85) and (2.1-86) may be rearranged by standard matrix operations and put into the form

$$S = s^D T + g^t D, \qquad (2.1\text{-}87)$$

$$E = -gT + \beta^T D, \qquad (2.1\text{-}88)$$

where

$$\beta^S = (\varepsilon^S)^{-1}, \qquad (2.1\text{-}89)$$

$$c^D = c^E + e^t \beta^S e, \qquad (2.1\text{-}90)$$

$$s^D = (c^D)^{-1}, \qquad (2.1\text{-}91)$$

$$h = \beta^S e, \qquad (2.1\text{-}92)$$

$$g = h s^D, \qquad (2.1\text{-}93)$$

$$\beta^t = \beta^S - h s^D h^t. \qquad (2.1\text{-}94)$$

Here h^t and g^t are the transposes (Hadley, 1961) of the h and g matrices. Equations (2.1-87) and (2.1-88) are the matrix expressions for the tensor expressions in Eqs. (2.1-10) and (2.1-11). The bar length is along the x_3 direction. The elastic stresses T_1, T_2, T_4, T_5, and T_6 and the electric displacements D_1 and D_2 are small (zero) everywhere. The exciting electric field

2 PIEZOELECTRIC RESONATORS AND WAVES

E_3 is along the bar length. With these assumptions the solution waves (one-dimensional but not plane) travel along the bar with a velocity

$$v_3^b = \sqrt{1/\rho s_{33}^D}. \tag{2.1-95}$$

The boundary conditions $T_3(x_3 = \pm l) = 0$ and the independence of D_3 from x_3 (since $D_{3,3} = 0$) lead to an expression for the elastic displacement as

$$u_3 = \frac{g_{33} D_3}{k_3 \cos k_3 l} \sin k_3 x_3, \tag{2.1-96}$$

where

$$k_3 = \omega \sqrt{\rho s_{33}^D} \tag{2.1-97}$$

and l is one-half the length of the bar. The bar is $2w$ wide and $2h$ thick. The following equations lead to an expression for the electrical impedance of the crystal resonator:

$$S_3 = s_{33}^D T_3 + g_{33} D_3, \tag{2.1-98}$$

$$T_3 = (1/s_{33}^D) S_3 - (g_{33}/s_{33}^D) D_3, \tag{2.1-99}$$

$$E_3 = -g_{33} T_3 + \beta_{33}^T D_3, \tag{2.1-100}$$

$$E_3 = -(g_{33}/s_{33}^D) S_3 + \beta_{33}^S D_3, \tag{2.1-101}$$

$$\beta_{33}^S = \beta_{33}^T + g_{33}^2/s_{33}^D, \tag{2.1-102}$$

$$V = \int_{-l}^{l} E_3 \, dx_3, \tag{2.1-103}$$

$$I = j\omega(2w)(2h) D_3, \tag{2.1-104}$$

$$S_3 = u_{3,3} = \frac{g_{33} D_3}{\cos k_3 l} \cos k_3 x_3, \tag{2.1-105}$$

$$k_3 = \omega/v_3^b, \tag{2.1-106}$$

$$V = \frac{g_{33}}{s_{33}^D} \int_{-l}^{l} -u_{3,3} \, dx_3 + \tilde{\beta}_{33}^S I \frac{(2l)}{j\omega(2w)(2h)}, \tag{2.1-107}$$

$$V = \frac{-g_{33}}{s_{33}^D} [u_3(l) - u_3(-l)] + \beta_{33}^S I \frac{(2l)}{j\omega(2w)(2h)}, \tag{2.1-108}$$

$$Z = \frac{1}{j\omega C_0} \left[1 - \frac{g_{33}^2}{s_{33}^D \beta_{33}^S} \frac{\tan k_3 l}{k_3 l} \right], \tag{2.1-109}$$

where
$$C_0 = \frac{(2w)(2h)}{(2l)\beta_{33}^S} \qquad (2.1\text{-}110)$$
is the clamped or high-frequency capacitance of the resonator.
The crystal impedance in Eq. (2.1-109) may also be written as
$$Z = \frac{1}{j\omega C_f}\left[1 + \frac{g_{33}^2}{s_{33}^D\beta_{33}^T}\left(1 - \frac{\tan k_3 l}{k_3 l}\right)\right], \qquad (2.1\text{-}111)$$
where
$$C_f = \frac{(2w)(2h)}{(2l)\beta_{33}^T} \qquad (2.1\text{-}112)$$
is the free or low-frequency capacitance of the resonator.

2.1.7.2 EXTENSIONAL-BAR MODE IN A THIN, NARROW BAR WITH ELECTRIC FIELD PERPENDICULAR TO BAR LENGTH

The case of the extensional-bar mode in a thin narrow bar with the electric field perpendicular to the bar length [discussed by Mason (1948, 1950), Berlincourt et al., (1964), and IEEE (1978)] has considerable technological importance. The E-element extensional-mode resonator is of this type.

The elastic stress conditions are the same as those for the bar with longitudinal electric field discussed in Section 2.1.7.1. The bar is $2l$ long, $2w$ wide, and $2h$ thick. The different electrical conditions change the frequency equation significantly. If the bar length is along x_3 and the electric field is along x_2, then the process described in detail in Section 2.1.7.1 gives the electrical admittance (I/V) as
$$Y = j\omega C_f\left[1 + \left(\frac{d_{23}^2}{s_{33}^E\varepsilon_{22}^T}\right)\left(\frac{\tan k_3 l}{k_3 l} - 1\right)\right], \qquad (2.1\text{-}113)$$
where
$$k_3 = \omega\sqrt{\rho s_{33}^E} \qquad (2.1\text{-}114)$$
and
$$C_f = \frac{(2w)(2l)}{(2h)}\varepsilon_{22}^T \qquad (2.1\text{-}115)$$
is the free or low-frequency capacitance of the crystal resonator. The electrical admittance may also be expressed as
$$Y = j\omega C_0\left[1 + \left(\frac{k_{23}^2}{1 - k_{23}^2}\right)\frac{\tan k_3 l}{k_3 l}\right], \qquad (2.1\text{-}116)$$
where
$$k_{23}^2 = \frac{d_{23}^2}{s_{33}^E\varepsilon_{22}^T} \qquad (2.1\text{-}117)$$

and

$$C_0 = \frac{(2\omega)(2l)}{(2k)} \frac{\varepsilon_{22}^T}{1 - k_{23}^2} \qquad (2.1\text{-}118)$$

is the clamped or high-frequency capacitance of the crystal resonator.

2.1.7.3 FACE-SHEAR MODES IN THIN, WIDE, LONG PLATES WITH ELECTRIC FIELD PERPENDICULAR TO PLATE SURFACE

A detailed analysis of the case of face-shear modes in thin, wide, long plates with the electric field perpendicular to the plate surface [presented by Mason (1950)] forms the basis for the following discussion. The plate normal is along the x_3 direction, the width is along x_2, and the length is along x_1. Appropriate boundary conditions for the thin plate are

$$T_3 = T_4 = T_5 = 0 \qquad (2.1\text{-}119)$$

everywhere. To derive the proper constants for this case, it is convenient to define a reduced set of constitutive equations as

$$S_i = s_{ij}^E T_j + d_{3i} E_3, \qquad (2.1\text{-}120)$$

$$D_3 = d_{3j} T_j + \varepsilon_{33}^T E_3, \qquad (2.1\text{-}121)$$

where $i = 1, 2, 6$ and $i = 1, 2, 6$. These equations can be written in matrix form as

$$S^p = s^{Ep} T^p + d^{pt} E_3, \qquad (2.1\text{-}122)$$

$$D_3 = d^p T^p + \varepsilon_{33}^T E_3. \qquad (2.1\text{-}123)$$

These equations are not in standard form because some of the stress terms are missing. This nonstandard form is emphasized by using the superscript p (for plate) on S, T, s^E, and d. The following matrix process will give expressions for the thin-plate material constants.

$$c^{c,E} = (s^c)^{-1}, \qquad (2.1\text{-}124)$$

$$e^c = d^{pt} c^{c,E}, \qquad (2.1\text{-}125)$$

$$\varepsilon_{33}^{c,S} = \varepsilon_{33}^T - e^c d^{pt}, \qquad (2.1\text{-}126)$$

$$T = c^{c,E} S - e^{ct} E_3, \qquad (2.1\text{-}127)$$

$$D_3 = e^c S + \varepsilon_{33}^{c,S} E_3. \qquad (2.1\text{-}128)$$

Equations (2.1-127) and (2.1-128) are the constitutive equations for the thin plate. The zero stresses everywhere in the plate (the surface stresses are zero and the plate is thin) have been incorporated into the elastic, piezoelectric, and dielectric constants in these equations. The three nonzero stresses (T_1, T_2, and T_6) in the plate indicate that the face shear (T_6) is coupled to the two extensions in the x_1 (T_1) and x_2 (T_2) directions. The dispersion relation [the relation between k_1, k_2, and ω so that waves travelling in the x_1–x_2 plane (normal to x_3) are solutions] has the following form:

$$[k_1^2 c_{11} + k_1 k_2 2c_{16} + k_2^2 c_{66} - \rho\omega^2]$$
$$\times [k_1^2 c_{66} + k_1 k_2 2c_{26} + k_2^2 c_{22} - \rho\omega^2]$$
$$= [k_1^2 c_{16} + k_1 k_2 (c_{12} + c_{66}) + k_2^2 c_{66}]^2, \quad (2.1\text{-}129)$$

where k_1 and k_2 are wavenumbers in the plate length and width directions and c_{ij} is shorthand for $c_{ij}^{c;E}$ as defined in the matrix inverse in Eq. (2.1-124). Equation (2.1-129) is fourth-order in k_1, k_2, and ω. For a plate very long in the x_1 direction, the variation with x_1 is removed and $k_1 = 0$. Then Eq. (2.1-129) becomes

$$(k_2^2 c_{66} - \rho\omega^2)(k_2^2 c_{22} - \rho\omega^2) = (k_2^2 c_{26})^2. \quad (2.1\text{-}130)$$

Equation (2.1-130) is in the standard form of two resonances coupled by c_{26}. When $c_{26} = 0$, the two values of k_2^2 are $\rho\omega^2/c_{66}$ (face shear) and $\rho\omega^2/c_{22}$ (plate extension). When c_{26} is not zero, the face shear and the extension are coupled together.

Even in this simplified case of a very long plate in which the length effects are suppressed, the admittance (Mason, 1950) is given by the long, involved expression

$$Y = \frac{j\omega(2l)(2w)}{(2h)} \varepsilon_{33}^{c,s}$$

$$\times \left\{ 1 - \frac{(e_{32}^c)^2}{\varepsilon_{33}^{C,s}} \left[\frac{\alpha^2\{\omega^2 \rho c_{66} - \beta^2(c_{22}c_{66} - c_{26}^2)\}}{(\beta^2 - \alpha^2)(c_{22}c_{66} - c_{26}^2)\rho\omega^2} \left(\frac{\tan \alpha w}{\alpha w}\right) \right. \right.$$

$$\left. - \frac{\beta^2\{\omega^2 \rho c_{66} - \alpha^2(c_{22}c_{66} - c_{26}^2)\}}{(\beta^2 - \alpha^2)(c_{22}c_{66} - c_{26}^2)\omega^2\rho} \left(\frac{\tan \beta w}{\beta w}\right) \right]$$

$$+ \frac{(e_{36}^c)^2}{\varepsilon_{33}^{c,s}} \left[\frac{(\omega^2 \rho - \alpha^2 c_{22})}{(\beta^2 - \alpha^2)(c_{22}c_{66} - c_{26}^2)} \left(\frac{\tan \alpha w}{\alpha w}\right) \right.$$

$$\left. - \frac{(\omega^2 \rho - \beta^2 c_{22})}{(\beta^2 - \alpha^2)(c_{22}c_{66} - c_{26}^2)} \left(\frac{\tan \beta w}{\beta w}\right) \right]$$

$$- \frac{e_{32}^c e_{36}^c}{\varepsilon_{33}^{c,s}} \left[\left(\frac{c_{66}(\omega^2 \rho - \alpha^2 c_{22})(\omega^2 \rho - \beta^2 c_{22}) - \alpha^2 \beta^2 c_{22} c_{26}^2}{(\beta^2 - \alpha^2)(c_{22}c_{66} - c_{26}^2)\omega^2 \rho c_{26}} \right) \right.$$

$$\left. \left. \times \left(\frac{\tan \alpha w}{\alpha w} - \frac{\tan \beta w}{\beta w} \right) + \left(\frac{c_{26}}{c_{22}c_{66} - c_{26}^2} \right)\left(\frac{\tan \alpha w}{\alpha w} + \frac{\tan \beta w}{\beta w} \right) \right] \right\}$$

$$(2.1\text{-}131)$$

where

$$\alpha = \omega A\sqrt{1 + B},$$
$$\beta = \omega A\sqrt{1 - B}.$$

In these expressions for α and β

$$A = \sqrt{0.5\rho(c_{22} + c_{66})/(c_{22}c_{66} - c_{26}^2)}$$

and

$$B = \sqrt{[(c_{22} - c_{66})^2 + 4c_{26}^2]/(c_{22} + c_{66})^2}.$$

If $c_{26} = 0$, then the admittance reduces to that for two uncoupled modes, that is,

$$Y = j\omega C_0 \left[1 + k^{(2)2} \frac{\tan x_{(2)}}{x_{(2)}} + k^{(6)2} \frac{\tan x_{(6)}}{x_{(6)}} \right], \quad (2.1\text{-}132)$$

where

$$C_0 = j\omega \frac{(2l)(2w)}{(2k)} \varepsilon_{33}^{c,s}, \quad (2.1\text{-}133)$$

$$x_{(2)} = \omega\sqrt{\rho/c_{22}}\, w, \quad (2.1\text{-}134)$$

$$x_{(6)} = \omega\sqrt{\rho/c_{66}}\, w, \quad (2.1\text{-}135)$$

$$k^{(2)2} = (e_{32}^c)^2/(c_{22}\varepsilon_{33}^{c,s}), \quad (2.1\text{-}136)$$

$$k^{(6)2} = (e_{36}^c)^2/(c_{66}\varepsilon_{33}^{c,s}). \quad (2.1\text{-}137)$$

These constants are defined by the matrix operations expressed in Eqs. (2.1-124) through (2.1-128).

2.1.7.4 FLEXURAL MODES IN THIN, WIDE, LONG PLATES WITH ELECTRIC FIELD PERPENDICULAR TO THE PLATE SURFACE AND PERPENDICULAR TO THE BENDING AXIS (ONE EDGE CLAMPED)

Since the flexural modes in a plate are antisymmetric in the plate-thickness direction, they cannot be excited strongly by simple surface electrodes. A common way of exciting these modes is to use a split crystal or *bimorph* arrangement. In the split crystal the two parts have piezoelectric constants with opposite signs so that the electric field associated with the positive strain in one part adds to the electric field associated with the negative strain in the other part.

A discussion of the detailed analysis of this structure (Mason, 1948) will illustrate the technique of developing an understanding of the flexural structure properties. The flexure analysis is complicated by the necessity of satisfying both stress and moment boundary conditions at the plate edges. This complication also appears in the basic differential equation, which is fourth-order in the elastic displacement. The simple thickness, extensional, face-shear, and torsional modes are all described by second-order differential equations with simple stress boundary conditions at the plate surfaces or bar ends. The fourth-order differential equation and the stress and moment boundary conditions required for flexure make it difficult to understand how all five mode types can arise from the same simple expressions listed on p. 53.

The large lateral dimensions of the plate (nearly infinite in the ideal case to be discussed here) make all of the lateral strains (S_1, S_2, S_4, S_5, and S_6) nearly equal to zero. Only S_3 is large. The electric field E_2 and the plate thickness $2h$ are along the x_2 direction. The plate length ($2l$) is along the x_3 direction. The bending axis is along the x_1 direction. Then, because the lateral strains are zero, a useful set of constitutive relations is

$$T_3 = c_{33}^D S_3 - h_{23} D_2, \qquad (2.1\text{-}138)$$

$$E_2 = -h_{23} S_3 + \beta_{22}^s D_2. \qquad (2.1\text{-}139)$$

Matrix expressions for the constants in Eqs. (2.1-138) and (2.1-139) may be derived as

$$\beta^s = (\varepsilon^s)^{-1}, \qquad (2.1\text{-}140)$$

$$c^D = c^E + e^t \beta^s e, \qquad (2.1\text{-}141)$$

$$h = \beta^s e. \qquad (2.1\text{-}142)$$

The matrix e^t is the transpose of the matrix e (Hadley, 1961). The desired moment or torque per unit area (M) is defined as

$$M = (2w) \int_{-h}^{h} T_3 x_2 \, dx_2. \qquad (2.1\text{-}143)$$

The plate is bent slightly as a result of the lateral end stress and moment and

$$u_{3,3} = x_2/R = S_3, \qquad (2.1\text{-}144)$$

where R is the radius of curvature of the bent plate. The radius of curvature is assumed not to depend on position in this analysis. The split plate idea is expressed as

$$h_{23}(+x_2) = -h_{23}(-x_2). \qquad (2.1\text{-}145)$$

2 PIEZOELECTRIC RESONATORS AND WAVES

Substituting T_3 [Eq. (2.1-138)] into Eq. (2.1-143), noting that R and D_2 do not depend on x_2, allows M to be written as

$$M = \frac{\bar{c}_{33}^D I_m}{R} - \frac{(2w)(2h)^2}{4} h_{23} D_2. \tag{2.1-146}$$

where

$$I_m = (2h)^3(2w)/12 \tag{2.1-147}$$

is the moment of inertia for the rectangular cross section of the plate (Gray, 1963). Since E_2 is also independent of x_2, integration of Eq. (2.1-139) along x_2 leads to an expression for E_2 as

$$E_2 = -h_{23}(2h)/4R + \beta_{22}^s D_2. \tag{2.1-148}$$

Since

$$u_{2,33} = 1/R, \tag{2.1-149}$$

M and E_2 may both be written in terms of $u_{2,33}$. Solving Eq. (2.1-148) for D_2 and substituting the result into Eq. (2.1-146) gives a second expression for M as

$$M = \bar{c}_{33}^D I_m u_{2,33} - \frac{(2w)(2h)^2}{4\beta_{22}^s} E_2. \tag{2.1-150}$$

where

$$\bar{c}_{33}^D = c_{33}^D(1 - \tfrac{3}{4} k_{23}^2) \tag{2.1-151}$$

and

$$h_{23}^2 = \frac{h_{23}^2}{c_{33}^D \beta_{22}^s}. \tag{2.1-152}$$

Both the moment M and the stress in the x_2 direction on the section (T_4) twist a thin x_3 section of the plate. Equating these two forces and substituting for I_m [Eq. (2.1-147)] in Eq. (2.1-150) gives the equation

$$T_{32,3} = -\frac{\bar{c}_{33}^D (2h)^2}{12} u_{2,3333}. \tag{2.1-153}$$

By the differential equation on p. 53,

$$T_{32,3} = -\rho \omega^2 u_2, \tag{2.1-154}$$

so that

$$u_{2,3333} = k^4 u_2, \tag{2.1-155}$$

where

$$k^4 = \frac{12\rho\omega^2}{(2h)^2 \bar{c}^D_{33}}. \tag{2.1-156}$$

The boundary conditions for this problem of the plate with one edge clamped and the other edge simply forced are

$$u_2(x_3 = 0) = 0, \tag{2.1-157}$$

$$u_{2,3}(x_3 = 0) = 0, \tag{2.1-158}$$

$$M(x_3 = 2l) = 0, \tag{2.1-159}$$

$$F_2(x_3 = 2l) = F_{23l} = -u_{2,333} I_m \bar{c}^D_{33}. \tag{2.1-160}$$

A general solution to Eq. (2.1-155) is

$$u_2 = A \cosh kx_3 + B \sinh kx_3 + C \cos kx_3 + D \sin kx_3. \tag{2.1-161}$$

For the boundary conditions in Eqs. (2.1-157) through (2.1-160), $C = -A$ and $D = -B$. Differentiating Eq. (2.1-161), substituting for C and D, and using the definition of voltage as

$$V = 0.5 E_2(2h) \tag{2.1-162}$$

for the split plate gives expressions for voltage V and transverse force F_{23l} at $x_3 = 2l$ as

$$V = A_1 A + B_1 B, \tag{2.1-163}$$

$$F_{23l} = A_2 A + B_2 B, \tag{2.1-164}$$

where

$$A_1 = k^2(\cosh) 2kl + \cos 2kl)(\beta^s_{22} \bar{c}^D_{33}(2h)^2/6), \tag{2.1-165}$$

$$B_1 = k^2(\sinh 2kl + \sin 2kl)(\beta^s_{22} \bar{c}^D_{33}(2w^2/6), \tag{2.1-166}$$

$$A_2 = -k^3(\sinh 2kl - \sin 2kl)\bar{c}^D_{33} I_m. \tag{2.1-167}$$

$$B_2 = -k^3(\cosh 2kl + \cos 2kl)\bar{c}^D_{33} I_m. \tag{2.1-168}$$

For this problem the electrical current I is

$$I = (2w) \int_0^{2l} \dot{D}_2 \, dx_3 \tag{2.1-169}$$

or

$$I = j\omega(2w) \int_0^{2l} D_2 \, dx_3. \tag{2.1-170}$$

Substituting for D_2 gives

$$I = \frac{j\omega(2w)(2l)V}{(2h)\beta_{22}^s} + \frac{kh_{23}(2h)(2w)}{4\beta_{22}^s}$$
$$\times [A(\sinh 2kl + \sin 2kl) + B(\cosh 2kl - \cos 2kl)]. \quad (2.1\text{-}171)$$

The elastic displacement at $x_3 = 2l$ can be obtained from Eq. (2.1-161) as

$$u_2 l = A(\cosh 2kl - \cos 2kl) + B(\sinh 2kl - \sin 2kl). \quad (2.1\text{-}172)$$

Equations (2.1-163) and (2.1-164) can be solved for the coefficients A and B as

$$A = \frac{B_2 V - B_1 F_{23l}}{A_1 B_2 - A_2 B_1}, \quad (2.1\text{-}173)$$

$$B = -\frac{A_2 V - A_1 F_{23l}}{A_1 B_2 - A_2 B_1}. \quad (2.1\text{-}174)$$

Equations (2.1-171) and (2.1-172) have the general form of the current–voltage equations for a two-port network. The electrical terminal is characterized by a voltage V and a current I; the mechanical or elastic terminal is characterized by a force F_{23l} and an elastic displacement u_{2l}. An exact equivalent electrical circuit for this case may be developed from these equations by the process described in Section 2.1-10. Equations (2.1-171) and (2.1-172) may also be solved for any desired combination of I, V, F_{23l}, and u_{2l}. It is important to keep in mind that the above discussion is only valid for the edge-clamped plate. Other edge conditions may be considered by reformulating the solution starting with Eq. (2.1-161). Other sets of boundary conditions for the plate with both edges free or both edges clamped may be used in Eqs. (2.1-157) through (2.1-160). The analysis proceeds in the same way; only the details are different.

2.1.7.5 FLEXURAL MODES IN THIN, NARROW BARS WITH ELECTRIC FIELD PERPENDICULAR TO THE BAR LENGTH AND PARALLEL TO THE BENDING AXIS

As in the plate discussed in Section 2.1.7.4, the flexural modes in a bar are also antisymmetric and cannot be excited strongly by a simple distribution of surface electrodes. The split crystal or bimorph arrangement can also be used to excite flexural modes in bars. The small cross-sectional area of the bar, however, allows the use of antisymmetrical electrode configurations (using the sides of the bar) to provide a convenient second way of exciting the flexural modes. Figure 2.1-4 shows a way of arranging the electrodes and their electrical connections so as to excite the flexure (Mason, 1950; Cady,

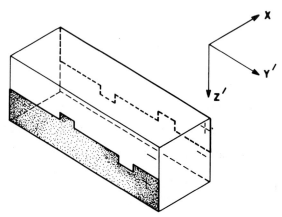

FIG. 2.1-4 Method for plating a longitudinal quartz crystal to excite a flexural mode.

1946). In Fig. 2.1-4 the two electrodes are arranged so that the transverse electric field along one side of the bar is opposite to that along the other side. By the transverse piezoelectric effect one side of the bar is elongated and the other side is shrunk. This elongation and shrinking causes the bar to bend along its length, thereby producing the flexure. Conversely, bending the bar along its length produces opposite electric fields and currents, which are added together by the antisymmetric electrode arrangement. The bar length is along the x_3 direction. In the very thin and very narrow bar all of the lateral stresses (T_1, T_2, T_4, T_5, and T_6) are zero everywhere in the bar. These conditions may be contrasted with the zero-lateral-strain conditions for the plate problem discussed in Section 2.1.7.4. The bar thickness $2h$, electric field E_1, and electric displacement D_1 are along the x_1 direction. The lateral electric displacements (D_2 and D_3) are also zero. In this case a convenient pair of constitutive relations [compare Eqs. (2.1-138) and (2.1-139)] are

$$S_3 = s_{33}^D T_3 + g_{13} D_1, \qquad (2.1\text{-}175)$$

$$E_1 = -g_{13} T_3 + \beta_{11}^T D_1. \qquad (2.1\text{-}176)$$

Equations (2.1-175) and (2.1-176) are equivalent to Eqs. (2.1-10) and (2.1-11) with the appropriate T_j and D_j set equal to zero, as indicated above. Matrix expressions for the constants in Eqs. (2.1-175) and (2.1-176) may be derived from the usually listed constants [c^E, ε^s, and e in Eqs. (2.1-8) and 2.1-9)] as

$$\beta^S = (\varepsilon^S)^{-1}, \qquad (2.1\text{-}177)$$

$$c^D = c^E + e^t \beta^S e, \qquad (2.1\text{-}178)$$

$$h = \beta^s e, \qquad (2.1\text{-}179)$$

$$s^D = (c^D)^{-1}, \qquad (2.1\text{-}180)$$

$$g = h s^D, \qquad (2.1\text{-}181)$$

$$\beta^T = \beta^S - h g^t. \qquad (2.1\text{-}182)$$

where e^t and g^t are the transposes of the e and g matrices (Hadley, 1961). Solving Eq. (2.1-175) for T_3 gives

$$T_3 = (1/s_{33}^D)S_3 - (g_{13}/s_{33}^D)D_1. \qquad (2.1\text{-}183)$$

Substituting Eq. (2.1-183) into Eq. (2.1-176) gives

$$E_1 = -(g_{13}/s_{33}^D)S_3 + \beta_{11}^B D_1, \qquad (2.1\text{-}184)$$

where

$$\beta_{11}^B = \beta_{11}^T - g_{13}^2/s_{33}^D.$$

Equations (2.1-183) and (2.1-184) have the same form as Eqs. (2.1-138) and (2.1-139); only the constant values and the electric-field and electric-displacement directions are different. Since the elastic-bending axes of the plate and bar are the same, the analysis follows the same path as Eqs. (2.1-143)–(2.1-174). Consequently, the electric current I for the bar with one end clamped and the other end simply forced becomes

$$I = \frac{j\omega(2w)(2l)V}{(2h)\beta_{11}^B} + \frac{k(g_{13}/s_{33}^D)(2h)(2w)}{4\beta_{11}^B} \qquad (2.1\text{-}185)$$
$$\times [A^B(\sinh 2kl + \sin 2kl) + B^B(\cosh 2kl - \cos 2kl)].$$

The elastic displacement at $x_3 = 2l$ may be written as

$$u_{2l} = A^B(\cosh 2kl - \cos 2kl) + B^B(\sinh 2kl - \sin 2kl), \quad (2.1\text{-}186)$$

where

$$A^B = \frac{B_2^B V - B_1^B F_{23l}}{B_2^B A_1^B - B_1^B A_2^B}, \qquad (2.1\text{-}187)$$

$$B^B = -\frac{A_2^B V - A_1^B F_{23l}}{B_2^B A_1^B - B_1^B A_2^B}, \qquad (2.1\text{-}188)$$

in which

$$A_1^B = k^2(\cosh 2kl + \cos 2kl)\left[\frac{\beta_{11}^B}{\bar{s}_{33}^D}\frac{(2h)^2}{6}\right], \qquad (2.1\text{-}189)$$

$$B_1^B = k^2(\sinh 2kl + \sin 2kl)\left[\frac{\beta_{11}^B}{\bar{s}_{33}^D}\frac{(2h)^2}{6}\right]. \qquad (2.1\text{-}190)$$

$$A_2^B = -k^3(\sinh 2kl - \sin 2kl)I_m/\bar{s}_{33}^D, \qquad (2.1\text{-}191)$$

$$B_2^B = -k^3(\cosh 2kl + \cos 2kl)I_m/\bar{s}_{33}^D, \qquad (2.1\text{-}192)$$

$$\bar{s}_{33}^D = s_{33}^D/(1 - \tfrac{3}{4}(k_{23}^B)^2), \qquad (2.1\text{-}193)$$

$$(k_{23}^B)^2 = g_{13}^2/(s_{33}^D \beta_{11}^B), \qquad (2.1\text{-}194)$$

$$k^4 = \frac{12\rho\omega^2 \bar{s}_{33}^D}{(2h)^2}. \qquad (2.1\text{-}195)$$

As in the example discussed in Section 2.1.7.4, Eqs. (2.1-185) and (2.1-186) may be solved for various desired relationships among the variables I, V, F_{231}, and u_{21}. Again it is important to keep in mind that these relations are only valid for the bar with one end clamped.

2.1.7.6 MODES NOT DISCUSSED IN DETAIL

Although their solutions will all be similar to those discussed in Section 2.1.7.3, detailed discussions of the following problems await careful study and publication:

(1) face-shear modes in thin, wide, long plates with electric field parallel to the plate surface;
(2) flexural modes in thin, wide, long plates with electric field parallel to the plate surface; and
(3) flexural modes in thin, narrow bars with electric field parallel to the bar length.

2.1.8 Theory for Combined Thickness and Contour Modes

Most theoretical work has been focused on various developments of the truncated two-dimensional equations for thin elastic plates (Mindlin, 1955, 1961) and for thin piezoelectric plates (Tiersten, 1969a). This approximation technique is so important that it needs to be discussed here. The elastic displacement components u_j are expanded in a power series in the plate thickness coordinate x_2. The coefficients of the powers of x_2 are called the *components of the displacement* $u_j^{(0)}$. After substituting the u_j expansions into the differential equations and boundary conditions, all quantities are integrated through the plate thickness. The integration removes the thickness direction from the problem and reduces the complex three-dimensional problem to a series of two-dimensional equations. The integrated quantities are

$$T_j^{(u)} = \int_{-h}^{h} x_2^n T_j \, dx_2 \qquad (2.1\text{-}196)$$

and

$$S_{ij}^{(n)} = \tfrac{1}{2}[u_{i,j}^{(n)} + u_{j,i}^{(n)} + (n+1)(\delta_{2j} u_i^{(n+1)} + \delta_{2i} u_j^{(n+1)})]. \qquad (2.1\text{-}197)$$

The simplification of this approach is realized when only the first one or two terms in the expansions are retained. Zeroth-, first-, and second-order theories have been constructed. At each level of approximation, the relevant enthalpy and energy densities are used to make the theory internally self-consistent. Correction factors modify the elastic, dielectric, and piezoelectric constants to make the results exact for a selected reference system (the infinite plate, for example). A similar approach was worked out for the piezoelectric plate (Tiersten and Mindlin, 1961; Tiersten, 1969a); the complete theory in terms of elastic and electrical nth-order quantities is summarized in the following lists. These lists are included here only for completeness and will not be discussed in detail.

The basic general definitions of the elastic, dielectric, and piezoelectric constants of the theory are as follows:

$$u_j = \sum_{n=0}^{g} x_2^n u_j^{(n)}, \qquad F_j^{(n)} = [x_2^n T_{2j}]_{-h}^{h},$$

$$S_{ij} = \sum_{n=0}^{g} x_2^n S_{ij}^{(n)}, \qquad \varphi^{(n)} = \int_{-h}^{h} x_2^n \phi \, dx_2,$$

$$D_j = \sum_{n=0}^{g} x_2^n D_j^{(n)}, \qquad E_i^{(n)} = \int_{-h}^{h} x_2^n E_i \, dx_2,$$

$$T_{ij}^{(n)} = \int_{-h}^{h} x_2^n T_{ij} \, dx_2.$$

The assumptions for the second-order theory are

$$u_j^{(n)} = D_j^{(n)} = 0 \qquad \text{for } n > 2$$

and

$$T_{22}^{(1)} = E_1^{(2)} = E_3^{(2)} = T_{21}^{(1)} = T_{23}^{(1)} = T_{22}^{(0)} = \ddot{u}_2^{(1)} = \ddot{u}_2^{(2)} = \delta u_1^{(2)} = \delta u_3^{(2)}$$
$$= \delta \varphi^{(2)} = 0.$$

The plate constants used in the second order theory are as follows:

$$c_{pq}^* = c_{pq}^E - (c_{p2}^E c_{2q}^E / c_{22}^E),$$

$$e_{pq}^* = e_{pq} - (e_{p2} c_{q2}^E / c_{22}^E),$$

$$\varepsilon_{ij}^* = \tfrac{9}{4}\varepsilon_{ij}^S + (e_{i2} e_{j2}/c_{22}^E),$$

$$\gamma_{rs} = c_{rs}^E - (c_{rw}^E c_{vs}^E / c_{vw}^E),$$

$$\psi_{ir} = e_{ir} - (e_{iv} c_{rw}^E / c_{vw}^E),$$

$$\zeta_{ij} = \varepsilon_{ij}^S + (e_{iv} e_{jw}/c_{vw}^E),$$

$$c_{pq}^{**} = x_p^\alpha x_q^\beta c_{pq}^*, \qquad \text{no sum},$$

$$e_{iq}^{**} = x_q^\beta e_{iq}, \qquad \text{no sum},$$

$$\alpha = \cos^2(p\pi/2),$$

$$\beta = \cos^2(q\pi/2).$$

The following are the potentials, fields, constitutive relations, differential equations, and boundary conditions used in the second order theory.

(1) Potentials
  (a) Elastic displacement: $u_j^{(0)}$, $u_j^{(1)}$
  (b) Electric potentials: $\varphi^{(0)}$, $\varphi^{(1)}$, $\Phi^{(0)}$, $\Phi^{(1)}$

(2) Fields
  (a) Elastic strain

  $$S_{ij}^{(0)} = \tfrac{1}{2}(u_{i,j}^{(0)} + u_{j,i}^{(0)} + \delta_{2j}u_i^{(1)} + \delta_{2i}u_j^{(1)})$$

  $$S_{ab}^{(1)} = \tfrac{1}{2}(u_{a,b}^{(1)} + u_{b,a}^{(1)})$$

  (b) Electric fields

  $$E_i^{(0)} = -\varphi_{,i}^{(0)} - \delta_{2i}\Phi^{(0)}$$

  $$E_i^{(1)} = -\varphi_{,i}^{(1)} + \delta_{2i}(\varphi^{(0)} - \Phi^{(1)})$$

  $$E_2^{(2)} = 2\varphi^{(1)} - h^2\Phi^{(0)}$$

(3) Constitutive equations

$$T_{ij}^{(0)} = 2hc_{ijkl}^{**}(u_{k,l}^{(0)} + \delta_{2l}u_k^{(1)}) + e_{kij}^{**}\varphi_{,k}^{(0)} + e_{2ij}^{**}\Phi^{(0)}$$

$$T_{ab}^{(1)} = \tfrac{2}{3}h^3\gamma_{abcd}u_{c,d}^{(1)} - \psi_{2ab}\varphi^{(0)} + \psi_{iab}\varphi_{,i}^{(1)} + \psi_{2ab}\Phi^{(1)}$$

$$D_i^{(0)} = e_{ikl}^{**}(u_{k,l}^{(0)} + \delta_{2l}u_k^{(1)}) - (\tfrac{1}{2}h)\varepsilon_{ij}^*\varphi_{,j}^{(0)} - (\tfrac{15}{4}h^3)\varepsilon_{i2}\varphi^{(1)} - (\tfrac{1}{8}h)\varepsilon_{i2}'\Phi^{(0)}$$

$$D_i^{(1)} = \psi_{iab}u_{a,b}^{(1)} + (\tfrac{3}{2}h^3)\zeta_{i2}\varphi^{(0)} - (\tfrac{3}{2}h^3)\zeta_{ij}\varphi_{,j}^{(1)} - (\tfrac{3}{2}h^3)\zeta_{i2}\Phi^{(1)}$$

$$D_2^{(2)} = (\tfrac{15}{8}h^5)(h^2\varepsilon_{2k}\varphi_{,k}^{(0)} + 6\varepsilon_{22}\varphi^{(1)} - 2h^2\varepsilon_{22}\Phi^{(0)})$$

$$\varepsilon_{i2}' = 4\varepsilon_{i2}^* - 15\varepsilon_{i2}$$

(4) Differential equations

$$2hc_{ijkl}^{**}(u_{k,li}^{(0)} + \delta_{2k}u_{l,i}^{(1)}) + e_{kij}^{**}\varphi_{,ki}^{(0)} + F_j^{(0)} + e_{2ij}^{**}\Phi_{,i}^{(0)} = 2\varrho h\ddot{u}_j^{(0)}$$

$$\tfrac{2}{3}h^3\gamma_{abcd}u_{c,da}^{(1)} - 2hc_{2bkl}^{**}(u_{k,l}^{(0)} + \delta_{2k}u_l^{(1)}) - e_{i2b}^{**}\varphi_{,i}^{(0)}$$
$$+ \psi_{iab}(\varphi_{,ia}^{(1)} - \delta_{2i}\varphi_{,a}^{(0)}) + F_b^{(1)} - e_{22b}^{**}\Phi^{(0)} + \psi_{2ab}\Phi_{,a}^{(1)} = \tfrac{2}{3}\varrho h^3 \ddot{u}_b^{(1)}$$

$$\tfrac{2}{3}h^3\psi_{2kl}u_{k,l}^{(1)} + \tfrac{2}{3}h^3 e_{ikl}^{**}(u_{k,li}^{(0)} + \delta_{2k}u_{l,i}^{(1)}) - \tfrac{1}{3}h^2\varepsilon_{ij}^*\varphi_{,ij}^{(0)}$$
$$- \zeta_{2j}(\varphi_{,j}^{(1)} - \delta_{2j}\varphi^{(0)}) - \tfrac{5}{2}\varepsilon_{k2}\varphi_{,k}^{(0)} - (h^2/12)\varepsilon_{i2}'\Phi_{,i}^{(0)} - \zeta_{22}\Phi^{(1)} = 0$$

$$\tfrac{2}{3}h^3\psi_{iab}u_{a,bi}^{(1)} - \zeta_{ij}(\varphi_{,ij}^{(1)} - \delta_{2j}\varphi_{,i}^{(0)}) + \tfrac{5}{2}\varepsilon_{2k}\varphi_{,k}^{(0)} + (15/h^2)\varepsilon_{22}\varphi^{(1)}$$
$$- 5\varepsilon_{22}\Phi^{(0)} - \zeta_{i2}\Phi_{,i}^{(1)} = 0.$$

(5) Boundary conditions

(a) Initial values: $u_i^{(0)}$, $u_a^{(1)}$, $\varphi^{(0)}$, $u_i^{(0)}$, $\ddot{u}_a^{(1)}$

(b) Edge conditions are one member of each:
  (i) $T_{nn}^{(0)}u_n^{(0)}$, $T_{ns}^{(0)}u_s^{(0)}$, $T_{n2}^{(0)}u_2^{(0)}$, $T_{nn}^{(1)}u_n^{(1)}$, $T_{ns}^{(1)}u_s^{(1)}$
  (ii) $\varphi^{(0)}D_n^{(0)}$, $\varphi^{(1)}D_n^{(1)}$

(c) Interior conditions are one member of each:
  (i) $F_2^{(0)}u_2^{(0)}$, $F_\alpha^{(0)}u_\alpha^{(0)}$, $F_\beta^{(0)}u_\beta^{(0)}$, $F_\alpha^{(1)}u_\alpha^{(1)}$, $F_\beta^{(1)}u_\beta^{(1)}$
  (ii) $\Phi^{(0)}(D_2^{(0)} + h^2 D_2^{(2)})$, $\Phi^{(1)}D_2^{(1)}$ where

$$D_2^{(0)} + h^2 D_2^{(2)} = \tfrac{1}{2}[D_2(h) + D_2(-h)]$$

and

$$D_2^{(1)} = (\tfrac{1}{2}h)[D_2(h) - D_2(-h)]$$

Although this theory seems to have increased the number of parameters, the solutions to the equations are tractable and often soluble in closed form.

Developments of the thin-plate approximation include a theory for extension (E), face shear (FS), flexure (F), thickness shear (TS), and thickness twist (TT) in each of two directions for both plates (Mindlin and Spencer, 1967) and bars (Lee, 1971a,b). Although these particular theories report the use of a power series expansion of elastic displacement through the plate or bar thickness, trigonometric and other orthogonal polynomial expansions have also been reported (Mindlin, 1956; Syngellakis and Lee, 1976). The trigonometric theory has not been widely used yet because it is somewhat more complicated than the power series. However, the trigonometric series may give a more accurate approximation for some problems because the assumed basis functions more nearly resemble what intuition suggests are the correct solutions to the differential equations and boundary conditions.

An important aspect of these theoretical developments is an emphasis on the dispersion relations for the system being studied. These relations give the dependence of lateral wave number (or propagation velocity) on frequency so that waves are solutions to the appropriate differential equations. Energy-trapped resonators and the related monolithic crystal filters can be understood in terms of these simple dispersion relations. In fact, since the wave number is a reciprocal wavelength, a simple wave-number substitution can convert the dispersion relations into approximate frequency equations (Meeker, 1977, 1979a). This development makes it easy to see how the very complex unwanted mode spectrum of thin plates and bars can arise from a theory containing only five basic mode types. Of course, it is necessary to satisfy the boundary conditions to obtain a more accurate understanding of these modes. Nevertheless, the simple wave-number model has important application in solving problems such as mode identification.

In all of these problems it is customary to include a consideration of the anisotropy of the system. It has proved very useful to rotate the coordinate axes of the plate away from the natural crystal axes. This rotation produces resonances with new dependences of frequency on various conditions external to the resonator. For many years the AT, BT, DT, GT, and CT crystal resonators have provided useful dependences of frequency, impedance, or unwanted mode spectra on temperature. The E and F modes have desireable properties for low-frequency applications. The dispersion relation for the F mode shows that the frequency approaches zero with zero slope so that the F mode has the lowest frequency of all the modes for the same bar or plate dimensions. In fact, the zero slope of the dispersion relation suggests that F-mode resonators could be designed to have almost zero frequency with realizable dimensions. F mode tuning forks have been popular resonators for low-frequency applications such as wrist watches. Some properties of these modes will be discussed in Section 2.2 of this chapter.

Figure 2.1-5 shows the dispersion relations for the five modes propagating along the x_1 or electric axis of the AT-cut quartz plate when x_2 is the direction of the plate normal. Figure 2.1-6 shows the dispersion relations for propagation along x_3. Figure 2.1-7 shows the displacement components associated with each of the five mode types in each direction. It is important to realize that these sets of dispersion relations are qualitatively the same for a plate of any material, whether piezoelectric or not. In fact, Figs. 2.1-5

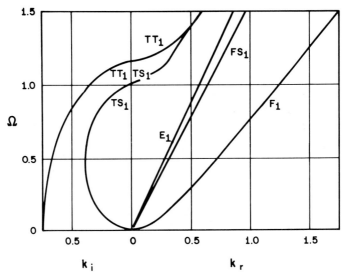

FIG. 2.1-5 Five-mode dispersion relations for x_1 propagation in a rotated Y-cut quartz plate, where $u = u_0 e^{j(\omega t - k_r x_1)} e^{-k_i x_1}$.

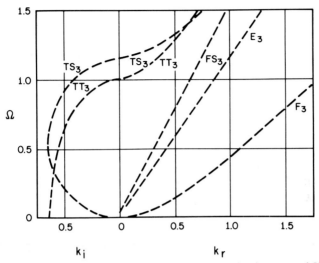

FIG. 2.1-6 Five-mode dispersion relation for x_3 propagation in a rotated Y-cut quartz plate, where $u = u_0 e^{j(\omega t - k_r x_3)} e^{-k_i x_3}$.

and 2.1-6 depict the dispersion curves for quartz with the small piezoelectric effect ignored. Quantitative details are likely to be important, but the main features of the mode spectrum will not depend very much on the exact values of the constants.

The thickness-shear and thickness-twist modes will always have cutoff frequencies that will separate frequencies with real lateral wave number from frequencies with imaginary lateral wave number. Flexure, face shear, and extension will always have zero wave number at zero frequency. The flexural wave number will always approach zero at zero frequency with zero slope. Boundary conditions on the ends of a bar or plate vibrating in flexure are more complicated than those for face shear or extension because an elastic moment or twist is involved. Corrections for the contribution of the moment may be applied to the wave-number model for low orders. The effect of neglecting the moment is likely to be small for higher-order flexural modes. For these reasons Figs. 2.1-5 and 2.1-6 are powerful tools in the development of a coordinated understanding of the resonance modes of a plate. Figure 2.1-8 shows the dispersion relations for a thin bar ($-5°$ X-Cut quartz), and Figure 2.1-9 shows the displacement components associated with each of the mode types. The similarity of Fig. 2.1-8 to Figs. 2.1-5 and 2.1-6 also shows the general power of the dispersion relations.

Figure 2.1-10 shows the dispersion relation for a plate with length L along the x direction. The plate has an infinite width along the direction perpendicular to the x direction, and the major surfaces are fully electroded. This

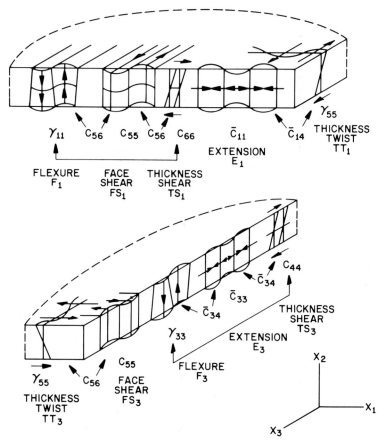

FIG. 2.1-7 Elastic displacement for five modes propagating along the x_1 and x_3 directions in a rotated Y-cut quartz plate.

dispersion relation represents the relationship between frequency and the complex lateral-propagation constant or wave number $(k_r + jk_i)$ so that waves are solutions to the differential equations. Only propagation in one lateral direction is considered here to simplify the discussion. In an actual resonator, both lateral-propagation constants must be considered, and the result is considerably more complicated (Tiersten, 1975d, 1976b). The next few paragraphs explain the significance of ω_c and $\omega_r^{(n)}$ in Figure 2.1-10. The use of only one lateral wave number means that the plate dimension in the other lateral direction is very large or infinite, so that the corresponding wave number is zero. When the wave number being considered is also zero, both lateral wave numbers are zero, and the plate is infinitely extended in both lateral directions. The frequency ω_c for this case is therefore the ordinary

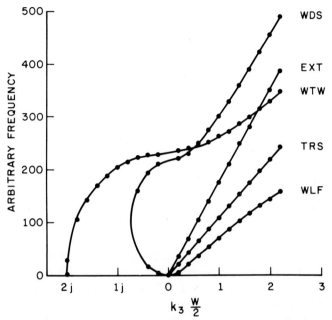

FIG. 2.1-8 Dispersion relations for five modes in a $-5°$ X-cut quartz bar. WDS is width shear, EXT is extension, WTW is width twist, TRS is transverse shear, and WLF is width–length flexure.

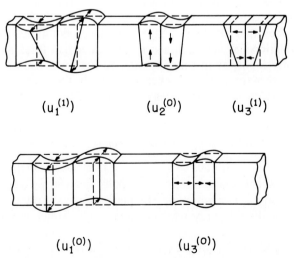

FIG. 2.1-9 Elastic displacements for five modes in a $-5°$ X-cut quartz bar.

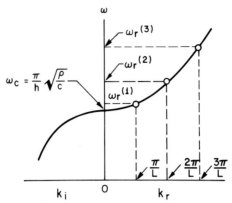

FIG. 2.1-10 Dispersion relation for elastic plate, where $u = u_0 \exp(j\omega t - jk_r x - k_i x)$.

infinite-plate frequency discussed in previous sections of this chapter. For any finite length, the edge boundary conditions are satisfied approximately at those frequencies for which the length is an integral number of half-lateral wavelengths. Since the frequency has a nearly parabolic dependence on wave number and is concave upward with positive curvature, a series of resonance frequencies (called *inharmonics*) occur above the infinite plate frequency. Each of these frequencies is characterized by the number of half-lateral wavelengths along the length direction. In Fig. 2.1-10 the three lowest inharmonic frequencies are labelled $\omega_r^{(2)}$, $\omega_r^{(2)}$, and $\omega_r^{(3)}$, corresponding to 1, 2, and 3 half-lateral wavelengths, respectively. The infinite-plate-thickness frequency is not observed because the infinite-plate-thickness mode does not satisfy the edge boundary conditions. Because the dispersion relation has zero slope at $k = 0$, the length fundamental mode is only slightly higher than the infinite-plate frequency. Since the lateral-propagation constant is a reciprocal wavelength, the integral number of half-lateral wavelength condition may be expressed as

$$L = m\lambda_m/2, \tag{2.1-198}$$

$$k_m = 2\pi/\lambda_m = m\pi/L. \tag{2.1-199}$$

The discussion above applies to each of the thickness branches, which are defined for different integral numbers of half-thickness wavelengths in the thickness direction. The resonance frequency depends on two numbers: One number signifies the approximate number of half-thickness wavelengths along the thickness direction, and the other number signifies the number of half-lateral wavelengths along the length direction. Frequencies for actual resonators depend on three numbers: thickness, length, and width.

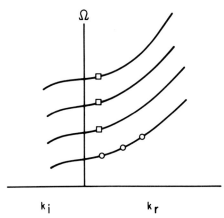

FIG. 2.1-11 Inharmonic length modes and thickness overtones.

In Fig. 2.1-11 the fundamental-length–thickness overtone frequencies are labelled with squares, and the inharmonic-length fundamental-thickness frequencies are labelled with circles. The value Ω is a normalized frequency, often normalized to the value $\Omega = 1$ for the lowest infinite-plate frequency. Charge cancellation in both directions reduces the activity of those modes that have antisymmetric displacements in either direction, at least as far as direct electrical excitation is concerned. Fringing electrical fields and elastic-mode coupling at plate edges may excite some other modes strongly. The inharmonic modes are relatively strongly excited and pose a significant design problem for filter crystals that need to be free of spurious transmission over a relatively wide band of frequencies. Frequency coincidences between overtone and inharmonic modes may add to the problem of spurious-mode suppression (flexure, face shear, and extension) in thickness-mode crystal resonators for narrow-band oscillator applications, especially if operation over a wide temperature range is desired.

Energy-trapped resonator designs that offer higher quality factor and reduced spurious transmission or excitation arose from some empirical observations (Bechmann, 1939) and some application of electromagnetic waveguide theory of modes with cut-off frequencies (Mortley, 1957; Shockley et al., 1963, 1967). This work suggested that proper conditions for energy-trapping a single mode could produce resonators with considerably better quality factor and spurious-mode rejection (Curran and Koneval, 1964, 1965; Lukaszek, 1965). A theoretical derivation for the plate $2h$ thick along x_2 with a strip electrode $2w$ wide along x and with an infinite length along x, gives the following relation (Shockley et al., 1967):

$$2w/2h \leq [M(q)/n]\sqrt{1/\Delta}, \qquad (2.1\text{-}200)$$

where $\Delta = (\omega_s - \omega_e)/\omega_e$ (ω_s being the resonance frequency of the unelectroded part of the plate and ω_e the resonance frequency of the electroded or active part of the plate), n is the thickness overtone number, and $M(q)$ is a different constant for each lateral mode with a different number g of half-wavelengths along the x_3 length direction. For the length along the x_3 direction of an AT-cut quartz plate, the first six values of $M(q)$ are

$$M(1) = 1.41, \ M(2) = 2.83, \ M(3) = 4.24, \ M(4) = 5.66, \ M(5) = 7.07$$

and

$$M(6) = 8.48.$$

Equation (2.1-200) is the condition for suppressing only modes with lateral wave number q or higher. To trap only the lowest-order mode, $M(2)$ is used in Equation (2.1-200), so that

$$2w/2h \leq (2.83/n)\sqrt{1/\Delta} \qquad (2.1\text{-}201)$$

is the proper condition for electrode length along the x_3 direction of the AT quartz plate. Other references (Shockley et al., 1967) give $M = 2.17$ for the same x_3 direction of the AT plate. A third reference for the same case (Mindlin, 1965, 1967) gives

$$M = \sqrt{2(c_{55}c_{66} - c_{56}^2)/c_{66}} \qquad (2.1\text{-}202)$$

for a plate with the same symmetry as the AT quartz plate. This expression gives the value $M = 2.17$ mentioned above for the x_3-propagating AT quartz plate. Since the energy-trapped theory (Mindlin, 1967; Shockley et al., 1967) is cast in terms of frequency lowering or plateback in the active resonator region, both mass loading and electrical conditions can be important. A contoured plate may also be considered as a plate with two regions, each with a different frequency. Figure 2.1-12 depicts a mass-loaded or thickness-contoured plate with the frequency difference and active resonator dimensions arranged to trap only the fundamental-inharmonic length mode (the thickness branch is not specified in this figure). Since the second-inharmonic mode as shown is charge-cancelled because of symmetry, the condition on the length could also be set so that the third-inharmonic frequency has a value just above the plate frequency. The condition shown results in a more conservative design. (The value Ω is a normalized frequency.) In quartz with low electromechanical coupling, the mass effects seem to dominate. In materials with high electromechanical coupling, the electrical effects may dominate, and it may be difficult to provide the proper conditions

2 PIEZOELECTRIC RESONATORS AND WAVES

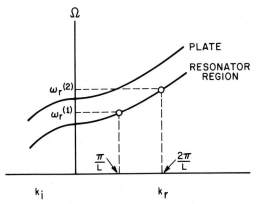

FIG. 2.1-12 Mass loading and thickness contouring—energy trapping.

with mass loading alone. Figure 2.1-13 depicts a plate with a high electromechanical coupling in which the frequency difference between the two regions is caused by the different electrical conditions (nearly open-circuited and nearly short-circuited) in the two regions. In this figure the electrode length has been improperly selected (too large), and the first three inharmonics are trapped. In such a resonator design, no ordinary mass-loading conditions will provide an optimized resonator. Mass-loading the open-circuited region would lower the frequency of the upper curve and produce less inharmonic trapping. (The value Ω is a normalized frequency.)

When the electrode thickness of a partially electroded quartz resonator is increased, the quality factor and the spurious rejection increase. At a critical value of frequency lowering that depends on the dimensions of the

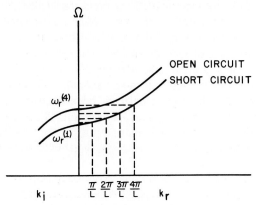

FIG. 2.1-13 Electrical conditions—energy trapping.

electrode, the quality factor begins to decrease, and the inharmonic spurious level increases (Lukaszek, 1965). The energy-trapped resonator allows plateback compensation for larger electrodes, and lower impedance, high-quality resonators are possible.

A monolithic crystal filter can be realized by putting more than one resonator on the same plate. The coupling between pairs of electroded regions can be controlled by the relative thickness of the electrode and plate and by the separation between the two electroded regions. Practical monolithic two- and multiresonator crystal filters have been reported (Sykes *et al.*, 1967; Spencer, 1972).

In all piezoelectric resonators, some of the elastic modes of vibration may not be driven or detected electrically because the electrode integrates the surface charge associated with that mode to zero. Consequently, modes with even-order elastic-displacement distributions in any direction (width, length, thickness) cannot be excited strongly. In practice, various nonideal effects cause these modes to be weakly excited, and they add to the mode spectrum of a real resonator. Under some conditions a mode not directly excited electrically couples elastically to a directly excited mode. This coupling may occur in the bulk material via the differential equation governing the motion or at the plate surfaces or edges via the boundary conditions. Since the frequency of the coupled mode may depend on temperature, electrode location or size or shape, or plate size, suppression of these unwanted modes often poses a challenging design problem for a particular resonator.

2.1.9 Electrical Effects in Piezoelectric Resonators

The most significant electrical effects in a piezoelectric resonator are the impedance level, the separation of resonance and antiresonance frequencies, and the quality factor. The impedance level is controlled principally by the electrode area and the dielectric constant. The impedance level depends on the active vibrating volume, which depends on the plate curvature and electrode thickness. The separation of resonance and antiresonance frequencies is controlled by the piezoelectric constant, the dielectric constant, and the elastic stiffness of the resonator material. The quality factor is controlled by dissipation in the resonator material, by the condition of the resonator surfaces, and by the location and type of electrical and mechanical lead attachments.

The electrical behavior of a piezoelectric resonator may be understood either directly in terms of the physical definitions of surface forces, particle velocities, electrical voltage, and current or in terms of various equivalent electrical circuits (Meeker, 1972). Equivalent electrical circuits for piezoelectric resonators are discussed in the next section.

2.1.10 Equivalent Electrical Circuits for Piezoelectric Resonators

In this section simple equivalent electrical circuits for crystal resonators are discussed. Equivalent electrical circuits for practical packaged resonators must include stray capacitances and stray inductances associated with the package and electrical leads. The more complicated equivalent circuit of a packaged crystal resonator is discussed in Chapter 7 of Volume 2.

Equivalent circuits are exact circuit representations of the crystal resonator if they have identical impedance, admittance, and transfer properties. Exact equivalent circuits for crystal resonators usually have elements that are not ordinary frequency-independent inductors, capacitors, and resistors. A particular equivalent circuit is generally useful only if it is cast in a form that is consistent with previous experience or if it otherwise leads to some new insight into the properties of the crystal resonator it represents. Approximate equivalent circuits are used to simplify the understanding of the resonator or to allow the crystal properties to be used in various circuit analysis techniques. Although new equivalent circuits are proposed from time to time, the lumped-element circuit (Butterworth, 1914; Cady, 1922; Van Dyke, 1925, 1928; Dye, 1926) is still used most often and is shown in Fig. 2.1-14. The most used exact equivalent electrical circuit for a simple thickness mode (Mason, 1948, 1950; Kossof, 1966; Meeker, 1972) is shown in Fig. 2.1-15. A useful description of the process of deriving an exact electrical equivalent circuit was given by Berlincourt et al., (1964). This derivation process may be summarized in the following way. First, the relationships between the elastic forces and velocities and the electrical displacements and potentials at the surfaces of the resonator are derived from the differential and constitutive equations as illustrated in the previous sections of this chapter. Then, a desired form for the equivalent electrical circuit is specified, and the circuit equations (nodal or mesh) are formulated. The coefficients in these two sets of equations are set equal to establish the equivalency. The

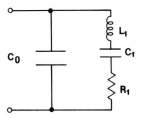

FIG. 2.1-14 Simple equivalent electrical circuit of the single-mode piezoelectric resonator.

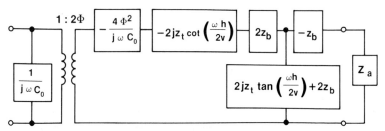

FIG. 2.1-15 Exact equivalent electrical circuit of the single-thickness mode with thickness excitation. Here the plate thickness is h, $c'_{rsrs} = c^E_{rsrs} + e^2_{rrs}/\varepsilon^S_{rr}$, $\Phi = c_{rrs}/h$, $z_t = \sqrt{\rho c'_{rsrs}}$, $C_0 = \varepsilon^S_{rr}/h$, $\omega h/2v = (\omega h/2)\sqrt{\rho/c'_{rsrs}}$, $4\Phi^2 = (4/\pi)(\omega_0 C_0)z_t k^2$, $k^2 = e_{rrs}e_{rrs}/c'_{rsrs}\varepsilon^S_{rr}$, $\omega_0 h/2v = \pi/2$, $Z_a = \sqrt{\rho_a c_a}$, and $Z_b = \sqrt{\rho_b c_b}$. The electric field is perpendicular to the plate. All quantities refer to unit area.

complexity of the relationships between the two sets of equations depends on the choice of the form of the specified equivalent circuit.

Impedance equations for a simple thickness mode piezoelectric transducer, derived directly from the differential equations, constitutive equations, and boundary conditions, are as follows (Meeker, 1972):

$$T_a = -jz_t \frac{C}{S} U_a - jz_t \frac{1}{S} U_b + \frac{\Phi A}{j\omega C_0} \frac{I}{A},$$

$$T_b = -jz_t \frac{1}{S} U_a - jz_t \frac{C}{S} U_b + \frac{\Phi A}{j\omega C_0} \frac{I}{A},$$

$$V = \frac{\Phi A}{j\omega C_0} U_a + \frac{\Phi A}{j\omega C_0} U_b + \frac{A}{j\omega C_0} \frac{I}{A},$$

where

$C = \cos k_r h,$ $S = \sin k_r h,$

$k_r = \omega \sqrt{\rho/c'_{rsrs}},$ $z_t = \sqrt{\rho c'_{rsrs}},$

$C_0 = \varepsilon^S_{rr} A/h,$ $\Phi A/C_0 = e_{rrs}/\varepsilon^S_{rr},$

$\Phi = e_{rss}/h,$ $c'_{rsrs} = c^E_{rsrs} + e_{rrs}e_{rrs}/\varepsilon^S_{rr},$

h being the plate thickness. A distributed equivalent circuit for thickness-mode resonators with arbitrary anisotropy was proposed by Ballato (1972a,b) and Ballato et al. (1974). The single-mode equivalent circuit is shown in Fig. 2.1-16. An equivalent circuit for a simple bar mode with transverse electric field is shown in Figure 2.1-17 (Mason, 1948, 1950; Berlincourt et al., 1964).

Crystal resonators are usually described or specified in terms of the lumped-element equivalent electrical circuit (Fig. 2.1-14) mentioned above.

2 PIEZOELECTRIC RESONATORS AND WAVES

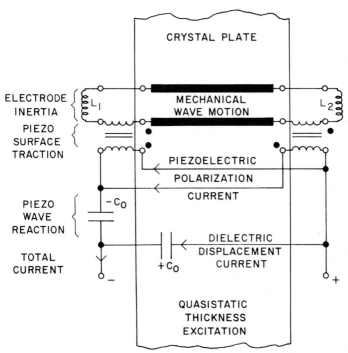

FIG. 2.1-16 Distributed exact equivalent electrical circuit for quasistatic thickness-excited thickness mode.

Since this circuit representation is not exact over any finite frequency range, the definitions of the circuit elements are not unique and depend on assumed secondary requirements that may not be clearly stated or understood. As an example, the mechanical or motional inductance L_1 can be defined in two ways. A crystal resonator specification should be very clear as to which inductance is meant.

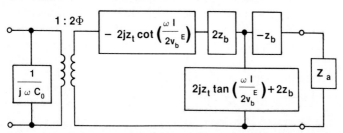

FIG. 2.1-17 Exact equivalent circuit for length mode—thickness excited. The plate length is l, $\Phi = wd_{31}/s_{11}^E$, $z_t = \rho w t v_b^E$, $C_0 = lw\varepsilon_{33}^T(1 - k_{31}^2)/t$, $v_b^E = \sqrt{1/\rho s_{11}^E}$, $4\Phi^2 = (4/\pi)(\omega_0 C_0)z_t k_{31}^3$, $k_{31}^2 = d_{31}^2/\varepsilon_{33}^T s_{11}^E$, $\omega_0 l/2v_b^E = \pi/2$, $Z_a = \rho_a wtv_b^a$, and $Z_b = e_b wtv_b^b$. The electric field is perpendicular to the bar length.

The first definition of the motional inductance uses the slope of the motional reactance at the resonance frequency as

$$L_1 = \tfrac{1}{2}\, dx_1/d\omega|_{\omega=\omega_r}. \qquad (2.1\text{-}203)$$

This equation is exact for a simple series circuit, even in the presence of loss. However, the motional reactance of even a simple crystal resonator is difficult to measure directly because of the shunt capacitance C_0. Combinations of fixture design and data manipulation are often used to find the motional reactance from measured electrical impedances over a frequency range. Sometimes a two-capacitance (series) or two-inductance (series) technique is used to convert the reactance slope to more easily measured frequency changes (Buist, 1961). The parasitic circuit elements in the resonator package and in the measurement fixture play an important role, and high-frequency measurements become difficult to interpret. The slope method of determining a crystal-resonator inductance is most useful for applications in which the resonator is used very near the resonance frequency, such as the crystal-oscillator application.

The second definition of the motional inductance uses the clamped capacitance C_0 and the resonance ($X = 0$) and antiresonance ($X = \infty$) frequencies of the resonator to determine the motional capacitance C_1, and then uses the resonance frequency condition to determine L_1 (Mason, 1950; IRE, 1957, 1961; IEEE, 1966) as

$$C_1 = 2C_0(f_a - f_r)/f_r, \qquad (2.1\text{-}204)$$

$$L_1 = 1/\omega_r^2 C_1. \qquad (2.1\text{-}205)$$

If the free (unrestrained) or low-frequency capacitance C_f is measured,

$$C_0 + C_1 = C_f. \qquad (2.1\text{-}206)$$

This definition of motional inductance is useful over a wide range of frequencies, from the resonance to the antiresonance frequencies of the resonator, and is often used to specify inductance for filter crystal resonators. This definition is accurate to about 1% if $C_0 > 25 C_1$. The assumptions that $f_a \approx f_p$ and $f_r \approx f_s$ are very well satisfied if $\omega_r L_1/R_1 \gg C_0/C_1$ (IEEE, 1966). For all cases, a good model of the packaged crystal resonator and the measuring circuit must be used in the interpretation of the impedance or admittance data.

For overtones, the motional inductance is approximately the same as that for the fundamental (Mason, 1950), so that the motional capacitance C_1 is reduced to

$$C_1 \text{ (overtone)} = (1/n^2) C_1 \text{ (fundamental)}, \qquad (2.1\text{-}207)$$

2 PIEZOELECTRIC RESONATORS AND WAVES

where n is the overtone number. This equation is only approximate for a trapped-energy resonator, since the effect of lateral dimensions is not included (Onoe and Jumonji, 1965; Burgess and Muir, 1975).

2.1.11 Properties of Modes in Crystal Resonators

Four parameters describe the useful properties of a single-mode resonator (Meeker, 1972); different sets of parameters are useful for particular applications, but only four are independent and all the others can be calculated from the basic four. Useful parameters include the free capacitance C_f, the clamped capacitance C_0, the motional inductance L_1, the motional capacitance C_1, the capacitance ratio $r = C_0/C_1$, the parallel resonance frequency f_p, the resonance frequency f_r, the motional resistance R_1, and the motional quality factor (Q_1).

Derivations of the electrical admittance and impedance (such as those illustrated in previous sections of this chapter) of various modes may all be used with the definitions in Eqs. (2.1-203) through (2.1-206) to obtain expressions for the desired equivalent electrical circuit parameters. These expressions serve only as estimates of these circuit parameters for actual physical resonator structures because the theoretical derivations are determined only for very idealized cases. These estimates may be used to determine material constants or to design resonators with desired properties.

The properties of various modes of quartz crystal resonators are discussed in Section 2.2. Reports on impedance levels (L_1 and C_1) of lithium tantalate plate resonators (Burgess et al., 1975) and quartz resonators (Onoe and Jumonji, 1965; Beaver, 1973; Burgess and Muir, 1975) include discussions of the effects of lateral dimensions.

2.1.12 Piezoelectric Materials

Some of the reports of work on piezoelectric materials other than quartz are referred to in this section. Reports of work on quartz are discussed in Section 2.2.

A very large number of materials are piezoelectric (Mason, 1950). For practical application, however, a piezoelectric material must be machinable into specific desired shapes and must be stable in ordinary processing environments. Brittleness, solubility, and hygroscopicity, for example, are difficult properties to manage in the fabrication of practical devices. These simple fabrication requirements reduce considerably the number of materials suitable for practical application. Another important restriction on the usefulness of a material is the possibility of using crystal plate or bar orientations in resonators with controlled (zero, linear, etc.) dependences of resonance frequency on temperature, stress, or other parameters of interest.

Some properties and material constants for the following materials have been reported by Mason (1950). This list is not intended to be complete and should only indicate that the piezoelectric properties of many materials have been studied. It should also be made clear that most of these materials have not been used in practical devices for one or more of the reasons mentioned above.

(1) Rochelle salt (sodium potassium tartrate tetrahydrate)
(2) Ethylene diamine tartrate (EDT)
(3) Dipotassium tartrate semihydrate (DKT)
(4) Ammonium dihydrogen phosphate (ADP)
(5) Potassium dihydrogen phosphate (KDP)
(6) Sodium chlorate
(7) Sodium bromate
(8) Dextrose sodium bromide
(9) Dextrose sodium chloride
(10) Dextrose sodium iodide
(11) Aluminum phosphate
(12) Tourmaline
(13) Lithium trisodium chromate hexahydrate
(14) Lithium trisodium molybdate hexahydrate
(15) Nickel sulfate hexahydrate
(16) Magnesium sulfate
(17) Lithium sulfate monohydrate
(18) Ammonium tartrate
(19) Lithium ammonium tartrate monohydrate
(20) Lithium potassium tartrate monohydrate
(21) Strontium formate dihydrate
(22) Barium formate
(23) Iodic acid
(24) Sodium ammonium tartrate tetrahydrate
(25) Lithium sulfate monohydrate
(26) Tartaric acid
(27) Ammonium tartrate

Thus far, quartz has had the best combination of properties for use in resonators for practical applications. A major disadvantage of quartz is a low piezoelectric coupling in all modes. The search for materials with higher electromechanical coupling factors has been a significant driving force for work on new materials. The following list refers to some of this work on various materials.

2 PIEZOELECTRIC RESONATORS AND WAVES

(1) Lithium niobate (Warner et al., 1967; Fukumoto and Watanabe, 1968; Lemanov et al., 1968; Schultz et al., 1970; Hannon et al., 1970; Kaliski, 1971; Smith and Welsh, 1971; Adachi and Kawabata, 1972; Burgess and Porter, 1973; Hales et al., 1974; Burgess and Hales, 1976; Klimenko et al., 1978; Watanabe and Yano, 1978; Nakazawa, 1979)

(2) Lithium tantalate (Smith, 1967; Warner et al., 1967; Sliker and Koneval, 1968; Onoe et al., 1969, 1973; Niizeki and Sawamoto, 1970; Ashida et al., 1970; Hannon et al., 1970; Kaliski, 1971; Smith and Welsh, 1971; Adachi and Kawabata, 1972; Burgess and Porter, 1973; Hales et al., 1974; Burgess et al., 1975; Burgess and Hales, 1976; Detaint and Lançon, 1976; Detaint, 1977; Uno, 1979; Nakazawa, 1979)

(3) Berlinite (orthoaluminum phosphate ; this material has considerable promise in practical devices; compare item (11) in the materials list on p. 108) (Stanley, 1954; Carr and O'Connell, 1976; Chang and Barsch, 1976; Jhunjhunwala et al., 1976a; Ozimet and Chai, 1977; Chai and Ozimet, 1979; Detaint et al., 1980)

(4) Lead zirconate titanate (polycrystalline; this type of material is often used in low-cost resonators for which low Q and wide frequency tolerances are acceptable) (Ikegami et al., 1974)

(5) Cadium sulfide (Sliker et al., 1969)

(6) Zinc oxide (Crisler et al., 1968)

(7) Zinc sulfide (Firsova, 1974)

(8) Bismuth germanate (Alekseev and Bondarenko, 1976; Sedlacek, 1977; Zelenka, 1978)

(9) Thallium vanadium sulfide (Weinert and Isaacs, 1975; Carr and O'Connell, 1976; Jhunjhunwala et al., 1976; Volluet, 1978)

(10) Rubidium biphthalate (Belikova et al., 1975)

(11) Triglycine sulfate (Pietrzak et al., 1976)

(12) Tellurium oxide (Carr and O'Connell, 1976)

(13) Lead potassium niobate (Yamada, 1973, 1975; Carr and O'Connell, 1976)

(14) Beta-eucryptite or lithium aluminum silicate ($LiAlSiO_4$) (Carr and O'Connell, 1976)

(15) Nepheline or potassium aluminum silicate–sodium aluminum silicate [($KAlSiO_4$) ($NaAlSiO_4$); this material has positive temperature coefficients of c_{11} and c_{66}, which make possible temperature compensated thickness-mode bulk-mode resonators and SAW devices] (Bonczar and Barsch, 1975; Carr and O'Connell, 1976).

(16) Lithium iodate (Jipson et al., 1976; Avdienko et al., 1977)

(17) Thallium tantalum selenide (Jhunjhunwala et al., 1976b)

(18) Barium germanium titanate (Kimura et al., 1973)

(19) Lithium gallate (Nanamatsu et al., 1973)

(20) Tourmaline (Kittinger et al., 1979)
(21) Lead titanate (Nagata et al., 1972)
(22) Potassium lithium niobate (Adachi and Kawabata, 1978)
(23) Calcium aluminate ($Ca_{12}Al_{14}O_{33}$) (Whatmore et al., 1979)

For all of these materials, application waits for low-cost, large-size crystals and information on material constants and engineering properties, such as hygroscopicity, hardness, and solubility.

The scope of this chapter does not allow further detail on the reports of work related to these materials. The references should guide the interested reader to the required information.

2.1.13 Conclusion

The need for higher-performance crystal oscillators makes the development of a more detailed understanding of the piezoelectric resonator more and more important. Work is presently underway on developing a more detailed understanding of the subtle and nonlinear properties of the quartz resonator. New crystalline materials (such as lithium niobate, lithium tantalate, and berlinite) are being used to provide useful devices with desired properties. The dependence of resonator properties on thermal and mechanical stress and acceleration are now being studied. The use of doubly rotated crystal plates and bars to control these effects has already begun. New improved resonator designs are merging as a result of this work. The need for even further improvements in crystal oscillator and filter performance is making further work in these directions more and more necessary.

2.2 PROPERTIES OF QUARTZ PIEZOELECTRIC RESONATORS

2.2.1 Temperature Coefficient of Resonance Frequency

Only a small sample of the reported work on the temperature coefficient of resonance frequency will be discussed in this section. After the analytical expressions for resonance frequency are derived (as in Section 2.1), the temperature coefficients of the relevant material constants can be used to calculate the dependence of the resonance frequency on temperature. This calculation can use the simple linear model involving homogeneous-equilibrium crystal states at each temperature (Section 2.1.3) or the nonlinear model using the higher-order constants and the reference temperature state (Section 2.1.4).

2 PIEZOELECTRIC RESONATORS AND WAVES

For quartz, resonance frequencies of the contour modes [bar extension (E), bar flexure (F), and plate face-shear (CT and DT)] have parabolic or linear dependences on temperature. The thickness dilatation (X) and thickness shear (Y) have linear dependences on temperature. The thickness shear (BT) has a parabolic dependence of resonance frequency on temperature. The plate-extensional (GT) and the thickness-shear (AT) resonance frequencies have cubic dependences on temperature. Changes in plate or bar orientation can be used to tailor these linear or cubic dependences to particular needs. Doubly rotated crystal plates (FC, SC, IT, RT, and LC) have been used to move the cubic turnover temperature into desired temperature ranges while changing the response of the resonator to external stresses. A more complete discussion of doubly rotated plates of quartz and other crystals, and in particular, of cuts insensitive to environmental disturbances of various sorts, leading to devices of highest precision, is given in Ballato (1977). See also Section 8.3 and Chapter 9 of Volume 2.

Figure 2.2-1 shows the loci of quartz crystal cuts with a zero linear temperature coefficient. The different dielectric and piezoelectric constants

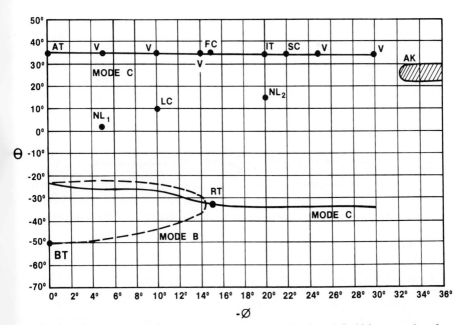

FIG. 2.2-1 Loci of zero linear temperature coefficient for B and C thickness modes of quartz plates as a function of crystallographic angles ϕ and θ. Mode B (faster) and mode C (slower) as quasi-shear modes. Orientations of several useful quartz crystal cuts are labelled by dots.

TABLE 2.2-1

Mode Designations and Properties of Quartz Resonators

ELEMENT DESIGNATION	REFERENCE	MODE OF VIBRATION	FREQUENCY RANGE
A	AT	THICKNESS SHEAR	0.5 – 100 MHz
B	BT	THICKNESS SHEAR	5 – 15 MHz
C	CT	PLATE SHEAR	300 – 1000 kHz
D	DT	PLATE SHEAR	200 – 500 kHz
E	$-5°$ X-CUT	LONGITUDINAL	60 – 300 kHz
F	$+18.5°$ X-CUT	LONGITUDINAL	60 – 300 kHz
G	GT	LONGITUDINAL	100 – 556 kHz
H	$-5°$ X-CUT	LENGTH–WIDTH FLEXURE	10 – 100 kHz
J	$-5°$ X-CUT 2 PLATES	DUPLEX LENGTH–THICKNESS FLEXURE	1.2 – 10 kHz
M	MT	LONGITUDINAL	10 – 100 kHz
N	NT	LENGTH–WIDTH FLEXURE	10 – 100 kHz

cause the ratio of capacitances and the motional inductance of each cut to be different.

The dependence of resonance-frequency temperature coefficient on crystal plate and bar orientation has been used to determine the temperature coefficients of the material constants (Bechmann, 1934, 1955b, 1961, 1962; Bechmann and Ayers, 1951; Bechmann et al., 1962, 1963).

2.2.2 Dependence of Crystal Inductance on Temperature

The motional inductance of an AT-quartz resonator decreases with temperature nearly linearly at about 240 ppm/K. (Holbeche and Morley, 1981; Bottom, 1982). This effect is primarily caused by the temperature coefficient of the e_{26} associated with the resonator plate orientation.

2.2.3 Tabulations of Properties of Quartz Resonators

Several tabulations of the properties of quartz resonators have been published. For reference, some of these are repeated here with some changes to bring them up-to-date. Properties reported recently for several kinds of resonator are also summarized in this section. Useful parameters that describe the single-mode resonator are discussed on p. 107.

2 PIEZOELECTRIC RESONATORS AND WAVES

TABLE 2.2-2

Typical Properties of Quartz Crystal Resonators[a]

Element designation	Resonance frequency f (kHz)	C_0/C_1	C_1 (pF)	L_1 (H)	R_1 (ohms)
$A^{(n)}$	$1.6n/t$	$250n^2$	$0.097wlf/n^3$	$2.62 \times 10^5 n^3/lwf^3$	100
B	$2.56/t$	650	$0.0242wlf$	$10.5 \times 10^5/lwf^3$	100
C	$3.07/l$	350	$1.08/tf^2$	$23300t$	1000
D	$2.07/l$	400	$0.43/tf^2$	$59000t$	1000
E	$2.82/l$	125	$0.383/tf^2$	$66000t$	1000
F	$2.56/l$	130	$0.301/tf^2$	$84000t$	1000
G	$3.37/l$	350	$1.52/tf^2$	$16700t$	100
H	$5.00w/l^2$	190	$0.0179/tf^2$	$1.42 \times 10^6 t$	10000
J	$5.60t/l^2$	200	$2.54 \times 10^{-4}/tf^2$	$1.0 \times 10^6 t$	10000
M	$2.80/l$	190	$0.664/tf^2$	$38200t$	1000
N	$5.60w/l^2$	900	$0.00242/tf^2$	$1.05 \times 10^7 t$	10000

[a] All linear dimensions are in meters; t is thickness, w width, and l length. n is the overtone number.

Table 2.2-1 shows most of the common mode types, with the simple letters that are used to refer to each type (Edson, 1953; Mason, 1964). Table 2.2-1 also shows the crystal orientation, the vibration-mode type, and the useful frequency range for each type of resonator. Table 2.2-2 shows various mode types with approximate expressions for f_r, C_0/C_1, C_1, L_1, and R_1 (Edson, 1953). Figure 2.2-2 shows motional inductance ranges (Mason, 1964), and Fig. 2.2-3 shows motional resistance ranges (Mason, 1964) for the various resonator types at different frequencies. Table 2.2-3 (see p. 118) shows a range of values of capacitance ratio for the various resonator types (Mason, 1964). Figures 2.2-4 and 2.2-5 show dependencies of resonance frequency on temperature for some of the more commonly used mode types (Edson, 1953; Mason, 1964). Figure 2.2-6 shows how the crystal plates are cut from the quartz stone (Mason, 1950; Mason, 1964). In Fig. 2.2-6 the crystal orientation angles have sign changes to make the figure consistent with the most recent standard (IEEE, 1978).

2.2.4 Conclusion

The brief review of quartz resonator work in this section shows that quartz has been used in practical crystal resonators for many years. Although other materials are being studied for use in resonators, quartz still has the best combination of properties. Present work on quartz is directed at the development of resonators with higher frequency precision and lower cost, as

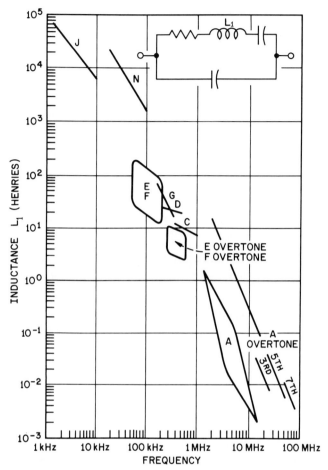

FIG. 2.2-2 Motional inductance ranges for quartz resonators.

2 PIEZOELECTRIC RESONATORS AND WAVES 115

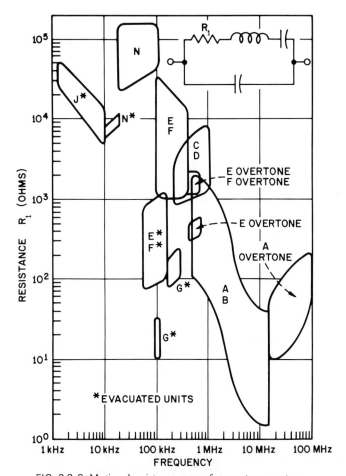

FIG. 2.2-3 Motional resistance ranges for quartz resonators.

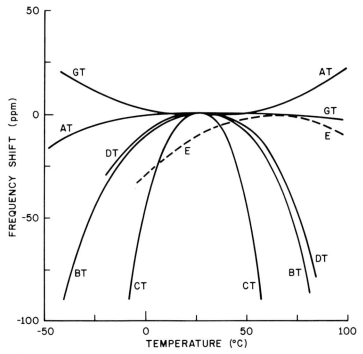

FIG. 2.2-4 Temperature coefficients of resonance frequency of quartz resonator types GT, AT, E, DT, CT and BT.

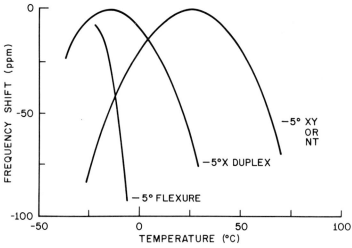

FIG. 2.2-5 Temperature coefficients of resonance frequency of quartz resonator types F, XD, and NT.

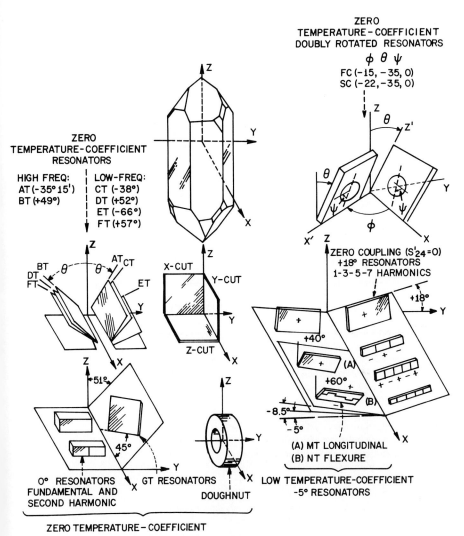

FIG. 2.2-6 Principal cuts of right-hand alpha quartz.

well as special dependences on temperature, acceleration, stress, and other forces. The current work on nonlinear effects and the design of stress-stable resonators is also of increasing importance. It appears that quartz will remain a useful material for crystal resonators for a very long time.

TABLE 2.2-3

Ratios of Capacitance (r) for Different Types of Quartz Crystal Resonator Elements

Resonator type[a]	$r = C_0/C_1$
A ($l/t > 5$)	250
B (l/t) > 5)	650
C ($w/l = 1$)	350
D ($w/l = 1$)	400
G ($w/l = 0.85$)	350
J ($t/l < 0.06$)	200
M ($w/l = 0.4$)	190
N ($w/l < 0.3$)	900

[a] Variable t is thickness, w width, and l length.

LIST OF SYMBOLS FOR SECTION 2.3

B	Stored energy susceptance in grating-equivalent circuit
C_0	Capacitance of interdigital transducer
C_1	Motional capacitance in resonator-equivalent circuit
f_t	Transducer center frequency
f_r	Frequency of resonance for SAWR
f_g	Frequency of maximum grating reflection
f_1	Frequency of first waveguide mode in grating
Δf_{long}	Longitudinal cavity mode resonance separation
Δf_{refl}	Grating reflection zero spacing
G_0	Interdigital-transducer conductance
$\hat{G}$	Maximum value of G_0
h	Groove depth
k^2	Piezoelectric coupling coefficient
L_1	Motional inductance in resonator-equivalent circuit
m	Effective cavity length in wavelengths
n	Distance between IDTS in wavelengths
N_g	Number of grooves in grating
N	Number of periods (wavelengths) in IDT
N_A	Electrode length in wavelengths (acoustic aperature)
p	IDT coupling parameter
Q_m	Material-limited resonator Q_u
Q_u	Unloaded resonator Q
Q_l	Loaded resonator Q
r	Reflection coefficient of one edge
R_1	Series resistance in resonator-equivalent circuit
R	Grating reflection coefficient
R_{peak}	Maximum value of R
R_Ω	Ohmic resistance

$R_\square$ Sheet resistivity
s_{max} Maximum grating separation
T Grating transmission coefficient
$T(f)$ Amplitude transmission factor of resonator
T_0 $T(f_r)$
v Acoustic velocity
v_0 Unperturbed acoustic velocity
w Groove width
x $\pi(f - f_t)N/f_t$
Y_0, Y_1 Admittances in transmission-line model of grating
Z_0 Source and load impedance
α Loss per grating period
δ Normalized deviation from stopband center frequency
ϵ Normalized admittance discontinuity from transmission-line model for grating
ϵ_0 Dielectric constant of free space (8.85×10^{-12} F/m)
ϵ_p Effective piezoelectric dielectric constant
γ Anisotropy parameter
θ Transmission phase for groove
κ Reflectivity per unit length
λ Wavelength
λ_t Wavelength in IDT
μ Power loss per transit of resonator cavity
τ Delay-line delay

2.3 SURFACE ACOUSTIC WAVES AND RESONATORS[§]

2.3.1 Introduction

The surface-acoustic-wave resonator (SAWR) is a recent addition to the group of components available for precision frequency control. In a SAWR, the properties of Rayleigh (surface-acoustic) waves are used to extend the range of fundamental-mode, piezoelectric resonators to frequencies well beyond 1 GHz. This capability for high-frequency operation eliminates the need for frequency multiplier chains and thereby improves the noise performance and pulling range of oscillators built with SAWRs. The small size and simplicity of SAWRs can lead to low-cost, UHF oscillators with spectral purity superior to that available using any other device for frequency control.

In this section, the role of SAWRs in frequency control applications is described through a comparison with the closely related and more familiar bulk-acoustic-wave resonators (BAWRs). The two types of resonators have identical equivalent electrical circuits but operate in nearly disjoint frequency regimes. Grating reflectors and interdigital transducers (IDTs), the subcomponents of SAWRs, are each discussed, and then the cavity design including losses is analyzed in detail. The process for SAWR fabrication is outlined, and the important packaging considerations required to ensure

[§] Section 2.3 was written by William R. Shreve and Peter S. Cross.

good long-term stability are discussed. Finally, the state-of-the-art performance characteristics of SAWRs are summarized including areas where significant improvement is anticipated.

2.3.1.1 BACKGROUND

Acoustic waves are propagating mechanical disturbances of a fluid or solid medium. A familiar example is an audible sound wave, but operating frequencies in the UHF and microwave regions are readily achievable. In hard materials like crystalline solids, acoustic waves cause elastic deformations with little frictional energy dissipation even at high frequency. Thus, these waves have low propagation loss with the lowest occuring in single-crystal materials. Because acoustic waves are mechanical rather than electromagnetic in nature, their velocity is low, typically 3000 m/sec or 10^{-5} times the velocity of light.

In crystals, acoustic waves are periodic deformations of the lattice that propagate through the crystal as the lattice relaxes toward its equilibrium position. Longitudinal waves cause displacements in the bulk of the crystal that are parallel to the direction of propagation and lead to alternating regions of compression and rarefaction. Transverse (shear) waves cause displacements in the bulk that are normal to the propagation direction. The most common surface wave, the Rayleigh wave, consists of both longitudinal and transverse displacements that can only propagate at a free boundary of the crystal. The displacement at this surface is elliptically polarized with the major axis of the ellipse normal to the surface. The motion near the surface is retrograde, but reverses at depths greater than approximately one-fifth of a wavelength (Auld, 1973). Essentially all of the energy in the Rayleigh wave is confined to a 1–2-wavelength-thick layer at the crystal surface. This energy confinement makes the wave accessible along the entire propagation path but also makes the propagation characteristics sensitive to surface loading or contamination.

In piezoelectric crystals the strains associated with acoustic waves generate electric fields, and conversely, electric fields applied to these crystals generate strains. Interdigitated sets of electrodes can be used to apply and detect these electric fields and thereby launch or detect surface waves. Typical interdigital structures of this type are shown in Fig. 2.3-1.

The stiffness, viscosity, and piezoelectric coefficients of a crystal and its symmetry determine properties like the SAW velocity, electromagnetic-to-acoustic coupling coefficient, temperature coefficient of delay, and attenuation coefficient. These quantities can be evaluated by solving the acoustic and electromagnetic field equations in conjunction with the constitutive relations between stress and strain. This calculation is quite involved for

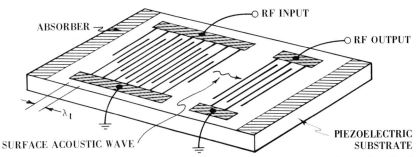

FIG. 2.3-1 Schematic of a SAW delay-line filter showing an overlap-weighted (apodized) input transducer, an unweighted output transducer, and acoustic absorbers on a piezoelectric substrate where $f_t = v/\lambda_t$.

anisotropic, piezoelectric crystals. The velocities of common crystals have been tabulated (Slobodnik *et al.*, 1973), but the complexity of the calculation and the need for accurate stiffness coefficients, especially when calculating temperature and stress sensitivities, make the discovery of new and better materials and orientations extremely difficult.

The accessibility of surface waves can be exploited to make simple delay lines, sophisticated bandpass filters, electroacoustic convolvers, and resonators. All of these devices depend on the excitation of waves with an electrical signal via the piezoelectric properties of the substrate or a layer on the substrate. These waves propagate nondispersively. They can be sampled with electrodes, reflected, modified by interaction with carriers in adjacent semiconductors, or effectively absorbed by lossy material on the surface. This flexibility has led to a wide variety of devices as evidenced by the large number of patents and publications in the SAW field.

The basic SAW device configuration, the delay-line filter, is shown in Fig. 2.3-1. The device consists of two IDTs fabricated on a piezoelectric substrate such as quartz or lithium niobate. The input IDT efficiently generates SAWs when an rf voltage is applied with a *temporal* period equal to the acoustic transit time across one *spatial* period of the electrode pattern. The IDTs are reciprocal devices so that acoustic waves are reconverted to an electrical signal by the output IDT. To first order, the filter frequency response is determined by the wave traveling from the input IDT to the output IDT. Waves traveling in the opposite direction or passing beyond the output IDT are absorbed. The impulse response of an IDT is simply a time replica of the overlap pattern of the electrodes. For example, in Fig. 2.3-1 the impulse response of the input IDT is a half-cycle cosine and that for the output IDT is a rectangle. If only one transducer has varying overlap, then the frequency response is the product of the Fourier transforms of the individual impulse responses of the IDTs. Tancrell and Holland (1971) showed that if both transducers are overlap weighted, then calculation of the frequency response is more complex. The group delay (or phase-slope) is determined by the center-to-center separation of the IDTs and can be made quite long. These filters are discussed in detail in Sections 5.3 and 5.4.

A SAW resonator differs from a SAW delay-line filter in that the response is determined by multiple passes of acoustic waves between reflectors. As shown in Fig. 2.3-2a, the device typically consists of two periodic arrays of shallow grooves that enclose two IDTs. Each periodic array is an efficient reflector of surface waves for a band of frequencies determined by the grating period and groove depth. When the two arrays are properly positioned near one another, a high Q, Fabry–Perot cavity is formed. Finally, coupling

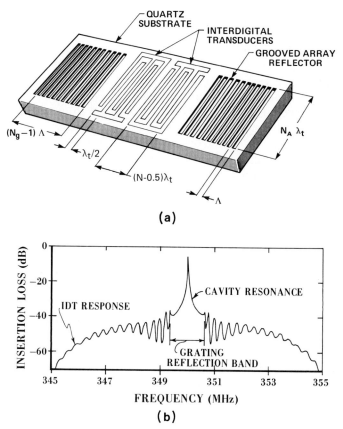

FIG. 2.3-2 (a) Schematic of a two-port SAW resonator showing two arrays of etched grooves that define the resonant cavity containing two IDTs. (b) Frequency response of a SAWR with two recessed, 120-electrode transducers, two 4.5-μm-period gratings, each consisting of 1000 grooves that are 800 Å deep, and an overall active width of 50 wavelengths.

TABLE 2.3-1

Comparison of ST-cut SAWRs and AT-cut BAWRs.

Property	SAWR	BAWR
Material	Quartz	Quartz
Orientation	ST cut	AT cut
Fundamental frequency (MHz)	50–1500	0.5–100
Q_m	$1.1 \times 10^{13}/f_r$	$1.5 \times 10^{13}/f_r$
Stability		
Temperature (0 to 50°C)	±11 ppm	±4 ppm
Acceleration	1 ppb/g	1 ppb/g
Drift	1 ppm/year	0.01 ppm/year
Configuration	1 or 2 port	1 port

to the electrical circuit is accomplished by the two relatively wideband IDTs. A typical SAWR frequency response is shown in Fig. 2.3-2b. The shape of the response near the center frequency is determined by the narrowband cavity resonance. In contrast to the delay-line filter, the overlap patterns of the IDTs are chosen to couple energy only to the fundamental cavity mode rather than to explicitly effect the filter impulse response.

2.3.1.2 COMPARISON OF SAWR AND BAWR

Surface-acoustic-wave resonators are closely related to bulk-acoustic-wave resonators, which have become the standard for instrument frequency control over the past 50 years. The major resonator characteristics for each type of device are summarized in Table 2.3-1. Both SAWRs and BAWRs are fabricated from α-quartz, although with different crystallographic orientations: 42.5°-rotated Y-cut (ST) for SAWR and 35.25°-rotated Y-cut (AT) for BAWR.

The significant difference between SAWRs and BAWRs is the range of frequencies for fundamental-mode operation. The fundamental resonance in bulk-wave resonators occurs when the plate is one-half wavelength thick. At frequencies above 100 MHz, the plate is less than 25 μm (1 mil) thick, making fabrication difficult and the resulting device very fragile. In SAWRs, however, the frequency of resonance is primarily determined by the periodicity of the grating reflectors and is independent of the substrate dimensions. Thus, the upper frequency limit is set by the achievable lithographic resolution. Tanski (1979a,b,c) used standard planar photolithography to fabricate SAWRs with linewidths as small as 0.55 μm, giving a resonance frequency of about 1.4 GHz. This upper limit has been extended to 2.6 GHz (0.3-μm lines–spaces) by Cross et al. (1980) with direct-write electron-beam lithography.

A lower frequency limit for practical SAWRs occurs because the devices require several hundred periods in each reflector resulting in an overall length of several centimeters at 50 MHz (where the period is about 30 μm).

Higher-frequency operation gives SAWRs several advantages in oscillators. To obtain an output at frequencies above about 100 MHz, a BAWR oscillator must be combined with a multiplier chain and filters to remove unwanted close-in harmonics. The multipliers not only add to the cost and complexity of the oscillator, but the noise floor increases as the square of the multiplication factor. Thus, fundamental-mode SAWR oscillators have a lower noise floor. Finally, the pulling range of an oscillator varies as the inverse of the multiplication factor, resulting in tighter initial tolerance on resonators (BAWRs) used in multiplied oscillators. A more detailed comparison of multiplied and fundamental-mode oscillators is given in Section 8.2 of Volume 2.

Viscous damping in the substrate places an upper limit on resonator quality factor Q_m, which is inversely proportional to frequency. This material-limited Q is dependent on crystallographic orientation and type of wave and is about 40% higher for AT-BAWRs than for ST-SAWRs.

The AT-cut for bulk waves has been widely used because of its excellent temperature characteristics, as shown in Fig. 2.3-3. The frequency varies as a *cubic* function of temperature, which minimizes frequency shifts over a wide temperature range. The first temperature-stable SAW cut, the ST-cut, has a *quadratic* dependence of frequency on temperature, which somewhat restricts the useful temperature range (Schulz *et al.*, 1970; Dias *et al.*, 1975).

The long-term stability of BAWRs is at present two orders of magnitude better than that of SAWRs. This is the result of the extensive development

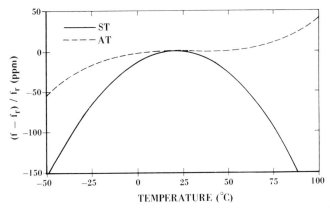

FIG. 2.3-3 Static frequency variation as a function of temperature for AT-cut BAWR and ST-cut SAWR.

that has been carried out on fabrication and packaging techniques over the past 50 years. Since SAWRs are relatively new devices, much less work on aging has been carried out. After Ash (1970) discussed SAWR structures, Staples (1974) was the first to publish results on the SAWR configuration discussed here. Despite this late start, by drawing on the experience gathered in BAWR work, rapid progress on SAWR long-term stability is now occurring.

SAWRs have the additional option of one- or two-port operation, which can simplify oscillator design. This difference shows up clearly when one examines the equivalent circuit for the resonator, shown in Fig. 2.3-4. The BAWR circuit shows the series RLC characteristic of a one-pole resonator shunted by the static capacitance of the electrodes. This static capacitance provides a relatively low-impedance path that can mask out the desired crystal resonance. As a result, an external inductor (shown dotted) is usually added to resonate out this capacitance. Additional filtering may also be required to remove the effects of the inductance and capacitance far from resonance. The shunt conductance $G_0(1/G_0 \gg R_1)$ is included to model the acoustic energy dissipated in the crystal in a relatively broad frequency band around resonance.

A SAWR with one transducer in the cavity (Fig. 2.3-4b) has an equivalent circuit identical to that of the BAWR. The shunt capacitance is due to the interdigital transducer itself. By using two transducers to couple energy into and out of the cavity, the problem of the shunt capacitance can be greatly reduced. As shown in Figure 2.3-4c, the capacitance now appears across

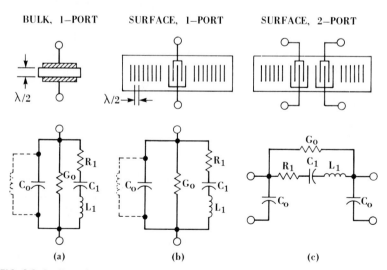

FIG. 2.3-4 Crystal resonator configurations and their associated equivalent circuits.

each port separately and does not shunt the resonant arm. Thus, there is no longer a path for spurious input–output coupling.

To briefly summarize, SAWRs and BAWRs are suited to different applications. At low frequencies, high Q and temperature stability make the BAWR the better choice for precision frequency control. At frequencies above 100 MHz, the lower noise floor, greater pulling range, and simplified circuitry make the SAWR oscillators an attractive alternative.

2.3.2 Resonator Design

The overall SAW resonator response is determined by the individual contributions of the grating reflectors, the cavity spacing and the IDTs. Each of these components and their aggregation into a resonator is discussed below followed by a description of the major loss mechanisms present in SAWRs. These losses are the ultimate limit of SAWR performance.

2.3.2.1 GRATING REFLECTORS

The key to the design of high-Q resonators is the choice of the proper reflector. Periodic gratings are nearly ideal SAW reflectors. They scatter primarily into a backward-traveling surface wave and couple only weakly to other acoustic modes. They reflect over a narrow band of frequencies so that a cavity that is many wavelengths long can be designed to support a single resonance. Furthermore, they can be fabricated in a straightforward manner using standard, planar processing technology much like that used to make integrated circuits.

Each grating is an array of weak acoustic perturbations spaced so that the small reflections from individual perturbations add in phase to produce nearly 100% reflection. Several types of perturbation have been used, including deposited metal (Staples, 1974) or dielectric (Staples *et al.*, 1974), etched grooves (Li *et al.*, 1975a; Miller *et al.*, 1975), localized material property changes introduced by diffusion (Schmidt, 1975), or ion implantation of impurities (Hartemann, 1975). Only etched grooves in quartz are discussed in detail here because they have been most thoroughly tested and have yielded the best results to date.

The basic physics of the reflective arrays has been studied by several groups and is now well characterized to second order in the perturbation magnitude (e.g., groove depth). There are two primary models used to analyze the gratings: repetitively mismatched transmission lines and a coupling-of-modes formalism. The transmission-line model provides good insight into the physics of the reflection mechanisms but leads to rather cumbersome analytical expressions. On the other hand, the coupling-of-modes

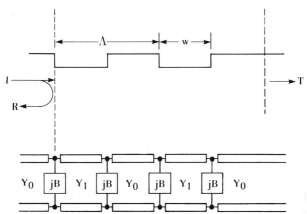

FIG. 2.3-5 Cross section of reflector grating consisting of grooves of width w with period Λ and the transmission-line equivalent circuit for a grating. R and T are the grating reflection and transmission coefficients, respectively. The groove edge acts as a transformer to change the transmission-line admittance from Y_0 to Y_1 and as an energy storage mechanism represented by the reactance B.

technique provides less intuitive understanding but yields simple, closed-form expressions. Over the frequency band of interest ($|f - f_g|/f_g < 0.1$, where f_g is the frequency of maximum grating reflection), the results from the two models are mathematically equivalent. Thus, we use here a hybrid of the two approaches to facilitate the presentation and hopefully clarify the notational ambiguities that now exist in the literature.

A cross-sectional view of a grooved array is shown in Fig. 2.3-5 along with its transmission-line equivalent circuit. Sittig and Coquin (1968) showed that groove and ridge regions in an acoustic substrate can be modeled by transmission lines with different characteristic impedances Y_0 and $Y_1 = Y_0(1 - \varepsilon)$. Li and Melngailis (1975) added the shunt elements jB to model stored reactive energy associated with each groove edge. Transmission-line theory predicts a reflection per groove $2r$, given by

$$2r = \varepsilon \sin \theta + \hat{B} \cos \theta, \qquad (2.3\text{-}1)$$

where $\theta = 2\pi w/\lambda$, w is the groove width, λ the free-surface acoustic wavelength, $\hat{B} = B/Y_0$, and the phase reference plane is the downstep of the groove. For arrays of grooves of depth h, the parameters ε and $\hat{B}$ are

$$\varepsilon = A_1(h/\lambda),$$
$$\hat{B} = A_2(h/\lambda)^2, \qquad (2.3\text{-}2)$$

representing the contributions due to impedance mismatch and stored energy, respectively. The multipliers A_1 and A_2 depend on the details of the groove shape (Shimizu and Takeuchi, 1979; Wright and Haus, 1980), but practical values for nearly rectangular grooves on ST-quartz are $A_1 \simeq 0.60$ and $A_2 \simeq 34$.

The reactive stored energy also has the effect of reducing the unperturbed wave velocity v_0 so that the frequency (f_g) of peak reflection for an array is

$$f_g = \frac{v_0}{2\Lambda}(1 - \hat{B}/\pi), \qquad (2.3\text{-}3)$$

where Λ is the grating period. Thus, f_g shifts down quadratically with the groove depth. This property can be exploited to provide fine control of the resonator frequency during fabrication.

In the coupled-mode formalism, Cross and Schmidt (1977) characterize the distributed reflection in a grating by the reflectivity per unit length (also called the coupling coefficient) κ which is related to r by

$$\kappa\Lambda = 2r. \qquad (2.3\text{-}4)$$

From the coupled-mode results, the amplitude reflection coefficient R of an array of N_g grooves is given by

$$R = \frac{2r}{\sqrt{4r^2 - \delta^2}\coth(N_g\sqrt{4r^2 - \delta^2}) + j\delta}, \qquad (2.3\text{-}5)$$

where Eq. (2.3-4) has been used and $\delta = \pi(f - f_g)/f_g$ is a measure of the deviation from the stopband center frequency. The effect of propagation loss can be included in Eq. (2.3-5) by replacing δ with $\delta - j\alpha$, where α is the loss per grating period.

The magnitude and phase of Eq. (2.3-5) are plotted in Fig. 2.3-6. The reflection magnitude in Fig. 2.3-6a exhibits high reflectivity over the central reflection bandwidth with a sidelobe structure away from the center frequency. The fractional reflection bandwidth is proportional to r (and therefore essentially proportional to groove depth), and the peak reflectivity R_{peak} is found from Eq. (2.3-5):

$$R_{\text{peak}} = \tanh(2N_g r). \qquad (2.3\text{-}6)$$

It is desirable to minimize the reflectivity outside the central reflection bandwidth to eliminate spurious longitudinal modes from the resonator response. The sidelobes arise because the local reflectivity abruptly starts and stops at the grating ends. The magnitude of these sidelobes can be substantially reduced by weighting the local reflectivity near the grating ends by varying the groove depth or groove length (Joseph and Lakin, 1975) or by selectively removing entire grooves (Tanski, 1979a,b,c) or parts of grooves (Cross, 1978).

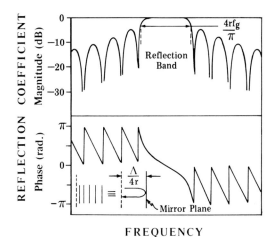

FIG. 2.3-6 Grating reflection coefficient. The inset shows the equivalence of the grating and a mirror at the phase center of reflection.

The reflection phase is plotted in Fig. 2.3-6b. Near the center frequency, the phase varies linearly with frequency, which leads to an interesting physical interpretation: The grating becomes equivalent to a mirror of reflectivity R_{peak} placed a distance $\Lambda/4r$ behind the actual position of the first reflector in the array, as indicated in the inset to the figure.

This simple, "equivalent mirror" model is very useful in understanding the behavior of a distributed Fabry–Perot cavity such as a SAWR despite its accuracy being limited to the center of the reflection band. For example, the frequency separation Δf_{long} between longitudinal cavity modes is

$$\Delta f_{long} = v/2m\lambda, \qquad (2.3\text{-}7)$$

where v is the acoustic propagation velocity and $m\lambda$ the effective mirror separation. Substituting $\lambda/4r$ (the minimum possible separation between the two equivalent mirrors) for $m\lambda$ we find

$$\Delta f_{long} = 2rf_g. \qquad (2.3\text{-}8)$$

From Eq. (2.3-5) one can prove that the separation of reflection zeros Δf_{refl} is

$$\Delta f_{refl} = (4/\pi)rf_g \qquad (2.3\text{-}9)$$

Thus, the longitudinal-mode spacing is *greater* than the reflection bandwidth thereby ensuring that the resonators can support only a single, longitudinal mode (at least for zero separation between the physical gratings).

2.3.2.2 TRANSDUCERS

The IDTs in the cavity in Fig. 2.3-2a largely determine the impedance level of the resonator. As a result, a prudent choice of IDT parameters can substantially reduce the complexity of the networks required to match the resonator and load impedances for a given application. As mentioned previously, the use of two transducers can simplify these networks by eliminating the capacitance shunting the resonant arm in the equivalent circuit and thereby can obviate the need for additional filtering. The matching circuitry can often be further simplified by designing the transducers so that the intrinsic resonator impedance is appropriate for the particular embedding network. In this section, the achievable impedance levels are related to design parameters.

Smith et al. (1969) showed that the admittance of an unapodized transducer consisting of N pairs of electrodes of length $N_A \lambda_t$ can be approximated by a capacitance C_0 in parallel with a conductance G_0 and a susceptance B_0 as follows:

$$C_0 = NN_A\lambda_t(\varepsilon_p + \varepsilon_0),$$

$$G_0 = \hat{G}[(\sin x)/x]^2,$$

$$B_0 = \hat{G}[(\sin 2x) - 2x]/2x^2, \qquad (2.3\text{-}10)$$

$$\hat{G} = 8k^2 v_0(\varepsilon_p + \varepsilon_0)N^2 N_A,$$

$$x = \pi(f - f_t)N/f_t,$$

where ε_p is the effective dielectric constant of the substrate, ε_0 that of air, k^2 the piezoelectric coupling coefficient of the substrate, f the frequency, f_t the synchronous frequency of the transducer (the frequency at which all the electrodes launch waves in synchronism), and $\lambda_t = v/f_t$ the wavelength at that frequency. The effects of mechanical reflections from the transducer electrodes are neglected in Eq. (2.3-10). For the purpose of SAWR analysis, the susceptance B_0 (which is zero at $f = f_t$) can also be neglected.

These element values can be related to the impedances of the other elements in the equivalent circuit (Fig. 2.3-4) for a resonator with a distance $m\lambda$ between reflection centers and total single-transit, fractional power loss μ (not including coupling losses) as follows (Shreve, 1975; Cross and Schmidt, 1977):

$$R_1 = \mu/p\hat{G}, \qquad (2.3\text{-}11\text{a})$$

$$L_1 = m/pf_r\hat{G}, \qquad (2.3\text{-}11\text{b})$$

$$C_1 = 1/(2\pi f_r)^2 L_1, \qquad (2.3\text{-}11\text{c})$$

where f_r is the frequency of resonance and p the IDT coupling parameter (see Section 2.3.2.3). With IDTs in the maximum coupling position, $p = 4$. In the derivation of these relationships, the cavity was assumed to resonate at the center of the grating reflection band.

In general, one would like to diminish the effect of the series resistance R_1 either by making it as small as possible or by using matching networks to transform the load impedances to high levels. From Eqs. (2.3-10) and (2.3-11a) one can see that R_1 can be reduced in three ways: (1) by choosing materials with a large piezoelectric coupling coefficient; (2) by increasing the length and number of electrodes in the transducer; or (3) by minimizing the total cavity losses. As noted above, quartz is the only suitable choice for frequency control applications due to its temperature stability. However, quartz has a low piezoelectric coupling coefficient. Therefore, assuming that cavity losses have been minimized, R_1 can only be reduced by increasing the transducer size (i.e., the length and number of electrodes). The various loss mechanisms present in SAWRs and techniques for their minimization are discussed in Section 2.3.2.4.

The maximum transducer size is set by the occurrence of spurious resonator modes and by second-order effects within the transducers. As noted at the end of Section 2.3.2.1, a grating resonator can support only one longitudinal mode if there is no distance between the gratings. In order to accommodate the IDTs, the gratings must be spaced apart causing the longitudinal-mode spacing to decrease. At some separation $s_{max} \lambda$, spurious longitudinal modes can occur within the reflection bandwidth of the gratings. For uniform highly reflective gratings, s_{max} is given by

$$s_{max} \approx 0.8 \lambda / h. \qquad (2.3\text{-}12)$$

The groove depth h is usually set at about 0.01λ to avoid excessive bulk-scattering loss from deep grooves or excessive radiation losses with a reasonable-length grating. This depth yields a value of 80 for s_{max}. Thus, a resonator can accomodate a single IDT with 80 electrode pairs or two transducers with 40 electrode pairs each. Special designs such as synchronous IDTs (Cross *et al.*, 1979), symmetric IDTs (Stevens *et al.*, 1977; Rosenberg and Coldren, 1980), or weighted gratings (Cross, 1978) can be employed to increase the numbers of electrodes somewhat, but even in these designs, the coupling cannot be increased without limit by increasing N. Further reductions of R_1 must be achieved by changing the resonator width.

Increasing the width of a resonator to reduce R_1 introduces the problem of spurious waveguide or transverse modes in the resonator. Haus (1977b) showed that grating arrays act as waveguides for Rayleigh waves. Since the gratings are many wavelengths wide, they are multimode waveguides with the mode spacing determined by the grating width. These grating-waveguide modes are commonly referred to as transverse modes because of their similarity to the transverse modes in electromagnetic waveguides. The amplitude profiles of the different modes have been calculated, and Shreve (1976a) showed that transducers can be designed to couple only to the desired mode. This technique, *transducer apodization*, requires varying the length of the electrodes so that the launched-wave amplitude matches the amplitude profile of the fundamental waveguide mode. The apodization also results in a reduction of the capacitance C_0 by a factor of approximately $2/\pi$ and a decrease in $\hat{G}$ by about 2. Most successful applications of this technique have been at frequencies below 300 MHz.

At higher frequencies it becomes increasingly difficult to control fabrication parameters such as groove profile, depth, and width to the degree required to match the waveguide-mode profile to the launched acoustic wave (Tanski, 1979c). As a result, unwanted modes cannot be effectively suppressed. The unwanted modes decrease resonator Q and significantly distort the resonator amplitude and phase response.

Transverse mode distortion can be removed from the frequency band around the fundamental resonance by reducing the grating width $N_A \lambda$. On ST-quartz, the frequency of the nearest transverse mode can be related to the fundamental frequency by (Haus, 1977b; Cross et al., 1980).

$$\Delta f_1/f_r = 1.035/N_A^2, \qquad (2.3\text{-}13)$$

where $\Delta f_1 = f_1 - f_r$ is the separation between the first transverse mode and the resonant frequency. If all distortion must be eliminated within a certain range around the peak response, then a limit on the maximum allowable width is effectively set by Eq. (2.3-13).

To briefly summarize, the transducers can be designed to control the impedance of a resonator within the limits on transducer size set by the occurrence of spurious longitudinal and transverse modes.

2.3.2.3 CAVITY DESIGN AND FREQUENCY RESPONSE

After the grating and transducer geometries have been selected, they must be combined to form a resonator. The positioning of the gratings and transducers determines the resonance frequency and the degree of coupling to the cavity. Proper positioning depends on a precise knowledge of the wave velocity and grating-reflection phase.

The grating-reflection phase has been determined for common substrate–reflector combinations (Dunnrowicz et al., 1976; Tanski and van de Vaart, 1976). The peak coupling position is either $\Lambda/4$ or $3\Lambda/4$ from the center of the last reflector, where Λ is the grating period, and the space between transducers is an integral number of periods. The maximum coupling position for an IDT on quartz with groove reflectors is shown in Fig. 2.3-7.

2 PIEZOELECTRIC RESONATORS AND WAVES

FIG. 2.3-7 Cross section of a quartz resonator showing transducers in the maximum coupling position.

If the grating-reflection-band center frequency f_g and the transducer synchronous frequency f_t are to correspond, the period of the grating and transducer must be slightly different to compensate for velocity differences in these regions. In the grating, the velocity is perturbed only by energy storage at the groove edges (Li and Melngailis, 1975). In the transducer region the velocity is perturbed by piezoelectric shorting and mass loading of the surface by the electrodes as well as by energy storage at electrode edges. Tanksi (1979a) characterized the velocity for recessed transducers on ST quartz.

Slobodnik et al. (1973) calculated the unperturbed velocity of common substrates from their material properties. An error in this velocity results in a shift from the desired frequency but does not change the response shape since the grating and transducer velocities are both calculated relative to the substrate velocity.

When the design is completed, it is instructive to check the design with an accurate model. Design trade-offs and constraints can be incorporated into a powerful computer-aided design tool. Cross and Schmidt (1977) employed coupled-mode analysis of gratings and a first-order transducer model to derive wave amplitude scattering matrices for gratings and IDTs. These matrices can be easily combined and computerized to give an accurate, numerically efficient model for SAWRs. The model allows one to "fine tune" a design and eliminate errors that would otherwise waste time and money in mask generation.

The primary performance specifications of a SAWR are its frequency response and the stability of that response. A typical plot of the electrical transmission of a two-port SAWR versus frequency is given in Fig. 2.3-8. Without the reflective gratings, the delay line formed by the two (unapodized) IDTs has the broad $[(\sin x)/x]^2$ response described by Eq. (2.3-10) and shown in Fig. 2.3-8a. The fractional bandwidth (between the zeros) of the central lobe is $2/N$, where N is the number of pairs of electrodes in the IDT.

When the gratings are included, the complete SAWR response is obtained as depicted in Fig. 2.3-8b. Inside the grating-reflection band, a high-Q cavity exists, and the sharp, resonance peak emerges from the broad IDT background. Outside the reflection band, the IDT response is modulated by numerous subsidiary peaks due to the sidelobes of the grating-reflection spectrum.

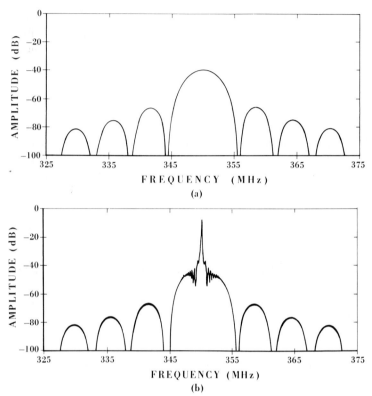

FIG. 2.3-8 (a) Frequency response of a delay line formed by two unweighted, nonreflecting, 120-electrode IDTs with 50 wavelength aperture. (b) Frequency response of the transducers in (a) flanked by 4.5 μm-period gratings each consisting of 1000 grooves that are 80 nm deep.

Near the center frequency f_r, the device behaves essentially as a one-pole resonator with an amplitude transmission factor $T(f)$ (the ratio of the output signal amplitude to the input amplitude) given by

$$T(f) = \frac{T_0}{1 + j2Q_l(f - f_r)/f_r}, \tag{2.3-14}$$

where T_0 is the amplitude transmission factor at resonance and Q_l is the loaded Q. When the SAWR is used in a circuit with source and load impedances of Z_0,

$$T_0 = (1 + R_1/2Z_0)^{-1}, \tag{2.3-15}$$

$$Q_l = Q_u(1 - T_0), \tag{2.3-16}$$

where Q_u is the unloaded Q of the resonator.

The resonance behavior of the SAWR can be characterized by the three parameters f_r, R_1, and Q_u. These parameters can be calculated from Eqs. (2.3-14) through (2.3-16) by measuring the frequency of resonance, bandwidth, and T_0.

2.3.2.4 LOSS MECHANISMS

It is important to minimize the losses in a SAWR cavity to optimize Q and R_1, and thereby achieve the best phase noise in resonator-stabilized oscillators. The losses in a resonator can be related to the measured electrical characteristics through the relation

$$Q_u = 2\pi m/\mu. \qquad (2.3\text{-}17)$$

The unloaded Q and cavity losses μ can be found by measuring T_0 and Q_1 and by calculating the cavity size m, the distance between the equivalent mirror planes defined in Section 2.3.2.1, from the resonator geometry.

The total loss can be subdivided into its constituent parts. Identifiable sources of loss include viscous damping, ohmic losses, bulk scattering, radiation through the gratings, diffraction, air loading, surface scattering, and geometrical losses. The individual losses and their frequency dependences are discussed below.

The *material loss* (viscous damping in the substrate) represents the fundamental limitation on device Q. This loss can be characterized by a constant a_{mat} that depends on the substrate orientation,

$$\mu_{\text{mat}} = a_{\text{mat}} m f. \qquad (2.3\text{-}18)$$

Budreau and Carr (1971) measured a value $a_{\text{mat}} = 6.0 \times 10^{-13}$ for ST-quartz that corresponds to a material-limited Q, Q_m, of 10,500 at 1 GHz. An additional viscous-damping loss can be introduced by the metal film used in the transducers. It is significantly less than the material losses of the quartz substrate, although no direct measurement has been reported.

Ohmic losses are a function of the current in the transducer. Specifically, the fractional loss μ_{ohm} in a single transit of the cavity is proportional to the product of the effective ohmic resistance R_Ω in an IDT times the transducer radiation conductance G_0 and is given approximately by

$$\mu_{\text{ohm}} = 2 p G_0 R_\Omega. \qquad (2.3\text{-}19)$$

The factor of 2 accounts for the two transducers in a two-port SAWR. The IDT resistance can be inferred from Lakin (1974) to be

$$R_\Omega = 4 c R_\square N_A / N, \qquad (2.3\text{-}20)$$

where $R_\square$ is the sheet resistivity of the metalization and c is a dimensionless parameter that measures the effective current-carrying length of the electrodes. This parameter falls into the range $\frac{2}{3} \leq c \leq 1$ depending on the apodization. From Eqs. (2.3-10), (2.3-19), and (2.3-20) it can be shown that the ohmic loss is related to the size of the transducer by

$$\mu_{\text{ohm}} \propto NN_A^2. \qquad (2.3-21)$$

Therefore even though R_Ω may decrease, the ohmic loss in the resonator increases as N is increased.

Scattering into bulk waves is a significant source of loss in most resonators and ultimately limits the size of the perturbation that can be used in the gratings. This scattering is related to the reactive energy storage discussed above and, like the stored energy reflection [Eqs. (2.3-1) and (2.3-2)], is proportional to $(h/\lambda)^2$ (Li et al., 1975a; Tanski, 1978). In general terms, the bulk waves scattered from individual reflectors in the center of an array cancel and result in evanescent nonzero strain fields (energy storage). At the ends of the array the scattered bulk waves do not cancel completely, and energy is carried away from the surface resulting in loss. Thus, the major losses occur at the edges of the cavity and

$$\mu_{\text{bulk}} = a_{\text{bulk}}(h/\lambda)^2, \qquad (2.3-22)$$

where a_{bulk} depends on material choice. From the work of Li and Tanski, for quartz grooves a_{bulk} is in the range of 17 to 20.

Radiation loss through the gratings is characterized by the grating transmission which can be found from Eq. (2.3-6):

$$\mu_{\text{rad}} = \text{sech}^2(2N_g r). \qquad (2.3-23)$$

This loss can usually be made negligibly small by increasing the number of reflectors. In cases where the substrate size is limited for technical or economic reasons, the reflector depth h can be adjusted to minimize the sum $\mu_{\text{rad}} + \mu_{\text{bulk}}$ with N_g set at the maximum allowable level.

Diffraction loss is not usually a major factor in resonators since most of the propagation takes place in periodic structures that guide the wave. Loss occurs only in the cavity region between the reflecting structures, be they gratings or transducers. A good estimate of this loss can be obtained by assuming that the beam in this region approximates a gaussian profile in the far field. Szabo and Slobodnik (1973) modeled diffraction on anisotropic substrates in the far field. Their work leads to a loss

$$\mu_{\text{diff}} = 0.4|1 + \gamma|D/N_A^2, \qquad (2.3-24)$$

where γ is the anisotropy parameter of the substrate and D is the number of wavelengths between the reflecting structures.

Acoustic energy can be coupled to the atmosphere surrounding the SAWR and thereby introduce loss. For air at atmospheric pressure loading the ST quartz, this loss is

$$\mu_{air} = (1.1 \times 10^{-4})m, \qquad (2.3\text{-}25)$$

which corresponds to a Q of 57,000. This loss can be reduced or eliminated by operating SAWRs in a rarefied atmosphere or vacuum.

Scattering from imperfections in the surface can also introduce significant losses for surface waves at high frequencies (Sabine, 1970), since scattering from a random distribution of small defects (Rayleigh scattering) increases as frequency to the fourth power. Scattering loss can be reduced to a negligible level at frequencies below 1 GHz by proper substrate preparation (Slobodnik, 1974).

Conversion of energy from the fundamental mode to higher-order transverse modes can occur if the higher-order modes are inadequately suppressed. Mode conversion is effectively a loss since it couples energy out of the desired mode and degrades resonator Q. The presence of significant conversion is often indicated by distortion of the resonator frequency response. The distortion is reduced by apodization and by decreasing the resonator width.

Geometrical nonuniformities in the resonator pattern can act as a "virtual loss" factor, especially at high frequencies (Cross *et al.*, 1980). Regular variations caused by pattern generator round-off, nonuniform groove depth, or varying metal thickness in transducers can cause the resonant frequency to vary as a function of position in the cavity. The variation in the resonant frequency in turn yields a broadened overall response. This Q degradation has the same effect as an additional cavity loss.

2.3.3 Fabrication

Section 4.8 describes the general fabrication method used for SAW resonators. Three advantages over BAW resonator fabrication result from planar processing on conventional silicon-wafer processing equipment. First, a large number of devices can be processed simultaneously on a wafer. When compared to individual device processing, wafer processing reduces the cost per device and the variations between devices. Secondly, critical device parameters are set by the mask, not by device fabrication. The resonance frequency of a SAWR is determined to within ± 250 ppm by the grating period and cavity size. This is to be compared to the tolerance of 1000 ppm achieved in BAWRs by lapping the crystals to thickness. The IDT apodization on the mask suppresses unwanted transverse modes in SAWRs just as contouring a BAWR to a particular shape results in energy trapping. Thirdly, the control over the frequency and uniformity achievable on a wafer makes it possible to trim entire wafers to frequency or to entirely avoid trimming and simply accept the yield of devices that fall on frequency.

A wide variety of processing techniques is possible for SAW devices, depending on the design details and personal preference. In all cases, the photolithography requires only a single critical masking step, so even though the line widths are often smaller than average integrated circuit dimensions, yields are typically high.

Figure 2.3-9 illustrates a typical lift-off process. The patterns of the IDTs and gratings are defined in a photoresist layer on the crystal surface. Aluminum is deposited on the entire wafer, and the photoresist is dissolved to lift-off the unwanted aluminum, thereby replicating the mask pattern for the transducers and grating in aluminum. (This lift-off process can be replaced by an etch process, but etching sacrifices some control of line width, a critical parameter at high frequency.) Next, more photoresist is applied and patterned to form a layer over only the transducers and cavity. The alignment of this pattern to the finer IDT-grating mask pattern is not critical to device performance since the IDT-grating spacing was determined by the first mask. The grooves are etched into the quartz by a CF_4-reactive ion-etch process that uses the aluminum in the grating as a mask. Finally, any aluminum in the grating region is etched away, the photoresist is stripped, and the devices are ready for testing. The wafer can be probed to determine the number of good devices and then diced up into individual devices.

The SAWRs can be trimmed either before or after dicing by a variety of methods. Since the velocity of the SAW is determined by the condition of the surface, any perturbation can be used for trimming. A nonconducting layer such as an oxide (Schoenwald *et al.*, 1975) or silicone (Parker, 1980) can be used, or if the aluminum mask is left in the grating region, then the groove depth can be changed to trim the frequency (Adams and Kusters, 1977). Penavaire *et al.* (1980) added aluminum strips to the cavity to trim resonators to ± 10 ppm of the desired frequency. Cross and Shreve (1981) invented a technique using reactive ion etching (the same process used to etch grooves) to trim the frequency of any SAW device. Proper selection of the gas mixture during etching gives a differential etch rate between the transducer metalization and substrate. This causes changes in the stored energy and reflectivity of the transducer electrodes, which in turn cause a shift in the resonance frequency. This technique introduces no additional materials to affect either electrical performance or stability.

The final processing steps and the techniques used to package SAWRs largely determine the SAWRs long-term stability. As discussed in detail

2 PIEZOELECTRIC RESONATORS AND WAVES 139

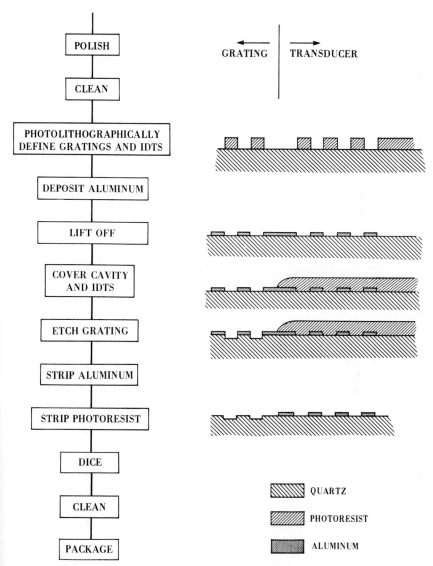

FIG. 2.3-9 Typical SAWR fabrication process flow diagram. The device cross section is pictured opposite the corresponding process step.

in Sections 4.8 and 6.3, the best aging results reported have been achieved by emulating packaging techniques used for precision BAWRs. The SAWRs are held in place by the metal straps used to make electrical connections or by separate straps connected to the packages. The packages are metal–ceramic and can be sealed with a cold-weld or thermocompression bond.

2.3.4 State-of-the-Art Performance

As stated in Section 2.3.2, resonator performance can be judged by the frequency response, which is characterized by f_r, Q_u, and R_1, and by frequency stability. Both Q_u and R_1 are in turn functions of the frequency. Realistic bounds for these parameters can be set by imposing a few arbitrary, but realistic limits on fabrication parameters and on the resulting SAWR response.

2.3.4.1 FREQUENCY RESPONSE

Consider a two-port resonator on ST quartz where the resonance occurs at the center of the reflection band and the transducers are recessed to eliminate reflections and placed in the maximum coupling position. The transducer size is chosen so that

(1) the transverse mode spacing defined in Eq. (2.3-13) is at least three times the material-limited (intrinsic) bandwidth f_r/Q_m and
(2) the grating separation is no more than $s_{max} \lambda$ (Eq. 2.3-12) so that the spurious longitudinal modes are adequately suppressed.

Using these postulates and neglecting all losses except viscous damping, bulk scattering, and radiation through the gratings, the maximum Q_u has been calculated by evaluating the losses in Eqs. (2.3-17) through (2.3-23) at the frequency of resonance. The result is plotted in Fig. 2.3-10a. At high frequencies, the maximum Q_u is determined by material losses alone, while at lower frequencies, bulk scattering and radiation losses predominate. This low-frequency limit is set by the substrate length ℓ, which is shown as a parameter in the figure. The groove depth was chosen to maximize the Q subject to the practical constraint that the depth is always at least 30 nm.

An analogous calculation was performed for SAW delay lines using the effective Q_u defined by Weglein and Otto (1977b) as the ratio of energy storage

---

FIG. 2.3-10 (a) Theoretical maximum value for resonator (solid curve) and delay-line (dashed curve) unloaded Q with total device length l as a parameter. Dots represent resonator experimental results. Squares represent delay-line experimental results. (b) Theoretical minimum resonator series resistance as a function of frequency. Points designated by letters are from the following sources: a, Lardat (1976); b, Tanski (1979a); c, Shreve (1976a); d, Tanski (1979b); e, Li (1977); f, Cross et al. (1979); g, Laker et al. (1977); h, Cross et al. (1980); i, Tanski (1980a); j, Coldren and Rosenberg (1976); k, Bale and Lewis (1974); l, Parker and Schulz (1975); m, Lee (1979); n, Weglein and Otto (1977b); o, Gilden et al. (1980).

2 PIEZOELECTRIC RESONATORS AND WAVES

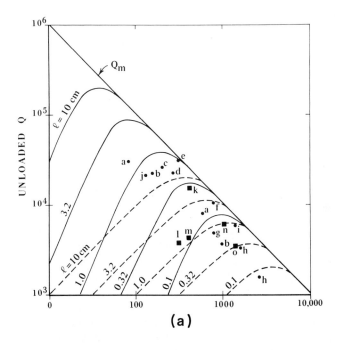

(a)

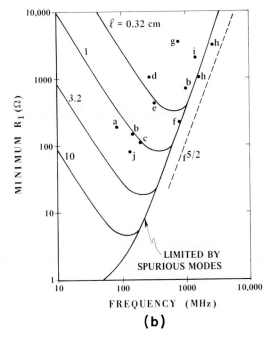

(b)

to dissipation. This effective Q can be expressed in terms of the center frequency f_t and the delay of the delay line τ as[§]

$$Q_u = 2\pi f_t \tau. \qquad (2.3\text{-}26)$$

Results obtained by neglecting the effects of bidirectional losses and losses within the transducer are plotted in Fig. 2.3-10a for comparison with the resonator calculation. A sampling of the best experimental results reported for resonators (dots) and delay lines (squares) is shown.

In Fig. 2.3-10b, the minimum value of R_1 is plotted as determined from Eqs. (2.3-10) and (2.3-11) with the transducer size limits described in Section 2.3.2.2. It should be noted that the design parameters for minimum R_1 are in general different from those for maximum Q_u. The increase in R_1 at low frequencies is caused by radiation losses and bulk scattering. At high frequencies, R_1 is roughly proportional to $f^{2.5}$ due to the transducer size limitation imposed by the 30 nm minimum groove-depth requirement.

2.3.4.2 STABILITY

The frequency of a SAWR-controlled oscillator is affected by four mechanisms: temperature variations, acceleration, phase noise, and aging.

Temperature stability is determined by the substrate material and can only be improved by placing the resonator in an oven to control the temperature or by adding external temperature sensing and electrical compensation. As mentioned above, the material most commonly used for stable operation is ST quartz. Dias *et al.* (1975) showed that the temperature at which the first-order coefficient vanishes, the turnover temperature, can be varied by changing the cut angle. In addition they showed that the temperature coefficient of external components shifts the effective turnover temperature downward. Adams and Kusters (1977) showed that the presence of aluminum IDTs in the SAWR cavity also causes a downward shift of the turnover temperature. Minowa (1978) calculated the magnitude of this shift for both aluminum and gold films on quartz. This film-thickness effect cannot be ignored in choosing a quartz cut.

For many frequency control applications, the ST temperature characteristic is not adequate. In recent years the search for new cuts of quartz, new materials, and layered substrates with better characteristics has intensified. Browning and Lewis (1978) discovered a cut of quartz in which the second-order temperature coefficient is reduced by about a factor of two from that of the ST cut. Subsequently Shimizu and Yamamoto (1980) and Williams

---

[§] Sometimes delay-line Q is defined as the Q of a parallel resonant circuit that has the same phase slope as the delay line. This definition leads to a Q that is one-half that in Eq. (2.3-26).

et al. (1980) found similar cuts. The search for better cuts is continuing, but Newton (1979) ruled out the possibility of finding a cut where both first- and second-order coefficients vanish.

Many temperature compensation approaches have been and are being tried to improve the temperature stability of SAW devices. Lewis (1979) surveyed the success achieved with new orientations, materials, layered substrates, ovening arrangements, and electrical compensation. His review shows that a great deal of effort has been expended in this area and that there is still a likelihood of advances in the state of the art from each approach.

The first study of the sensitivity of SAWRs to external forces was performed by Dias et al. (1976). These results and other early work made it appear that SAWRs would make good force, pressure, or acceleration sensors. Weglein and Otto (1977a) demonstrated that the sensitivity of SAWRs to external forces did not necessarily degrade their performance in frequency control applications. They showed that the noise spectrum of a SAW oscillator was essentially immune to random vibrations that caused a 10–20 dB increase in the noise of a bulk crystal oscillator. Subsequently, Levesque et al. (1979) reported frequency sensitivity of $4.3 \times 10^{-8}/g$, in agreement with their theory. The work was extended by Hauden et al. (1980) who predicted and measured no shift from properly applied diametrically opposed forces on circular plates. These results can be applied to desensitize SAWRs to the influence of mounting forces. Without describing their mounting technique, Tanski et al. (1980a) report an acceleration sensitivity of $3 \times 10^{-10}/g$.

The short-term stability of resonator-controlled oscillators is affected by the resonator Q, R_1, and drive level. As mentioned previously, fundamental-mode SAWR oscillators have a lower noise floor than multiplied BAWR oscillators. Early work by Lewis (1973) on the phase noise of SAW delay-line oscillators has been carried over to resonators. Parker (1979) has studied phase noise in detail. He showed a phase noise of -97 dBc/Hz at 300 Hz from the carrier for a resonator oscillator compared to -85 dBc/Hz for a delay-line oscillator. Phase noise is discussed in detail in Section 8.2.3.

Perhaps the most important characteristic of SAWRs for frequency control is their long-term stability. The first SAWRs were fabricated and packaged using the conventional techniques for SAW filters. Contamination that outgassed from the package and mounting materials limited the stability to around 20 ppm/year. As described more fully in Section 6.3, the emulation of BAWR cleaning and packaging techniques has led to aging rates as low as 0.1 ppm/year. The aging results for one of these devices are shown in Fig. 2.3-11.

The ultimate limit of SAWR stability is not known. Organic contamination has been identified as a major contributor to SAW aging, but little is known

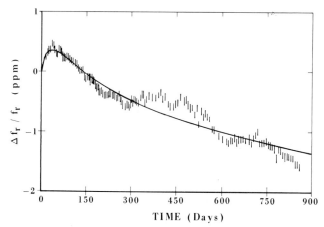

FIG. 2.3-11 State-of-the-art aging of a SAWR oscillator.

about the magnitude of other mechanisms such as stress relief, metalization variations, outgassing, and surface adsorption.

2.3.5 Conclusion

The SAWRs are becoming viable components for precision frequency control. Their major advantages over conventional BAWRs are small size and higher-frequency operation. They can be used in low-cost UHF oscillators to yield high spectral purity.

SAWRs can be modeled by a simple equivalent circuit analogous to that used for BAWRs. They can be made in one-port or two-port configurations to provide added design flexibility. Fabrication consists of conventional integrated circuit processing techniques. Critical dimensions that give coarse frequency control and apodization (the SAWR equivalent of energy trapping) are set by the mask. Wafer processing reduces the cost per device and variations between devices from the levels achievable in individual fabrication of BAWR devices. Because of their higher frequency of operation, clean packaging of SAWRs should be at least as critical as it is for BAWRs. The areas of mounting and packaging of SAWRs may be fruitful areas for future reasearch. State-of-the-art quartz SAWRs have the following characteristics: (1) an unloaded Q greater than $9 \times 10^{12}/f_r$ (80% of the material Q); (2) quadratic temperature stability of $-15 \times 10^{-9}/(°C)^2$; (3) acceleration sensitivity of about $1 \times 10^{-9}/g$; and (4) long-term stability of 0.1 ppm/year.

SAWR technology is still developing. It is likely that the temperature characteristics of SAWRs will improve as new quartz cuts are fully characterized. Careful examination of fabrication and packaging techniques

2 PIEZOELECTRIC RESONATORS AND WAVES   145

may improve the aging characteristics of SAWRs. It is also probable that new mounting techniques and the discovery of stress-insensitive cuts (similar to the SC cut for BAWRs) will lead to reductions in the acceleration sensitivity of SAWRs. As these improvements are made, SAWRs will be used in an increasing number of areas for precision frequency control.

3 Radiation Effects on Resonators

James C. King
Dale R. Koehler

Sandia National Laboratories
Albuquerque, New Mexico

3.1	Introduction	147
3.2	Radiation Effects and Modeling	148
	3.2.1 Substitutional Al^{3+} Defect Center	148
	3.2.2 Frequency Changes	149
	3.2.3 Optical Effects	153
	3.2.4 Elastic Modulus Changes	153
3.3	Dynamics of Radiation Effects	154
	3.3.1 Hydrogen and Transient Effects	154
	3.3.2 ESR and IR Studies	155
	3.3.3 Trap Characterization	156
	3.3.4 Material Quality and Anelastic Losses	157
	3.3.5 Thermal Effects	158

3.1 INTRODUCTION

The study of radiation effects in quartz crystal resonators has proved to be a useful investigative technique in describing modifications in the basic structure of quartz caused by impurities. The success of this technique is due primarily to the remarkable sensitivity of certain resonator characteristics, such as its elastic modulus, to the presence of trace amounts of some contaminants commonly found in natural and synthetic quartz. Work in this field has therefore been devoted almost exclusively to analysis of the manifold effects of ionizing radiation (x rays, gamma rays, and electrons) on point defects. The following discussion reflects this emphasis. The displacement damage caused by collision processes of neutrons (King and Fraser, 1962) in quartz, although of less technological importance, is also of interest in a complete perspective of radiation effects, as are the changes caused by alpha particles (Aoki *et al.*, 1976c).

3.2 RADIATION EFFECTS AND MODELING

3.2.1 Substitutional Al^{3+} Defect Center

Except for the purest quartz, in which radiation heating of the resonator causes both static and dynamic frequency changes and is the only measurable effect, the induced effects of ionizing radiation on quartz crystal resonators can be discussed in terms of a model of one of the primary impurity defects in quartz. This defect is the substitutional Al^{3+} defect with an associated interstitial charge compensator, either a H^+, Li^+, Na^+ ion or a hole. A paper by Weil (1975a) reviewed the literature dealing with the role of aluminum centers in α-quartz for the 20-year period before 1975. The radiation-effects model has been developed over the years as experimenters have accumulated data from many measurements. In this model the ionizing radiation produces electron–hole pairs. The holes migrate to, and are trapped by, the impurity Al sites, and the original compensating cation is then released. The hole-compensated Al site, which is active both optically and paramagnetically, is also the cause of a specific acoustic loss. The freed cations, in the case of Li and Na, are believed to be loosely trapped in the relatively large channels along the optic axis of the crystal lattice. These cations contribute to the acoustic loss in a resonator if the alternating stress field in the resonator is in a direction to couple mechanically with the associated lattice disturbance. However, such acoustic losses are not observed in air-swept quartz, in which hydrogen has replaced the heavier Li and Na cations. A more detailed discussion of the sweeping process (high-temperature electrolysis) and crystal growth characterization is contained in the literature (King, 1959; Chapter 1, Sections 1.2.2 and 1.2.3 of this volume).

Throughout this irradiation scenario, the lattice near the point defect is altered, resulting in a change in the elastic constant of the structure and hence a shift in the resonance frequency of the crystal. In terms of the Al defect model, the nature and amount of the impurity content of the crystal is of the utmost importance in describing the radiation effects. The charge compensator at Al centers in natural quartz seems to be predominantly Li or Na; in synthetic material it is primarily Na. In electrolyzed quartz, hydrogen replaces these compensators, and in vacuum-swept material the Al sites become predominantly hole-compensated.

Besides affecting the optical, paramagnetic, and acoustic characteristics of a resonator, the radiation-modified defect sites also cause changes in the infrared and dielectric absorption parameters of the quartz. These experimentally measured effects are examined in this chapter in the context of the previously mentioned model.

3.2.2 Frequency Changes

To a great degree the measurement of the quartz resonator's frequency change has constituted the most extensive experimental effort (Aoki et al., 1975, 1976a,c; Aoki and Wada, 1978; Bahadur and Parshad, 1979, 1980; Berg and Erickson, 1969; Capone et al., 1970; Esquivel and Sagara, 1974; Euler et al., 1978; Flanagan and Wrobel, 1969; Koehler, 1979; Lipson et al., 1979; Lobanov et al., 1968; Ludanov et al., 1976; Pellegrini et al., 1978) in part because of the relative ease of obtaining precision frequency data. Much of the work on radiation-induced changes in acoustic parameters during the period preceding the last decade was reviewed by Fraser (1968). More recent work has also been reported (King and Sander, 1972, 1973a,b, 1975).

Early in the decade, radiation work (Capone et al., 1970) showed that for an exposure of 1Mrad (Si) of 10-MeV electrons, the accumulated frequency changes for selected natural quartz (Fig. 3-1) are as large as 4 ppm and negative. Offsets as large as 10 ppm negative for unselected natural quartz were also seen. Western Electric fast-growth, lithium-doped quartz displayed negative frequency shifts of about 8 ppm; Sawyer Research Products (SARP) high-Q quartz exhibited a positive offset of about 4 ppm. The best material, electrolytically swept cultured quartz, showed a positive frequency change of only 0.02 ppm. This behavior was interpreted (King, 1958) and also treated in a more general fashion (King and Sander, 1972). Because of

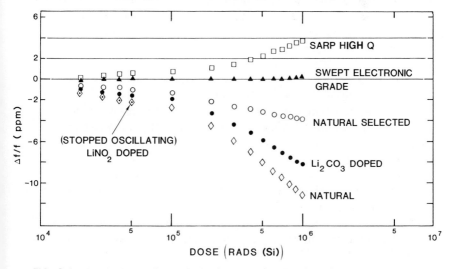

FIG. 3-1 Accumulated offset oscillator frequency as a function of 10-MeV electron dose. [From Capone et al. (1970), © 1970 IEEE.]

the nonlinear and saturation behavior of radiation-induced frequency changes, these effects should not be characterized on the basis of a per-unit dose.

As outline above, steady-state frequency offsets in crystal resonators have been accounted for by a basic mechanism (King, 1958, 1959; O'Brien and Pryce, 1954) involving specific crystal defects that cause stress relaxation at some temperature below the operating temperature of the crystal. These defects are modified by the ionizing radiation so as to vary the mechanical coupling of the resonator with the defect-associated stress field. Acoustic relaxation processes occurring below room temperature are usually caused by small changes in the deformation of the lattice about substitutional or interstitial point defects. These processes have small activation energies because of the small thermal energy that is available. Anelastic processes are characterized in terms of a relaxation strength that is a measure of the degree of "coupling" between the acoustically active defect and the alternating stress field in the resonator. The frequency-determining elastic modulus of the resonator is affected by the defects so that if radiation causes stress relaxation (i.e., a decrease in resonator stiffness), the resonance frequency is reduced at temperatures above the onset of stress relaxation. Conversely, an accompanying increase in resonance frequency occurs if the radiation modifies an acoustically active defect so that it ceases to contribute to stress relaxation.

At any temperature, the net change in crystal frequency after irradiation results from the net radiation-induced change in the defects that contribute to stress relaxation below that temperature. It has been shown that the net frequency change is given by

$$\sum_{i=1}^{n} \Delta f_i/f = \sum_{i=1}^{n} (Q_{max}^{-1})_i,$$

where Δf_i is the positive or negative frequency offset resulting from the decrease or increase in the relaxation strength of the ith defect and $(Q_{max}^{-1})_i$ is the peak value of the relaxation absorption associated with the stress relaxation of the same defect.

A specific defect giving rise to an acoustic loss is usually identified by the temperature at which the acoustic loss is greatest. For a relaxation process, that is the temperature at which the relaxation frequency equals the frequency of the crystal resonator. A good example is the so-called 50-K defect. The 50-K defect is commonly accepted as the substitutional Al^{3+} defect compensated by interstitial Na, which together cause a deformation-relaxation absorption at 50 K in a 5-MHz resonator. After irradiation, the Na defect is removed from the site, and a positive frequency offset occurs at temperatures above 50 K (Fig. 3-2). The radiation-induced 100-K defect has been

3 RADIATION EFFECTS ON RESONATORS 151

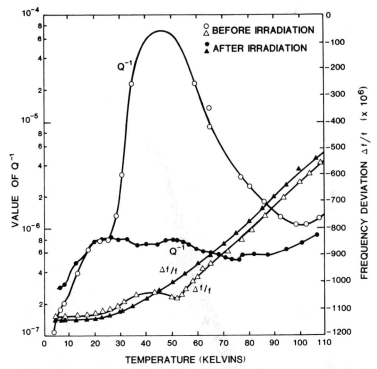

FIG. 3-2 Internal friction and frequency deviation at low temperatures for 5-MHz thickness-shear vibration in Z-growth synthetic quartz before and after x irradiation. [From King (1959), © 1959 AT&T. Reprinted from *The Bell System Technical Journal* by permission.]

characterized as a relaxation mechanism involving a substitutional Al site stripped of an electron on one of the neighboring oxygen atoms. In other words, it is a hole-compensated Al center. During irradiation, the production of this defect causes a frequency decrease at temperatures above 100 K (Fig. 3-3). This defect is the well-known paramagnetic center (O'Brien and Pryce, 1954; Martin *et al.*, 1979) that imparts a smoky color to irradiated quartz and is associated with A-band absorption. As Fig. 3-1 shows, swept Z-growth synthetic quartz is the quartz most tolerant to radiation. This occurs as a result of the removal (by electrolysis) of the Na^+ and consequent reduction of the 50-K defect, as well as because of the removal of Li^+ and K^+, potential sources of similar defects. In natural quartz, production of the 100-K defect dominates, and the frequency changes produce a negative offset. In Z-growth synthetic quartz, reduction of the 50-K defect dominates and frequency changes result in a positive offset. However, only a negligible offset ensues in swept Z-growth synthetic quartz.

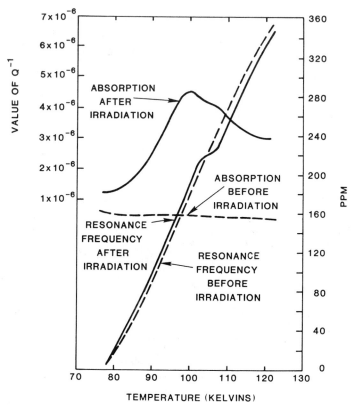

FIG. 3-3 Plot of Q^{-1} and frequency of vibration at low temperatures for an AT-cut natural quartz resonator before and after x-irradiation. [From King and Sander (1972), © 1972 IEEE.]

More recent measurements (Aoki et al., 1976a,b; Aoki and Wada, 1978) of frequency changes in synthetic and natural quartz irradiated by 1-MeV electrons are in good agreement with the above characterization, but at higher radiation doses (as much as 10^{17} electrons/cm^2 or $\sim 4 \times 10^9$ rad) a progressively greater positive frequency offset accrues for all quartz materials tested. This monotonically increasing behavior at very high dose levels in AT-cut resonators has been attributed to displacement effects produced by radiation, probably the removal of oxygen from the lattice. Displacement damage is also the primary effect of neutron irradiation, and at doses up to 10^{19} n/cm^2 the nature of the damage is localized disordering.

With neutron irradiation above this level ($\sim 10^{20}$ n/cm^2), displacement damage throughout the crystal causes large nonannealable (at temperatures below the α–β inversion point) density increases of the order of 4%. Associated with the more localized disordering damage is a monotonically increasing frequency shift (as a function of neutron dose) in AT-cut quartz resonators as observed by King and Fraser (1962) and others (Flanagan and Wrobel, 1969; EerNisse, 1971). Locally disordered structural defects, similar to those found in glassy materials, have also been reported in hypersonic attenuation studies (Laermans, 1979) in neutron-irradiated crystalline quartz. In addition, neutron irradiation has produced amorphous or glasslike thermal properties in quartz (Saint-Paul and Lasjaunias, 1981; Gardner and Anderson, 1981.)

3.2.3 Optical Effects

The concomitant observation of c-band optical absorption, persistent to 400°C, has also been seen earlier. This phenomenon is associated with an oxygen vacancy center at which an electron is trapped (Mitchell and Paige, 1954; Nelson and Crawford, 1958; Weeks, 1956). Optical absorption measurements in synthetic quartz of differing impurity concentrations showed no color changes in the purer material; the less-pure quartz colored readily. Classification of the optical absorption into A_1, A_2, and c-bands follows an earlier characterization (Mitchell and Paige, 1954) and correlates closely with associated frequency changes in the resonators. Annealing experiments showed that the A_1- and A_2 bands are extinguished near 250°C, which is in good agreement with the annealing behavior of the radiation-produced hole centers and H centers described below. The c-band centers arise with the A_1 and A_2 bands. Insofar as excess hydrogen can diffuse to the radiation-produced color centers, the number of hole-compensated centers will be reduced. Also, because the hydrogen–aluminum center does not absorb in the visible region, coloration will be reduced. It is therefore seen that an additional critical index for describing radiation-induced colorability is the concentration of diffusable hydrogen.

3.2.4 Elastic Modulus Changes

Measurements have been made to determine the explicit effect of radiation on the individual elastic moduli (i.e., c_{66}, c_{14}, and c_{44} for AT, BT, and Y cuts) (Aoki et al., 1976a; Ludanov et al., 1976). However, because there are (1) different impurity defect centers in quartz, (2) varying relative amounts of impurities from crystal to crystal, and (3) different anelastic coupling strengths of each defect center on the elastic modulus, it is clear that a universal statement of the relative radiation-changed moduli ratios for different cuts is of questionable utility.

Changes in the elastic constants of the crystalline structure, besides causing obvious frequency changes, will also cause changes in the frequency–temperature characteristics of the quartz resonator. Recent work (Aoki and

Wada, 1978; Benedikter et al., 1974) on natural and synthetic quartz shows that the materials display characteristics similar to those reported earlier (King, 1959). These changes, although expected as a result of the frequency behavior induced by the radiation, require detailed determination of in-individual modulus changes in the quartz (Aoki et al., 1976a; Ludanov et al., 1976). Only then can predictions be made of effects on the frequency-temperature characteristic.

3.3 DYNAMICS OF RADIATION EFFECTS

3.3.1 Hydrogen and Transient Effects

The role of hydrogen (the most abundant impurity in quartz) as (1) a preirradiation charge compensator at Al sites, (2) a reservoir for postirradiation charge compensators, and (3) a primary constituent in the radiation-induced dynamic charge rearrangement process in quartz is becoming better understood. This is a result of transient frequency measurements (King and Sander, 1972, 1973a,b, 1975; Koehler et al., 1977; Koehler, 1979; Pellegrini et al., 1978; Young et al., 1978) and low-temperature ESR and IR experiments (Markes and Halliburton, 1979). After pulsed irradiation involving exposure-time intervals from nanoseconds to microseconds, observation of the quartz shows a significant annealable negative frequency offset at room temperature. Some interesting transient thermal effects have also been observed (Hartman and King, 1973, 1975; Koehler et al., 1977; Koehler, 1979; Young et al., 1978). These will be discussed later in the text. The transient frequency change has been attributed (King and Sander, 1972, 1973a,b, 1975) to a relaxation process, which anneals above 165 K. The kinetics of the annealing process obeys a $t^{-1/2}$ relationship and is theoretically (Sosin, 1975) interpreted in terms of a one-dimensional diffusion-limited annealing of uncorrelated defects. More specifically, the monovalent cation H^+ is trapped at substitutional Al sites. Sosin's model, and calculations from it, should constitute the correct approach. Many investigators (King and Sander, 1975 and references cited therein) have demonstrated that monovalent cations such as Li^+, Na^+, and H^+, generally found as interstitial impurities in quartz, diffuse most readily along a single crystallographic direction, the optic axis. The experiments showed that the resonance frequency of all the crystal units tested, after exposure to a gamma burst at room temperature, exhibited a negative offset of several parts per million, which annealed out to a relatively stable value within 10 to 15 minutes after exposure. The acoustic relaxation process believed responsible for this frequency offset involves the 100-K defect mentioned above, part of which is annealable at room temperature. If the quartz is of very high purity (Young et al., 1978) (i.e., high Q as determined by IR characterization of the hydrogen content), then the rapid annealing of the hole-compensated centers is not observed, presumably because of a smaller concentration of hydrogen-compensated Al precursors. This situation also obtains in vacuum-swept quartz where the cation compensators have been replaced by holes instead of by hydrogen. Further experimental evidence (Krefft, 1975) shows that the concentration of hydrogen in vacuum-swept quartz is significantly reduced, verifying that a reduction of the transient $\Delta f/f$ is associated with a reduction of hydrogen.

The optical analog to this annealable 100-K defect is the short time observation of coloration, or A-band absorption, after pulse exposure. The validity of this facet of the model has been supported by transient optical absorption (within the A band) data taken after irradiation of various quartz specimens (Flanagan and Wrobel, 1969; Spitsyn et al., 1978). Still further support came from A-band absorption measurements at 77 K (Mattern, 1973; Mattern et al., 1975), followed by room-temperature measurements in which a reduction in A-band intensity occurred as the irradiated sample warmed. This situation would be expected to follow from the suggestion that the H^+-compensated fraction of the substitutional Al centers would remain colored, or hole-compensated, if irradiated at temperatures low enough to prevent the proton from migrating back to the hole-compensated Al site after it is freed by irradiation. The earlier work cited above (Markes and Halliburton, 1979; Weil, 1975a,b) indeed suggests that the freed proton constitutes an electron trap at low temperatures. Viewed another way, atomic hydrogen is freed from the Al site.

3.3.2 ESR and IR Studies

If electronic-grade quartz is irradiated at room temperature (Markes and Halliburton, 1979), an increase in the number of hole-compensated centers occurs, as determined by electron spin resonance (ESR) measurements. After a second irradiation at 77 K, the number of these centers is further increased. In terms of the model, the cation (Na^+) compensated Al sites would release their charge compensators and become hole- or hydrogen-compensated under irradiation at room temperatures. At low temperatures only the hydrogen-compensated sites could do so. The annealing studies further show that a measurable component of the hole-compensated Al center persists up to room temperature, indicating that the available hydrogen need not or cannot compensate all of the Al centers.

Complementary IR studies (Martin et al., 1979) substantiate the dynamics of the defect rearrangement processes after irradiation by revealing that an enhancement of IR absorption associated with the Al-OH center occurs after irradiation at room temperature. This is additional evidence that hydrogen, at sites other than the substitutional Al sites, has migrated to the originally cation (Na^+) compensated center and replaced it as a postirradiation Al^{3+} charge compensator. Later irradiation at 77 K, which is able to free the hydrogen from such sites, would therefore show an increase in the number of hole-compensated centers (in agreement with experimental observation).

In comparative experiments that were part of the studies of Martin et al. (1979), both air-swept and unswept samples from the same bar of electronic-grade synthetic quartz were irradiated. An initial 77-K irradiation showed a factor of 25 more Al-hole centers in the swept specimen than in the unswept specimen. This sweeping process removed the cation compensators from the Al centers and replaced them with hydrogen ions that are readily freed under the 77-K irradiation, leaving the centers hole-compensated.

As noted in earlier work, radiation-produced effects like those described here can be removed by annealing at high temperatures (Bahadur and Parshad, 1979). In a study (Martin et al., 1979) of the temperature region from 500 to 650 K, both the ESR measurements of hole-compensated Al centers and the IR measurements of the Al-OH centers showed a destruction of these defects and a return to preirradiation conditions. In contrast to the unswept material, no significant changes were observed in swept quartz. For the swept samples this is expected because of the absence of heavier cation compensators in the swept quartz and the substitution of hydrogen as the preirradiation equilibrium Al compensator.

3.3.3 Trap Characterization

Earlier workers (Freymuth and Sauerbrey, 1963) were able to establish a two-component activation energy fit to the experimental data in detailed annealing measurements of radiation-induced frequency changes (negative) in natural quartz. The temperature dependence of the annealing curves was best reproduced by traps with activation energies of $E_1 = 0.3 \pm 0.1$ eV and $E_2 = 1.3 \pm 0.3$ eV. Current interpretation of these data pictures the preirradiation Al compensating cations as loosely trapped (0.3 eV) in the c channels after they are freed by irradiation. Later heating allows them to return more readily to Al sites, thereby replacing the holes or hydrogen ions that had taken their places. A reasonable interpretation is that the 1.3-eV trap is the coulombic potential well surrounding the substitutional Al defect.

Other recent conductivity and dielectric relaxation experiments (Jain and Nowick, 1982a,b) on synthetic and natural quartz resonators have been tentatively analyzed in terms of single defect centers (i.e., trapping sites). For the synthetic quartz resonators, the motional energy (thermally activated mobility) as determined from the conductivity measurements over the temperature range from 230 to 280 K was 0.14 eV. This value is significantly lower than that for the natural material. The result for the natural quartz resonator was a motional energy of 0.45 to 0.50 eV. Prompt-radiation-induced photoconductivity and transient increase of acoustic losses, discussed in Section 3.3.4 are understood as two manifestations of the temporarily freed impurity cations in quartz. To support this notion, the independently determined activation energies should agree. Additional work needs to be done to measure this parameter more extensively for natural and synthetic quartz materials shortly after irradiation and at a later time for the more persistent contributors to conductivity and acoustic losses.

3.3.4 Material Quality and Anelastic Losses

From another perspective, the observed absence or reduction of an ionic current in swept quartz (Hughes, 1975), as well as the elimination of radiation-produced Q changes by sweeping (King and Sander, 1973b, 1975), arises from the removal of cation charge compensators. Sweeping therefore removes the source of the ionic current as well as the transient 290-K (at 5 MHz) acoustic loss mechanism. The effects of sweeping are therefore seen to be consistent with the model.

The importance of material quality has been stressed (King and Sander, 1973b) because in some quartz the magnitude of the annealable acoustic loss increase at room temperature has been large enough to cause the oscillator to stop for long periods of time. Oscillator gain margin (Paradyz and Smith, 1973, 1975) determines the capacity for sustaining vibration. The observed change in resonator resistance, or acoustic loss, is a function of radiation dose and impurity concentration in the quartz.

More recent anelastic absorption (Q^{-1}) measurements (Aoki and Wada, 1978; Capone *et al.*, 1970; Martin *et al.*, 1979) are consistent with earlier results. These measurements demonstrate qualitative agreement for radiation-induced reduction of the Na loss peak, for an increase in the hole-compensated Al loss peak, and for substantiation of the expected dependence of frequency on anelastic loss processes (Aoki and Wada, 1978). The Q^{-1} versus temperature data (Martin *et al.*, 1979) taken from Na-swept Premium-Q quartz also agree with the ESR results (Markes and Halliburton, 1979) (that the alkali ion becomes mobile under irradiation at temperatures above 200 K). These latter results showed that resonators irradiated at 77 K exhibited no change in the 50-K loss peak but that irradiation at 300 K completely removed this loss mechanism. Sweeping, by removing the Na charge-compensating cations, should therefore also eliminate the 50-K loss peak (and it does). The loss mechanisms described thus far have been fairly well established, but the origins of smaller anelastic absorption peaks in such Q^{-1} spectra have not been identified.

3.3.5 Thermal Effects

The response to pulsed irradiation is no longer impurity-related in high-purity material that has been electrolyzed in vacuum, where the source of changes in resistance and transient frequency (namely H^+, Li^+, and Na^+ ions) has been removed from the crystal by the sweeping process. Studies (Koehler et al., 1977; Koehler, 1979; Young et al., 1978) with 5-MHz, fifth-overtone, AT-cut resonators have shown transient frequency changes that have been interpreted in terms of dynamic and static thermal effects resulting from the deposition of radiation energy and from heating the resonator structure. Thermal modeling of the crystal resonator and associated oven environment, in conjunction with earlier (Anderson and Merrill, 1960) empirically derived frequency dependencies on the rate of change of

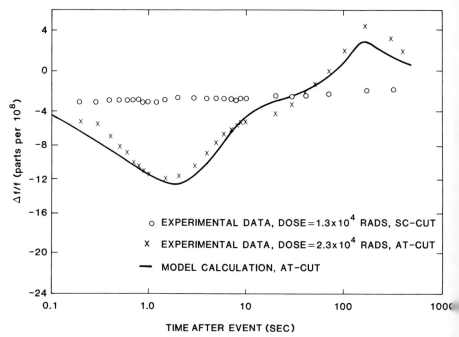

FIG. 3-4 Transient frequency change after pulsed gamma irradiation for an AT-cut and SC-cut resonator. The frequency offset for the SC-cut air-swept Premium-Q quartz resonator has been attributed to the presence of some residual postsweeping impurities in the quartz.

3 RADIATION EFFECTS ON RESONATORS

temperature of the quartz, has led to good agreement between the calculations from the model and the experimentally measured frequency transients (Fig. 3-4). Considerations of temperature gradients in quartz crystals, caused by thermal transients, led Holland (1974) to design the thermal-transient compensated (TTC) cut. EerNisse (1975) was led to the equivalent orientation (the stress-compensated or SC-cut) from a consideration of stress effects. In irradiation studies on these SC- (as well as BT- and AT-) cut swept Premium-Q quartz units, the resonators displayed the frequency changes expected and calculated from the resonator thermal model mentioned above (i.e., negative transients in AT resonators, positive transients in BT resonators, and negligible changes in SC-cut resonators). A typical "thermal-signature" frequency transient ensues from pulsed irradiation because of a radiation deposition throughout the resonator structure that is a function of radiation energy and material and because of later changes in temperature equilibration that are a function of the material's thermal properties. The negligible radiation-induced thermal-frequency transient in the SC- (or TTC-) cut resonator of course, stems from the explicitly designed insensitivity of this crystal orientation to such thermal effects.

In conclusion, studies of the radiation-induced effects in quartz over the past decade have led to a significantly increased understanding of the relationship between radiation sensitivity and crystal defects. This knowledge has been used to modify and control the concentration of defects in the raw material, which in turn permits the fabrication of precision resonators that are little affected by radiation environments.

4 Resonator and Device Technology

John A. Kusters

Hewlett-Packard Company
Santa Clara, California

4.1	Resonator Material Selection	161
4.2	Sawing	163
	4.2.1 Natural Quartz	165
	4.2.2 Cultured Quartz	166
4.3	X-Ray Orientation	166
4.4	Mechanical Operations	168
4.5	Cleaning	170
4.6	Vacuum Deposition	170
4.7	Mounting and Sealing	174
4.8	Special Fabrication Considerations for SAW Devices	178
4.9	Novel Resonator Techniques	180
4.10	Environmental Effects	182

Of primary importance in precision frequency control devices is the manufacturing technology employed to fabricate these units. For crystal resonators, the choice of technology used is governed by the desired end use of the device. Quartz resonator applications range from high-volume low-cost resonators in color TVs and quartz watches to low-volume very-high-cost resonators in precision frequency standards. The price range for finished devices ranges from under one dollar to more than several thousand dollars for high-precision quartz transducer designs. Each fabrication facility differs in the type of processes used (Piwonski, 1971; Wasshausen, 1971; Metcalf, 1972; Royer, 1973).

4.1 RESONATOR MATERIAL SELECTION

Quartz is the material of choice in the majority of resonator devices made today (refer to Sections 1.1 and 1.2). The recent expansion of surface-acoustic-wave (SAW) activity also has introduced a variety of other materials, such as lithium niobate, lithium tantalate, berlinite (aluminum phosphate), zinc oxide, and others (refer to Sections 2.3, 5.3, and 5.4).

In terms of the total volume of devices made, quartz dominates all other materials. Quartz is easily obtained, has been extensively studied for years, and exhibits properties that make it suitable for a wide variety of devices. Initially, quartz devices were fabricated only from naturally occuring quartz crystals obtained primarily from Brazil, with minor deposits in other parts of the world. Since the 1940s, extensive efforts have led to the development of cultured quartz, which exhibits greater uniformity and better utilization of material than natural quartz. As a result, virtually all devices today use cultured quartz. Natural quartz is used primarily in devices where the problems of available size, lattice defects, or inclusions in cultured quartz outweigh the much lower overall material yield in natural quartz. Endpoint material yield in natural quartz can be as low as 1 to 2% compared to greater than 20% in cultured material. This is primarily due to excessive twinning, veils, and fractures in the natural material.

Cultured material is available in a wide variety of sizes, orientations, and grades. A particularly useful property of cultured material is that the quartz bars can be selectively grown in different crystallographic directions to maximize yield in producing different resonator devices and orientations. Size and quality of the material can be also controlled by the grower to meet almost any requirement of the user.

Major problems in currently available cultured material are inclusions, lattice defects that result in etch channels, and susceptibility to radiation damage. Extensive research is being done to eliminate these problems.

Cultured quartz can be loosely and imprecisely divided into five categories in order of increasing price. *Commercial grade*, also known as *electronic grade*, is usually fast grown and suitable for a wide variety of low-precision devices where the device performance is dominated by other than material factors. The Q, or quality factor, of this material is about 1.8×10^6.

Premium Q material is suitable for a number of medium- to high-precision devices. It's Q is typically 2.2×10^6.

Special premium Q material, with Q values in the vicinity of 2.6×10^6 and higher, is used for precision resonators where material loss dominates device performance.

Optical-grade material is used for special optical devices and other components requiring material with the lowest possible internal strain birefringence. These include acoustooptic filters, deflectors, and optical modulators.

Swept quartz is basically a high-Q material where, using a variety of different techniques, impurities are "swept" out of the quartz material using a strong electric field at elevated temperatures. This material is used primarily

4 RESONATOR AND DEVICE TECHNOLOGY

in devices where susceptibility to ionizing radiation is an important consideration.

The Q of the material is defined in terms of a 5-MHz, fifth overtone, AT cut, designed and mounted so that material losses are the dominant effect. Q is defined in the usual electronic sense, that is, the device frequency divided by the half-height resonance line width. In practice, in the unprocessed raw material the Q is measured indirectly using optical absorption techniques at different wavelengths, which generally correlate with the acoustic losses seen in a finished device.

Generally, SAW resonators and shallow bulk acoustic wave (SBAW) resonators are also fabricated on quartz substrates. For most of these devices the electromechanical coupling factor in quartz is quite small. In addition, although SAW orientations have been identified with usable frequency versus temperature characteristics, these do not have the performance of common bulk-wave devices such as the AT and SC cuts (refer to Section 2.2). The general SAW technology permits a much wider variety of acoustic devices than conventional bulk-wave resonators (refer to Sections 5.3 and 5.4).

SAW filters on quartz are usually restricted to bandwidths of less than 1%. Other materials such as lithium niobate permit bandwidths up to 10% because of much higher electromechanical coupling factors. However, the higher-coupling-factor materials generally have undesireable frequency–temperature characteristics. Recent work in berlinite holds promise for higher coupling factors with better temperature performance than current quartz SAW devices.

Material technology for SAW devices is one of the more active areas of crystal areas of crystal research. Some of this effort may ultimately result in improved conventional bulk-wave devices.

The final choice of the material used will depend on the ultimate use and the desired performance of the device.

4.2 SAWING

The first step in device fabrication is the sawing of raw material to the correct orientation and size. Because piezoelectric materials are anisotropic, the final device performance depends on the exact crystallographic orientation of each finished surface of the device. For most devices, this orientation is determined during the cutting operation.

Crystal cutting is done by one of two methods, sawing with diamond or other abrasive-coated rotary blades or lapping with an abrasive compound. Cutting may be done with a plunge saw where the cutting blade makes a single cut through the entire surface starting at the top surface. Also used are reciprocating saws where either the crystal or the saw blade traverses horizontally. Internal diameter (ID) cutting saws are also used where the saw blade is a thin material tensioned on its outer circumference with an abrasive-coated hole in the center of the blade material.

Lapping saws (see Fig. 4-1) depend on wearing a path through the crystal by traversing a string, wire, or tensioned blade over the surface of the crystal while flooding the area of contact between crystal and blade with a slurry compound consisting of a carrier fluid and an abrasive. Typical abrasives used include alumnium oxide, silicon carbide, and diamond dust. Commercially available lapping saws contain 20–100 cutting blades, depending on the width of the cut and the desired thickness of the finished piece.

The type of saw used depends on the size, shape, and type of crystal material used. For example, natural quartz crystals, due to their irregular shape, are almost always cut with a rotating blade. Cultured quartz crystal bars

FIG. 4-1 Modern, multiblade lapping saw. Work to be cut is positioned below the blade pack. Blade pack consists of 10 to 100 steel blades under high tension. (Photo courtesy of Varian Industrial Equipment Group.)

4 RESONATOR AND DEVICE TECHNOLOGY

obtained from the supplier precut to a particular size and orientation are more easily cut on a lapping saw.

4.2.1 Natural Quartz

The first step in sawing a crystal is to determine where to cut. Natural quartz crystals grow with a number of well-defined facets on the surface. The quality of material available today is rather poor. It is rare that faces other than the prism or m faces can be easily determined. The problem is further complicated in that both right- and left-handed quartz exist in roughly equal proportions. Natural quartz may also exhibit twinning where both left- and right-handed quartz exist in the same piece.

An accepted method is to mount the crystal on an m face. Usually the crystal is fastened to a glass, ceramic, or other easily cut material using wax, casting resin, or plaster of Paris. Cutting perpendicular to this face, parallel to the optic or z axis, will generate an X-face on the crystal. The direction of the z axis can be determined either from natural features on the crystal or by use of a polariscope or an immersion iconoscope. In practice, a test cut is usually taken first and the saw (or crystal) mount corrected based on the results of x-ray measurements of the test cut (Merigoux *et al.*, 1980)

After cutting X-faces on two sides of the crystal, the cut surfaces can be heavily etched in a commerical quartz etch. The etching clearly shows regions of electrical and optical twinning in the crystal (Heising, 1946). Also, the asymmetry of etch pits on the surface allows the use of an oriascope to determine uniquely the handedness of the crystal and the direction of the +x-axis. It is also possible to determine this information using optical and electrical tests, but the etch method is simpler.

Using the etching information, singly rotated Y cuts can be made by mounting the crystal on a previously cut X face and rotating the optic axis about the x axis to the proper angle. In practice, this is almost always done using transfer fixturing and x rays. The crystal is mounted on appropriate tooling in an x-ray system and the proper rotation set by x-ray Bragg diffraction from chosen crystallographic planes (Heising, 1946). Additional test cuts might also be used to refine the actual cutting angle.

To cut doubly rotated cuts such as the SC, IT, RT, and LC, intermediate cutting operations are needed (Bond and Kusters, 1977).

This process, using diamond-cutting tools, usually results in a slab of quartz material with the desired thickness but with irregular edge dimensions. Further etching and inspection determines twinned regions and usable areas in the slab. Final cutting of the crystal blank uses a saw for square or rectangular blanks or a core drill for circular blanks.

4.2.2 Cultured Quartz

The difficulty and labor involved in cutting natural quartz has led to a rapid growth in the use of cultured quartz for the majority of quartz-resonator devices. Cultured material usually contains no twinning and is grown uniformly as either right- or left-handed material. Quartz suppliers provide precut sections of almost any desired size, with major faces oriented crystallographically to within 15′ of arc. In addition, in recent years a number of small speciality cutting shops have provided precut crystal blanks of the necessary size and angle to meet most of the crystal industry's requirements.

If the material is obtained in bar form, the bars are mounted on mechanically indexing fixtures for sawing. If further precision is needed, transfer fixturing and x-rays can be used, the proper rotation determined from x-ray diffraction and transferred to the saw. Because of the uniformity of the bars, cutting is usually done with lapping saws. Occasionally a conventional rotary cutting saw might be used.

Doubly rotated cuts are more easily handled in cultured material. Bars already cut to the first rotation can be obtained from the quartz supplier. This makes cutting of the second rotation similar to cutting AT and BT cuts.

Again, test cuts might be used to further refine the cutting angle.

4.3 X-RAY ORIENTATION

A further step is necessary, once cut blanks have been obtained, before the blanks can be classified as usable. Because of minor variations and mechanical tolerances in sawing and possible lattice variations in the parent material, each blank produced is not at exactly the same orientation. To obtain blanks of the necessary precision, two methods, both using x-ray diffraction, are used. Blanks may be either sorted according to a pre-established specification, or may be angle-corrected so that the major surfaces have the desired orientation.

The orientation of a major surface can be accurately determined using x-ray Bragg diffraction from known crystal planes (Bond, 1976). Usually a twin-crystal diffractometer is used with the reference crystal precisely adjusted to the desired orientation, although this is not a necessary condition. Perhaps the greatest difficulty lies in ensuring that the crystal surface is properly mounted on the goniometer reference surface. A recently introduced laser-assisted method (Vig, 1975) permits maximum accuracy to be obtained.

For singly rotated cuts such as the AT or BT, the rotation angle about the X axis is the most important since this angle governs the temperature performance of the unit. The usual plane for AT cuts to measure this rotation is the 01·1 crystallographic plane. For BT cuts, either the 10·1 or the 10·2 planes give satisfactory results. For this class of cuts, the rotation angle about the Z axis is not critical and is usually ignored except for the highest-precision units.

Doubly rotated cuts, except for the IT cut, pose a special problem since there is no suitable plane to directly determine the angle with respect to the optic axis or the rotation angle about the z–z′ axis in the blank. For these cuts, special methods have been determined using either multiple planes or pretilted blanks so that planes such as the 01·1 can be used (Bond and Kusters, 1977; Clastre et al., 1978; Asanuma and Asahara, 1980). Recent improvements include the development of fully automated goniometers capable of accurately measuring a wide variety of orientations (Darces and Merigoux, 1978; Kobayashi, 1978; Birrel et al., 1980).

If the yield due to sorting is not sufficiently high, a more difficult process may be used. Angle correction is a process whereby the actual surface of the blank can be changed slightly in orientation angle to provide the correct crystallographic orientation in the final blank. Several methods have been either used or proposed to change the orientation of a sawn blank.

In one method, the crystal is mounted on a lapping fixture with adjustable diamond feet which rest against a reference surface on the x-ray system. Instead of actually measuring the surface orientation, the diamond feet are adjusted until the crystal is at the desired orientation. The lapping fixture with the crystal blank attached is placed on an abrasive lapping machine. Material removal continues until the diamond tipped feet prevent any further lapping. The resulting crystal surface is now at the desired orientation (Hammond, 1961; Kusters and Adams, 1980).

Another accepted method is to etch a step on one half of the crystal blank. The depth of the step is directly related to the necessary correction required (Husgen and Calmes, 1976). Another proposed method involves using a laser under computer control to burn small pits on the surface of the crystal (Birrell et al., 1980). For both of these methods, during subsequent parallel lapping the etched, or laser-damaged, areas will provide an asymmetry in the lapping operation that will ultimately result in a correctly oriented surface.

Another method takes advantage of the fact that the temperature performance of BT cuts and fundamental-mode SC cuts are highly sensitive to surface contour. Instead of correcting the orientation of the surface, the proper contour is chosen to provide a blank whose temperature–frequency characteristics are essentially the same as if the blank had the correct surface orientation (Vig et al., 1981).

Significant improvement can be seen during x-ray measurement if the quartz blank is given a heavy etch prior to irradiation. This tends to remove the outer damaged layer formed on the quartz blank during previous mechanical operations. Heavy etch will also stop any further propagation of microcracks that might have started as a result of surface damage.

4.4 MECHANICAL OPERATIONS

The next steps in the manufacturing process are the various mechanical and chemical processing steps that turn a sawn blank into a finished blank ready for final processing. The actual sequence depends on the type of resonator unit and its final intended use. Many of the steps discussed may not be used for low-precision units but might be essential for the proper performance of high-precision devices.

Such a step is parallel lapping. Either a pin lap or, more conventionally, a planetary lap is used (Miller, 1970). This process produces a blank that has both sides parallel and also is the first step in determining the final blank thickness. The blanks are placed in a carrier of the proper thickness between two lapping plates. An abrasive grit in a fluid carrier provides the lapping agent. Parallelism is obtained by frequently alternating the position of the crystal blanks in the carrier. Final thickness is determined either by use of a thickness gauge or for thickness-mode devices, since the crystal blank is piezoelectric and generates a small radio signal at its resonance frequency during a lapping operation, by the use of an HF or VHF radio receiver placed near the parallel lap and tuned to the desired frequency.

Circular blanks not already at the proper diameter at this time are stacked one on top of another into a cylinder. The blanks may be waxed into a stack, or held by pressure if suitable tooling is available. A cylindrical grinder removes the excess crystal material and rounds the stack to the proper diameter. If diametric control is important for the device, diamond honing and edge polishing can be used to set the blank diameter as precisely as necessary.

Rectangular or square blanks, contour-mode resonators, flexure bars, and extensional-mode bars are cut from the paralleled blanks if necessary and trimmed to their final dimensions using sawing, grinding, lapping, and polishing operations.

From this point, the sequence followed is determined by the device. Possible steps include polishing, contouring, and final mechanical trim to frequency.

Polishing has traditionally been a mechanical process used on resonator devices where attainment of the highest possible Q and long-term stability are important (Miller, 1970; Vig *et al.*, 1973). It typically is not done to low-precision crystal units. Recent developments in crystal processing techniques have led to the use of chemical polishing with ammonium bifluoride and other etchant compounds. This technique, which must be tailored to the specific orientation of the device, leads to an acceptable surface polish and a greatly increased resistance to crystal damage, especially in high-shock environments (Vig *et al.*, 1977a,b; Brandmayr *et al.*, 1979; Suda *et al.*, 1979). Material-limited Q and good long-term aging performance have also been obtained with a 3-μm lapping step followed with a heavy chemical etch (Castellano et al., 1977).

Surface contouring is normally restricted to thickness-mode devices. It produces a surface that is a section of a sphere. It provides a means of confining acoustic energy to the center of the crystal and minimizes acoustic leakage through the crystal mount. These acoustic losses decrease Q and increase the equivalent series resistance of the device. Contouring is done either to one side (plano-convex) or to both sides (biconvex). The amount of contouring is customarily expressed in diopters, the inverse of the equivalent focal length in meters of a glass lens of the same radius of curvature, or by the actual radius of curvature of the surface. Actual contouring can be performed on a large scale by gently tumbling the blanks in a contouring drum partially filled with an abrasive slurry. Final contour achieved closely matches the radius of the drum. Contouring is also done using cylindrical grinding machines and using "diopter cups" from the optical industry. In the latter method the blanks are mounted in a holder and placed on the surface of a spherical cup of the proper radius of curvature. The cup is spun rapidly while slurry runs into the cup. Pressure applied to the crystal holder forces it against the lapping surface and generates a spherical surface on the blank.

Final mechanical trim-to-frequency can be done only when a method exists at this stage that permits the resonator to be excited in its resonant mode. Thickness-mode devices can be easily excited using air-gap electrodes that are designed to hold the blank on its periphery and to provide an electric field normal to the surface through a small air gap. The device can then be driven into resonance by electronic means. The most popular driver is a crystal impedance (CI) meter whose output frequency is monitored by a frequency counter.

In this manner, the current resonance frequency of the device can be measured and a small additional amount of lapping, mechanical polishing, or chemical etching can be done to remove additional material. The process is continued until the device is within frequency tolerance for this stage of the process. If the frequency becomes too high, the unit must be rejected.

Similarly, contour-mode devices, and flexure- and extensional-mode units may be adjusted to frequency by proper electrical excitation and small amounts of material removal. A useful material removal technique for these devices is the use of an air-abrasive unit where abrasive powder is driven onto the surface of the resonator to be trimmed by air pressure (Kulischenko, 1975). This is usually cleaner and faster than fluid-carrier lapping compounds.

Following the mechanical processes, a final etch may be given to the device. This may be the chemical polishing step, a light etch after mechanical lap or polish, or the final mechanical trim-to-frequency. This final etch removes a thin damaged layer produced on any crystal during mechanical processing (Fukuyo and Oura, 1976) and is essential for maximizing Q and for proper long-term aging of the resonator device. Similar results have been obtained using rf back-sputtering, plasma etching, and ion-milling (Castellano and Hokanson, 1975).

4.5 CLEANING

Cleanliness is absolutely essential for proper long-term aging performance of crystal-resonator devices (refer to Chapter 6). At this stage in its manufacturing process, the crystal blank has been exposed to virtually every known contaminant harmful to proper long-term performance: it has been immersed in various oils, soaps, and noxious liquids; ground into its surface have been lapping and polishing compounds of a wide variety of chemical compositions; it has been attacked by etchants; it has been handled many times by humans so its surfaces are also loaded with unknown organics.

Each crystal-processing facility tends to develop its own cleaning technology (Simpson, 1970; Vig et al., 1973, 1974; Hart et al., 1974; Hart, 1974). Cleaning may involve combinations of acidic and caustic baths, washing in polar and nonpolar solvents, ultrasonic cleaning, boiling in solvents, vapor degreasing, and vapor drying (White, 1973). Recent work indicates that whatever method of chemical cleaning is used, residuals from all previous work can be detected on the crystal surface if sufficiently sensitive surface analysis, such as ESCA or AUGER, is used (Bryson et al., 1979).

Further cleaning, which appears to remove all of the detectable residues, involves exposure of the crystal blank to intense ultraviolet (UV) radiation. Since the UV also creates ozone when oxygen is present in the cleaning system, this method is termed "UV–ozone" cleaning (Vig et al., 1974, 1975).

Effective removal of residue may also be done using various vacuum cleaning methods. Of greatest importance are ion bombardment (Vig et al., 1973; Hart et al., 1974; Hart 1974), plasma scrubbing, and electron bombardment. Each of these require a high degree of cleanliness prior to exposing the blank to vacuum.

4.6 VACUUM DEPOSITION

Inherent in the design of resonator devices is some form of electrode structure that creates the proper electric field distribution in the crystal unit. The most commonly used method involves vacuum deposition of metallic films through properly designed evaporation masks. The masking used will define the final electrode size and shape and influence the final parametric performance of the finished resonator (Mindlin, 1968; Werner and Dyer, 1976).

4 RESONATOR AND DEVICE TECHNOLOGY

Deposition must be done in a vacuum sufficiently high so that the mean free path of an atom is significantly longer than the distance from the deposition source to the substrate. This implies a vacuum of 10^{-6} Torr (millimeters of mercury) or greater (Rankin, 1972). Pumps capable of achieving this level are of four generic types; diffusion, turbomolecular, ionization, and cryogenic.

Oil (or earlier, mercury) diffusion pumps depend on a momentum exchange between hot oil molecules that have a flow path directed to the bottom of the pump and the residual gas molecules to be pumped. This pump must be backed, or operated in series, with a mechanical roughing pump capable of achieving vacuum in the low micron range. A major problem with diffusion pumps is their tendency to "back-stream," a phenomenon by which some of the oil molecules tend to diffuse upward into the vacuum chamber and contaminate the surfaces being plated. Proper arrangement of optically dense, cooled baffle plates can reduce this to acceptable levels. Maximum achievable vacuum is 2–5×10^{-8} Torr, primarily depending on the vapor pressure of the pump oil.

Diffusion pumps are attractive because of relatively low cost and very high pumping speeds. However, the possibility of surface contamination generally eliminates their use for high-precision resonators where long-term aging is an important factor.

Turbomolecular pumps also require backing by an external mechanical pump. The turbo pump is essentially a multistage turbine that is motor driven. The pump's rotors and stators are arranged to provide a net momentum exchange to the residual gas molecules to direct them from the vacuum system through the turbo pump to the roughing pump. Maximum achievable vacuum is in the 10^{-8}-Torr range. A common fault with earlier pumps of this design allowed oil vapor from either the turbo pump bearings or the mechanical backing pump to be drawn back into the vacuum chamber in case of pump malfunction or power failure. Modern pumps, backed with fast-acting solenoid valves, have virtually eliminated this problem.

The ionization pump is essentially a small sputtering cell where residual gasses are ionized and driven into an electrode. At the same time, titanium atoms are sputtered from the system and either combine chemically with the gas atoms or drive them to the opposite wall and bury them. Gas atoms that cannot be ionized are driven to an electrode surface through momentum exchange. Modern pumps are constructed from a large array of basic sputtering cells. Maximum achievable vacuum approaches 10^{-11} Torr. The major difficulty with this pump is its poor pumping speed for noble gases. Special configurations of the sputtering cell have been developed that enhance noble gas pumping. Pumping under high gas loads can cause heating of the electrode surfaces. Under this condition, heavy out-gassing of previously pumped gasses can occur. For this reason, the vacuum system must

FIG. 4-2 Cryogenic pumping head. Vacuum shroud has been removed to show the head structure and the optical baffle. (Photo courtesy of Varian Industrial Components Operation.)

first be pumped down to the $1-5 \times 10^{-3}$ Torr region before the ionization pump can be used. Ionization pumps are essentially contamination free.

Reliable cryogenic pumps (see Fig. 4-2) are a recent development. This pump operates by cooling an activated charcoal surface (see Fig. 4-3) down to 10 to 15 K using a closed-cycle helium refrigeration system. All gasses, except helium, simply condense on the cold surface. An appreciable amount of helium will also be adsorbed on the cold area. Since the gasses do not combine chemically and are not removed from the pump using an external roughing pump, periodic regeneration of the cold surface is necessary. Regeneration involves flowing dry nitrogen through the area of the cold surface while the refrigeration system is turned off. Several hours of off-time are usually sufficient to allow the cold surface to heat up and previously condensed gasses to be removed from the system. Vacuum in the 10^{-8}-Torr region is easily achieved. Cryogenic pumps are essentially contamination

4 RESONATOR AND DEVICE TECHNOLOGY

FIG. 4-3 Cross section of a modern cryo pump showing the inside of the optical baffle and the location and configuration of the activated charcoal adsorption surface. (Photo courtesy of Varian Industrial Components Operation.)

free but have a limited heat capacity. Optical baffling must be used to prevent the pump cold surface from seeing vacuum system heat sources such as evaporation filaments or system heaters.

Regardless of the type of final vacuum pump used, the vacuum system must first be preroughed with external pumps to at least the 10^{-1} Torr region. A widely used roughing pump that is not mechanical depends on the cooling of a molecular sieve material such as zeolite to liquid nitrogen temperatures. Modern vacuum systems may use a series of these units to attain the necessary vacuum levels during roughing. The major advantage of this pump is that it avoids the possibility of contamination of the vacuum system with oils that are present in any form of mechanical roughing pump.

Actual deposition of the electrode material may be done through thermal evaporation, where the electrode material is heated to its vaporization point in the vacuum, or by sputtering, where bombardment of a target of electrode material by ionized gas atoms causes some of the target atoms to be driven off and captured by the substrate to be plated. Sputtering generally results in better adherence between metallization and the substrate.

Thermal evaporation may be done by electron bombardment of the plating material or by vaporization from an electrically heated filament, boat, or specially designed source (Andres, 1976).

The electrode materials most widely used are gold, aluminum (Bottom, 1976; Ang, 1979, 1980), silver (Fukuyo et al., 1979), and combinations such as chrome–gold, molybdinum–gold, or titanium–palladium–gold (Dybwad, 1978). The choice of electrode material depends on the processes being used and the final intended use of the device. Best long-term aging rates have been seen with gold and copper (refer to Chapter 6).

Final frequency trimming of thickness-mode resonators can be done during the deposition process. If the crystal being plated can be electrically driven by an external system during deposition, the actual instantaneous resonator frequency can be used to control the plating process (Snell, 1975a,b). This process is especially useful for gold and copper electrodes.

Final trim-to-frequency conventionally involves spot-plating on one side of the resonator. With the proper mounting configuration, both sides can be plated simultaneously, which can lead to better control of the resonator motional parameters (Fischer and Schulzke, 1976).

Aluminum electrodes almost always require an additional step. Since aluminum is readily oxidized, either thermal treating or a final anodization is required to passivate the aluminum surface. Anodization can also be used to trim the device to the final frequency (Bottom, 1976; Reche, 1978). Both plasma and liquid anodization have been used successfully. Anodization can also be controlled automatically to control the final trim-to-frequency (Ang, 1979, 1980).

Other methods of trimming to the desired final frequency involve laser removal of electrode material (Hokanson, 1969; Smagin, 1974; Caruso 1977), exposure of silver electrodes to halides such as iodine vapor, air-abrasive units to remove small amounts of metallization, rf back-sputtering, and galvanic plating of additional electrode material (Kosecki, 1970).

4.7 MOUNTING AND SEALING

To be a useful device, the resonator must be mounted in some form of holder with appropriate electrical connections to the crystal electrode structure.

4 RESONATOR AND DEVICE TECHNOLOGY

Crystal holders are generally fabricated from a header that contains the electrical leads for external connection and a can, or outer enclosure, which will eventually be fastened to the header to form a complete package. Speciality holders have been developed for unique applications that combine the functions of header and enclosure in a single unit.

Crystal headers are made with two or more electrical feedthroughs that are isolated from each other, other possible metal portions of the header, and the enclosure with ceramic, glass, or organic insulators. The header itself may be made of glass, metal, ceramic brazed to metal, or plastic. Electric feedthroughs are terminated inside the package with wire or shaped metal fittings that attach to the crystal electrode pattern. The number, size, shape, and location of the electrical feedthroughs and their connection to the crystal are dictated by the desired response of the resonator to shock, vibration, and acceleration (Bernstein, 1970, 1971; Filler and Vig, 1976a,b; Lee and Wu, 1977).

Methods of mounting the crystal resonator to the electrical feedthroughs can be divided into two general categories, metallic and adhesive. Metallic methods include brazing with indium or gold–germanium alloys (Grzegorzewicz, 1975; Kusters et al., 1977), acoustic bonding (Nickols and Fay, 1978), thermocompression bonding, nickel electrobonding (Vig et al., 1975), and soldering (Fyfe, 1972b). Adhesive bonding uses either conductive epoxies, conductive polyimides (Filler et al., 1978), or metallic-loaded pastes that are fired to drive off organic binders.

Special crystal holders have also been used in which the crystal is held captive between pressure plates or where the electrode contact is made using coiled-spring clips. Ceramic flat-pack versions have also been fabricated that combine a low-profile holder with an integral outer enclosure (Wilcox et al., 1975; Peters, 1976; Filler et al., 1980).

Of the various methods used, best long-term aging for high-precision units have been obtained using thermocompression bonding, gold–germanium brazing, or conductive polyimides (refer to Chapter 6).

Of equal importance to the resonator performance is the outer enclosure. The type of header and mounting method used and the method of sealing the enclosure to the header dictate the design and material used in the enclosure. Typical enclosures are glass (Wolfskill, 1968), copper, nickel, and ceramic. Methods of sealing the enclosure to the header include solder, epoxy, melting, capacitive discharge welding (Fuchs, 1978), resistance heating (Fuchs, 1979), thermocompression bonding, and cold welding (Jamiolkowski and Sobocinski, 1974; Kusters et al., 1977). The latter two generally give the best long-term aging results.

Solder and epoxy sealing usually contain volatile compounds that can degrade aging characteristics of the device. Melting to seal all-glass enclosures

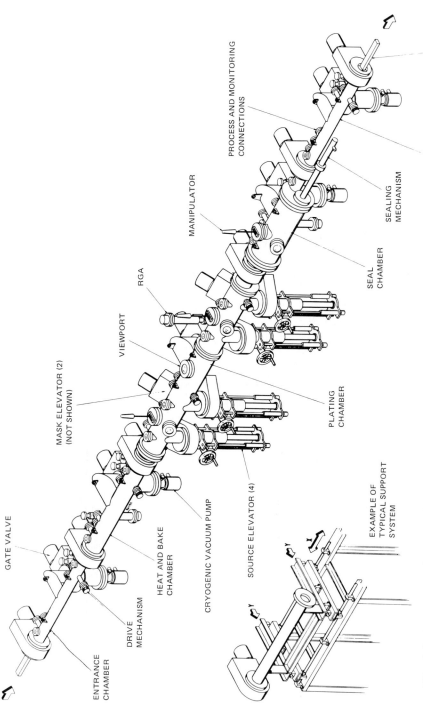

FIG. 4-4 Quartz crystal fabrication facility. This is a modern, in-line system that permits crystal cleaning, baking, frequency plating, and sealing without breaking system vacuum. Actual facility length is approximately 27 ft. [From Ney and Hafner, (1979).]

and resistance heating or welding for metal packages can liberate residual gasses and other contaminants present on the header and enclosure wall.

Final processing of the resonator unit may take place either before or after the enclosure is sealed to the header.

The highest-precision units are processed in a single system where the crystal blank, attached to its mount, receives a final cleaning using UV-ozone, ion etching, or plasma scrubbing, then is baked at an elevated temperature in high vacuum, is frequency plated, and is sealed in the final enclosure without breaking vacuum in the plating system (Ney and Hafner, 1979) (see Fig. 4-4). Resonator units designed for high-precision, oven-controlled oscillators may also receive a partial-pressure backfill with hydrogen or

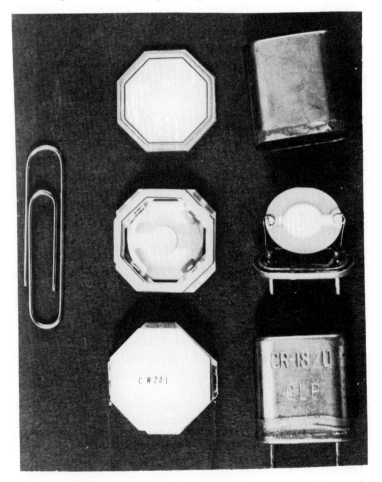

FIG. 4-5 Low-precision and high-precision crystal units, with and without enclosure.

helium prior to final sealing. A slight backfill atmosphere, 100 μm to several torr, helps to improve the thermal coupling between resonator blank and oven mass.

Lower-precision units may receive a vacuum bake, may be exposed to room atmosphere during the final processing steps, and may be packaged in a partially evacuated enclosure. The choice of process steps depends on the final use of the unit. Low-cost, low-precision units usually receive no further processing after the crystal package is sealed.

The finished resonator is now ready for final testing and use as a precision frequency control device.

Figure 4-5 shows a low-cost, low-precision unit and a high-precision crystal unit, with and without enclosure.

4.8 SPECIAL FABRICATION CONSIDERATIONS FOR SAW DEVICES

The SAW resonators present a special challenge. Complete surface preparation is usually only necessary on one side of the resonator blank. The degree of surface perfection required almost always exceeds that of a bulk-wave resonator. The SAW devices have their greatest utility at frequencies above that obtained in the usual bulk-wave device. The usual range of interest for SAW devices is 100 MHz to several gigahertz. At these frequencies, any surface imperfection or microcrack becomes an acoustic scattering center that can result in wave distortion or conversion to other acoustic modes. In SAW resonators this leads to reduced Q and increased device resistance.

An acceptable method, which has consistently produced material-limited Q values in quartz SAW resonators, is to first lap the substrate to the desired orientation. This followed by a heavy etch, a second lapping with 3-μm abrasive, heavy etch, and a final lap with 1-μm abrasive. All lapping is done on a metal lap. Following another light etch, the substrate is polished using a soft-pitch lap and a suitable polishing compound such as cerium oxide in water. The substrate is hand-polished until no visible surface imperfections are found. After a light etch, the substrate surface is examined under 600× dark field in a microscope. The sequence of polish, etch, examine is continued until no further imperfections are found using the microscope.

The type of treatment that a SAW substrate receives after surface preparation differs significantly from a bulk-wave device (refer to Section 2.3). Whereas simple metal masking is usually adequate for a bulk-wave resonator, metalization on a SAW device is done using techniques and equipment developed for integrated circuit processing. These include metallization, application of photoresist, exposure of the photoresist through precision

4 RESONATOR AND DEVICE TECHNOLOGY

masks, development of the photoresist, chemical etching, and photoresist removal (Smith, 1977). The pattern definition process defines the performance and utility of the SAW device (Field and Chen, 1976). Pattern definition may be done using visible light (Adams and Kusters, 1977), UV light, or electron-beam exposure (Hartemann, 1978; MacDonald et al., 1979; Cross et al., 1980).

For SAW resonators, the active area is usually metallized using aluminum (Adams and Kusters, 1977) or an aluminum alloy (Latham et al., 1979). After pattern definition, the SAW resonator may have grooves etched into its surface using either ion milling (Castellano and Hokanson, 1975) or a combination of sputter etching and plasma etching in a fluorine atmosphere (Adams and Kusters, 1977).

Final trim-to-frequency may be done during the milling or plasma etching stage. Similar to bulk-wave devices, the actual frequency can be monitored during final processing. Processing continues until the final frequency tolerance is reached. Some success has also been achieved using thin-film overlays to trim-to-final frequency (Urabe et al., 1979) and using argon-ion bombardment (James and Wilson, 1979).

Because SAW devices are confined to a single surface of the substrate, have a performance that is not limited by lateral surface boundary conditions, and are small in size, a common practice is to produce multiple resonator devices on a single substrate. This tends to complicate final trim-to-frequency since all devices on a substrate receive the same processing. Minor perturbations may result in a distribution of final frequencies across the substrate. Following resonator fabrication on the substrate, the substrate is sectioned into individual devices.

In certain applications, packaging considerations may dictate that several devices, perhaps at different frequencies, be mounted in the same package. If proper consideration is given to possible acoustic coupling between resonator sections, the substrate layout and sectioning can be simplified.

Sectioning in conventional IC devices is made possible by natural cleavage planes in silicon that lend themselves readily to a score-and-break technique. Natural cleavage planes do not exist in quartz. Individual sections must be sawn from the substrate.

Long-term aging in a SAW device is controlled by essentially the same effects that plague bulk-wave devices. Surface stress and contamination are perhaps the leading contributers (Dolochycki et al., 1979). Acoustic energy distribution in a SAW resonator penetrates about one acoustic wavelength. This is roughly equivalent to a bulk-mode fundamental or third-overtone resonator operating at the same frequency. SAW devices, however, typically operate at frequency ranges considerably above that of bulk-wave devices, so the problem becomes more severe.

Complicating the problem is that for devices that are ion milled or plasma etched, reaction products may be driven into the substrate. Early results with plasma etching in a fluorine atmosphere showed that significant frequency changes were observed during a vacuum bakeout subsequent to final substrate processing (Adams and Kusters, 1978). Vacuum outgassing showed traces of fluorine that were apparently trapped within the quartz substrate.

Further aging effects may be the result of the type of mount chosen for the SAW device. Since energy is confined primarily to one surface, the obvious choice is to mount the device, using some form of adhesive, solder, or brazing, to the inactive surface. For a given substrate thickness, however, a SAW device is more sensitive to external stress than an equivalent bulk-wave device (Dias *et al.*, 1976). A hard mount such as that achieved by brazing allows thermally induced stresses to affect frequency stability. Soft, compliant mounts do not have this problem. Problems arise because usual materials such as the room-temperature vulcanizing compounds (RTV) have severe outgassing and temperature problems. The resultant contamination of the resonator surface may result in severe degradation of long-term stability.

The best results reported to date incorporate all of the techniques developed for bulk-wave resonators. This includes mounting by brazing to compliant supports, thorough baking in vacuum, and sealing in cold-welded enclosures (refer to Chapter 6).

4.9 NOVEL RESONATOR TECHNIQUES

Resonator processing is usually characterized as involving techniques that have been developed through many years of use. Fundamental improvements to conventional technology occur rarely. This condition seems to be changing. Recently introduced techniques promise a considerable improvement in yield, efficiency, and cost reduction. Primary among these are chemical polishing (refer to Section 4.4), UV–ozone cleaning (refer to Section 4.5), low-profile, rugged crystal mounts (refer to Section 4.6), and in-line processing systems (refer to Section 4.6).

In addition, a recent development in resonator processing techniques has its foundation in the integrated circuit industry. This method uses photolithographic and etching techniques to produce multiple resonators from a single substrate. In this process, a large substrate is oriented, lapped, and perhaps polished to the desired thickness. Photolithographic methods using metallization, photoresist, and precision masking, similar to SAW device processing, are used to define the electrode pattern on the substrate and also the final resonator outline. A multistep process may be used with

4 RESONATOR AND DEVICE TECHNOLOGY

several different metallization layers. A typical process is to define final electrode configuration with a first layer of metallization and define the final outside dimensions of the device with a second metallization layer of a different material. The substrate is then etched using chemical or plasma etching techniques. This produces a substrate with many individual resonators defined on the blank still attached to the substrate by small break-away tabs. Chemical removal of the second layer of metallization leaves the individual resonators with a properly defined electrode pattern.

The choices of metallization used, for both device definition and electrode pattern, and the various crystal and metal etchants are a result of a proper chemical analysis of the entire process.

The advantage of this technique is that multiple resonator devices can be made on a single substrate with only a minor increase in process complexity. Final trim-to-frequency cannot be controlled in the usual manner of monitoring during a process step but is easily done using laser trimming after completion of fabrication.

Currently, the use of this technique is restricted to ultraminiature resonators for watch applications and other special purpose applications (Staudte, 1968, 1973; Oguchi and Momosaki, 1978; Hatschek, 1980) (See Fig. 4-6).

FIG. 4-6 32.768-kHz quartz tuning fork produced using photolithographic techniques to define the pattern and reactive plasma etching to define the quartz resonator shape. (Photo courtesy Hewlett-Packard Co.)

Another fabrication technique that shows great promise in reduced aging rates and improved short-term stability is to remove the influence of the electrode material on crystal performance through the use of a new electrodeless resonator design (Cutler and Hammond, 1969; Besson, 1976, 1977). This device is a conventional resonator blank without any surface electrode metallization. Instead, the device is excited by air-gap electrodes deposited on additional surfaces that are precisely spaced a very small distance away from the acoustically active surface of the resonator blank.

4.10 ENVIRONMENTAL EFFECTS

Crystal resonators are generally carefully designed to minimize the effect of any environmental changes. The choice of crystal orientation determines the gross performance of the device in a changing temperature environment. The choice of mounting location and method of mounting determine the acceleration response of the unit. A number of other factors can also limit crystal performance.

Static compensation to temperature change is exhibited by the conventional AT and BT cuts. These cuts exhibit rather large dynamic changes in frequency in any situation where the temperature changes rapidly. Recent development of elasticity theory has led to the development of cuts where both static and dynamic compensation under rapid temperature change is achieved. The most notable of these is the SC cut (Holland, 1974; EerNisse, 1975; Kusters, 1976).

Acceleration effects can also be minimized by a proper choice of mount and mounting location (Filler and Vig, 1976a,b; Ballato *et al.*, 1977; Lee and Wu, 1977). Several novel schemes have been proposed and tested that have led to approximately an order of magnitude reduction in the acceleration sensitivity of resonator devices (Warner *et al.*, 1979). One item of importance is that the proper mount design also strongly influences the vibrational response of the resonator device. Mechanical resonances in the support structure can couple strongly to the resonator device. Under mechanical excitation, this can show up as a spur in the phase noise response of the oscillator system using that resonator.

Of equal importance in precision frequency applications is the device sensitivity to electric and magnetic fields. To first order, thickness-mode resonators such as the AT and BT cuts are not affected by applied dc voltages. All of the doubly-rotated cuts are quite sensitive to applied dc biases. For example, an LC cut (a doubly rotated cut with only a first-order frequency-temperature coefficient used for thermometry) changes its frequency by 3 ppm when 100 V dc is applied to the resonator (Kusters, 1970). Proper oscillator design must take this possibility into account.

4 RESONATOR AND DEVICE TEQNOLOGY   183

Resonator crystals appear capacitive at low frequencies, with very low leakage. Static charges can accumulate on the crystal with unpredictable results. While it may look promising to use dc biasing as a method of oscillator frequency control, static charges applied to the resonator with biasing potentials tend to be compensated on the resonator surface by mobile ions in the blank. The amount of compensation is dependent on the impurities present in the blank. The time constant of the ion mobility is dependent on the resonator crystallographic orientation and the blank temperature. Time constants range from several seconds at 80°C. to several minutes at room temperature (Kusters, 1970).

Magnetic field sensitivity is usually not of importance in precision resonators. Quartz is inherently magnetically insensitive. A wrong choice of material for electrode patterns, mounting structure, or header may make the precision resonator device sensitive to applied magnetic fields.

Sensitivity to ionizing radiation has been shown to be related to impurities in the quartz material. The effect seems also to be dependent on crystallographic orientation. Vacuum-swept quartz has shown the best results for minimizing permanent frequency changes due to radiation (refer to Section 4.1).

Sensitivity to applied pressure is a fundamental property of any piezoelectric device. This has been used to advantage in special devices designed for metrology applications (Karrer and Leach, 1969). While this is not a problem in conventional resonators that are vacuum encapsulated, it may pose problems either for units not sealed under vacuum or special units where the resonator design permits external pressure changes to be applied to the resonator.

In general, a crystal resonator is sensitive to a variety of external stimuli. Proper design of the resonator and of the driving circuitry that will use the resonator is necessary to obtain precision frequency control.

5 Piezoelectric and Electromechanical Filters

Robert C. Smythe
Piezo Technology Incorporated
Orlando, Florida

Robert S. Wagers
Central Research Laboratories
Texas Instruments Incorporated
Dallas, Texas

	List of Symbols for Sections 5.1 and 5.2	186
5.1	General	187
	by Robert C. Smythe	
5.2	Bulk-Acoustic-Wave Filters	188
	by Robert C. Smythe	
	5.2.1 Introduction	188
	5.2.2 Crystal Filters	189
	5.2.2.1 Discrete-Resonator Crystal Filters	192
	5.2.2.2 Monolithic Crystal Filters	199
	5.2.2.3 Nonlinear Effects	216
	5.2.2.4 Crystal Filters Using Other Materials	219
	5.2.3 Electromechanical Filters	221
	5.2.3.1 Flexure-Mode Bars and Plates	223
	5.2.3.2 Extensional-Mode Filters	224
	5.2.3.3 Disk–Wire Filters	225
	5.2.3.4 Torsional-Mode Filters	226
	5.2.3.5 Nonlinear Effects	227
	List of Symbols for Sections 5.3 and 5.4	228
5.3	Surface-Acoustic-Wave Filters	230
	by Robert S. Wagers	
	5.3.1 Introduction	230
	5.3.2 Interdigital Transducer Admittance	233
	5.3.2.1 Normal-Mode Representation of Acoustic Admittance	233
	5.3.2.2 Interdigital Transducer Capacitance	236
	5.3.2.3 Evaluation of Interdigital Transducer Admittance	237
	5.3.3 Relation of Normal-Mode Theory Admittance to the Impulse Model	239
	5.3.4 Limitations on the Use of Electrostatic Fields	240
	5.3.5 Electromechanical Coupling Constant k	241

5.3.6	Electrical Q and Insertion Loss	242
5.3.7	Bulk-Wave Modeling of Interdigital Transducers	244
5.3.8	Advanced Bulk-Wave Models	249
5.4	SAW Bandpass and Bandstop Filters	257
	by Robert S. Wagers	
5.4.1	Introduction	257
5.4.2	Impulse-Response Realizations	260
5.4.3	SAW Bandpass Filter Capabilities	263
5.4.4	SAW Bandstop Filters	266

LIST OF SYMBOLS FOR SECTIONS 5.1 AND 5.2

A_e	Resonator electrode area
BW_3	3-dB bandwidth (of a filter)
C_0	Static capacitance (of a resonator or monolithic filter)
C_1	Motional capacitance (of a resonator or monolithic filter)
c_{22}, c_{55}	Unstiffened elastic constants of AT-cut quartz
$\bar{c}_{66}$	Stiffened elastic constant of AT-cut quartz
Δf	$(f_p - f_e)$
f	Frequency
f_e, f_p	Cutoff frequencies of a trapped-energy resonator or monolithic filter
f_s, f_a	Principal symmetric and antisymmetric mode frequencies of a two-resonator monolithic filter
g	Normalized gap width of a monolithic filter
h	Electrode height
I	Current
k	A dimensionless trapping constant, either k_{ts} or k_{tt}
k_e, k'_e, k_p, k'_p	Wave numbers of a trapped-energy resonator
k_{ts}, k_{tt}	Dimensionless trapping constants, Eqs. (5-9) and (5-10)
k_{26}	Electromechanical coupling constant of AT-cut quartz
m, p, q	Mode indices
M_n	Tiersten's effective elastic constant
n	Overtone number
N	Frequency-thickness constant of AT-cut quartz
P	Power
r	Ratio of static to motional capacitance, C_0/C_1
t	Wafer thickness
t'	Electrode thickness
V	Voltage
w	Electrode width
γ	An effective nonlinear elastic constant
δf	$f_a - f_s$
$\delta_m, \delta_p, \delta_q$	Frequency offsets
ρ	Mass density of quartz
ρ'	Mass density of electrode film
ω	Circular frequency, $2\pi f$

5 PIEZOELECTRIC AND ELECTROMECHANICAL FILTERS   187

5.1 GENERAL[§]

This chapter treats bulk-acoustic-wave (BAW) filters and surface-acoustic-wave (SAW) filters. The BAW filters include electro-mechanical filters, discrete-resonator crystal filters, monolithic crystal filters, and ceramic filters. Because the scope of this book is limited to precision frequency selection and control, ceramic filters will not be treated. For similar reasons, the treatment of SAW filters will be restricted to topics relevant to their use as frequency-selective devices. The more general signal-processing applications, such as convolution and correlation, though important, would take us beyond the present scope of frequency control.

By design, the depth of treatment differs for the various filter categories. Discrete-resonator crystal filters and electromechanical filters, because they are well-established technologies, receive rather limited consideration, with the emphasis being on recent developments. (However, using the references cited, the interested reader can study these fields in greater detail.) Monolithic filters and SAW filters, on the other hand, are fairly new technologies on which there has been a very great deal of recent work published, work that this chapter attempts to summarize and to which it attempts to serve as an introduction and guide. This is particularly important in the case of SAW filters, since this technology has essentially developed since the mid-1960s, beginning with the demonstration of the piezoelectric SAW transducer by White and Voltmer (1965). Monolithic filter technology, on the other hand, has grown out of quartz resonator technology.

While SAW and BAW technologies have developed separately, there is much common ground. Lukaszek and Ballato (1980) discussed some ways in which SAW technology might benefit from bulk-wave experience, pointing out related problem areas and solutions. The SAW and BAW filters are alike insofar as they both employ acoustic waves and some means of converting electrical energy to acoustic energy and vice versa. In addition, both are primarily used as bandpass filters.

There are also basic differences. The means by which SAW and BAW filters perform the bandpass function are quite different. The BAW filters are made up of (acoustic) resonators, coupled or interconnected in various ways. Most SAW filters on the other hand, are transversal filters [tapped delay-line filters (Kallman, 1940)], although SAW resonators and SAW resonator filters (Sec. 2.3) are also of importance. Consequently, BAW filters may be described in the frequency domain by rational functions, while SAW filters are most easily described in the time domain (to first order) by a sum of impulse functions. Many BAW filters are minimum-phase networks, while for SAW filters, amplitude and phase response can be

[§] Sections 5.1 and 5.2 were written by Robert C. Smythe.

controlled separately. The BAW filters make use of classical filter theory. Transversal filter theory, on the other hand, was not highly developed prior to the advent of SAW filters.

As a further guide to the field the reader may refer to a number of survey papers, collected papers, and texts. Matthews (1977) edited a valuable text on SAW filter design and applications. Sheahan and Johnson (1977) edited a very useful collection of papers on crystal and mechanical filters (many of which are referred to in this chapter) and provided helpful introductory remarks. The text on filter design edited by Temes and Mitra (1973) includes chapters on crystal and mechanical filter design. Another useful collection of papers on crystal, mechanical, and SAW filters is the January, 1979 issue of the *Proceedings of the IEEE*, a special issue on miniaturized filters. Progress in SAW filter technology was recorded in the May, 1976 issue of the *Proceedings of the IEEE* and in three special issues of the *IEEE Transactions* (November, 1969; April, 1973; May, 1981) published jointly by the Sonics and Ultrasonics Group and the Microwave Theory and Techniques Society.

For both SAW and BAW filters, this is a particularly appropriate time for review. The SAW filters are just now entering a period of serious commercial development. As evidence of this new maturity, increasing attention is being given to such matters as cost, manufacturing methods, and secondary performance characteristics such as aging and reliability. The BAW filters, on the other hand, seem ripe for new levels of sophistication as major areas of application expand and new ones open up.

5.2 BULK-ACOUSTIC-WAVE FILTERS

5.2.1 Introduction

Bulk-acoustic-wave filters include piezoelectric crystal filters, piezoelectric ceramic filters, and electromechanical filters. Piezoelectric ceramic filters, like piezoelectric ceramic resonators, though of importance in a number of applications, are beyond the scope of this book.

Piezoelectric crystal filters and electromechanical filters, which from now on will be referred to simply as crystal filters and mechanical filters, have much in common conceptually and, moreover, share some applications. Yet the two technologies are essentially separate, chiefly because of differences in manufacturing methods.[§]

An important difference between crystal and mechanical filters is that in the former each resonator is also an electromechanical transducer, while

---

[§] An illuminating comparison written by Sheahan and Johnson (1975) was reprinted in a collection of papers in the field edited by Sheahan and Johnson (1977).

5 PIEZOELECTRIC AND ELECTROMECHANICAL FILTERS

in a mechanical filter the transducers are formed separately and, with few exceptions, are associated with the first and last resonators of the filter, whose topology is that of a ladder or bridged-ladder network. Consequently, the topology of crystal filters, especially those using discrete resonators, is much more varied than that of mechanical filters. However, as Sheahan and Johnson (1975) point out, monolithic crystal filters are essentially mechanical filters, have similar topologies, and hence are designed by the same network synthesis methods. At the same time, the fact that each resonator has electrical terminals makes it practical to use monolithic filter elements as sections of larger filter networks.

Both crystal and mechanical filters can be realized at very low frequencies, although in practice few crystal filters are made below 60 kHz, with the majority being above 1 MHz. Monolithic crystal filters are usually impractical or uneconomical below 4 to 5 MHz. Mechanical filters are useful from below 1 kHz up to about 500 to 700 kHz, while the frequency range of crystal filters extends to about 300 MHz. Figure 5.2-1 shows the frequency-bandwidth domains in more detail.

Applications for both crystal and mechanical filters are primarily in communication and navigation systems. The earliest applications of crystal filters (in the 1930s) were in telephone frequency-division multiplex (FDM) equipment (Lane, 1938; Simmonds, 1979). This has remained an important area of application for crystal filters, especially in North America, while in Europe and Japan many FDM systems use mechanical filters (Guenther et al., 1979; Onoe, 1979; Yakuwa et al., 1979).

More important uses of crystal filters are in all classes of mobile two-way radio and paging equipment, as well as in point-to-point radio communications, electronic navigation systems, and frequency synthesizers (Smythe, 1979a,b). In addition to FDM systems, mechanical filters are used in hf radio communication applications, in low-frequency electronic navigation systems (Johnson, 1977), and in a variety of special applications such as automatic train control systems.

5.2.2 Crystal Filters

If we consider any frequency-selective network incorporating one or more crystal resonators to be a crystal filter, then we may say that crystal filters can be used to obtain all the common types of filter functions. Nevertheless, most crystal filters are bandpass networks, a few are band-reject filters, and only rarely are high-pass or low-pass functions realized using crystal resonators. Accordingly, the discussion that follows will be limited to bandpass crystal filters.

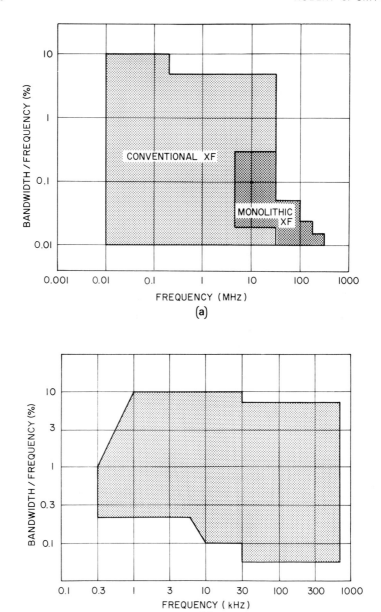

FIG. 5.2-1 Bandwidth and frequency capabilities of (a) quartz crystal filters and (b) mechanical filters.

5 PIEZOELECTRIC AND ELECTROMECHANICAL FILTERS

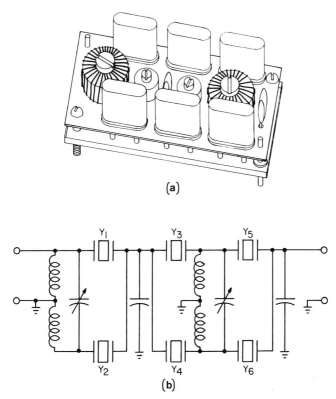

FIG. 5.2-2 Typical discrete resonator crystal filter. A six-pole narrow-band filter is shown. (a) Layout of the three half-lattice sections; (b) circuit diagram.

For reasons noted in earlier chapters, the primary resonator material is quartz; hence, we will be concerned chiefly with filters using quartz. Lithium tantalate crystal filters are treated in Section 5.2.2.4. Because of space limitations, a number of important topics have been omitted. These include frequency discriminators (Smith, 1968), stacked crystal filters (Ballato and Lukaszek, 1973a,b; Stearns et al., 1977), and active network crystal filters (Means and Ghausi, 1972; Waddington, 1975).

Bandpass crystal filters may be divided into discrete-resonator filters, in which each resonator is electrically and, most often, physically a separate device (Fig. 5.2-2) and acoustically-coupled or monolithic crystal filters, in which at least some of the resonators are coupled acoustically (Fig. 5.2-3). Since their introduction in the 1960s, acoustically-coupled crystal filters have developed rapidly, and much of the growth in crystal filter applications has been associated with the monolithic filter technology. Nevertheless,

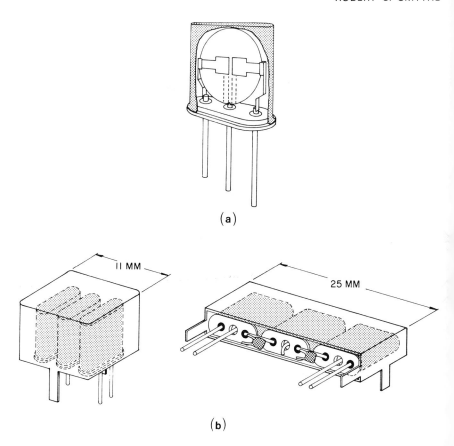

(a)

(b)

discrete-resonator filters continue to be of importance, particularly at frequencies below 5 MHz. The wide range of filter requirements makes it likely that both technologies will continue to develop.

5.2.2.1 DISCRETE-RESONATOR CRYSTAL FILTERS

Although for purposes of discussion we consider discrete-resonator crystal filters separately from monolithic filters, from the standpoint of circuit design theory the two are more alike than different. It follows that much of the material in this section is useful background for the following one. More detailed design information can be found in numerous references, including Kosowsky (1955, 1958), which gives an image-parameter treatment and Zverev (1967) and Temes and Mitra (1973), which treat aspects of insertion loss synthesis methods.

5 PIEZOELECTRIC AND ELECTROMECHANICAL FILTERS   193

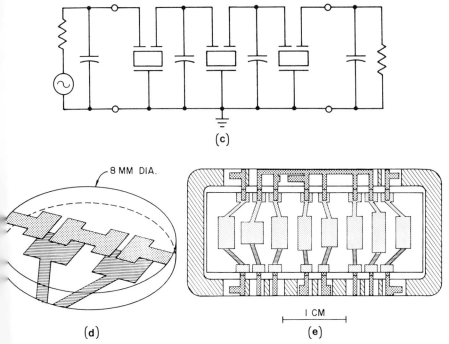

(d)   (e)

FIG. 5.2-3 Monolithic crystal filters. (a) Typical two-pole MCF; (b) miniature tandem monolithic filters; (c) circuit diagram, six-pole tandem monolithic filter; (d) electrode configuration, four-pole VHF monolithic filter (typical of construction, 30–180 MHz) [from Smythe (1979), © 1979 IEEE]; (e) eight-resonator FDM channel filter [from Pearman and Rennick (1977)].

A. Structures. Crystal-filter networks may take a variety of forms too numerous to list completely. The most important are those related to the symmetrical lattice (Fig. 5.2-4). Usually, the symmetrical lattice is replaced by its half-lattice (Jaumann network) equivalent (Fig. 5.2-5) to reduce the

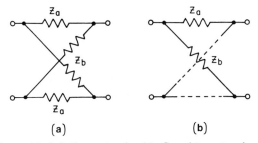

FIG. 5.2-4 Symmetrical lattice network. (a) Complete network; (b) Drafting representation.

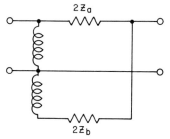

FIG. 5.2-5 Half-lattice equivalent of symmetrical lattice network, using ideal transformer.

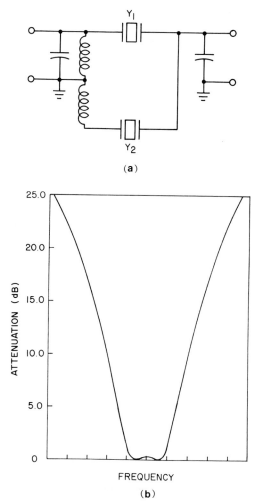

FIG. 5.2-6 Two-pole narrow-band crystal filter. (a) Circuit diagram; (b) calculated attenuation characteristic (frequency units arbitrary).

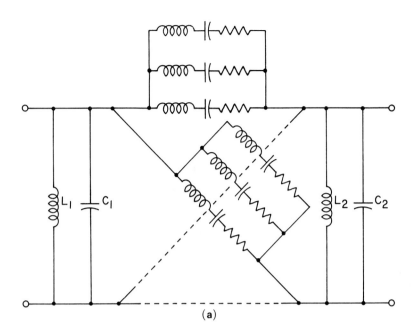

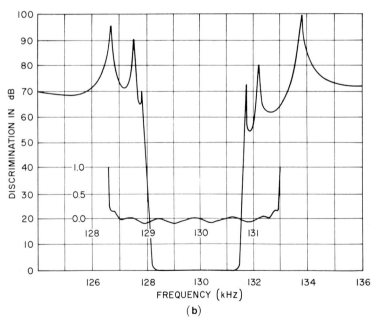

FIG. 5.2-7 Eight-pole wide-band symmetrical-lattice filter. (a) Simplified circuit diagram; (b) attenuation characteristic [from McLean et al. (1979)].

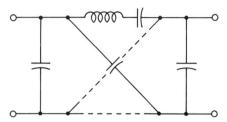

FIG. 5.2-8 Equivalent circuit, 5° X-cut divided-electrode crystal resonator. Six of these resonators are used in the filter shown in Fig. 5.2-7.

number of components and obtain a grounded network. Synthesis, however, may still be carried out assuming the full lattice.

Figure 5.2-6 shows a simple two-pole, narrow-band (see below) half-lattice crystal filter and its theoretical attenuation characteristic. To obtain greater selectivity more resonators can be added, as in Fig. 5.2-7 (McLean et al., 1979), which shows the circuit of a 128-kHz eight-pole wide-band symmetrical-lattice filter and its response. In this frequency range, two-port resonators[§] having the symmetrical-lattice-equivalent circuit of Fig. 5.2-8 are easily realized so that the full lattice can be realized with the same number of crystal elements as the half-lattice.

The lattice and half-lattice achieve attenuation by a balance of the series and diagonal arm impedances, and for high stopband attenuation a near-perfect balance is required. In the filter just described, a stopband attenuation of 70 dB is attained by careful construction, taking advantage of the inherent balance of the two-port resonators used.[¶] Most often, such high attenuation is not achievable in a single lattice, and it becomes more practical to divide the filters into two or more lattice or half-lattice sections (Fig. 5.2-2). This tandem lattice configuration is by far the most common discrete-resonator crystal filter structure. Intermediate-band (see below) and wide-band tandem-lattice synthesis have been treated by Blinchikoff (1975) and Szentirmai (Temes and Mitra, 1973, ch. 4).

For narrow-band tandem-lattice filters, the synthesis for all-pole response has been known for many years. The realization from the low-pass ladder prototype of symmetrical response with $j\omega$-axis transmission zeros was given in a particularly simple form by Holt and Gray (1968). As will be seen in Sec. 5.2.2.2, each two-pole lattice section may be replaced by a symmetrical two-resonator monolithic filter plus, perhaps, additional reactive elements.

[§] These use a single mode of vibration and should not be confused with the acoustically coupled resonators to be discussed in Sec. 5.2.2.2, which use two or more modes.
[¶] Also, in this example, the first and last resonators are realized by tank circuits L_1, C_1 and L_2, C_2, so that the balance requirement of the lattice itself is less than 70 dB.

5 PIEZOELECTRIC AND ELECTROMECHANICAL FILTERS

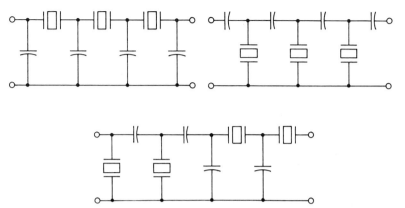

FIG. 5.2-9 Typical ladder network configurations suitable for narrow-band applications.

Dillon and Lind (1976) showed that such a structure can realize all the classic filter sections.

The tandem-lattice structure combines the stopband attenuation advantage of a multisection filter with the ability of the lattice to allow resonators of similar impedance level to be used throughout. For very narrow bandwidths, these advantages may sometimes be obtained by a ladder network. Typical forms are shown in Fig. 5.2-9. Dishal (1958, 1965) gave a simple method for the synthesis of a class of SSB filters in ladder form that was refined by Haine (1977). One may also combine lattice or half-lattice sections with ladder sections.

B. *Design Types.* Bandpass crystal filters may be classified by design types, which are summarized in Table 5.2-1. The corresponding forms of

TABLE 5.2-1

Crystal Filter Design Types[a]

Design type	Abbreviation	Maximum bandwidth	Design assumptions
Very narrow band	VNB	$0.05f/r$, typical	C_0 neglected
Narrow band	NB	$0.7f/r$	None
Intermediate band	IB-1	$2f/r$–$4f/r$, typical	Reactance of L_0 is constant
	IB-2		
Wide band	WB	$f\sqrt{2/r}$	None

[a] For type IB-1, L_0 is used to cancel a portion of C_0; for type IB-2, C_0 is completely cancelled. $r = C_0/C_1$.

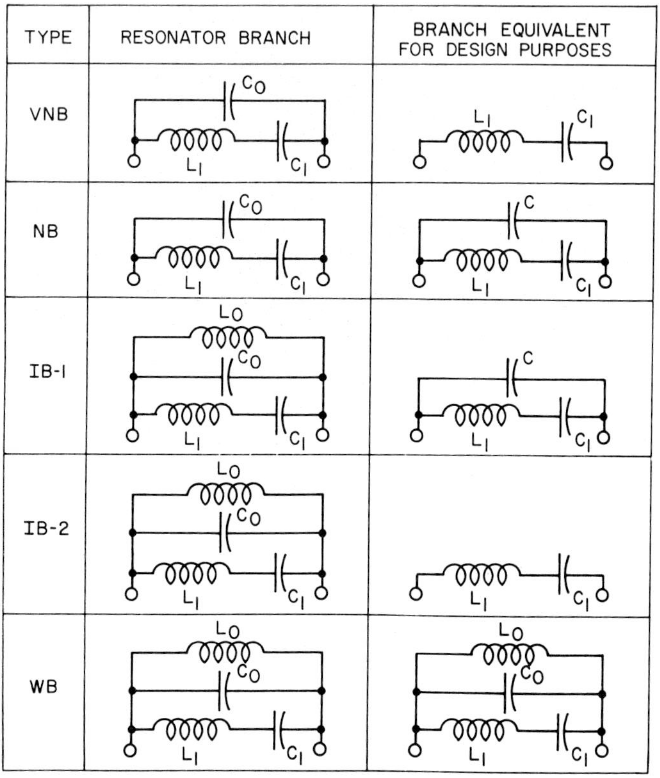

TYPE	RESONATOR BRANCH	BRANCH EQUIVALENT FOR DESIGN PURPOSES
VNB		
NB		
IB-1		
IB-2		
WB		

FIG. 5.2-10 Typical resonator branches for bandpass crystal-filter networks.

the resonator branches of the filter network are given in Fig. 5.2-10. Although these classifications are described for discrete filters, they also apply to monolithic filters. Note also that in Fig. 5.2-10, the loss associated with the inductors has been omitted from the circuit representation but must be taken into account in actual practice. Very-narrow-band and narrow-band (VNB and NB) designs employ crystals and capacitors only. In intermediate-band (IB) designs inductors are used to overcome the maximum bandwidth limitations imposed by resonator capacitance ratios by cancelling or "tuning out" all (type IB-2) or a portion (type IB-1) of the resonator shunt capacitance. In wide-band (WB) designs, such as Fig. 5.2-7, the inductors are used to form resonators, so that wide-band filters are in effect partly LC filters.

Of these types, the most commonly used are NB designs, followed by VNB and IB. The WB designs are used mostly at low frequencies. While single- and tandem-lattice and half-lattice filters of all types are produced, ladder crystal filters are generally of types VNB or NB.

5 PIEZOELECTRIC AND ELECTROMECHANICAL FILTERS

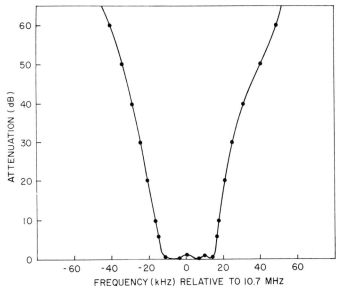

FIG. 5.2-11 Attenuation characteristic of early six-pole tandem monolithic filter. [From Nakazawa (1962).]

5.2.2.2 MONOLITHIC CRYSTAL FILTERS

By far the most important class of present-day crystal filters are those using acoustical coupling—monolithic crystal filters (MCFs) in common parlance, but including a variety of nonmonolithic structures.

The first successful filters using acoustical coupling were demonstrated by Nakazawa (1962). Nakazawa's devices consisted of two acoustically-coupled resonators. These could be connected in tandem to obtain higher-order filters, up to six poles being demonstrated in his 1962 paper. Figure 5.2-11 shows the attenuation characteristic of one of these early filters.

The development, or rather rediscovery (Shockley et al., 1963), shortly thereafter of the trapped-energy theory of thickness-shear resonators (Mortley, 1951, 1957) provided a basis for the understanding of acoustical coupling, as was soon recognized by Curran's group (Gerber et al., 1965) and others. Onoe and Jumonji (1965) gave a particularly clear, simplified analysis of coupling for an isotropic material. Coupled-resonator devices were demonstrated by Onoe et al., (1966), Sykes and Beaver (1966), and Mailer and Beuerle (1966), all of whom made use of trapped-energy concepts. It was also soon recognized that any number of resonators could be acoustically coupled. Onoe et al. (1966)[§] demonstrated a three-pole filter, while Sykes and Beaver (1966) demonstrated a six-pole acoustically coupled filter.

[§] See also earlier publications in Japanese by Onoe and co-workers, some of which are referred to in Onoe (1979).

Further development by a number of organizations quickly followed. The most important are

(1) development of multiresonator MCFs,
(2) development of tandem-connected two-pole MCFs,
(3) development of hybrid monolithic filters, which use both single resonators and two-pole MCFs,
(4) development of linear and nonlinear analysis methods for MCFs and trapped-energy resonators, and
(5) development of VHF MCFs.

Some aspects of these will now be treated.

A. Acoustical Coupling. The modern "energy-trapping" theory of thickness-shear resonators, first proposed by Mortley (1951, 1957), was independently put forward a decade later by Shockley et al. (1963), who were developing multiple, uncoupled resonator devices for use in conventional piezoelectric ceramic- and crystal-filter networks. Although they were interested in coupling between resonators—or rather, how to reduce or eliminate such coupling—their analysis was limited to single-resonator models, as was Mortley's.

The earliest analytical treatment of inter-resonator coupling is that of Onoe and Jumonji (1965), who treated thickness-twist waves in an isotropic, nonpiezoelectric plate. Horton and Smythe (1967) showed how Onoe and Jumonji's analysis could be applied to AT-cut quartz by using results obtained by Mindlin (1966) for thickness-twist waves and by Mindlin and Lee (1966) for fundamental thickness-shear waves.

Beaver (1967a,b, 1968) applied the methods of Mindlin and Lee (1966) and Tiersten and Mindlin (1962) to the analysis of an arbitrary number of

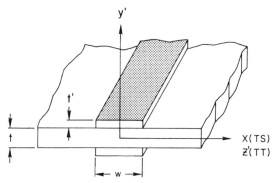

FIG. 5.2-12 Strip-electrode model of trapped-energy resonator. For thickness-twist (TT) propagation, the right-hand axis is Z'; for thickness-shear (TS) propagation, the right-hand axis is x.

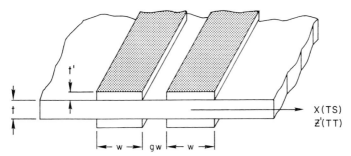

FIG. 5.2-13 Strip-electrode model of two acoustically coupled resonators. The ratio of electrode separation to electrode width is the gap ratio g.

coupled resonators, giving specific results for 2-, 3-, and 6-resonator structures with either thickness-twist or thickness-shear coupling.

These early analyses were for two-dimensional strip electrode models (Figs. 5.2-12 and 5.2-13). Further, the thickness-shear approximation of Mindlin and Lee was restricted to the fundamental mode. These limitations were removed by Tiersten. In an important series of papers (Tiersten, 1974a, 1975a, 1976a,b, 1977), he treated fundamental- and overtone-mode rectangular-electrode resonators and coupled-resonator pairs, the latter with either thickness-twist or thickness-shear coupling.

Tiersten's analysis of two coupled resonators gives a transcendental equation that can readily be solved numerically to obtain the natural frequencies of the symmetric and antisymmetric modes of the device, including the (unwanted) anharmonic modes. These frequencies correspond to short-circuit frequencies of the device. The remaining short-circuit admittance parameters are obtained by evaluation of closed-form solutions of the corresponding integrals, so that a complete lumped-element equivalent-circuit model, including unwanted modes, is obtained) (Fig. 5.2-14a). An example of the calculated and measured attenuation of a third-overtone two-resonator monolithic filter is given in Fig. 5.2-15, which shows good agreement between the two. Tiersten's analysis treats two identical resonators. Extension to more than two resonators, or to nonidentical resonators, is straightforward, at least with regard to the solution for the natural frequencies.

Other aspects of coupled-resonator theory were treated by many authors, who cannot all be given recognition here. Mason (1969b) used transmission-line models, while Ashida (1971, 1974) used transmission matrices to analyze acoustical coupling. Glowinski et al. (1973) and Lançon (1973) considered the effects of asymmetry. Dworsky (1981) used a Rayleigh–Ritz variational technique to calculate coupling in a variety of two-resonator configurations.

In what follows, some important results concerning acoustically coupled resonators are presented. It should be made clear that the emphasis is on

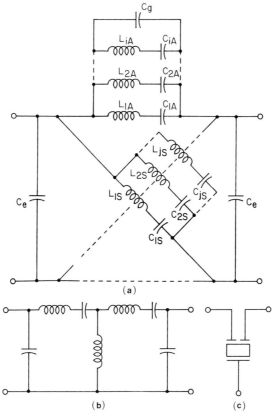

FIG. 5.2-14 Symmetrical two-pole monolithic-filter equivalent circuits. (a) Lattice equivalent circuit. Branches with subscripts $2S, \ldots, jS$ and $2A, \ldots, iA$ represent unwanted modes of vibration. C_g represents fringing (gap) capacitance. (b) Simplified ladder equivalent network, omitting unwanted modes and fringing capacitance. (c) Drafting symbol.

presenting results in a relatively simple form suitable for ready interpretation. For the underlying theory, the interested reader is referred to the previously cited works of Tiersten, Mindlin, and Onoe, which form the basis for this exposition, and the treatment by Spencer (Mason and Thurston, 1972, ch. 4). For simplicity, the treatment is restricted to thickness-shear vibrations in in AT-cut quartz.

Trapped-energy analysis makes use of wave-propagation concepts. An infinite plate of uniform thickness possesses, in the absence of an electric field, thickness-shear mode resonances at frequencies f_p at which the plate thickness t is n acoustic half-wavelengths:

$$f_p = nN/t, \quad n = 1, 2, 3, \ldots, \tag{5.2-1}$$

5 PIEZOELECTRIC AND ELECTROMECHANICAL FILTERS

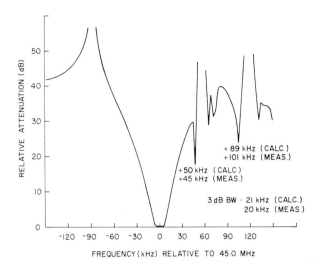

FIG. 5.2-15 Attenuation characteristic of a 45-MHz, third-overtone, two-pole monolithic filter, showing agreement between measured and calculated anharmonic mode frequencies [from Smythe (1979).]

where N is one-half the shear-wave velocity, is called the frequency-thickness constant, and is given by

$$N = \tfrac{1}{2}(\bar{c}_{66}/\rho)^{1/2}, \tag{5.2-2}$$

$\bar{c}_{66}$ being a stiffened elastic constant and ρ the mass density. For AT-cut quartz,

$$\bar{c}_{66} = 29.24 \times 10^9 \quad \text{N/m}^2,$$
$$\rho = 2649 \quad \text{kg/m}^3,$$
$$N = 1661 \quad \text{Hz m}.$$

The uniform plate may be considered as an acoustical waveguide, loosely analogous to a parallel-plate electromagnetic waveguide, with the frequencies f_p corresponding to waveguide cutoff frequencies. That is, for a given n, at frequencies above f_p the shear wave can propagate laterally in the plate, or waveguide, while below f_p the wave is evanescent. As in the electromagnetic case, the cutoff frequencies are those at which the thickness is n half-wavelengths.

If now the faces of the plate are coated with thin, perfectly conducting electrodes of thickness t' and mass density ρ', then the cutoff frequency, which we will now call f_e, is lowered by an amount

$$\Delta f = f_p - f_e, \tag{5.2-3}$$

where

$$\Delta f = (2\rho' t'/\rho t + 4k_{26}^2/n^2\pi^2)f_p. \qquad (5.2\text{-}4)$$

In Eq. (5.2-4) the first term is the lowering due to the electrode mass and is called the *mass loading*. The second term is called by analogy the *piezoelectric loading* and represents the reduction in stiffness that occurs in the quartz when the shorted electrodes allow displaced charge to equalize. The k_{26} term is the electromechanical coupling coefficient. For AT-cut quartz,

$$k_{26}^2 = 7.752 \times 10^{-3}.$$

We will consider waves that travel parallel to the crystallographic x axis (thickness-shear, or TS, waves) and waves that travel parallel to the z' axis (thickness-twist, or TT, waves). For TS waves the wave numbers for electroded and unelectroded plates, k_e and k_p, respectively, are given by

$$\begin{aligned}k_e^2 &= (4\pi^2\rho/M_n)(f^2 - f_e^2),\\ k_p^2 &= (4\pi^2\rho/M_n)(f^2 - f_p^2),\end{aligned} \qquad (5.2\text{-}5)$$

where M_n (Table 5.2-2) is a quantity dependent on n, obtained by Tiersten (1974a).

For TT waves, M_n is replaced by c_{55}. For AT-cut quartz,

$$c_{55} = 68.81 \times 10^9 \quad \text{N/m}^2.$$

In a "real" trapped-energy resonator or monolithic filter, both TS and TT wave prepagation must be dealt with simultaneously, but for infinite-strip-electrode models, only one or the other need be considered. Figure 5.2-12 shows a strip-electrode model of a single resonator in which the electrodes are perpendicular to either the x-axis (the TS case) or the z'axis (the TT case). At frequencies below f_e, k_e is imaginary, corresponding to an evanescent wave in the electroded region, while above f_e, k_e is real and the wave can propagate. Similarly, for the unelectroded region the shear wave can propagate only for frequencies greater than f_p. Hence, for frequencies lying between the two cutoff frequencies,

$$f_e < f < f_p, \qquad (5.2\text{-}6)$$

a wave can propagate freely in the electroded region, but the surrounding plate acts like a waveguide below cutoff. Under such conditions, at one or more frequencies in the interval of Eq. (5.2-6) a standing wave will exist in the electroded region, while in the plate regions the wave will decay expon-

5 PIEZOELECTRIC AND ELECTROMECHANICAL FILTERS

TABLE 5.2-2

Constants for Thickness-Shear (TS) and Thickness-Twist (TT) Wave Propagation

Wave type	Overtone	M_n $(10^9 N/m^2)^a$	k^b
TS	1	109.94	0.730
	3	75.80	0.878
	5	90.09	0.806
	7	80.44	0.852
TT	All	—	0.924

[a] M_n is Tiersten's elastic constant.
[b] k is defined in Eqs. (5.2-9) and (5.2-10).

entially. The standing waves occur at the natural frequencies (resonances) of the structure, given by the roots of

$$k_e \tan(k_e w - m\pi/2) = k'_p, \quad m = 0, 1, 2, \ldots, M - 1, \quad (5.2\text{-}7)$$

where

$$k'_p = -jk_p.$$

Equation (5.2-7) may readily be solved by iteration, using Eq. (5.2-5). Even values of m correspond to symmetric modes, for which the amplitudes in the electroded region are cosinusoidal functions of the propagation distance from the origin, while odd values correspond to antisymmetric modes and sinusoidal mode shapes. Outside the electroded region, the wave amplitude decays exponentially.

If we call the frequency of the mth mode f_{nm} ($m = 0, 1, 2, \ldots, M - 1$), then only the lowest mode f_{n0} is wanted. Given n, f_p, and w, the number of trapped modes M depends on Δf. Let $\Delta f_{\max}$ be the maximum value of Δf of which only one mode is trapped ($M = 1$). Then $\Delta f_{\max}$ can be shown to be

$$\Delta f_{\max}/f_p = (t/knw)^2,$$

$$\Delta f_{\max}/f_p = (N/kf_p w)^2 \quad (5.2\text{-}8)$$

where k is a dimensionless constant and N is given by Eq. (5.2-2).

For TS waves,

$$k = k_{ts} = (2\bar{c}_{66}/M_n)^{1/2}. \quad (5.2\text{-}9)$$

For TT waves,

$$k = k_{tt} \simeq (2\bar{c}_{66}/c_{55})^{1/2}. \quad (5.2\text{-}10)$$

Values of k are given in Table 5.2-2.

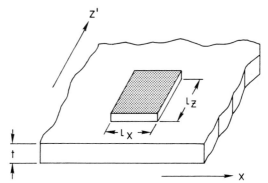

FIG. 5.2-16 Rectangular-electrode trapped energy resonator.

The strip electrode natural frequencies given by Eq. (5.2-7) can be written

$$f_{nm} = f_e + \delta_m, \quad m = 0, 1, 2, \ldots, M - 1. \quad (5.2\text{-}11)$$

Then for a rectangular-electrode resonator (Fig. 5.2-16) it can be shown that the natural frequencies are

$$f_{npq} = f_e + \delta_p + \delta_q, \quad (5.2\text{-}12)$$

where p and q are the x- and z'-direction mode indices, respectively. Thus, to obtain the natural frequencies for the rectangular-electrode case, the strip electrode model is solved twice, once for TS waves and once for TT waves; the frequency offsets δ_p and δ_q are thus obtained to give the total offset from f_e.

Now consider two identical coupled strip resonators (Fig. 5.2-13). The following transcendental equations are obtained for the frequencies of the symmetric and antisymmetric modes.

For the symmetric modes,

$$k_e w = \tan^{-1}(k'_p/k_e) + \tan^{-1}[(k'_p/k_e) \tanh(k'_p gw/2)] + m\pi,$$

$$m = 0, 1, 2, \ldots, M - 1. \quad (5.2\text{-}13)$$

For the antisymmetric modes,

$$k_e w = \tan^{-1}(k'_p/k_e) + \tan^{-1}[(k'_p/k_e) \coth(k'_p gw/2)] + m\pi,$$

$$m = 0, 1, 2, \ldots, M - 1. \quad (5.2\text{-}14)$$

The principal symmetric and antisymmetric mode frequencies, f_s and f_a, correspond, of course, to $m = 0$. A quantity of primary interest is the mode spacing, defined as

$$\delta f = f_a - f_s. \quad (5.2\text{-}15)$$

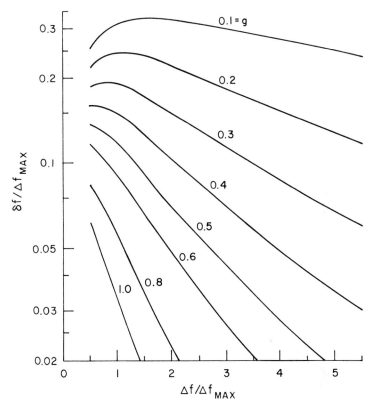

FIG. 5.2-17 Normalized mode spacing, $\delta f/\Delta f_{max}$ versus normalized frequency lowering, $\Delta f/\Delta f_{max}$, for two identical acoustically coupled resonators. The parameter g is the gap ratio, the ratio of electrode separation to electrode width.

Figure 5.2-17 plots, in normalized form, the mode spacing versus the electrode cutoff frequency lowering Δf for various values of the gap ratio g, the ratio of electrode separation to electrode width.

Finally, consider two identical coupled resonators having rectangular electrodes (Fig. 5.2-18). It can be shown from the work of Tiersten that the mode spacing is the same as for the strip electrode model. The actual frequencies of all modes are obtained by adding to the frequencies from Eqs. (5.2-13) and (5.2-14) the offset corresponding to the electrode height h obtained from Eqs. (5.2-7) and (5.2-11).

Let us illustrate the results just presented by calculating the frequency and mode spacing for a two-resonator, third-overtone, AT-cut quartz monolithic filter. For this example, let $f_p = 60$ MHz, $n = 3$, with thickness-twist coupling so that from Eqs. (5.2-1) and (5.2-2), the wafer thickness is $t = 83.05$ μm.

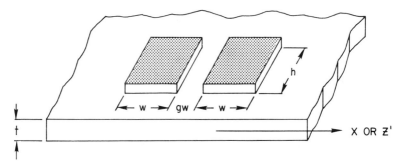

FIG. 5.2-18 Two acoustically coupled resonators.

Let the electrode dimensions (Fig. 5.2-18) be $w = h = 0.6$ mm, $g = 0.2$, $gw = 0.12$ mm, and let the thickness of the aluminum electrodes be 0.2 μm. The frequency lowering Δf [Eq. (5.2-4)] is then

$$\Delta f = 295 + 21 = 316 \quad \text{kHz},$$

where the first term is the mass loading and the second term is the piezoelectric loading.

At this point, we may proceed in at least two different ways. In normal practice Eqs. (5.2-13) and (5.2-14) are next solved numerically to obtain the mode frequencies. For the principal modes these are

$$f_s = 59737.6 \quad \text{kHz},$$
$$f_a = 59770.2 \quad \text{kHz},$$

while the mode spacing is $\delta f = 32.6$ kHz.

These frequencies are for the strip electrode model (Fig. 5.2-13). To obtain the actual frequencies for the rectangular electrodes, solve Eq. (5.2-7) numerically for the given electrode height h and TS waves, and calculate the offset δ from Eq. (5.2-11). For the principal mode this offset is $\delta_0 = 75.5$ kHz. Adding this offset to f_s and f_a gives the final values:

$$f_s = 59813.1 \quad \text{kHz},$$
$$f_a = 59845.7 \quad \text{kHz}.$$

The mode spacing is unchanged.

As an alternative to the procedure just outlined, the mode spacing δf of the principal modes can be obtained directly from the normalized curves of Fig. 5.2-17, without resorting to numerical methods, in the following manner:

From the second equality of Eq. (5.2-8) and Table 5.2-2, $\Delta f_{\max} = 150$ kHz; then $\Delta f/\Delta f_{\max} = 2.11$. Entering Fig. 5.2-17 with this value, for $g = 0.2$

5 PIEZOELECTRIC AND ELECTROMECHANICAL FILTERS   209

$\delta f/\Delta f_{max} \simeq 0.22$, and the mode spacing is $\delta f \simeq 33$ kHz, which compares favorably with the value obtained numerically, the difference being attributable to error in reading from the curves.

B. *Configurations.* Filters using MCFs can be realized in very many different configurations. The most elegant of these is the full monolithic, of which the most important example is the eight-pole, 8.14-MHz filter (Fig. 5.2-3e) used in the Western Electric A-6 channel bank. This filter was described by Pearman and Rennick (1977). A prime difficulty with multiresonator monolithic filters is suppression of unwanted modes of vibration (Werner *et al.*, 1969). By making the length-to-width ratio of each resonator different, Pearman and Rennick were able to obtain adequate inharmonic mode control. An earlier version of this filter using identical resonators required two plates to attenuate the unwanted modes (Olster *et al.*, 1975; Cawley *et al.*, 1975). Over the course of the development of these filters a large number of publications have described various aspects of design, performance, and fabrication. Among these Byrne (1970), Grenier (1974), Hokanson (1969a), Miller (1970), and Lloyd (1971) described fabrication methods, Lloyd and Haruta (1969) and Haruta *et al.* (1969) treated acoustical design, and Rennick (1973, 1975) treated circuit design and tuning procedures.

The in-line arrangement of resonators results in coupling only between adjacent resonators. The resulting transfer function is of the all-pole type, having all transmission zeros at zero and infinity, although Braun (1972, 1973) has shown that by suitable interconnection of resonators transmission zeros can be produced. More general resonator configurations allow additional couplings to be realized, producing finite transmission zeros. Masuda *et al.* (1973; 1974a,b) studied several such arrangements, and Onoe and Spassov (1974) studied a space-saving arrangement of four resonators. None of these techniques seems to have found practical application.

The simplicity of the multiresonator monolithic filter must be balanced against difficulties of manufacture and of suppressing unwanted modes. Because of these difficulties, for many applications the dominant MCF configuration is the tandem two-pole configuration (Fig. 5.2-3b,c). Dividing a multipole filter into tandem sections divides the out-of-band attenuation requirements (including spurious mode performance) among the several sections and at the same time increases manufacturing yield with respect to passband requirements, since an out-of-tolerance parameter can be remedied by replacing a single section. In general, two-pole sections are preferable to, say, three- or four-pole sections because the filter can usually be designed so that the two-pole sections are physically symmetrical, while this is not the case for three- or four-pole sections. Nevertheless, the use of higher-order sections reduces the number of components. Eight-pole filters composed

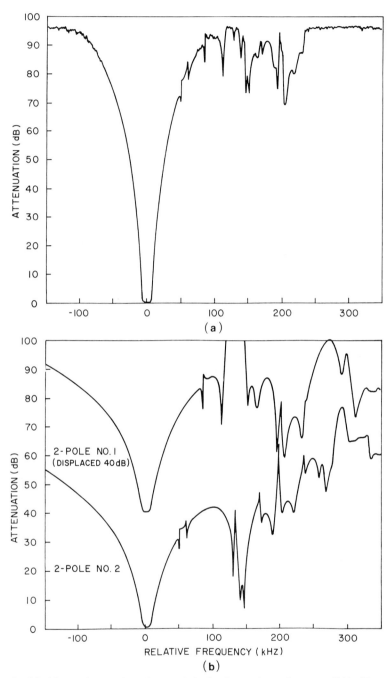

FIG. 5.2-19 (a) Attenuation characteristic for four-pole tandem monolithic filter. (b) Attenuation of the individual two-pole monolithic sections showing staggering of unwanted mode frequencies.

of two four-poles have been described by Werner *et al.* (1969), Kohlbacher (1972), as well as Olster *et al.*, (1975).

An important feature of the tandem arrangement is that by designing so that the unwanted modes associated with different sections occur at different frequencies, excellent stopband attenuation can be achieved. This is illustrated in Fig. 5.2-19, which shows the attenuation of each two-pole section of a four-pole tandem monolithic and of the complete filter.

A limitation of the tandem monolithic configuration is the maximum bandwidth that can be achieved. As the bandwidth of a tandem monolithic filter increases, the termination capacitances and the capacitances at the junctions between monolithics decrease. The maximum inductorless bandwidth is defined (Smythe, 1969, 1972) as the bandwidth at which one or more of the junction capacitances vanishes. The exact limit depends on a number of design parameters. Figure 5.2-20 shows approximate limits. Note that at the inductorless limit the termination capacitances may already be negative. In many practical applications, impedance-matching networks containing inductors are required, and the fact that the terminations are not capacitive is then inconsequential.

To extend the range of practical bandwidths, several approaches may be followed. By using low-loss, temperature-stable inductors to "tune out" excess nodal capacitance (Class IB-1 design, Table 5.2-1) the inductorless limit can be exceeded. Figure 5.2-21 shows the response of a 143-MHz four-pole tandem monolithic filter using this technique.

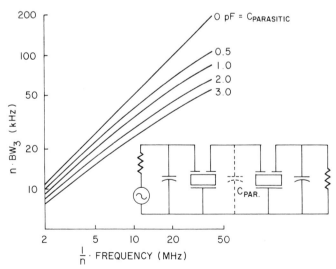

FIG. 5.2-20 Maximum inductorless bandwidth BW_3, for a tandem monolithic filter. n is the overtone, $C_{parastic}$ is the stray node-to-ground capacitance, including holder capacitance.

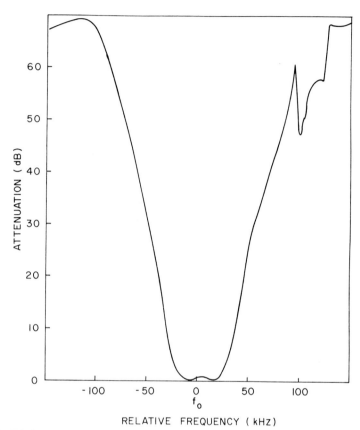

FIG. 5.2-21 Attenuation characteristic of a 143-MHz four-pole tandem monolithic filter. The 3-dB bandwidth is 46 kHz and loss is 7 dB.

Figure 5.2-20 shows that because filter impedances are high, the maximum inductorless bandwidth is a sensitive function of the stray nodal capacitance. By fabricating two two-poles on a single wafer (Fig. 5.2-3d) (Smythe, 1972), the parasitic capacitance at the junction between them can be made very small, allowing the bandwidth to approach the uppermost curve (Fig. 5.2-20). The response of a 167-MHz four-pole filter made in this manner is shown in Fig. 5.2-22. This configuration, while utilized primarily to eliminate the use of an inductor, retains most of the advantages of other tandem configurations and is at the same time fully monolithic.

Figure 5.2-20 also shows that the maximum inductorless bandwidth decreases approximately as the square of the overtone. Thus, the development of techniques for fabricating higher fundamental-frequency plates such as

5 PIEZOELECTRIC AND ELECTROMECHANICAL FILTERS

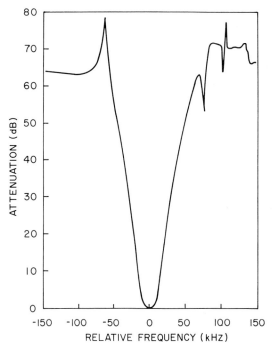

FIG. 5.2-22 Attenuation characteristic of a 167-MHz four-pole MCF constructed as shown in Fig. 5.2-3d. The 6-dB bandwidth is 24 kHz.

ion milling (Berté, 1977) and chemical etching (Vig et al., 1977) serve to increase the practical range of bandwidths. Berté employed ion-beam milling to fabricate fundamental-mode monolithic filters and resonators at frequencies up to 275 MHz using AT-cut quartz and lithium tantalate. In Berté's devices a central diaphragm is milled to the required thickness, leaving an integral outer support ring.

The maximum inductorless bandwidth increases approximately as the square of the piezoelectric coupling coefficient. Studies of high-coupling materials are discussed in Sec. 5.2.2.4. Limitations on the characteristics of such materials have led a number of workers to investigate composite structures. Mason (1969a,b) analyzed monolithic filters having a thin film of high-coupling-constant piezoelectric material on each face of a quartz plate. Roberts (1971) fabricated 190-MHz and 335-MHz two-pole monolithic filters having a single film of cadmium sulfide on an AT-cut quartz plate. By placing one electrode between quartz and CdS and the other on the top surface of the film, a filter impedance of 50 Ω was obtained for a bandwidth of 135 kHz at 190 MHz, using the seventh overtone. To obtain a higher fundamental frequency. Grudkowski et al. (1980) etched a thin diaphragm in a silicon wafer and used the electroded ZnO film to form a two-pole monolithic filter at 425 MHz.

The filters just discussed have all-pole responses. However, a two-pole monolithic filter, which may in turn be part of a hybrid or tandem monolithic filter, can realize a pair of real-frequency ($j\omega$-axis) or complex-frequency transmission zeros. This may be accomplished by either circuit means or charge cancellation. The latter was treated by Yoda et al. (1969). In the former, by capacitively bridging a two-pole monolithic (Fig. 5.2-23a), a pair of real-frequency transmission zeros is produced, symmetrically disposed about ω_m, the mean of the two resonator frequencies. In some instances the bridging capacitor may be realized by electrodes on the quartz wafer (Smythe, 1979a,b). If the bridging capacitor is replaced by an inductor, then a complex pair of transmission zeros at $\sigma \pm j\omega_m$ is realized. This realization is usually impractical, but an equivalent one is obtained by retaining the capacitive bridging while reversing the connections to the electrodes of one resonator (Fig. 5.2-23b). The reversal is represented in the two-pole equivalent circuit (Fig. 5.2-14b) by changing the sign of L_{12}.

A more flexible method of realizing transmission zeros incorporates, in addition to the bridging capacitance, a common lead reactance, usually a capacitance (Fig. 5.2-23c), which is equivalent, in the symmetrical case, to the use of unequal motional inductances in the two-pole narrow-band lattice section.

Realization methods for tandem two-pole monolithic filters with transmission zeros are closely related to those for narrow-band tandem-lattice discrete crystal filters, since each lattice section of the latter is equivalent to a symmetrical two-pole MCF with, possibly, bridging and common lead reactances. Hence, for example, Holt and Gray's (1968) method may be used. For real or complex transmission zeros, the generalization by Dillon and Lind (1976) is available. Herzig and Swanson (1978a,b) employed the latter, using transfer functions put forward by Rhodes (1970) to realize MCFs with uniform group delay.

Additional design freedom in tandem monolithic filter design can be obtained by dropping the requirement of physical symmetry. Lee (1974, 1975) gave a useful network equivalence and applied it to the synthesis of an upper-sideband MCF.

An alternate configuration for producing real-frequency transmission zeros by using both single resonators and two-poles—the hybrid monolithic or polylithic filter—was suggested by McLean (1967a,b), introduced into the production of channel filters by Sheahan (1971, 1975), and is particularly suited to the realization of highly asymmetrical response characteristics. Figure 5.2-24 shows the circuit diagram and response characteristic of a 4.896-MHz hybrid monolithic employing three two-pole and two single resonators.

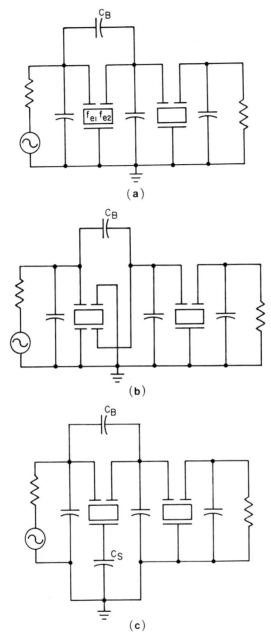

FIG. 5.2-23 Addition of circuit elements to obtain transmission zeros. (a) Bridging capacitance C_B produces real zeros, $(f_{\infty 1} + f_{\infty 2})/2 = (f_{e1} + f_{e2})/2$; (b) the phase reversal obtained by interchanging the connections to one electrode pair causes the transmission zeros to be complex; (c) adding a common lead reactance such as capacitance C_s, provides another degree of freedom. Common lead reactance may be used alone or in conjunction with bridging capacitance and/or phase reversal.

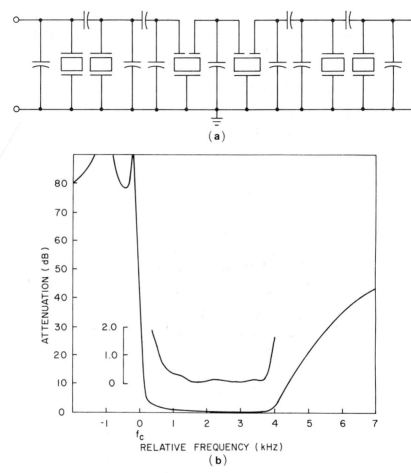

FIG. 5.2-24 A 4.896-MHz hybrid monolithic filter. (a) Circuit diagram; (b) attenuation characteristic.

5.2.2.3 NONLINEAR EFFECTS

While nonlinearity in crystal filters may be due to any of the circuit components, such as powdered iron or ferrite inductive devices, the following discussion will be limited to nonlinearities in the piezoelectric elements.

Important nonlinear effects, which may occur in resonators or circuits that contain them, include nonlinear resonance, anomalous resistance at low current levels (sometimes accompanied by resonance-frequency anomalies), nonlinear mode coupling, intermodulation, excess phase noise, and level dependence of amplitude and phase response. In addition, crossmodulation may no doubt occur but has not been reliably reported.

5 PIEZOELECTRIC AND ELECTROMECHANICAL FILTERS   217

In discussing these nonlinear effects it will be useful to consider three mechanisms that are dominant in three different regions of strain amplitude, which will be loosely referred to as low, high, and very high. In reality, of course, all three mechanisms are present at all strain levels, and additional mechanisms can no doubt exist.

At very high strain levels, acoustic losses (and for very thin electrode films, ohmic losses) cause thermal gradients and consequent changes in frequency and resistance. Such high-level effects will not be treated here and are of limited interest for filter applications. By high-level nonlinearity we mean that which is due to the imperfect elasticity of the piezoelectric material used. Such inelasticity gives rise to nonlinear resonance (Tiersten, 1975c, 1976c) and intermodulation. The latter is discussed following the discussion of low-level nonlinearity.

While not perfectly understood, nonlinearity at low strain appears to be associated with surface defects including loose particles, surface damage (caused, for example, by lapping), contamination (especially viscous contamination such as oil), and electrode film defects including flakes, nodules and poorly adhering films.

The most important low-level effects are anomalous resonator resistance (sometimes accompanied by resonance-frequency anomalies) and, in filter applications, intermodulation. Both effects may be caused by the surface defects just named.

Bernstein (1967) observed that small particles of quartz or other material in combination with a thin film of oil on the surface of AT-cut resonators produce high starting resistance (high resistance at low current). He also found that unetched or lightly etched quartz surfaces produce high starting resistance.

Nonaka et al. (1971) made a detailed study of the effects of nodules and scratches in gold electrode films on starting resistance and frequency. He found that by driving resonators at high strain levels, nodules could be removed. Resonators so treated showed excellent resistance linearity and did not revert to their previous behavior after a period of one year. Bernstein (1967), on the other hand, observed that resistance nonlinearity associated with particles made adherent by an oily film was only temporarily improved by high levels of strain. These results are not inconsistent with Nonaka's but are probably due to the oily film.

In filters, resonator resistance and frequency anomalies cause level-dependent changes in amplitude and phase such as that reported by Swanson (1978). In frequency- and phase-shift keyed (FSK and PSK) data transmission systems he observed that these anomalies can produce AM to PM conversion, that is, a change in phase due to a change in signal amplitude. Since the signals are phase or frequency modulated, transmission errors can result.

Intermodulation (IM) at low strain levels was reported by Rider (1970) and Malinowski and Smith (1972). Horton and Smythe (1973) showed experimentally that at low levels of strain, IM was caused primarily by surface defects.

A characteristic of low-level nonlinearity is its frequent erratic nature. Resonator resistance and IM may exhibit hysteresis with respect to current or strain and temperature and may exhibit temporal instability. Third-order IM may not exhibit cubic dependence on input power. Such variability seems consistent with the postulated mechanisms.

At higher levels of strain it has long been recognized that the elastic nonlinearity of the piezoelectric material must be considered. From an analysis by Tiersten (1974b, 1975b), it can be shown that the third-order IM occurring at a frequency ω_{IM} in a trapped-energy resonator can be represented by a dependent voltage source in the motional impedance branch of the equivalent circuit (Fig. 5.2-25) with rms amplitude

$$V_{IM} = j\gamma K I_1^2 I_2^*(\omega_{IM} C_1)^{-3}. \qquad (5.2\text{-}16)$$

In Eq. (5.2-16), I_1 and I_2 are the rms currents through the resonator at the test-tone frequencies ω_1 and ω_2. The frequency of the third-order IM product is

$$\omega_{IM} = 2\omega_1 - \omega_2, \qquad 5.2\text{-}17)$$

γ is an effective nonlinear elastic constant

$$\gamma = \tfrac{1}{2}c_{22} + c_{266} + \tfrac{1}{6}c_{6666}, \qquad (5.2\text{-}18)$$

and K is a complex combination of material constants, wafer thickness, and wave number.

For AT-cut quartz, γ was determined by Smythe (1976) to be approximately 8.1×10^{11} N/m² at 11.7 MHz and 8.3×10^{11} N/m² at 111 MHz, based on intermodulation measurements, while nonlinear resonance measurements

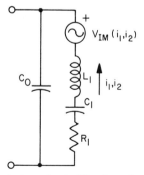

FIG. 5.2-25 Resonator-equivalent circuit. The dependent voltage source V represents the intermodulation of i_1 and i_2.

gave a value of $1.3 \times 10^{12} N/m^2$ at 11.7 MHz. Planat et al. (1980) obtained values of $7.9 \times 10^{11} N/m^2$ at 10 MHz and $6.6 \times 10^{11} N/m^2$ at 100 MHz using intermodulation measurements. For X-cut lithium tantalate resonators, Planat et al. (1980) obtained a γ value of $1.7 \times 10^{10} N/m^2$ using intermodulation measurements.

For a device or system excited by two "test tones" at ω_1 and ω_2, the third-order intermodulation ratio (IMR) may be defined as

$$\text{IMR} = (P_1^2 P_2)^{1/3}/P_{\text{IM}},$$

where P_1 and P_2 are the available power at the test frequencies ω_1 and ω_2, respectively, and P_{IM} is the IM power delivered to the load at ω_{IM}. For a single-resonator filter, the IMR is approximately

$$\text{IMR} = 2.4 \times 10^{13} n^2 A_e^2 (\text{BW}_3/f_0)^4 (P_1^2 P_2)^{-2/3} \quad (5.2\text{-}19)$$

for AT-cut quartz, where A_e is the electrode area in square millimeters, n is the overtone, and BW_3 and f_0 are the 3-dB bandwidth and center frequency in hertz. The values P_1 and P_2 are in watts, and γ has been taken as $8.2 \times 10^{11} N/m^2$. Equation (5.2-19) can also be shown to hold for the two-pole Butterworth case with in-band test tones and may be a useful estimator for in-band IM higher-order filters.

While Tiersten's analysis was restricted to single resonators, with reasonable accuracy it may be applied to acoustically coupled resonators by neglecting the change in mode shape. In the corresponding equivalent circuit representation, a dependent voltage source is associated with each resonator.

5.2.2.4 CRYSTAL FILTERS USING OTHER MATERIALS

The maximum frequency and bandwidth limitations associated with the use of quartz may be overcome by techniques which allow thinner wafers to be fabricated. Alternatively, or in conjunction, the use of piezoelectric materials with higher electromechanical coupling or higher bulk wave velocity may be considered. Motivated in part by SAW filter requirements, new materials are being actively investigated.

Berlinite (aluminum metaphosphate, $AlPO_4$) holds much promise for the future (Chang and Barsch, 1976; Detaint et al., 1979, 1980; Kolb 1979; Ozimek and Chai, 1979), but techniques for its growth require further development, and large crystals are not found in nature.

Lithium niobate, widely used for wide-band SAW devices, has been investigated for monolithic filter applications by Burgess and Porter (1973) and (in conjunction with high-fundamental-mode fabrication techniques) by Berté (1977). Because lithium niobate does not possess a plate orientation with a low frequency–temperature coefficient, typical coefficients for thickness modes being -60 to $-90 \times 10^{-6}\,°\mathrm{C}$, it has not found practical application to bulk-mode filters.

For lithium tantalate, on the other hand, temperature-compensated bulk-wave modes have been found. Warner and Ballman (1967) found that the strongly coupled fast shear wave of X-cut lithium tantalate has a parabolic frequency–temperature behavior for the fundamental thickness-shear resonance. On the other hand, overtone resonance frequencies and antiresonance frequencies of all orders possess a large negative temperature coefficient. Sawamoto (1971) made an experimental study of energy trapping in X-cut resonators and calculated inter-resonator coupling. Siffert (1981) and co-workers employed fundamental mode X-cut resonators in commercially produced filters.

Burgess and Porter (1973) and Hales and Burgess (1976) fabricated rotated Y-cut lithium tantalate resonators and monolithic filters with frequency–temperature coefficient of $-22 \times 10^{-6}/°\mathrm{C}$. Detaint and Lançon (1976) calculated first-order temperature coefficients and fundamental thickness-shear modes of singly and doubly rotated lithium tantalate plates. The calculations were extended to the third and fifth overtones and supplemented by experimental measurements by Detaint (1977), who found temperature-compensated orientations.

A third-overtone, 200-MHz monolithic filter having a bandwidth of 45 kHz and a frequency–temperature coefficient of $-36 \times 10^{-6}/°\mathrm{C}$ was reported by Uno (1975), who used Z-cut lithium tantalate and the thickness-extensional mode. This mode has the advantage of a high frequency–thickness constant, 3020 Hz m for the Z-cut and 3080 Hz m for the 47°-rotated Y-cut. Energy trapping occurs for the harmonic overtones but not for the fundamental mode.

Using lateral-field excitation, Uno (1978, 1979) obtained quadratic frequency-temperature behavior in rotated Y-cut lithium tantalate for thickness-shear motion and constructed a 199-MHz monolithic filter with 33-kHz bandwidth.

Lithium niobate and lithium tantalate have also been evaluated for low-frequency applications. Sawamoto and Niizeki (1970) investigated length-extensional modes in singly rotated cuts of lithium tantalate and found orientations having quadratic frequency–temperature behavior. Hannon *et al.* (1970) investigated length-extensional bar resonators of both materials at frequencies from 150 to 800 kHz. For lithium niobate resonators the temperature coefficient of frequency was typically $-75 \times 10^{-6}/°\mathrm{C}$, limiting their usefulness. For lithium tantalate, doubly rotated cuts having quadratic frequency–temperature behavior of the series resonance frequency with a second-order coefficient of approximately $0.11 \times 10^{-6}/(°\mathrm{C})^2$ and a capacitance ratio of around 20 were exhibited. Flexural-mode resonators were

5 PIEZOELECTRIC AND ELECTROMECHANICAL FILTERS

studied by Onoe (1973) and co-workers. Arranz (1977) developed a nine-pole, 128-kHz ladder filter using extensional-mode lithium tantalate resonators for use as a FDM voice channel filter.

5.2.3 Electromechanical Filters

Electromechanical filters, usually referred to simply as mechanical filters, occupy an important place in the filter spectrum at frequencies between 300 Hz and 700 kHz, with the largest usage being in the range from 3 to 500 kHz. Bandwidths range from 0.1 to 10% of center frequency. The variety of mechanical filter structures that have been manufactured is quite large, and no attempt will be made here to catalog them. Instead, we will attempt to summarize recent developments and the current state of the art. However, as an introduction, a brief generalized description may be helpful. In addition, excellent tutorial material may be found, for example, in Chapter 5 of Temes and Mitra (1973) and in Johnson and Guenther (1974). Also, a number of survey papers may be consulted, including Onoe (1979), Johnson and Yakuwa (1978), Konno et al. (1978), Sawamoto et al. (1978), and Kunemund (1975).

Most mechanical filters consist of an input transducer, a number of mechanical resonators (usually coupled by welded wires), and an output transducer. The resonators are made of one of a number of proprietary nickel–iron alloys, with small quantities of other metals, such as chromium or molybdenum, added to improve the frequency–temperature characteristics. Temperature coefficients of frequency may be as low as 1 to $2 \times 10^{-6}/°C$. Values of Q typically range from 10^4 to 3×10^4.

Resonators may be in the form of rods or bars, plates, disks, or tuning forks. Commonly used modes of vibration include flexure (of disks, rods, bars, forks, and plates) and torsion and extension (of rods and bars). All of these modes have also been employed for coupling wires.

The earliest transducers used the magnetostrictive properties of ferrous–nickel alloys. Modern transducers use primarily magnetostrictive ferrites or piezoelectric ceramics. Frequently, the transducers are composite structures. In either case the transducers may also serve as resonators.

In analyzing and designing mechanical filters it is extremely helpful to develop electrical circuit analogs. In the modern literature the mobility analogy is commonly employed (Temes and Mitra, 1973, pp. 163–165) in which the analog of mechanical force is electrical current ("through" variables) and the analog of velocity is voltage ("across" variables). While mechanical filters, like crystal resonators and monolithic filters, are distributed systems, it is usually possible to obtain adequate lumped-element representations.

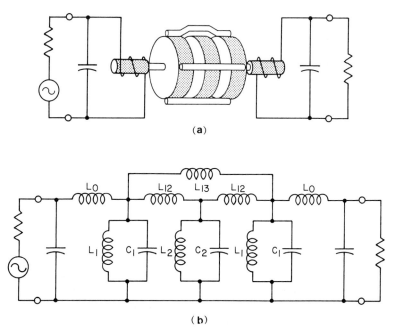

FIG. 5.2-26 Disk-wire mechanical filters. (a) Pictorial representation showing three resonators, ferrite transducers, and coupling wires; (b) approximate equivalent electrical circuit. [From Johnson and Winget (1974), © 1974 IEEE.]

A simplified example is shown in Fig. 5.2-26. Here L_0 represents the inductance of the magnetostrictive transducer coil, and disk resonators are represented by L_1, C_1 and L_2, C_2, while L_{12} represents the adjacent-resonator coupling wires. For narrow bandwidths these representations are usually adequate and permit straightforward application of well-known filter-network synthesis procedures. For wider bandwidths, more elaborate representations are necessary. Guenther (1973b) and Guenther and Traub (1980b) developed and applied the single-mode resonator circuit models shown in Fig. 5.2-27 as well as more general multimode models. Coupling wires are usually modeled with sufficient accuracy by pi networks (Johnson, 1968; Johnson and Guenther, 1974).

Mechanical filters are basically coupled-resonator structures with all-pole bandpass responses. As for other types of coupled-resonator filters, the introduction of coupling between nonadjacent resonators introduces transmission zeros at real or complex frequencies. In Fig. 5.2-26 the first and third resonators are coupled by a bridging wire, represented by L_{13}, producing an upper stopband attenuation pole.

5 PIEZOELECTRIC AND ELECTROMECHANICAL FILTERS 223

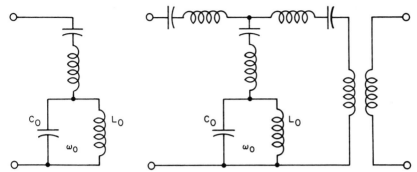

FIG. 5.2-27 One-port and two-port mechanical resonator wideband equivalent circuits. [From Guenther and Traub (1980b). © 1980 IEEE.]

5.2.3.1 FLEXURE-MODE BARS AND PLATES

For most mechanical bodies, the lowest resonance frequencies are associated with flexure modes. Figure 5.2-28 shows a modern, two-resonator mechanical filter (Johnson, 1977). The two resonator plates are coupled along their nodal axes by torsion forces transmitted through the two coupling wires. Piezoelectric ceramic transducers are bonded to the resonator

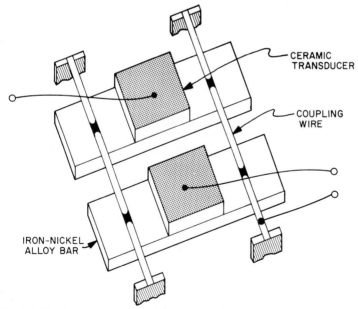

FIG. 5.2-28 Two-pole flexure-mode mechanical filters. [From Johnson (1977); reprinted from *Electronics*, October 13, 1977; © 1982; McGraw-Hill, Inc.; all rights reserved.]

plates, the material of which has been heat-treated so that its temperature coefficient compensates for that of the transducers.

A typical application of filters of this type is in the very-low-frequency radio navigation system, Omega. Here mechanical filters with bandwidths of 25 Hz serve as receiver front-end filters at the Omega operating frequencies in the 10 to 14 kHz range.

Other applications ranging from 3 kHz to nearly 100 kHz include frequency-shift-keyed modems and telephone signaling and pilot tone filters (Pfleiderer and Wollmershauser, 1976). Albsmeier, Guenther, and their associates (Albsmeier *et al.*, 1974; Guenther *et al.*, 1979) presented details of a 48-kHz channel filter using 12 bending-mode cylindrical rod resonators and piezoelectric ceramic transducers that has been in use since 1973. A separate signaling filter is used. A later design at 128 kHz incorporates transmission zeros, making it possible to reduce the number of resonators to 10 and improve the group delay characteristic. Related papers by Guenther and Thiele (1980a,b) and Guenther and Traub (1980a,b) describe the optimization techniques used in design. Because the fractional bandwidth of the 48-kHz filter is approximately 7% and because of the stringent requirements on the attenuation characteristic, an improved equivalent-circuit resonator model was developed by Guenther (1973b), as noted earlier.

In addition to bars and plates, flexure-mode tuning fork filters are used at frequencies down to 300 Hz, mostly as single resonators (Konno *et al.*, 1978), but they may also be used as elements of higher-order filters in the same manner as crystal resonators. In addition, a novel three-prong fork provides a two-pole response (Johnson and Yakuwa, 1978; Konno *et al.*, 1978).

Other low-frequency coupled-mode resonators were described by Konno and Tomikawa (1967). In one type, either one edge or two diagonally opposite edges of a square bar are chamfered to remove degeneracy of the two flexural modes used (Johnson and Yakuwa, 1978). In another type, flexural modes of a dumbbell configuration are used. Similar configurations have employed extensional and torsional modes (Johnson and Yakuwa, 1978).

5.2.3.2 EXTENSIONAL-MODE FILTERS

A number of mechanical filters at frequencies from 40 to 500 kHz use length-extensional-mode resonators. A representative configuration is shown in Fig. 5.2-29. In Japan, 60 to 108 kHz channel filters have been used in applications outside the public telephone system for many years (Onoe, 1979). Coupling is by means of wires vibrating in extension.

A 128-kHz channel filter using 13 extensional-mode cylindrical rod

5 PIEZOELECTRIC AND ELECTROMECHANICAL FILTERS 225

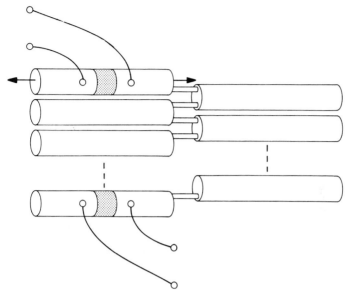

FIG. 5.2-29 Typical extensional-mode filter construction using a folded arrangement of half-wavelength elements. The resonators are held by silicone rubber supports (not shown) at center nodal planes. The Langevin resonant transducers are of composite construction: nickel–iron alloy with piezoelectric ceramic center sections. [From Johnson and Guenther (1974); © 1974 IEEE.]

resonators with quarter-wavelength extensional-mode couplers was developed at CNET in France (Bosc and Loyez, 1974; Bon et al., 1976, 1977). In connection with this program, Carru et al. (1977) investigated the use of lithium niobate in both Langevin and sandwich-type transducers.

A 455-kHz filter produced in Germany uses length-extensional rod resonators with flexural wire coupling and ceramic transducers, resulting in a very compact assembly (Johnson et al., 1971). Ernyei (1978) illustrated how the resonator array of a filter of this type can be folded to facilitate the introduction of bridging couplers.

5.2.3.3 DISK-WIRE FILTERS

Mechanical filters using flexural-mode discs coupled by wires welded to the circumference have been used for many years at frequencies from 60 to 600 kHz and fractional bandwidths from 0.1 to 10%. Figure 5.2-26 shows a simplified example. At the lower frequencies, the fundamental (one-nodal-circle) mode is used, but for higher frequencies the two-modal-circle overtone is preferred. Although multimode disk resonators have been proposed (Johnson, 1966), they have not been widely used in production.

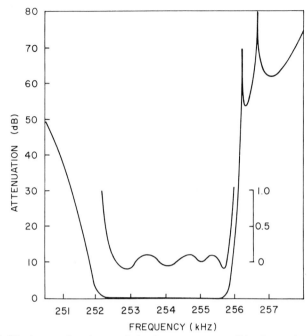

FIG. 5.2-30 Attenuation characteristic of seven-resonator disk-wire mechanical channel filter. [From Johnson and Winget (1974). © 1974 IEEE.]

The earliest disk–wire filters used magnetostrictive alloy extensional-mode transducers; present-day ones use magnetostrictive ferrites. Other improvements include refinements in design made possible by more accurate circuit models of resonators and coupling wires (Johnson, 1968) and the introduction of real and complex transmission zeros by the use of bridging couplers (Johnson, 1975), as well as by improved manufacturing methods. Important examples of modern disk–wire mechanical filters are the 256-kHz six- and seven-resonator channel filters described by Johnson and Winget (1974) (Fig. 5.2-30).

5.2.3.4 TORSIONAL-MODE FILTERS

Torsional-mode resonators are employed, usually with extensional coupling, in a number of recently developed FDM channel filters, as well as in other applications. Representative construction is shown in Fig. 5.2-31. Kohlhammer and Schuessler (1971; Schuessler, 1971) developed 200-kHz channel and signaling filters using eight and five resonators, respectively, with bridging wires to produce attenuation poles.

5 PIEZOELECTRIC AND ELECTROMECHANICAL FILTERS

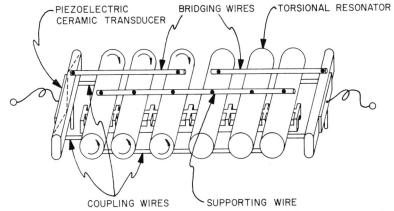

FIG. 5.2-31 Torsional-mode mechanical filter using second-flexure-mode transducers. [From Yakuwa et al. (1977).]

Sawamoto et al. (1975, 1976) developed filters in the 120-kHz frequency range, using torsional-mode piezoelectric ceramic-composite transducers longitudinally coupled and having torsional-mode iron–nickel alloy resonators. The 1976 paper gives detailed design information for a nine-pole, four-zero filter.

Other channel filters were developed and put into production by two organizations. Yano et al. (1974, 1975) developed 128-kHz band-separation filters using torsional-mode resonators, both with and without transmission zeros (Onoe, 1979). These consist of a channel filter and a signaling filter, both driven by the same input transducer but having separate output transducers. Production techniques were discussed by Watanabe et al. (1979).

Yakuwa et al. (1977, 1979) described the design and performance of a 128-kHz channel filter using six torsional-mode cylindrical rod resonators and two second-overtone flexural-mode piezoelectric ceramic–metal composite transducers that also serve as resonators. Manufacturing methods for this filter were described by Tsuchida et al. (1979).

5.2.3.5 NONLINEAR EFFECTS

The resonance frequency and loss of a mechanical resonator change as the amplitude of vibration increases. Yano et al. (1975) measured the nonlinearity of flexure- and torsional-mode resonators. In both cases, resonance frequency change was 0.3 Hz at 50 Hz (flexure mode) and at 110 kHz (torsional mode) for amplitudes corresponding to 0 dBm filter input power.

Yakuwa and Okuda (1976) measured nonlinear elastic and loss parameters for a variety of materials. Since the energy stored in each resonator is related to the filter bandwidth and input power, they were able to determine the minimum resonator volume for a specified degree of nonlinearity.

LIST OF SYMBOLS FOR SECTIONS 5.3 AND 5.4

a	Fraction of surface that is metallized
a_0	Reference value of a
a_-	Amplitude of normal mode propagating in negative direction
a_+	Amplitude of normal mode propagating in positive direction
A	Permittivity ratio, cross-sectional area
B_1, B_a	Acoustic susceptance
B_2	Inverter susceptance
$BB(f)$	Baseband frequency response
c_{ij}	Elements of stiffness matrix
$\bar{c}_{44}$	Piezoelectrically stiffened c_{44}
C, C_i, C_0	Capacitance
C_s	Capacitance per finger pair (per unit width sometimes) of single electrode IDT with 50% metallization
C_T	Capacitance of IDT
d_{31}	Element of piezoelectric strain matrix
$D_-(0)$	Normal component of electric displacement, evaluated at the surface, of Rayleigh wave propagating in the negative direction
$D_+(0)$	Normal component of electric displacement, evaluated at the surface, of Rayleigh wave propagating in the positive direction
D_i, D_y	Electric displacement in the i direction, electric displacement in the y direction
D_{max}	Maximum electric displacement
D_3^n	Normalized electric displacement in the z direction
e_{ij}	Elements of piezoelectric stress matrix
$E(t)$	Envelope of impulse response
E_i	Component of electric field in i direction
E_{max}	Maximum electric field
f	Frequency
$f(x)$	Spatial excitation function
f_0	Center frequency, carrier frequency
f_i	Frequency at which $\lambda/2 = 2L_i$
f_{notch}	Notch frequency
Δf	3-dB bandwidth
F	Spatial Fourier transform of electric potential, electric field, or electric displacement
F_e, F_0	Even and odd parts of F
F_a, F_b, F_i	Mechanical "voltage"
G_a	Acoustic radiation conductance
G_0	Acoustic radiation conductance at mid-band
H	Hilbert transform
i	Integer
I_n	Integral of electric potential over nth half wavelength of IDT
I_c	Electric current into port C
$I(t)$	Impulse response
j	Square root of -1

5 PIEZOELECTRIC AND ELECTROMECHANICAL FILTERS

J	Inverter matrix
k^2, k_0^2, k_{ij}^2	Square of the electromechanical coupling constant
k_{eff}^2	Square of effective electromechanical coupling constant
K, K'	Complete elliptic integrals of the first kind
$K_1, K_2,$	Arbitrary constants
l	Value of z coordinate at the $+z$ edge of the IDT
	Length of one section in bulk-wave model
$-l$	Value of z coordinate at the $-z$ edge of the IDT
L	Inductance
L_i	Electrode width plus adjacent gap width
m	Term within argument of Legendre polynomials; parameter of complete elliptic integrals
n	Integer, index of summation
N	Number of electrode pairs in IDT, transformer turns ratio
p	Integer, harmonic number
P	Magnitude of Rayleigh-wave power flow per meter
P_n	nth Legendre polynomial
Q	ω_0 (average energy stored)/(energy dissipation rate)
R_B	Balance resistance
R_G	Generator resistance
R_L	Load resistance
s_{ij}	Elastic compliance element
$S(f)$	Spectrum function
t	Thickness, time
t_{peak}	Time at peaks of impulse response
t_d	Time delay
T	Temperature
T_i	ith component of stress
u_i	Mechanical displacement
U_a, U_b, U_i	Mechanical "current"
v	Velocity
$\Delta v/v$	Frational change in Rayleigh wave velocity when top surface of crystal is short-circuited
V, V_0, V_c	Voltage applied to transducer
w, w_i	Width of transducer, aperture of ith overlap
x	Space coordinate
X	Notation reduction term equal to $N\pi(f - f_0)/f_0$
y	Space Coordinate
Y	Admittance of IDT
Y_0	Admittance of transmission line
z	Space coordinate
Z_0	Mechanical impedance, characteristic impedance of transmission line
Z_1, Z_2	Equivalent circuit impedance elements
γ	Propagation constant
γ_0	Propagation constant at midband, reference wave number equal to π/L_i
Γ_1	Notation reduction term
Γ_2	Notation reduction term
ε_0	Permittivity of free space
$\varepsilon_{ij}, \varepsilon_{ij}^S, \varepsilon_{ij}^T$	Element of permittivity matrix
ε_p	$\sqrt{\varepsilon_{22}\varepsilon_{33} - \varepsilon_{23}^2}$

η	z/λ_0, z coordinate measured in center frequency wavelengths
λ	Wavelength
λ_0	Wavelength at midband, periodicity of transducer
ξ	Dummy variable of integration through IDT
ρ	Mass density
ϕ	Electric potential, Rayleigh wave potential, transformer turns ratio
$\bar{\phi}$	Relative electric potential
Φ	Phase, phase of baseband impulse response
ψ	Notation reduction term
ω	Radian frequency
ω_0	Radian frequency at midband

5.3 SURFACE-ACOUSTIC-WAVE FILTERS[§]

5.3.1 Introduction

Central to the emergence of surface-acoustic-wave (SAW) filters was the development of the interdigital transducer (IDT). In the earliest filters, the IDTs (Fig. 5.3-1) were merely quarter-wavelength wide stripes of alternating polarity with quarter-wavelength wide gaps between the electrodes. These gave a $[(\sin X)/X]^2$ type of frequency response with -27-dB sidelobes. Since that time, multiple transducer configurations have been developed, multiple electrical phases employed, and complex weighting applied to the IDTs to obtain greater than 60-dB sidelobe suppression and asymmetric bandpass characteristics.

The IDT weighting techniques are such that phase and amplitude can be independently specified. This feature is certainly key to the sustained interest that SAW devices have experienced over the last decade. Because of this ability, two major technological needs, radar pulse-compression filters and television IF filters, have been met. The radar pulse-compression filter typically has a time delay that is a nonlinear function of frequency, and the television IF filter has both nonlinear group delay and asymmetric amplitude response. Today most modern radar systems have SAW pulse compressors in them and several of the major television manufacturers have SAW filters in their IF stage.

Exemplified in Fig. 5.3-2 are two of the parameters of bandpass filters, center frequency and fractional bandwidth. Not only are SAW transversal filter results shown, but for comparison SAW resonator filter data (Coldren and Rosenberg, 1978b) and bulk-acoustic-wave (BAW) filter data (Zverev, 1967) are also given. Immediately one can see that the SAW transversal filters serve those applications requiring larger fractional bandwidths and

---

[§] Sections 5.3 and 5.4 were written by Robert S. Wagers.

5 PIEZOELECTRIC AND ELECTROMECHANICAL FILTERS

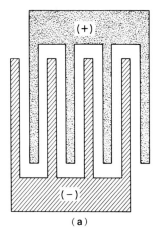
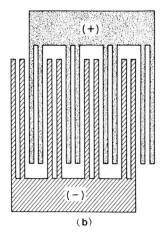

(a)   (b)

FIG. 5.3-1 Metallization in two types of SAW two-phase interdigital transducers. (a) Single-electrode IDT with alternating polarity electrodes, each of nominal width $\lambda_0/4$. The gaps between electrodes are also nominally $\lambda_0/4$. (b) Double electrode IDT with electrodes and gaps of nominal width $\lambda_0/8$. The double-electrode transducer has lower acoustic reflections than the single electrode transducer does.

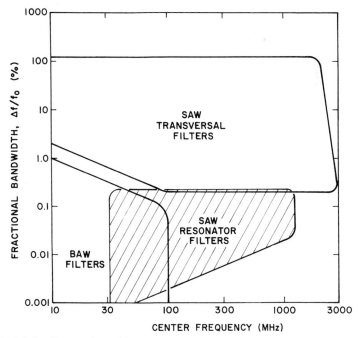

FIG. 5.3-2 Frequencies and fractional bandwidths served by three crystal filter technologies. The BAW data are from Zverev (1967), and the SAW resonator filter data are from Coldren and Rosenberg (1978b).

higher carrier frequencies. The upper limit of $\Delta f/f_0 \simeq 125\%$ defines the range free of harmonic interference. The right-hand limit is set by photolithographic capabilities (the line shown was obtained by assuming 0.5-μm feature definition capability). The lower limit on $\Delta f/f_0$ arises from design limitations imposed when the Rayleigh wave potential becomes comparable to the source potential; and finally the boundary paralleling the BAW region is defined by physical length constraints. In Fig. 5.3-2 the length of a SAW transversal filter was restricted to 2 cm.

Other features bearing on SAW filter application are the insertion loss, sidelobe ratio, phase error, and temperature sensitivity. The first three parameters are interrelated, and compromises must be made among the three according to the system specifications. However, insertion losses can range from a few tenths of a decibel to greater than 40 dB; sidelobes may vary from as poor as -27 dB to as good as -60 dB; phase errors can be better than a few tenths of a degree.

Temperature sensitivity is set primarily by substrate choice. Materials commonly chosen for SAW transversal filters include ST-cut quartz and several cuts of $LiNbO_3$ and $LiTaO_3$. ST-cut quartz has no first-order temperature coefficient but has very weak coupling, whereas the other substrates have strong coupling but first-order temperature coefficients of 40 to 90 ppm. The choice of substrate is determined by weighing bandwidth, insertion loss, and crystal heater requirements against one another.

The performance of SAW transversal filters is dominated by the transduction processes at the IDT. To be sure, the filter characteristics are also modulated by physical processes such as wave attenuation and diffraction, but those effects generally are a perturbation to the response dictated by transduction physics. It is the transduction process that must be understood and controlled to design SAW filters.

In the following sections the operation of SAW IDTs will be developed. The three most common physical interpretations of IDT operation will be considered: the bulk-wave model, the impulse model, and the normal-mode model. These three schools overlap in the IDT properties they describe much of the time. However, each has its own advantage in some aspect of analysis or synthesis.

Bulk-wave modeling (Smith et al., 1969; Smith and Pedler, 1975, 1976) requires interpreting each electrode or gap between electrodes (Fig. 5.3-1) as a bulk-wave transmission line. A description of the entire IDT is developed as a cascade of bulk-wave transducers. This approach has been extremely successful in analyzing IDT operation. It has quantitative accuracy in describing not only electric-to-acoustic conversions but acoustic-to-acoustic conversions as well. It tends to become numerically overbearing if the electrode overlap of the IDT is varied; it does not contain synthesis concepts distinct from those of the impulse model.

5 PIEZOELECTRIC AND ELECTROMECHANICAL FILTERS   233

The impulse response model (Tancrell and Holland, 1971); Hartmann et al., 1973a; Tancrell, 1974) represents the IDT as a sequence of half sinusoids placed at each electrode position or a sequence of delta functions placed at the electrode edges. It then draws on known Fourier transform relations to relate the frequency response of the filter to the metallization pattern. This approach has its greatest power in performing synthesis of electric-to-acoustic transfer functions. It tends to have less quantitative accuracy than the bulk-wave model approach in doing analysis of terminal transfer relations, and it has virtually no capability for the analysis of acoustic-to-acoustic transfer relations.

The normal-mode technique (Auld and Kino, 1971; Auld, 1973; Wagers, 1978) gives greatest emphasis to the wave nature of the SAW, treating the disturbance as two SAW eigenmodes propagating in opposite directions under the IDT synchronously with the applied traveling waves of the IDT. Again this is an analysis technique that does not contain synthesis elements distinct from those contained in the impulse response model. The normal-mode approach considers the entire IDT, without dividing it into segments as is done in the bulk-wave-modeling approach, and provides closed-form expressions for wave amplitudes and IDT admittances. Though the formulation is potentially capable of providing approximations to acoustic-to-acoustic conversion processes, the most significant contribution from this analysis has been in clarifying electric-to-acoustic conversion processes.

5.3.2 Interdigital Transducer Admittance

5.3.2.1 NORMAL-MODE REPRESENTATION OF ACOUSTIC ADMITTANCE

Almost all transduction to SAW is accomplished with IDTs. The transducers consist of alternating-polarity electrodes as illustrated in Fig. 5.3-1. Typically the gaps between electrodes are the same size as the electrodes. At midband of the IDT response, the frequency and spatial periodicity of the applied voltage is such that $\omega_0/\gamma_0 = v$, where $\gamma_0 = 2\pi/\lambda_0$ (λ_0 being the periodicity of the transducer) and v is the phase velocity of the wave. Thus an observer (or wave) traveling along the surface at velocity ω_0/γ_0 sees a constant voltage and can extract power from the applied voltage.

In the transducer impedance analysis presented here, it is assumed that the transducer electrodes have constant overlap. In obtaining numerical values for the transducer admittance it is also ultimately assumed that the potential on the electrodes and in the gaps is essentially the same as would exist if the substrate were nonpiezoelectric, that is, the weak-coupling approximation is made. We shall see that this approximation sets limits on the range of validity of the analysis and hence the model of the transducer.

Figure 5.3-1 illustrates the types of IDT structures considered. The polarity of the voltage applied to the electrodes reverses approximately every half wavelength of the excited Rayleigh wave in the most common case. While this is the most common transducer configuration, the results presented are not limited to such cases alone. Nor does it require that all the electrodes be present as in Fig. 5.3-1.

Figure 5.3-3 illustrates the orientation of coordinates with respect to the substrate and Rayleigh wave propagation direction. In Fig. 5.3-3, the transducer is shown as simply a wide distribution of potential $\phi(0, z)$. In fact, any transducer will be finite in the x direction. But it is assumed that the width dimension is great enough that any variation with x can be neglected. Two normal-mode amplitudes (Rayleigh waves) are excited by the transducer. They emerge as $a_-(z)$ and $a_+(z)$ traveling in the negative and positive z directions, respectively. Each mode starts with zero amplitude at one edge of the transducer and grows in amplitude as it travels through the transducer and emerges at the other end.

The normal-mode analysis begins by finding integral equations for the mode amplitudes $a_-(z)$ and $a_+(z)$. Both Auld (1973) and Wagers (1978) illustrated how the mode-amplitude differential equations are derived. Auld (1973) used open-circuited modes for the basis of his expansion whereas Wagers (1978) used short-circuited modes. The choice of basis determines whether one obtains integral equations dependent on IDT electric potential or IDT electric displacement.

Wagers (1978) showed that the Rayleigh wave amplitudes are given by

$$a_+(z) = e^{-j\gamma z}\Gamma_1 \int_{-\infty}^{z} e^{j\gamma\xi}\bar{\phi}(0, \xi)\, d\xi, \qquad (5.3\text{-}1)$$

$$a_-(z) = -e^{j\gamma z}\Gamma_2 \int_{z}^{\infty} e^{-j\gamma\xi}\bar{\phi}(0, \xi)\, d\xi, \qquad (5.3\text{-}2)$$

where ξ is a dummy variable of integration along the z-axis of Fig. 5.3-3,

$$\Gamma_1 = j\omega V_0 D_+^*(0)/4P, \qquad \Gamma_2 = -j\omega V_0 D_-^*(0)/4P, \qquad \phi(0, z) = V_0 \bar{\phi}(0, z),$$

and V_0 is the peak-to-peak voltage applied across the transducer.

In deriving Eqs. (5.3-1) and (5.3-2), propagation of the waves as $\exp[j(\omega t \pm \gamma z)]$ was assumed, ω and γ are the radian frequency and propagation constant, respectively, of the waves ($\omega = 2\pi f = \gamma v$, $\gamma = 2\pi/\lambda$), λ is the Rayleigh wavelength, P the magnitude of power flow per meter of acoustic beamwidth, and $\phi(0, z)$ the electric potential at the interface between the transducer and the substrate. The terms $D_+(0)$ and $D_-(0)$ are the values at $y = 0$ of the y components of electric displacement of the forward and backward freely propagating Rayleigh modes. Superscript * specifies complex conjugate of the designated quantity.

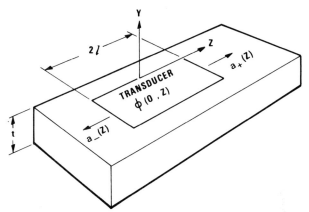

FIG. 5.3-3 Orientation of coordinates relative to the crystal and the propagation direction of straight-crested Rayleigh waves excited by the IDT.

Equations (5.3-1) and (5.3-2) give the amplitudes of the Rayleigh waves throughout the transducer region. The limits of $\pm\infty$ in the integrals are merely symbolic. The integrations need be extended only to the points beyond which $\bar{\phi}(0, \xi) = 0$, that is, at the edges of the transducer.

The electric displacement existing at the interface between the IDT and the crystal can be expressed as a superposition of that arising from all physical processes active at the surface of the crystal. The contribution due to Rayleigh wave generation can be written as

$$D_y = \{a_+(z)D_+(0) + a_-(z)D_-(0)\}. \tag{5.3-3}$$

Using Eqs. (5.3-1) and (5.3-2) for the mode amplitudes and Eq. (5.3-3) for the electric displacement, one can evaluate the complex Poynting theorem over a boundary enclosing the IDT (Wagers, 1978) to obtain for the transducer admittance:

$$Y = \psi \int_{-l}^{l} dz\, \bar{\phi}(0, z) \int_{-l}^{l} \cos[\gamma(\xi - z)]\bar{\phi}(0, \xi)\, d\xi$$

$$+ j\psi \int_{-l}^{l} dz\, \bar{\phi}(0, z) \left\{ \int_{-l}^{z} \sin[\gamma(\xi - z)]\bar{\phi}(0, \xi)\, d\xi \right.$$

$$\left. - \int_{z}^{l} \sin[\gamma(\xi - z)]\bar{\phi}(0, \xi)\, d\xi \right\}, \tag{5.3-4}$$

where

$$\psi = \omega^2 w |D_+(0)|^2 / 4P,$$

w is the overlap width of the IDT, and $2l$ the length of the IDT in the propagation direction (Fig. 5.3-3). The portion of the input admittance of the transducer due to Rayleigh wave transduction is completely specified by Eq. (5.3-4) in terms of the self-consistent electric potential existing at the plane between the crystal and the transducer. It should also be noted that the real and imaginary parts of Eq. (5.3-4) are Hilbert transform pairs for all electric potentials. This is a general property, true of all admittances, and in Wagers (1978) explicit satisfaction of the Hilbert transform pair relation is demonstrated without restriction of the transducer structure.

5.3.2.2 INTERDIGITAL TRANSDUCER CAPACITANCE

The admittance expression of Eq. (5.3-4) represents only that portion of the IDT terminal properties arising from acoustic effects. In parallel with the acoustic admittance is the electrostatic capacitance of the IDT that exists even in the absence of piezoelectricity. A complete IDT model requires representation of the static capacitance. Engan (1969) was the first to solve the electrostatic boundary value problem illustrated in Fig. 5.3-1a. He assumed the structure properties had no variation across the width (as we have assumed) and that the IDT pattern extended to $\pm \infty$ along the z direction. While he assumed uniform periodicity to $z = \pm \infty$ he did allow the metallizations to vary in width. He found that the potential of the transducer could be represented by the series,

$$\phi(0, z) = V_0 \sum_{n=0}^{\infty} \frac{P_n(2m - 1)}{(2n + 1)K'(m)} \sin\left[(2n + 1)\frac{2\pi z}{\lambda_0}\right], \quad (5.3\text{-}5)$$

where λ_0 is the periodicity of the transducer, $m = \sin^2(\pi a/2)$, a is the portion of the surface that is metallized, $a = a_0 = 0.5$ for 50% metallization, P_n are Legendre polynominals of the first kind, and K' is the complete elliptic integral of the first kind to the complementary parameter $1 - m$. By integrating the electric displacement to find the charge on the IDT electrodes, he was able to express the capacitance of the transducers as

$$C_T = Nw(\varepsilon_p + \varepsilon_0)K(m)/K'(m), \quad (5.3\text{-}6)$$

where N is the number of electrode pairs in the IDT, $\varepsilon_p = \sqrt{\varepsilon_{22}\varepsilon_{33} - \varepsilon_{23}^2}$, ε_{ij} are elements of the permittivity matrix, ε_0 is the permittivity of free space, and K is the complete elliptic integral of the first kind.

Equation (5.3-6) shows that the static capacitance of the IDT is a function of the electrode width. In Fig. 5.3-4 the relative capacitance variation is shown for both single-electrode IDTs and double-electrode IDTs. For

5 PIEZOELECTRIC AND ELECTROMECHANICAL FILTERS

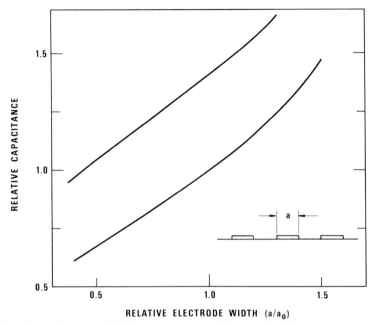

FIG. 5.3-4 Variation of IDT static capacitance with electrode width. For single-electrode transducers, $a/a_0 = 1$ corresponds to electrodes of width $\lambda_0/4$, whereas for double electrodes, $a/a_0 = 1$ corresponds to electrodes of width $\lambda_0/8$. [From Wagers (1976). © 1976 IEEE.]

single-electrode IDTs, $a/a_0 = 1$ corresponds to electrodes of width $\lambda_0/4$, whereas for double-electrode IDTs, $a/a_0 = 1$ corresponds to electrodes of width $\lambda_0/8$. One can see that for 50% metallization IDTs, the double-electrode IDT has approximately 40% more capacitance than the single-electrode IDT. The greater capacitance of the double-electrode IDT increases its Q (susceptance/conductance at midband). However, its coupling strength relative to the single-electrode IDT (as measured by the IDT radiation conductance) is 7% larger, which helps to decrease its Q.

5.3.2.3 EVALUATION OF INTERDIGITAL TRANSDUCER ADMITTANCE

To evaluate Eq. (5.3-4), the electric potential must be known throughout the transducer. Often the contribution to the total potential from the acoustic waves is neglected, and only the electrostatic solution is used for $\bar{\phi}(0, \xi)$. This is generally a very good approximation, but still, obtaining the fields of the transducer structure in the absence of piezoelectricity is a formidable task in the general case. It is because of this difficulty that the admittance expression was developed in the form of Eq. (5.3-4). The admittance, as specified, has a dependence only on the electric potential distribution [the electric displacement of the IDT does not appear in Eq. (5.3-4)]. This facilitates analysis since the potential is the specified quantity in the transducer boundary value problem. In most transducers, $\bar{\phi}(0, \xi)$ is known exactly over half the range of ξ since it is specified on the electrodes. In the regions between the electrodes, the exact value of the potential is unknown, but because the potential must make a transition from one electrode to the adjacent one in a short distance, the variation can be guessed with sufficient accuracy to yield admittance values adequate for many applications.

In some cases, the potentials within the transducers are known from numerical analysis. Thus, Eq. (5.3-4) can be evaluated numerically and the results tabulated as a function of γ. Electrode withdrawal transducers, for example, employ highly nonuniform electrode sequences (Hartmann, 1973; Laker et al., 1978; Wagers, 1978). In these transducers, bandpass shaping is accomplished by removing electrodes from what would otherwise be a regularly periodic array of electrodes. The synthesis procedures for electrode withdrawal transducers limit the allowed electrode sequences to only a few possibilities. Thus, the exact integrations for the allowed electrode sequences can be performed numerically and tabulated. Wagers (1978) discusses how this is accomplished; it is shown that at midband, the radiation conductance of Eq. (5.3-4) can be written in the form

$$G_a = \omega w(\varepsilon_p + \varepsilon_0)(2\pi)^2(\Delta v/v)\left|\sum_n I_n\right|^2, \quad (5.3\text{-}7)$$

where

$$I_n = \int_{n/2}^{(n+1)/2} d\eta \, \bar{\phi}(0, \eta)e^{j2\pi n}, \quad (5.3\text{-}8)$$

$\eta = z/\lambda_0$, G_a is the radiation conductance, and $\Delta v/v$ is a measure of the electromechanical coupling (Section 5.3.4). The term I_n of Eq. (5.3-8) is the integral of the electric potential over the nth half wavelength of the transducer. For periodic, *double-electrode* IDTs, as in Fig. 5.3-1b, $I_n = 0.0 + j0.13945$. Eq. (5.3-7) shows that the IDT radiation conductance increases linearly with frequency, IDT width, substrate permittivity, and coupling; it increases as the square of the number of coupling sections.

If analytic forms for $\bar{\phi}(0, z)$ of the interdigital capacitor can be found, then analytic expressions can be obtained for Y. For an N-electrode-pair periodic transducer, if the electric potential $\bar{\phi}(0, z)$ of the transducer is approximated by that of an interdigital structure on a nonpiezoelectric substrate, then the results of Engan (1969) [Eq. (5.3-5)] can be used. Using only the fundamental

# 5 PIEZOELECTRIC AND ELECTROMECHANICAL FILTERS

term of Eq. (5.3-5), one can substitute for $\bar{\phi}(0, z)$ in Eq. (5.3-4) to obtain for the IDT admittance $Y = G_a + jB_a$, where

$$G_a = G_0 \left(\frac{\sin X}{X}\right)^2, \qquad (5.3\text{-}9)$$

$$B_a = G_0 \frac{1}{X}\left(\frac{\sin 2X}{2X} - 1\right), \qquad (5.3\text{-}10)$$

$$G_0 = w[N\omega(\varepsilon_p + \varepsilon_0)]^2 \frac{\pi^2}{[K'(m)]^2}\left(\frac{\gamma}{\gamma_0}\right)^2 |\phi|^2/4P, \qquad (5.3\text{-}11)$$

$\gamma_0 = 2\pi/\lambda_0$, $\omega_0 = 2\pi f_0 = \gamma_0 v$, $X = (\gamma - \gamma_0)l = N\pi(\omega - \omega_0)/\omega_0$, G_0 is the radiation conductance at midband, and ϕ the Rayleigh wave potential associated with a power flow of magnitude P.

5.3.3 Relation of Normal-Mode Theory Admittance to the Impulse Model

If in Eq. (5.3-4) $\cos[\gamma(\xi - z)]$ is replaced by $\cos(\gamma\xi)\cos(\gamma z) + \sin(\gamma\xi)\sin(\gamma z)$, then one can readily show that

$$G_a = \psi |F(\gamma)|^2, \qquad (5.3\text{-}12)$$

where

$$F(\gamma) = \int_{-l}^{l} dz\, \bar{\phi}(0, z)e^{j\gamma z}. \qquad (5.3\text{-}13)$$

Equation (5.3-12) specifies that the radiation conductance is obtained from a frequency response $F(\gamma)$. Equation (5.3-13) shows that $F(\gamma)$ is obtained from an integral over $\bar{\phi}(0, z)$. The form of the integral is the same as a Fourier transform. In fact extending the integration limits in Eq. (5.3-13) to $\pm\infty$ and prescribing $\bar{\phi}(0, z) = 0$ for $|z| > l$, the integral becomes the Fourier transform of $\bar{\phi}$. Since $F(\gamma)$ is the frequency domain response, then $\bar{\phi}(0, z)$ is the spatial impulse response of the transducer. When a suitable method for prescribing $\bar{\phi}$ is chosen, then one can bring to bear on the design question all previous experience with Fourier transform pairs.

Consider substituting Eq. (5.3-5) for the *electrostatic* potential in the integral of Eq. (5.3-13) and assume that $2l = N\lambda_0$. When $\gamma = \gamma_0$ all terms in the series will integrate to zero except the $n = 0$ term because of the orthogonality of $\exp[j\gamma_0 z]$ and $\sin(2n + 1)\gamma_0 z$. The result is that at midband, $F(\gamma)$ equals the first Fourier coefficient of the series of Eq. (5.3-5). Physically this is the same result as would have been obtained if we had placed a half wave of sinusoid at each electrode with the same polarity as the electrode. For values of γ near γ_0, the first term of Eq. (5.3-5) would still dominate the Fourier transform of Eq. (5.3-13). Thus one can see that G_a can be physically related to a transform of half waves of sinusoid positioned at the electrodes. This physical interpretation was first elaborated by Hartmann et al. (1973).

5.3.4 Limitations on the Use of Electrostatic Fields

In the preceding discussion we used the electrostatic fields [Eq. (5.3-5)] as an approximation to the total electric potential of the transducer. Obviously this cannot always be a reasonable approximation, for if the transducer is made long enough, then eventually the potential of the wave will equal that being applied to the transducer. Certainly the transducer length must be limited if the use of electrostatic potentials in Eq. (5.3-4) is to be a good approximation.

Under steady-state operation of an IDT at frequency ω, the power input to the transducer is $V_0^2 G_a/2$ where V_0 *is now the peak applied voltage* and G_a the radiation conductance of the IDT. This power is carried away by the forward and backward traveling Rayleigh waves. Thus

$$2Pw = V_0^2 G_a/2. \qquad (5.3\text{-}14)$$

A fundamental relation (Auld, 1973) between Rayleigh wave potential, acoustic power flow, and coupling $\Delta v/v$ is

$$\frac{|\phi|^2}{4P} = \frac{1}{\omega(\varepsilon_p + \varepsilon_0)} \frac{\Delta v}{v}, \qquad (5.3\text{-}15)$$

where $\Delta v/v$ is the fractional change in Rayleigh wave velocity that occurs when a massless perfect short circuit is applied to the top of the substrate. If Eqs. (5.3-7) and (5.3-15) are substituted in Eq. (5.3-14), then we can write the ratio of the Rayleigh wave potential to the applied potential as

$$|\phi|/V_0 = 2\pi |\sum_n I_n| \Delta v/v, \qquad (5.3\text{-}16)$$

where I_n is defined in Eq. (5.3-8). The term I_n in Eq. (5.3-16) can be evaluated for an N-electrode pair double-electrode transducer (shown in Fig. 5.3-1b) by using the value (quoted above) $0.0 + j0.13945$. We obtain

$$|\phi|/V_0 = 1.75N \, \Delta v/v. \qquad (5.3\text{-}17)$$

Equation (5.3-17) is a completely general expression relating the magnitude of the Rayleigh wave potential to the applied potential as a function of the length of the transducer in the propagation direction. Having neglected the Rayleigh wave amplitude in the source terms for mode excitation, we find

5 PIEZOELECTRIC AND ELECTROMECHANICAL FILTERS 241

that the wave is predicted to grow linearly with the number of electrodes in the IDT. Obviously this type of growth cannot continue without bound. A limit to the applicability of the analysis can be estimated by calculating the IDT parameters at which the Rayleigh wave potential equals the applied potential. This is obtained by setting Eq. (5.3-17) equal to unity. For (YZ) $LiNbO_3$ with $\Delta v/v = 0.024$ we obtain $N = 24$, and for ST-cut quartz with $\Delta v/v = 0.000581$ we obtain $N = 984$.

While one can calculate at what point the transducer length forces the wave potential to be a significant part of the total potential, one cannot with similar ease quantify the effect on filter response of neglecting the Rayleigh potential and using only the electrostatic solution. As a practical matter, the Rayleigh wave potential is neglected in the design of transducers hundreds of wavelengths long for operation on ST-cut quartz, and 50–60-dB sidelobes are obtained. Emtage (1972) and Zuliani et al. (1975) deal with the question of including the electric potential of the piezoelectric substrate.

5.3.5 Electromechanical Coupling Constant k

Commonly in the analysis of BAW devices a very simple relationship is found between the acoustic radiation conductance at synchronism and the static capacitance of the transducer. It is not surprising, then, that in the bulk-wave modeling approach to IDT analysis (Sec. 5.3.7) a simple relation like that of bulk-wave transducers was found to relate $G_a(\omega_0)$ and C_T. It has been shown (Smith et al., 1969) that, consistent with the physical assumptions,

$$G_a(\omega_0) = G_0 = (4/\pi)k^2 N \omega_0 C_T, \quad (5.3\text{-}18)$$

where k^2 is the square of the electromechanical coupling constant. While the form of Eq. (5.3-18) is appealing because of its close relation to rigorously correct relations in bulk-wave acoustics, the equation when applied to IDTs is an approximation because of the semiempirical manner in which it was derived. Nevertheless, the bulk-wave modeling approach to the analysis of IDT excitation and detection of SAWs is so successful that Eq. (5.3-18) has been accepted worldwide, not only as essentially rigorous but in effect as the definition of k^2 for IDTs.

It is of interest to relate the k^2 defined by Eq. (5.3-18) to material constants through the fundamental relations for G_a and C_T found earlier. Thus, if G_0 of Eq. (5.3-11) is used along with Eq. (5.3-15) and the relation for static IDT capacitance [Eq. (5.3-6)], then the electromechanical coupling constant for IDTs can be written as

$$k^2 = \frac{\pi^3}{4K(m)K'(m)} \frac{\Delta v}{v}. \quad (5.3\text{-}19)$$

We see from this equation that the efficiency of IDT coupling to the Rayleigh wave can be related to two factors:

(1) the magnitude of the velocity perturbation that arises from shorting the top surface of the crystal (this is totally a material relation and provides a basis for comparing one crystal cut to another); and
(2) the fraction of the surface covered by electrodes.

The product of the two elliptic integrals, $K'(m)K(m)$, reaches a minimum at $a = \frac{1}{2}$ (50% metallization) and increases monotonically to infinity as a varies from $\frac{1}{2}$ to 0 (no electrodes) or from $\frac{1}{2}$ to 1 (continuous metal). Thus, maximum IDT coupling is achieved with electrodes $\lambda_0/4$ wide.

When $\lambda_0/4$ electrodes are used, $K(m) = K'(m) = 1.854$. Substituting for the elliptic integrals in Eq. (5.3-19) gives $k^2 = 2.26\,\Delta v/v$, a result very similar to that obtained for bulk-wave crystals. In bulk-wave analysis, one commonly obtains $k^2 = 2\,\Delta v/v$, and in SAW literature one often sees $k^2 = 2\,\Delta v/v$. The latter relation, $k^2 = 2\,\Delta v/v$, is not consistent with k^2 defined by Eq. (5.3-18). Obviously, as a practical matter, the difference between the two definitions is only $\sim 10\%$, and practical results will not be much affected by the choice of k^2.

5.3.6 Electrical Q and Insertion Loss

In resonant circuits of constant-valued inductors and capacitors, the circuit Q and the 3-dB fractional bandwidth are reciprocals of one another. Thus, one can view the bandwidth as determined by the Q. In narrow-band SAW IDTs, however, the bandwidth is controlled by phasing between the constant-length IDT and a variable-wavelength Rayleigh wave. As a consequence the traditional concepts of Q do not give the half-power bandwidth. (However, the Q of the IDT is still a relevant parameter for gauging matching difficulties.)

The definition of Q considered is the traditional one for parallel connected circuits: $Q = \omega_0$(average energy stored)/(energy dissipation rate). Thus, for the IDT at synchronism this becomes $Q = \omega_0 C_T/G_a(\omega_0)$. If, as in the previous section, we substitute for $G_a(\omega_0)$ from Eq. (5.3-11), for $|\phi|^2/4P$ from Eq. (5.3-15), and for the static capacitance from Eq. (5.3-6) then we obtain

$$Q = \frac{K(m)K'(m)}{N\pi^2\,\Delta v/v}. \tag{5.3-20}$$

Now, referring to Eq. (5.3-9) one can see that the radiation conductance varies as $[(\sin X)/X)]^2$, where $X = N\pi(\omega - \omega_0)/\omega_0$. The 3-dB points for one transducer occur when $N(\omega - \omega_0)/\omega_0 = N(f - f_0)/f_0 = 0.44295$. Thus,

5 PIEZOELECTRIC AND ELECTROMECHANICAL FILTERS

the fractional 3-dB bandwidth $\Delta f/f_0$ of the acoustic radiation conductance is related to the number of electrode pairs by

$$N = 0.8859/(\Delta f/f_0). \quad (5.3\text{-}21)$$

Substituting N from Eq. (5.3-21) into Eq. (5.3-20) and replacing the elliptic integrals by their value for $\lambda_0/4$-wide electrodes, (1.854) we obtain

$$Q \simeq \frac{\pi}{8 \, \Delta v/v} \frac{\Delta f}{f_0}. \quad (5.3\text{-}22)$$

This equation shows that for very narrow-band IDTs, the Q of the IDT is lower for materials with large $\Delta v/v$. For example, with (YZ) $LiNbO_3$ ($\Delta v/v = 0.024$) a 10%-bandwidth IDT has a Q of 1.6. Note also that the Q is independent of IDT width. In principle the aperture can be chosen to match the IDT at midband to the characteristic admittance Y_0 of the drive transmission line. Thus, narrowband SAW filters composed of two constant-aperture IDTs have a minimum insertion loss of 6 dB. This loss arises solely from the bidirectionality of the IDTs.

Equation (5.3-22) is very similar to expressions found in other works. Indeed if one replaces $\Delta v/v$ by $k^2/2$, then Eq. (5.3-22) is functionally identical to Eq. (5.9) of Snow (1977). However, an examination of the definition of terms in the paper by Snow (1977) reveals that the $\Delta f/f$ in that expression is the 4-dB bandwidth of the transducer instead of the 3-dB bandwidth assumed above. This difference can be traced to the different definitions of k^2 discussed in the previous section. If one uses $k^2 = 2.26 \, \Delta v/v$ instead of $k^2 = 2 \, \Delta v/v$, then the expressions for the IDT Q obtained here and in Snow (1977) are consistent.

If the filter bandwidth is required to be greater than that of the IDTs, then an insertion loss penalty must be paid. The loss–bandwidth relation can be derived by a variety of approaches. Commonly one assumes that the IDT is shunt-tuned with an inductor, loaded with a resistor, and driven through a transformer. However, the transformer is not essential to the analysis; direct circuit analysis assuming a transmission line driving a loaded, inductor-tuned IDT yields the same loss–bandwidth relation.

Using a theorem of Bode and Fano, Gerard (1978) derived and discussed the loss–bandwidth relation. Snow (1977) also presented a very simple derivation. In those cases, or by considering the IDT to be driven directly by a transmission line, one obtains a relation of the form

$$\frac{\Delta f}{f_0} \leq \sqrt{\frac{8 \, \Delta v/v}{\pi}} \quad (5.3\text{-}23)$$

for an upper limit on the 3-dB bandwidth of an N-wavelength-long, unapodized IDT that can be tuned to have only bidirectionality loss (3 dB) at

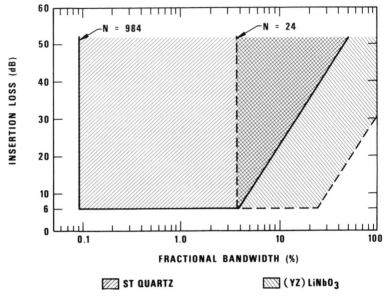

FIG. 5.3-5 Relation between SAW filter insertion loss and 6-dB fractional bandwidth. The bounds shown indicate approximately what can be achieved with two bidirectional IDTs. The lower limit of 6 dB comes from bidirectional power loss. The left limit is where the weak-coupling approximation is grossly violated. The right limit follows from interplay of the SAW IDT with the matching network.

midband. Because two IDTs are used for a filter, Eq. (5.3-22) gives the maximum 6-dB bandwidth of a filter. Snow (1977) shows that for bandwidths greater than that specified by equality in Eq. (5.3-23), the insertion loss increases at 12 dB/octave.

The asymptotic relation on insertion loss of 12 dB/octave, the corner point given by equality in Eq. (5.3-23), and the physical limits of analysis accuracy given by $1.75N \, \Delta v/v = 1$ [Eq. (5.3-17)] can be combined in one plot to show the insertion loss versus fractional bandwidth ranges covered by various SAW filters. Figure 5.3-5 shows results for substrates of (yz) $LiNbO_3$ and ST-cut quartz.

5.3.7 Bulk-Wave Modeling of Interdigital Transducers

Rayleigh wave propagation on anisotropic substrates is inherently a three-dimensional problem. The degree of complexity can be reduced by modeling real devices in terms of infinitely wide ones. In that case the boundary value problem becomes two-dimensional; there is variation in the field variables in both the propagation direction and with distance into the substrate. Yet

5 PIEZOELECTRIC AND ELECTROMECHANICAL FILTERS 245

on examination of Eqs. (5.3-1) and (5.3-2), one finds that the excited fields of the IDT are *rigorously* described by a function with only one-dimensional variation. An integration along the top surface of the crystal is all that is required. The nature of the fields within the crystal is not in evidence in Eqs. (5.3-1) and (5.3-2). Indeed it does not influence the form of the excitation mathematics.

For years, one-dimensional bulk-wave analysis has been used to describe Rayleigh wave processes. The approximations have met with remarkable success in describing interactions with IDTs. Particularly good results have been achieved with the "crossed-field" bulk-wave model. It can be shown that the crossed-field model can yield excitation mathematics identical in form to that obtained rigorously from the normal-mode theory [Eqs. (5.3-1) and (5.3-2)].

Figure 5.3-6 shows the orientation of coordinates and the assumed physical nature of the exciting electric field in the substrate. It is assumed that the direction of the electric field in any section l long is constant throughout the section. At the boundaries of each segment, the field makes abrupt changes in direction. For the following development, each section is assumed to be a hexagonal 6-mm crystal with its polar axis aligned along the z coordinate. The physical attributes of the crystal sections (mass, stiffness, etc.) are assumed to be identical.

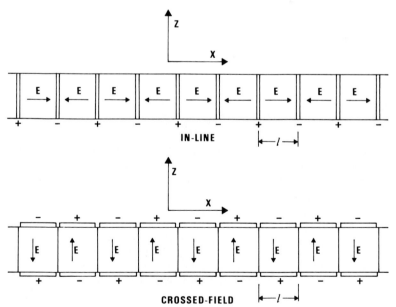

FIG. 5.3-6 Assumed physical configuration of one-dimensional bulk-wave structures having either an in-line or crossed-field orientation of the electric field.

We assume that the structures of Fig. 5.3-6 are infinite in the y direction. Thus, there is no variation with y. Additionally we assume that the disturbance within the structure is independent of z. Therefore $\partial/\partial z$ is also required to vanish. With only $\partial/\partial x$ nonzero, the equations of motion (Auld, 1973) take on a particularly simple form.

By assuming that the electric field has either one polarity or the other, different sets of equations become relevant. If one considers the in-line model, then only the electric field in the x direction, E_1, is nonzero. On the other hand, in considering the crossed-field model one assumes that only the electric field in the z direction, E_3, is nonzero. Quite different motions of the plate are produced by the two models. The crossed-field model describes a pure compressional wave, whereas the in-line model represents pure shearing in the x–z plane. Obviously neither of these acoustic disturbances is Rayleigh-like. The Rayleigh wave has some shear motion, some compressional content, and varies rapidly in the z direction. All bulk-wave components of the Rayleigh wave are evanescent in the z direction.

If we define mechanical voltages and currents by $F_i = -T_i A$ and $U_i = j\omega u_i$, then the constitutive equations (refer to Chapter 2) for the two models can be cast in transmission-line form (u_i is particle displacement in the i direction, T_i the ith component of stress, and A the cross-sectional area of the plate). For the crossed-field model we obtain

$$F_1 = -\frac{c_{11} A}{j\omega} \frac{\partial U_1}{\partial x} + e_{31} A E_3, \tag{5.3-24}$$

$$U_1 = \frac{-1}{j\omega \rho A} \frac{\partial F_1}{\partial x}, \tag{5.3-25}$$

$$D_3 = \varepsilon_{33}^s E_3 + \frac{e_{31}}{j\omega} \frac{\partial U_1}{\partial x}, \tag{5.3-26}$$

where c_{ij} are the stiffness constants, e_{ij} the piezoelectric stress constants, ρ is the mass density of the medium, D_i the electric displacement in the i direction, and the superscript s on the permittivity element ε_{33}^s signifies that the permittivity is measured at contract strain.

The in-line model equations become

$$F_5 = -\frac{c_{44} A}{j\omega} \frac{\partial U_3}{\partial x} + e_{15} A E_1, \tag{5.3-27}$$

$$U_3 = \frac{-1}{j\omega \rho A} \frac{\partial F_5}{\partial x}, \tag{5.3-28}$$

$$D_1 = \varepsilon_{11}^s E_1 + \frac{e_{15}}{j\omega} \frac{\partial U_3}{\partial x}. \tag{5.3-29}$$

The transmission-line equations for the two models are identical in form. Quite different physical assumptions about the nature of the fields internal to the medium have produced coupled equations with no apparent difference. Differences do arise however when boundary conditions are applied.

In solving Eqs. (5.3-24) and (5.3-25), one assumes that the disturbance propagates as $\exp[j(\omega t - \gamma x)]$. The plate thickness is t, and the width in the y direction is w. The differential equations produce dispersion relations of the form $\gamma = \omega\sqrt{\rho/c_{ii}}$, where $i = 1$ for the crossed-field model and $i = 4$ for the in-line model. Two boundary conditions are required for each model. For the in-line model, it is assumed that the interfaces between segments are infinitesimally thin, perfectly conducting electrodes with acoustic properties identical to the hexagonal crystal. Under these conditions, D_1 is a constant in each segment. The electric current into each segment comes from integrating D_1 over the cross-sectional area, and the voltage across a segment is found by integrating the electric field E_1.

For the crossed-field model, one assumes that each segment is electroded on the outer surfaces. This sets $\partial E_3/\partial x = 0$ at the surfaces. It is further assumed that the plate is thin enough that $\partial E_3/\partial x = 0$ everywhere in each segment and that the voltage across each segment is $V = -E_3 t$. The electric current into each electrode is obtained by integrating D_3 over the length of a segment.

Solving Eqs. (5.3-24)–(5.3-26) and applying the above boundary conditions, the crossed-field model yields

$$F_a = Z_0 \left\{ \frac{U_a}{j \tan \gamma l} + \frac{U_b}{j \sin \gamma l} \right\} - e_{31} w V_c, \qquad (5.3\text{-}30)$$

$$F_b = Z_0 \left\{ \frac{U_a}{j \sin \gamma l} + \frac{U_b}{j \tan \gamma l} \right\} - e_{31} w V_c, \qquad (5.3\text{-}31)$$

$$I_c = j\omega C_0 V_c + e_{31} w (U_a + U_b), \qquad (5.3\text{-}32)$$

where $Z_0 = c_{11} w t \gamma/\omega$ is a mechanical impedance, $C_0 = w l \varepsilon_{33}^s/t$ the capacitance of a section, l the length of a section, subscripts a and b denote the mechanical ports, subscript c denotes the electric port, V_c is voltage applied to port c, and I_c current into port c.

For the in-line model, Eqs. (5.3-27)–(5.3-29) yield

$$F_a = Z_0 \left\{ \frac{U_a}{j \tan \gamma l} + \frac{U_b}{j \sin \gamma l} \right\} - \frac{e_{15}}{j\omega \varepsilon_{11}^s} I_c, \qquad (5.3\text{-}33)$$

$$F_b = Z_0 \left\{ \frac{U_a}{j \sin \gamma l} + \frac{U_b}{j \tan \gamma l} \right\} - \frac{e_{15}}{j\omega \varepsilon_{11}^s} I_c, \qquad (5.3\text{-}34)$$

$$V_c = \frac{1}{j\omega C_0} I_c - \frac{e_{15}}{j\omega \varepsilon_{11}^s} (U_a + U_b), \qquad (5.3\text{-}35)$$

where $Z_0 = \bar{c}_{44} wt\gamma/\omega$, $\bar{c}_{44} = c_{44}(1 + k_{15}^2)$, $k_{15}^2 = e_{15}^2/(\varepsilon_{11}^s c_{44})$, and $C_0 = wt\varepsilon_{11}^s/l$. If impedance elements Z_1 and Z_2 and transformer ratio N are defined as in Fig. 5.3-7, then it can be shown that the two circuits of Fig. 5.3-7 have exactly the same terminal property relations as Eqs. (5.3-30)–(5.3-32) and Eqs. (5.3-33)–(5.3-35). Thus, these circuits are referred to as the equivalent circuits of the acoustic problems of Fig. 5.3-6.

Now differences between the two models are apparent. Equations (5.3-30) and (5.3-31) show that the mechanical quantities in the crossed-field model are driven by the electric voltage, whereas Eqs. (5.3-33) and (5.3-34) for the in-line model indicate that the mechanical quantities are driven by the electric current. Small differences exist in the equivalent circuits as well, with the in-line model requiring the presence of a negative capacitance.

When considering SAW devices, one cannot easily calculate numerical values for the parameters of the circuits of Fig. 5.3-7. The thickness applicable to Rayleigh waves is not defined. Also when considering practical materials such as (YZ) $LiNbO_3$ and ST-cut quartz, the equations of motion do not separate neatly into a few mechanical variables with a single piezoelectric coupling term as was found above for a hexagonal 6-mm crystal. Instead, several piezoelectric constants are involved with coupled systems of equations. Simple models such as in Fig. 5.3-7 do not emerge from the mathematics.

When applied to Rayleigh waves, the models of Fig. 5.3-7 are used as representative of the underlying physical processes; the characteristic impedance Z_0, segment capacitance C_0, and turns ratio N are chosen to give best agreement with experimental results. Smith et al. (1969) were the first to provide extensive development of these models. Gerard (1969) provided further clarification of the model. Quantitative application of the model to broadband devices was described by Smith et al. (1972), and the issue of

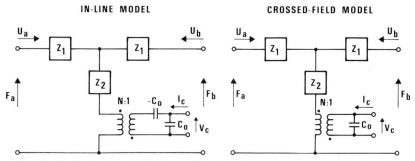

FIG. 5.3-7 Equivalent circuits for one-dimensional acoustic-wave generation by an in-line [where $Z_1 = j\bar{Z}_0 \tan \gamma l/2, Z_2 = -j\bar{Z}_0 \csc \gamma l, N = e_{15} wt/l, C_0 = \varepsilon_{11}^s wt/l, \bar{Z}_0 = \bar{c}_{44} wt\gamma/\omega$, and $\bar{c}_{44} = c_{44}(1 + e_{15}^2/\varepsilon_{11}^s c_{44})$] or crossed-field [where $Z_1 = j\bar{Z}_0 \tan \gamma l/2, Z_2 = -j\bar{Z}_0 \csc \gamma l, N = e_{31}w, C_0 = \varepsilon_{33}^s wl/t$, and $Z_0 = c_{11} wt\gamma/\omega$] electric intensity.

5   PIEZOELECTRIC AND ELECTROMECHANICAL FILTERS   249

which model was better was decided (Smith, 1973) on the basis of triple transit measurements. Smith (1973) found that for both $LiNbO_3$ and quartz substrates, the crossed-field model gave good approximations to the actual self-consistent acoustic reflections from IDTs. The in-line model predicted completely wrong trends for acoustic reflections as a function of IDT load impedance.

5.3.8   Advanced Bulk-Wave Models

SAW IDTs usually have many sections of the form shown in Fig. 5.3-6. The application of the equivalent circuits of Fig. 5.3-7 to a description of transduction processes requires cascading many such circuits, one for each electrode of the IDT, and computing the composite result. This is an exceedingly difficult task. It is virtually impossible to carry out analytic evaluation; numerical methods must be employed. In resorting to the computer, insight into the device physics is lost. The factors dominating device performance are merged with less significant factors to become part of the "bottom line" prediction.

In 1971 a new method of analysis and an equivalent circuit were presented that expressed the essential features of cascaded excitation sections without requiring laborious numerical procedures. The models put forth by Leedom *et al.* (1971) had only three elements no matter how many sections, or electrodes, existed in the overall transducer. The essential difference between the development of Leedom *et al.* (1971) and that of previous bulk-wave models was that they considered the entire transducer (instead of just one section) to be one acoustic region and accounted for the alternating electrical excitation by introducing a spatially varying source term. Starting from one-dimensional acoustic equations, they developed a Green's function for the acoustic transmission line and integrated it over the total length of the transducer. The result was that their model predicted transducer performance in terms of a Fourier transform over the electric potential of the transducer. A major simplification in the complex model accrued from their approach, because all the frequency dependences of the device were contained in three elements defined in terms of the Fourier transform of potential.

Subsequent to the Leedhom *et al.* (1971) paper describing new equivalent circuits for disturbances characterizable by one-dimensional analysis, Krimholtz (1972) applied the new circuit approach to SAW IDTs. More than just applying the explicit circuits reported by Leedhom *et al.* (1971) that were applicable to either even- or odd-symmetry excitation functions, Krimholtz extended the concepts to derive new equivalent circuits that were valid for arbitrary-symmetry, *real* excitation. The circuit he found is illustrated in Fig. 5.3-8. Note that the circuit is no more complicated than one

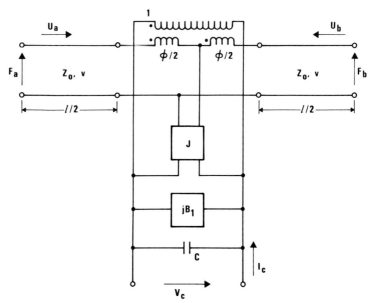

FIG 5.3-8 One-dimensional bulk-wave equivalent circuit for arbitrary-symmetry excitation functions. [From Krimholtz (1972).]

of the bulk-wave models of Fig. 5.3-7. However, the circuit in Fig. 5.3-8 describes the entire transducer. The complex interaction between sections of the transducer is taken into account in the definition of the elements.

Consider the circuit of Fig. 5.3-8 to describe a physical situation like the crossed-field model of Fig. 5.3-6. Then with a voltage V_c applied across a section, the maximum electric field inside a section is $E_{max} = |V_c|/t$. The maximum electric displacement is $D_{max} = \varepsilon_{33} E_{max}$. Krimholtz (1972) defined an excitation function

$$f(x) = d_{31} D_3^n(x)/s_{11}, \qquad (5.3\text{-}36)$$

where d_{31} is the active piezoelectric strain constant, s_{11} the active compliance element, and $D_3^n(x) = D_3(x)/D_{max}$. The associated form of the Fourier transform of this function was

$$F(\gamma) = \int_{-\infty}^{\infty} f(x) e^{-j\gamma x}\, dx. \qquad (5.3\text{-}37)$$

The only restriction on $f(x)$ was that it be a real function. It was explicitly intended to be a function that described the entire transducer, not just a single electrode. The form of $f(x)$ considered by Krimholtz (1972) was basically a pulse sequence with $f(x) = 0$ in the gaps between electrodes and $f(x)$

5 PIEZOELECTRIC AND ELECTROMECHANICAL FILTERS

equal to a constant underneath an electrode. Of course, the sign of $f(x)$ changed according to whether the electrode was connected to the positive or negative side of the source.

The shunt susceptance B_1 of Fig. 5.3-8 is defined as

$$B_1 = \frac{w}{2}\frac{\omega^2}{v}\frac{s_{11}}{t} H\{|F(\gamma)|^2\}, \tag{5.3-38}$$

where $v = (s_{11}\rho)^{-1/2}$, $\gamma = \omega/v$, $Z_0 = \rho v w t$, s_{ij} is an elastic compliance element, and $H\{\ \}$ is the Hilbert transform. Coupling of the electrical and mechanical ports is provided through the J inverter and transformer defined by

$$J = \begin{bmatrix} 0 & j/B_2 \\ jB_2 & 0 \end{bmatrix}, \tag{5.3-39}$$

$$B_2 = -\frac{\omega s_{11}}{t} F_e(\gamma), \tag{5.3-40}$$

$$\phi = -jw\gamma F_o(\gamma), \tag{5.3-41}$$

where the subscripts o and e refer to the odd and even parts of $f(x)$. The capacitance C is obtained in the usual way as charge divided by voltage,

$$C = (w/V_e) \int_{-\infty}^{\infty} |E_3(x)|\varepsilon_{33}^T(x)\,dx, \tag{5.3-42}$$

where the superscript T signifies the use of stress-free permittivity.

The array of mathematics from Eqs. (5.3-36)–(5.3-42) may not seem like a clarification of the transduction physics when viewed as a collection of symbols. If one considers a few examples, though, then the power of the new circuit becomes evident. For example, consider a transducer with a spatially even excitation function radiating on to an infinite half-space. The transformer ϕ would disappear from the circuit being replaced by shorts in the secondary and an open in the primary. The transmission line of Fig. 5.3-8, being terminated in its characteristic impedance Z_0, would present an impedance of $Z_0/2$ at the center of the line. The inverter J would transform that impedance into the admittance $B_2^2 Z_0/2$, and the input admittance of the transducer becomes

$$Y = j\omega C + jB_1 + B_2^2 Z_0/2. \tag{5.3-43}$$

Thus, the transducer admittance is a capacitance shunted by a conductance $B_2^2 Z_0/2$ and a frequency dependent susceptance B_1. By the relation of Eq. (5.3-38), B_1 satisfies the physical consistency requirement by being the Hilbert transform of the real part of the circuit.

Equation (5.3-43) shows at a glance that energy conversion from the electric terminals to the acoustic form follows the variation of $B_2^2 Z_0/2$. By reference to Eqs. (5.3-40) and (5.3-37), we see once again that this is determined by the Fourier transform over the electric potential (through $D = \varepsilon E$) of the transducer. Perhaps the most important feature of the Leedom et al. (1971) and Krimholtz (1972) models is that they make clear the steps that need to be followed in performing synthesis. In the above case, a procedure might take the following form.

(1) Specify the circuit (complete with matching elements) that will drive Y in Eq. (5.3-43). For example, the admittance Y may be series- or shunt-tuned by an inductor chosen to resonate with C at midband and the tuned admittance driven by a source with 50-Ω impedance.

(2) Specify the desired frequency response; make a guess at C; ignore B_1 and calculate the required $B_2^2 Z_0/2$.

(3) Take the inverse Fourier transform to find $f(x)$, the electrode excitation function.

(4) Modify $f(x)$ to obtain a physically achievable electrode pattern. For example, $f(x)$ must be truncated to a finite length; very small electrode overlaps for which overlap and excitation strength are not proportional may be adjusted according to the experience of the designer; and electrodes occurring near zeros of the baseband impulse response that are required to have extremely narrow widths (relative to the rest of the array) due to a rapid phase change may be omitted.

(5) Then calculate B_2, B_1, C, and the transfer function for comparison to the required frequency response.

Iteration and looping between the five steps is then required to optimize the transducer design.

As with the bulk-wave models of Fig. 5.3-7, the quantitative application of the Krimholtz (1972) model of Fig. 5.3-8 to IDTs is not straightforward. Again, the thickness t to use for Rayleigh waves is unknown. The IDT capacitance is most certainly not given by Eq. (5.3-42). (The correct expression for IDT capacitance of a single-electrode transducer is given by Eq. (5.3-6).] Rayleigh wave velocity is much more complicated than the formula $v = (s_{11}\rho)^{-1/2}$, and Rayleigh wave coupling to an IDT cannot even begin to be approximated by a single piezoelectric constant d_{31}, as was done here.

However, if we consider the formalism described by the model to be representative of the processes underlying Rayleigh wave transduction, then remarkably good results can be obtained. For example, if in the expression for input conductance, $B_2^2 Z_0/2$, we replace $d_{31}^2/(\varepsilon_{33}^T s_{11}^E) = k_{31}^2$ by 2 $\Delta v/v$, take an effective thickness of $t \simeq 2\lambda/3$, and assume $\varepsilon_p + \varepsilon_0 \simeq \varepsilon_{33}^T$, then $B_2^2 Z_0/2$ gives the same numerical value for radiation conductance as was obtained rigorously in Eq. (5.3-12).

5 PIEZOELECTRIC AND ELECTROMECHANICAL FILTERS

Krimholtz (1972) developed the model of Fig. 5.3-8 for application to dispersive SAW IDTs. With the bulk-wave definition of elements chosen by him, the circuit performance would be quantitatively different from the actual SAW device results; only relative device responses could be calculated. Any results critically dependent on the interplay between a matching network and the SAW impedance would not be well approximated. While Krimholtz (1972) presented no experimental results for SAW devices, the significance of his new circuit was noticed by Bahr and Lee (1973). What they saw as a significant new feature of the Krimholtz (1972) circuit was that the model explicitly allowed for variation of the electric field under an electrode along the propagation direction. It was well known (Engan, 1969) that the electric field was not constant under each electrode, yet all the preceding SAW excitation analyses based on bulk-wave models had treated it as a constant. It was also well known that while the constant-field models gave good results at the fundamental frequency, their accuracy at harmonics was considerably worse. Bahr and Lee (1973) proposed that the excitation function $E_3(x)$ in Eq. (5.3-36) be explicitly represented by the actual fields. To prove the merit of this approach they carried out an analysis for N-pair single-electrode IDTs in which $E_3(x)$ was represented by the exact calculation of Engan (1969) for a nonpiezoelectric, anisotropic dielectric. They derived the radiation conductance Q and effective coupling factor k_{eff}^2, for the odd harmonics of the transducer. From their bulk-wave analysis they found that

$$k_{\text{eff}}^2 = \frac{\pi^3}{4K(m)K'(m)} \frac{\Delta v}{v} P_n^2(\cos \pi a), \qquad (5.3\text{-}44)$$

where $n = (p - 1)/2$ and p refers to the harmonic number 1, 3, 5, ...

Consider Eq. (5.3-44) evaluated at the fundamental. In this case, $p = 1$ and $P_0 = 1$. The functional definition of k_{eff}^2 is then identical to that of Eq. (5.3-19) in Section 5.3-5. [If in the derivation of Eq. (5.3-9) for G_a we had retained those portions of the analysis responsible for harmonic operation, then Eq. (5.3-19) would have the same form as Eq. (5.3-44) at harmonics as well.] The effective coupling factor, defined in Eq. (5.3-44), was compared to experimental results for the first three harmonics as a function of metallization ratio (Bahr and Lee, 1973). Agreement between experimental results for (YZ) LiNbO$_3$ delay lines and theoretical predictions of Eq. (5.3-44) was quite good.

In a subsequent paper on the application of the Krimholtz (1972) circuit to SAW IDTs, Matthaei et al. (1975) recast the notation to be directly applicable to SAW device analysis. They maintained the main thrusts of Leedhom et al. (1971) and Krimholtz (1972); that is, they sought a simplification of transducer analysis in which the aggregate effect of all source elements was combined into a few frequency-dependent elements. They showed how to scale the source terms for apodized transducers and added an approximation to the effects caused by different transmission-line impedances in the electrode and gap regions. Their efforts continued to focus on fundamental-frequency operation with idealizations to the IDT electric field distribution, that is, flat-topped pulse sequences. They showed directly that transfer functions from electric to acoustic ports were dependent on the Fourier transform not of the pulse sequence but of the derivative of the pulse sequence. Of course this is the transform of a sequence of impulses positioned at the edges of the electrodes. Thus, by a quite different initial approach they had obtain the impulse response model of Tancrell and Holland (1971) and Tancrell (1974).

It is fair to say that the major objective of Leedhom et al. (1971), Krimholtz (1972), and Matthaei et al. (1975) was the simplification of the laborious analysis technique of cascading tens or hundreds of equivalent circuits to obtain a transducer response. This objective was generally coupled with approximations that limited accuracy. The use of idealized pulse-sequence source terms meant poor harmonic frequency modeling. The omission of impedance discontinuities between the electrode and gap regions meant inaccurate modeling of the reflected and regenerated acoustic waves.

The steps remaining in the development of bulk-wave models for SAW IDT operation were to add actual source field distributions for arbitrary metallization ratios and sequences as well as to quantify the element values. This step was taken by Smith and Pedler (1975, 1976). In contrast to the previous users of the Krimholtz model, Smith and Pedler (1975, 1976) had as their major objective the creation of a model capable of *quantitative* analysis even if it was numerically complex. Their concept was simple:

(1) use the Krimholtz (1972) model for each electrode and for each gap;
(2) in the excitation term [Eq. (5.3-36)], use a function that well approximated the actual normal electric field $E_3(x)$;
(3) develop appropriate frequency dependence and magnitude scaling such that the circuits gave quantitatively correct values for the IDT properties.

The electric fields were found by solving multiple electrostatic boundary value problems for the metallization ratios and sequences that are encountered in IDTs. The normal component of the electric field, $E_3(x)$, was expanded as a series of Chebyshev polynominals, and the coefficients of the expansion were tabulated as a function of the electrode environment in which a given electrode was embedded. Following Krimholtz (1972), Smith and Pedler (1975, 1976) could then define circuit elements for Fig. 5.3-8 as

$$\phi = j\gamma Z_0^{1/2} F_o(\gamma), \qquad (5.3\text{-}45)$$

$$B_2 = -\frac{\gamma}{Z_0^{1/2}} F_e(\gamma), \qquad (5.3\text{-}46)$$

5 PIEZOELECTRIC AND ELECTROMECHANICAL FILTERS

with

$$F(\gamma) \propto \int_{-\infty}^{\infty} E_3(x) e^{j\gamma x}\, dx. \qquad (5.3\text{-}47)$$

To make the model quantitatively accurate the correct proportionality function in Eq. (5.3-47) needed to be constructed.

This construction is carried out by reference to results derived from the normal-mode theory, which does have the correct frequency dependence. After substituting for the higher-level expressions contained in the formula for radiation conductance [Eq. (5.3-12)] one obtains

$$G_a = \omega \frac{\Delta v}{v}(\varepsilon_p + \varepsilon_o)\frac{2\pi}{A}\left|\int_{-\infty}^{\infty} e^{j\gamma x}\frac{E_3(x)}{V_c}\,dx\right|^2, \qquad (5.3\text{-}48)$$

where $A = [\varepsilon_{11}/\varepsilon_{33} - \varepsilon_{13}^2/\varepsilon_{33}^2]^{1/2}$ and V_c is the voltage applied across the IDT. Equation (5.3-48) is a completely general, quantitatively accurate expression for the radiation conductance due to a single metal stripe. To obtain a numerical representation for G_a it is necessary to take the Fourier transform of the actual self-consistent normal electric field that exists under an electrode.

For a single metal stripe radiating into a half space, the equivalent circuit of Fig. 5.3-8 would have acoustic terminations of Z_0. Then the electric port would exhibit a radiation conductance of

$$G_a = (\gamma^2/2)|F(\gamma)|^2. \qquad (5.3\text{-}49)$$

To obtain the proper proportionality for Eq. (5.3-47), it is necessary that Eqs. (5.3-48) and (5.3-49) yield the same answer. Thus, equating Eqs. (5.3-48) and (5.3-49), defining terms consistent with Smith and Pedler (1975, 1976), taking $E_3(x)$ to be the field under the ith electrode, we obtain

$$|F(\gamma)| = \left[\frac{4}{\pi}\frac{\gamma_0}{\gamma}f_i k_0^2 C_i\right]^{1/2}\left|\frac{2\pi}{A}\int_{-\infty}^{\infty}\frac{e^{j\gamma x}E_3(x)}{V_c/L_i}\,dx\right|, \qquad (5.3\text{-}50)$$

where the following conditions hold.

(1) L_i is a distance equal to the width of the ith electrode and the adjacent gap. (Thus for single-electrode transducers $L_i = \lambda_0/2$; for double electrode transducers $L_i = \lambda_0/4$.)
(2) γ_0 is a reference wave number equal to π/L_i.
(3) f_i is the frequency at which $2L_i$ equals one-half wavelength.
(4) k_0^2 is the electromechanical coupling constant (which equals $2\,\Delta v/v$ in this derivation).
(5) $C_i = w_i C_s/2$.
(6) w_i is the aperture dimension of the ith electrode.
(7) C_s is the static capacitance per finger pair per unit width of a single electrode transducer with 50% metallization.

Equation (5.3-50) is identical in form to that quoted by Smith and Pedler (1975, 1976) in their Eq. 13. The capacitance C_s used in Eq. (5.3-50) is the same whether single or double electrodes are considered. In addition to the Fourier transform, two terms that do change their numerical values as a function of single or double electrodes are γ_0 and f_i. Both f_i and γ_0 for double electrodes are twice as large as for single electrodes in a transducer with the same fundamental frequency.

The power of the Smith and Pedler (1975, 1976) model comes not only in treating harmonic responses quantitatively but in handling subtleties in the fundamental responses of IDT designs that by the nature of the weighting technique have significant end effects. Figures 5.3-9 and 5.3-10 illustrate the effectiveness of their model in predicting actual filter responses. In Fig. 5.3-9, results for a filter on (YZ) LiNbO$_3$ are shown. The filter had 10 double electrodes with an experimental metallization ratio of $\sim 56\%$. (The theoretical metallization ratio was 58%.) Note the quantitative agreement between experiment and theory in the insertion loss out to the 11th harmonic.

Even more impressive in illustrating the predictive power of the Fig. 5.3-8 model over the Fig. 5.3-7 models is the experiment–theory comparison of Fig. 5.3-10. Shown there are results for a phase-reversal transducer having 13 single electrodes. The dashed curve labeled "crossed-field model" (which used the Fig. 5.3-7 model) shows a response at the third harmonic, 90 MHz;

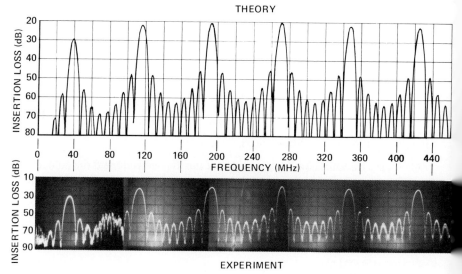

FIG. 5.3-9 Theoretical and experimental response of a (YZ) LiNbO$_3$ delay line with IDTs composed of 10 double electrodes. The theoretical model of Fig. 5.3-8 was used for each electrode and each gap. [From Smith and Pedler (1975).© 1975 IEEE.]

5 PIEZOELECTRIC AND ELECTROMECHANICAL FILTERS

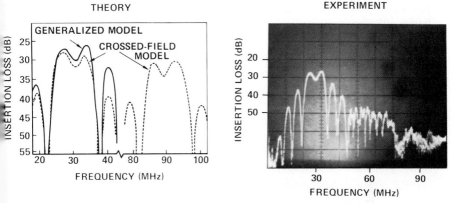

FIG. 5.3-10 Comparison of experimental insertion loss with theoretical predictions using the Fig. 5.3-8 element model and the Fig. 5.3-7 crossed-field model. The phase reversal transducers, on (YZ) LiNbO$_3$, had 13 single electrodes. The solid curve is for the Fig. 5.3-8 model while the dashed curve shows predictions from the crossed-field model. [From Smith and Pedler (1975). © 1975 IEEE.]

the experimental fact is that there was no third harmonic response. As Fig. 5.3-10 shows (curve labeled "generalized model"), the Fig. 5.3-8 model correctly predicts the filter response.

5.4 SAW BANDPASS AND BANDSTOP FILTERS

5.4.1 Introduction

An enormous volume of literature has been written on the design of SAW bandpass filters. Most of the literature addresses the uses of a bulk-wave model, the delta-function impulse model, and the sine-wave impulse model. Combinations, perturbations, and refinements of these approaches have been published, and each of these approaches has been made to work for specific classes of designs. However, every design procedure is not equally accurate in synthesizing all possible filter responses (Szabo et al., 1979). Generally, the more physical effects comprehended by the design procedure, the more accuracy can be achieved in the filter performance (and the more complex the design task). For example, in designing the electrical phase of a filter, electrically generated acoustic reflection from an IDT can usually be ignored if the insertion loss is set to a large value, say 25 dB. If large insertion losses are not tolerable, then more complicated design procedures must be employed.

Most SAW band pass filter synthesis begins in the frequency domain (rather than the time domain) with a specification of the amplitude and phase characteristics. The specification can be made analytically, but commonly it is specified by defining numerically or piecewise linearly the ranges within which the resultant filter characteristics must lie. Thus, there is not a unique frequency-response specification and not a unique impulse response. The problem for the SAW filter designer is to produce a *finite* impulse response, which can be implemented and which has a Fourier transform satisfying the boundary conditions specified on the frequency domain.

In fact, while the designer starts with the overall filter specification he must separate the response into two or more subresponses, the product of which gives the desired terminal properties of the filter. Each of these responses must then be translated into an IDT design. Approaches to subdividing the overall characteristics into constituent responses involve art, ingenuity, and experience. A simple approach is to let one IDT be an unapodized N-electrode-pair transducer and to apodize the other transducer. The frequency response required of the apodized transducer is obtained by dividing the overall filter response by that of the N-electrode-pair IDT, $[(\sin X)/X]^2$, where $X = N\pi(f - f_0)/f_0$.

Figure 5.4-1 shows a baseband[§] frequency response, $BB(f)$, which we take to be the numerically specified requirements for one IDT. The response shown is complex. What is specified is the amplitude ($= [(\text{real})^2 + (\text{imaginary})^2]^{1/2}$) and the deviation from constant time delay. Phase requirements for the filter are obtained by integrating the deviation from constant time delay:

$$\text{Phase} = \Phi(\omega) + K_1\omega + K_2. \quad (5.4\text{-}1)$$

Thus $t_d = K_1 + d\Phi/d\omega$, where K_1 is the constant time delay associated with wave propagation from one IDT to the other; $d\Phi/d\omega$ represents variation in the time delay and is a function almost exclusively of IDT design. The value K_1 is a function not only of the IDT design but of the absolute positions of the IDTs on the substrate; K_2, which is usually not specified, sets the absolute phase of the filters. It is a function of IDT design and placement, temperature, fabrication tolerances, and all second-order effects.

In describing the baseband phase, one commonly considers only $\Phi(\omega)$ and neglects the K_1 and K_2 terms, which are not solely a function of IDT design. Thus, with the phase term $\Phi(\omega)$ from Eq. (5.4-1) and the amplitude specification, a complex baseband frequency response $BB(f)$ can be defined. That characteristic is illustrated in Fig. 5.4-1a. The Fourier transform of

---

[§] The baseband is the spectral range occupied by the information containing waveform alone without an rf carrier.

5 PIEZOELECTRIC AND ELECTROMECHANICAL FILTERS

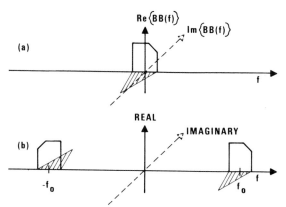

FIG. 5.4-1 (a) Asymmetric, and possibly dispersive, baseband frequency response specification for a SAW IDT. (b) Hermitian frequency response of a SAW IDT using the baseband specification of Fig. 5.4-1a.

BB(f) is given by

$$\int_{-\infty}^{\infty} BB(f)e^{-j2\pi ft}\,df = \frac{E(t)}{2}e^{-j\Phi(t)}, \qquad (5.4\text{-}2)$$

where $E(t)$ and $\Phi(t)$ are the time domain envelope and phase functions, respectively, defined by the Fourier transform [Eq. (5.4-2)]. Unless BB(f) is hermitian, the baseband impulse response will be complex, as Eq. (5.4-2) shows. Often the characteristics of BB(f) are far from hermitian. It is in realizing such complicated frequency responses that SAW technology demonstrates its power.

Consider now creating an IDT frequency specification such as is illustrated in Fig. 5.4-1b The baseband response is translated to the carrier frequency f_0, and its mirror-imaged complex conjugate is translated to $-f_0$. This frequency response, specified for all $-\infty \leq f \leq \infty$, is hermitian, and its Fourier transform is real. Thus

$$\int_{-\infty}^{\infty} S(f)e^{-j2\pi ft}\,df = E(t)\cos[2\pi f_0 t + \Phi(t)], \qquad (5.4\text{-}3)$$

where the following definitions hold.

(1) $S(f)$ is the spectrum of Fig. 5.4-1b constructed from the baseband response illustrated in Fig. 5.4-1a and provided as the initial IDT specifications.

(2) $E(t)$ is the real envelope function of the impulse response and is the same $E(t)$ appearing in Eq. (5.4-2).

(3) f_0 is a constant equal to the desired carrier frequency.

(4) $\Phi(t)$ is a real phase modulation of the impulse response. It is equal to the $\Phi(t)$ appearing in Eq. (5.4-2).

Thus, by producing a baseband specification for the filter and by carrying out Fourier transforms at baseband, the functions for creating a *real* impulse response at the carrier frequency can be realized.

It is important to note that the impulse response of a SAW IDT is a real function always. The construction process described above did not make the impulse response real. Instead the construction process is merely mathematically consistent with the physical reality that when an impulse of acoustic energy traverses an IDT, the voltage appearing on the terminals of the IDT has the real representation given by the right-hand side of Eq. (5.4-3).

The value of describing the bandpass characteristics in terms of the parameters of a baseband response is that the functions relating to the carrier-frequency impulse response can be obtained by fast Fourier transform (FFT) operations on the baseband specifications. This greatly reduces the number of points required to satisfy the sampling theorem.

5.4.2 Impulse-Response Realizations

The impulse response of Eq. (5.4-3) has an envelope function $E(t)$ that is slowly varying relative to $\cos(2\pi f_0 t)$. It also has a phase modulation term $\Phi(t)$ that is normally slowly varying relative to the carrier phase $2\pi f_0 t$. The phase term $\Phi(t)$ modulates the temporal positions of the peaks of the impulse response. With Φ the peaks occur at

$$t_{\text{peak}} = \frac{n}{2f_0} - \frac{\Phi(t_{\text{peak}})}{2\pi f_0}. \tag{5.4-4}$$

When designing IDTs according to impulse-response modeling, one would adjust the electrode positions to correspond to the impulse-response peaks given by Eq. (5.4-4). This type of placement is illustrated in Fig. 5.4-2.

The type of electrode positioning described above, where the electrodes are placed at the peaks of the rf impulse response, derives its analytic basis from the development of Section 5.3.3. There it was shown that the IDT radiation conductance was related to the Fourier transform of the *electrostatic* potential of an IDT metallization. After representing the electrostatic potential by the series solution [Eq. (5.3-5)] of Engan (1969) and evaluating the Fourier integral for $\omega \simeq \omega_0$, it was shown that only the fundamental component of the potential contributed significantly to the radiation conductance. The fundamental component of IDT potential is identified with the impulse response of Eq. (5.4-3). When the impulse response has phase modulation, that is, when $\Phi(t) \neq$ constant, one makes a physical correlation to Eq. (5.3-5) by considering λ_0 of Eq. (5.3-5) to be a function of position.

5 PIEZOELECTRIC AND ELECTROMECHANICAL FILTERS 261

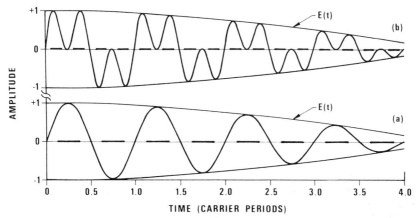

FIG. 5.4-2   Envelope function of Fig. 5.4-1a baseband response suitably sampled for (a) single-electrode transducers and (b) double-electrode transducers.

Some designers have been concerned about nonperiodic electrode placement. Boege *et al.* (1976) noted that automatic pattern generators require discrete step sizes, and Mitchell and Parker (1974) questioned whether or not the IDT response is rigorously correct at frequencies off synchronism. In an effort to obtain periodically sampled real impulse responses that are consistent with, and exploit, double-electrode embodiments, various decomposition and replication techniques have been developed. For example, Fig. 5.4-3 shows two replication techniques, both of which produce hermitian responses, which lead to impulse responses with twice the sampling rate of

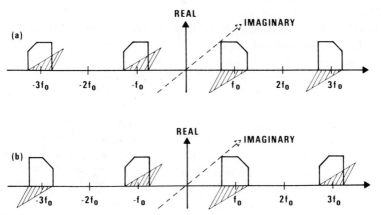

FIG. 5.4-3   Two baseband replication schemes that produce hermitian frequency responses and have impulse responses with two peaks per half cycle at the carrier frequency f_0.

the response in Eq. (5.4-3). The Fig. 5.4-3a spectrum has an impulse response

$$I(t) = E(t)\{\cos(\omega_0 t + \Phi)[1 + \cos(2\omega_0 t)] - \sin(\omega_0 t + \Phi) \sin(2\omega_0 t)\},$$

(5.4-5)

where $\omega_0 = 2\pi f_0$. Similarly, transformation of the Fig. 5.4-3b spectrum leads to the impulse response

$$I(t) = E(t)\{\cos(\omega_0 t + \Phi)[1 + \cos(4\omega_0 t)] + \sin(\omega_0 t + \Phi) \sin(4\omega_0 t)\}.$$

(5.4-6)

For the slowly varying functions $E(t)$ and $\Phi(t)$, both Eqs. (5.4-5) and (5.4-6) have peaks at twice the rate of Eq. (5.4-3). This is readily seen in Fig. 5.4-2b where Eq. (5.4-5) is shown for $\Phi = -\pi/2$. Note that there are two peaks per half cycle of the rf carrier ω_0 and that they are in phase with the carrier. The impulse response of Eq. (5.4-6) has a very similar appearance to that of Eq. (5.4-5). It can be shown that for $\Phi = -\pi/2$ in Eq. (5.4-5) and $\Phi = 5\pi/4$ in Eq. (5.4-6), the portions of the two equations in braces are identically equal if Eq. (5.4-5) is evaluated at $\omega_0 t$ while Eq. (5.4-6) is evaluated at $\omega_0 t + \pi/4$. Thus, the sampling peaks of Eq. (5.4-6) are shifted 45° relative to those of Eq. (5.4-5).

Since Eqs. (5.4-5) and (5.4-6) indicate sampling at four samples per wavelength, they are more physically consistent with double-electrode IDTs. One is able to weight each metallization of a double electrode independently, and the two impulse responses, Eqs. (5.4-5) and (5.4-6), give the required values.

Both impulse responses, Eq. (5.4-5) and (5.4-6), show that if a dispersive filter is required and the baseband response is not hermitian (i.e., $\phi \neq$ constant), then the impulse-response peaks will not be periodic. Nondispersive filters (Mitchell and Parker, 1974; Chao et al., 1975) and dispersive filters with hermitian baseband responses (Reilly et al., 1977) have been realized with uniform double-electrode placements. Boege et al. (1976) presented a discussion for dispersive filters that are not hermitian at baseband and yet have uniform double-electrode sampling. They used an imaging technique of the Fig. 5.4-3b type in which the carrier was taken as $2f_0$ and the response from $f = 0$ to $f = 4f_0$ was considered to be the desired baseband response. One can see from Fig. 5.4-3b that this larger baseband characteristic is hermitian. While one may uniformly sample an impulse response, such as Eq. (5.4-6), for a dispersive filter this will always lead to some electrode overlaps that are less than would have occurred if the electrode positions had been modulated to place the sample at an impulse-response peak. A greater difficulty with diffraction from the smaller time-domain sidelobes will likely result.

5 PIEZOELECTRIC AND ELECTROMECHANICAL FILTERS

Impulse-response concepts have been used to illustrate where the electrodes should be positioned along the crystal surface. Additionally, those electrodes so positioned must provide weighted samples of voltage that are proportional to the amplitude of the impulse-response peaks. Most commonly, this is accomplished in one IDT by apodization (Tancrell, 1974). It has also been achieved with capacitive voltage dividers (Malocha and Wilkus, 1978; Sato et al., 1974) and with binary tap-weight approximations. The latter technique is referred to in the literature as electrode withdrawal (Hartmann, 1973; Laker et al., 1978; Wagers, 1978).

Once a design has been synthesized from the bandpass characteristics, the next step is not to fabricate the device but to analyze the structure. Second-order effects are much easier to comprehend in analysis than in synthesis. Thus, an analysis procedure such as the bulk-wave models described in Sections 5.3.7 and 5.3.8 is applied to the IDT design including terminating impedances. Departures in the predicted response from the required characteristics are corrected either by designer action or automatically within the analysis loop. When the analysis procedure indicates the design is acceptable, photomasks for the filter are generated and filter tests conducted. If the tested filter does not meet the desired specifications, then one approach to design iteration is to take the measured bandpass data, FFT the data to obtain the effective impulse response of the filter IDTs, correlate the desired impulse response (and metallization) to the achieved impulse response, and make perturbations to the design. This can be a convergent procedure (Savage and Matthaei, 1979; Savage, 1980).

5.4.3 SAW Bandpass Filter Capabilities

Three types of SAW filters well illustrate the capabilities of this technology. They are

(1) hermitian-baseband, extreme-rejection, two-phase[§] SAW filters;
(2) hermitian-baseband, low-loss, three-phase[¶] SAW filters; and
(3) asymmetric-amplitude, nonlinear-group-delay, television IF SAW filters.

An example of the first category of filters is shown in Fig. 5.4-4a (Hays and Hartmann, 1976). This filter employed three two-phase transducers on ST-cut quartz. The packaged filter was smaller than a dime. Having only 10-dB

[§] Two-phase filters employ IDTs requiring two voltage phases. For example, $V = 0$ and $V = V_0$ can be applied to the two bond pads. Also $V = V_0 \exp[j\Phi]$ and $V = -V_0 \exp[j\Phi]$ can be applied.
[¶] Three-phase filters employ IDTs requiring three voltage phases. Commonly these IDTs are driven by a voltage set $V_0 \exp[j\Phi]$, $V_0 \exp[j(\Phi + 120°)]$, and $V_0 \exp[j(\Phi + 240°)]$.

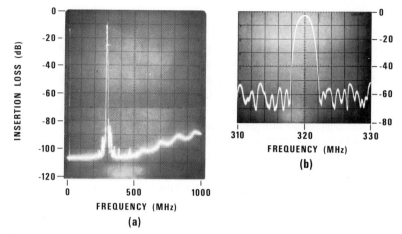

FIG. 5.4-4 SAW filter frequency responses illustrating exceptional capabilities. (a) The sidelobes are more than 70 dB down from dc to 1 GHz. [Adapted from Hays and Hartmann (1976). © 1976 IEEE.] (b) The insertion loss is only 2.3 dB. [Adapted from Potter and Hartmann (1977). © 1977 IEEE.]

insertion loss at 287 MHz, the filter had both electrode withdrawal and apodization weighting. A central IDT had constant-overlap electrode withdrawal weighting applied to it; output IDTs on either side of the central IDT were apodized. The output IDTs were bonded together to form a common output port. As can be seen from Fig. 5.4-4a, out-of-band rejection of greater than 70 dB was obtained from dc to >1 GHz.

By employing a three-transducer configuration, the 3-dB bidirectionality loss of the central transducer could be avoided. Another reason for the three-transducer configuration was to achieve phase linearity in a relatively low-loss two-phase filter. In a two-transducer two-phase filter, if the insertion loss is as low as 10 dB, acoustic reflections between the IDTs cause phase deviation in the output signal as the direct signal beats with the signal that experiences acoustic reflection at each IDT and transverses the delay line three times. The reflected signal, referred to as *triple transit echo*, has three times the phase slope of the direct signal. Phase linearity in a three-transducer filter is achieved by balancing against one another different signals each of which has three times the phase slope of the direct signal. The Fig. 5.4-4a filter exhibited a midband phase response within $\pm 2°$ of linear over the range $0°C \leq T \leq 50°C$ (Hays and Hartmann, 1976).

For the filter of Fig. 5.4-4a, insertion loss, phase linearity, and out-of-band rejection were all key issues in determining the filter configuration. In that particular application, relatively large insertion loss was acceptable, while the ultimate performance in out-of-band rejection was essential. In many applications, however, insertion loss must be maintained at the lowest possible level. In such cases multiphase SAW IDTs are employed. Three-phase unidirectional IDTs are often used (Potter and Hartmann, 1977). While these filters require the added system complexity of three voltage phases (as opposed to one phase relative to ground for the two-phase IDTs), they provide the lowest insertion losses possible in SAW transversal filters. This capability is illustrated in Fig. 5.4-4b for a filter with only 2.3 dB of loss at 320 MHz (Potter and Hartmann, 1977). This is a phenomenal performance. No other electronic component could have achieved the same insertion loss with the same out-of-band rejection in as small a component size.

Note that while the out-of-band rejection in Fig. 5.4-4b is excellent ($>$ 50 dB), it is less than has been demonstrated with single-filter two-phase transducer technology. On the other hand, it is clear that one could cascade two filters of the Fig. 5.4-4b type and achieve better out-of-band rejection than obtained in Fig. 5.4-4a filter while still incurring only $\sim$ 5-dB insertion loss. Again the question of whether to build one two-phase filter, one three-phase filter, or to cascade several, etc., is determined by the system-dictated rank ordering of out-of-band rejection, insertion loss, and phase linearity.

One filter that combines the most difficult of all parameters is the television IF filter. Not only must the passband amplitude response be asymmetric with prescribed shelves, but the phase response must produce the correct nonlinear time delay, and deep rejection notches must be imbedded adjacent to the general out-of-band rejection regions. If possible it would also be desirable to achieve zero insertion loss at midband and require no matching components. No SAW filter can meet all these objectives, but the technology has met the essential one. Many different solutions have been found. Substrates in commercial use for television IF filters now range from zinc oxide on glass (Fujishima *et al.*, 1979) to $LiTaO_3$ (Takahashi *et al.*, 1978) to several cuts of $LiNbO_3$ (Hazama *et al.*, 1978; Komatsu and Yanagisawa, 1977; DeVries and Adler, 1976). All the commercial metallization schemes use two-phase transducers, and apodization weighting is generally employed. In general those filters employing multistrip couplers (MSC) did so to eliminate BAW interference with high-frequency traps. They also tend to be found on substrates of (YZ) $LiNbO_3$ (DeVries and Adler, 1976), although Komatsu and Yanagisawa (1977) achieved acceptable bulk-mode levels for their purposes with (YZ) $LiNbO_3$ and no MSCs by attaching the substrate to the header with a metallic bond.

Figure 5.4-5 shows the IF characteristics of a SAW filter and associated drive electronics developed by Hitachi (Hazama *et al.*, 1978). In this case the group delay is essentially flat across the band. Their SAW filters had one unapodized IDT and one apodized IDT in line with one another. Like

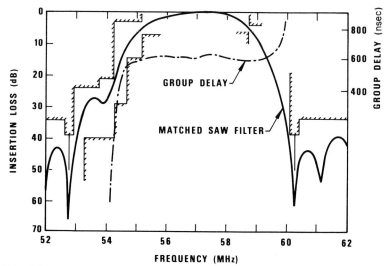

FIG. 5.4-5 SAW television IF filter response. [Adapted from Hazama et al. (1978). © 1978 IEEE.] Note the extreme asymmetry possible in the amplitude response while the group delay is held constant.

most of the successful developers of these filters they chose to eliminate the bulk waves by selecting a superior material, 128° Y-cut LiNbO$_3$ (Shibayama et al., 1976) for the substrate rather than using an MSC. Those designs not requiring MSCs achieve more than twice as many filters per 2-in. substrate wafer. Hazama et al. (1978) quote a wafer density of 100 filters per 2-in. substrate and a production volume in 1978 of 100,000 filters per month.

5.4.4 SAW Bandstop Filters

Much less effort has gone into the development of SAW-based notch filter technology. Perhaps this is because most receiver architectures are designed to avoid the need for notch filters. The few notch filter efforts found in the literature can be divided into two groups: those based on acoustic resonators and those based on transversal acoustic devices. Representative of possible resonator-dependent circuits is the configuration shown in Fig. 5.4-6a, where a SAW resonator is placed in the transformer secondary with a load resistor R_L and a balance resistor R_B. If R_B is equal to the series resistance of the SAW resonator at resonance, then the current through the load resistor will vanish producing the desired transmission notch. Also, by designing the SAW resonator (Section 2.3) such that the resistance at resonance is

$2R_L$, the attenuation outside the stopband notch can be as little as 6 dB (theoretical minimum). Using (yz) $LiNbO_3$ as a substrate, a 50-dB notch has been obtained at 70 MHz by Akitt (1976). At 10 kHz from the notch frequency the transmission was back up to within ~ 3 dB of the pass-band response. The passband insertion loss of 7 dB was within 1 dB of the theoretical value. By loading the resonator with shunt capacitance they found that the notch frequency could be trimmed downward as much as 40 kHz. While no information was provided on temperature sensitivity of the filter, (yz) $LiNbO_3$ is known to be highly temperature sensitive. Section 8.2.3.3 (Volume 2) discusses temperature sensitivity of quartz-based SAW resonators.

Notch filter techniques based on SAW IDTs can be divided into two groups: either the IDT has been treated as an impedance element, or various versions of SAW delay-line coupling to other signal paths have been attempted. Two kinds of interferometers employing SAW delay lines are possible:

(1) the SAW delay line signal is allowed to beat with another delay line signal (Dieulesaint and Hartemann, 1973), or

(2) the SAW delay line signal is allowed to beat with a signal transmitted through a conventional circuit path (Plass, 1973).

In the former approach both delay lines can be fabricated on the same substrate and the design compensated such that balanced outputs from the two delay lines phase to produce strong nulls. The broadband response of such of an interferometer has many transmission zeros. The separation between zeros equals the reciprocal of the time-delay difference of the two lines. With this approach the insertion loss away from the rejection frequencies is large, being equal to the SAW delay-line insertion loss. Dieulesaint and Hartemann (1973) reported results at 100 MHz with a zero separation of 3.3 MHz.

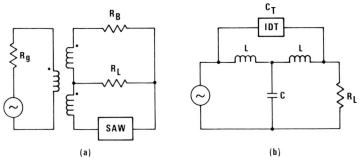

FIG. 5.4-6 Two circuits used for bandstop filters. Circuit (a) employs a SAW resonator. Circuit (b) uses a single IDT as an impedance element, where $C_T \gg C$ and $(2\pi f_{\text{notch}})^2 = 1/LC$.

An interferometer in which the reference path is derived from a conventional RLC circuit operates on the same principles as the two-delay-line approach. Depending on one's requirements it could be viewed as having either a deficiency or an advantage over the two-delay-line approach. If fixed component RLCs were used for the reference path, then each interferometer built would have to be hand tuned to position the desired notches. This, of course, is a costly fabrication process. On the other hand, if the RLC path is adaptive with say a variable capacitance, then perhaps a lower frequency notch could be locked to a more temperature-stable frequency standard. Also, if the desired rejection frequency is not constant, then the interferometer can be made adaptive or be operator controlled. A consideration of interferometers of the RLC reference path type as applied to European car telephones (455–470 MHz) was presented by Plass (1973).

By far the best results obtained for SAW bandstop filters are those where the SAW IDT is used as an impedance element. In these cases (Lakin et al., 1974; Koyamada et al., 1975; Ishihara et al., 1975) the number of electrodes of the IDT is made large enough to make the IDT become self-resonant at frequencies just above synchronous radiation of Rayleigh waves.

Remarkable results have been obtained with the self-resonant IDT as an impedance element replacing conventional capacitors in RLC circuits. This is illustrated in Fig. 5.4-7 (Koyamada et al., 1975) where an 80-dB-deep, 60-kHz-wide notch is shown at 153 MHz. Outside the stop band, the insertion loss was < 1 dB from dc to 500 MHz. The circuit employed four sections of the type illustrated in Fig. 5.4-6b with LiNbO$_3$ as the SAW substrate. The circuit of Fig. 5.4-6b is basically an all-pass circuit except when the IDT impedance becomes $1/j\omega C$ (real part equal to zero). This condition obtains at a frequency above synchronism where the radiation conductance has gone to zero and the acoustic reactance has swung sufficiently inductive

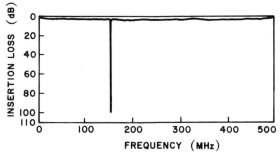

FIG. 5.4-7 Bandstop filter characteristics of a circuit made by cascading four of the Fig. 5.4-6b circuits. The crystal was 128° Y-cut, X-propagation LiNbO$_3$. [Adapted from Koyamada et al. (1975).]

5 PIEZOELECTRIC AND ELECTROMECHANICAL FILTERS   269

to lower the effective capacitance of the IDT to C. The susceptance of the IDT can also be reduced to ωC at a lower frequency that is just above synchronism, but the real part of the radiation conductance is not zero at that frequency.

The approach illustrated in Fig. 5.4-6b was extended by Ishihara *et al.* (1975) to SAW IDTs on ST-cut quartz. A 45-dB notch (15 kHz wide) was obtained at 153 MHz with a pass-band loss of less than 1.2 dB. The transition bandwidth for this filter was ~ 1 MHz. While the frequency response characteristics of the Ishihara *et al.* (1975) filters are somewhat less impressive than those obtained with LiNbO$_3$ substrates (Koyamada *et al.*, 1975), the temperature characteristics are outstanding. The ST-cut quartz-based filter exhibited total frequency deviation of ± 125 ppm over the range $-20°C \leq T \leq 80°C$, whereas the LiNbO$_3$-based filters exhibited a temperature sensitivity of ~ 70 ppm/°C.

6 Long-Term Stability and Aging of Resonators

Eduard A. Gerber

U.S. Army Electronics Technology and Devices Laboratory
Fort Monmouth, New Jersey

6.1	Low-Frequency Bulk-Wave Devices	271
6.2	High-Frequency Bulk-Wave Devices	273
	6.2.1 Causes of Aging	273
	6.2.2 Progress through Holder Design	274
	6.2.3 Progress through Mounting and Crystal Plate Design	275
	6.2.4 Isolation of Aging Causes	277
	6.2.5 Influence of Temperature	279
	6.2.6 Influence of Radiation	279
6.3	Surface-Wave Devices	279
	6.3.1 SAW Resonators	280
	6.3.2 SAW Delay Lines	283

The change in frequency of quartz crystal units with time, called aging or long-term drift, has received much attention and accounts for a great deal of the development effort on improving stability. It therefore merits treatment in a separate chapter. However, short-term frequency changes (with a sampling time of a few seconds or less) are generally caused by crystal and oscillator (Gapnepain, 1976) and are discussed in Section 8.3. Great strides have been made in the past years to isolate various physical and mechanical processes that contribute to aging of high-frequency thickness-shear resonators and to develop crystal units with improved frequency stability.

6.1 LOW-FREQUENCY BULK-WAVE DEVICES

Wire-mounted low-frequency types, however, received relatively little attention until recently, when the advent of quartz resonators for wrist watches provided impetus for new research into low-frequency resonators having low aging rates. The first improvements in low-frequency types were noted in measurements of width-shear resonators (Gerber and Sykes, 1966). The apparent reason for improved long-term stability of this type of wire-mounted crystal unit is that the support and electrical connection cover only a part of the nodal area of the plate; consequently, less dissipation and influence is produced. In other types of extensional modes and square face-shear types, the node is a point, and thus the energy loss and long-term strain relaxation at the connection is greater. This probably helps to explain why some flexure types of crystal units have low aging rates. The suspension system connected to the nodal points is subject to rotary motion instead of compression and extension. While none of the present low-frequency types, including flexure or width-shear types, possess the low drift rates obtained in high-frequency thickness-shear types, it is probable that many improvements can be made by employing mounting systems that result in less dissipation and lower strain. Figure 6-1 shows typical aging rates for low-frequency-type crystal units produced under careful process control and enclosed in the cold-weld holders described elsewhere in this book. It is apparent from these data that of the low-frequency types, the width-shear resonator exhibits the lowest aging rate. Flexure resonators are next, and the square face-shear as well as extensional units have the highest aging rates. Lower frequency or more massive plates result in lower aging

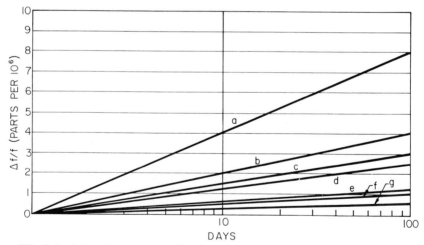

FIG. 6-1 Aging characteristics of low-frequency wire-mounted crystal units. a, 200 kHz E type, extensional mode; b, 200 kHz C type, face-shear mode; c, 100 kHz E type, extensional mode; d, 990 kHz D type, $w/e = 0.4$, width-shear mode; e, 8 kHz N or K type, flexure mode; f, 550 kHz D type, $w/e = 0.4$, width-shear mode; g, 230 kHz D type, $w/e = 0.4$, width-shear mode.

6 LONG-TERM STABILITY AND AGING OF RESONATORS

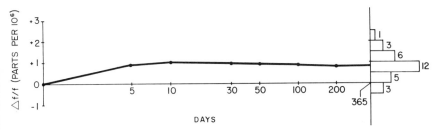

FIG. 6-2 Typical aging curve and histogram of 32.768-kHz tuning fork crystal units, number of samples: 30.

rates. This is as would be expected: A given mounting system will have less effect if it represents a smaller amount of the total vibrating system. The figure shows the aging of crystal units to be linear with logarithmic time. One would expect the rate to decrease over long periods of continuous operation (Gerber and Sykes, 1966).

Tuning fork and other low-frequency watch types now being developed using numerical techniques, such as finite element analysis, promise to have low aging rates similar to N and K types shown in Fig. 6-1. Figure 6-2 shows a typical aging curve of a 32.768-kHz quartz tuning fork unit, together with a histogram of frequency change distribution of 30 specimens measured at room temperature for one year (Kanbayashi *et al.*, 1976). Similar results are reported by other authors (Engdahl and Matthey 1975; Yoda *et al.*, 1972; Forrer, 1969).

6.2 HIGH-FREQUENCY BULK-WAVE DEVICES

The aging rates of high-frequency thickness-shear resonators, such as the AT and BT types (refer to Section 2.2 and Chapter 4) and particularly those of the so-called precision type, have been reduced during the past few years to exceptionally low rates.

Their frequency–time performance seems to be divided into two distinct parts: (1) an initial stabilization period during which there may be frequency changes of as much as several parts per 10^8 for a period of one to five weeks and (2) a much slower drift rate in which the total frequency change may be the order of 1 to 5 parts per 10^{10} per month. The initial aging particularly is highly process dependent.

6.2.1 Causes of Aging

Aging of thickness-shear crystal units is caused mainly by four processes (Gerber and Sykes, 1966; Vig, 1977):

(1) changes in strains due to temperature gradients and to stress relief in the mounting clips, bonding agents, electrodes, and quartz; the electrode stresses are functions of the metal used, method of deposition (e.g., evaporation, sputtering, electroplating), substrate cleanliness, and temperature during deposition, as well as electrode thickness;

(2) changes in mass loading due to adsorption and desorption of contamination; it is interesting to observe that for bulk-wave resonators, if contamination equal in mass to $1\frac{1}{2}$ monolayers of quartz is adsorbed or desorbed from the surfaces, then the frequency change in parts per million is equal to the resonator's frequency in megahertz. For example, the frequency of a 5 MHz crystal changes by 5 ppm;

(3) changes in materials, such as electrode diffusion and recrystallization, reactions at the electrode–quartz interface, diffusion of impurities, structural changes in quartz due to imperfections in the crystal lattice, and radiation effects; and

(4) other effects, such as changes in hydrostatic pressure due to leaks, and static charge decay.

6.2.2 Progress through Holder Design

Some general statements may be made, first about the aging of the general-purpose, high-frequency crystal units, governed by the type of construction. In solder-sealed metal holders, aging rates could be as high as 5 ppm per month for the first year and as low as 1 to 2 ppm per year for the first two years. The high value represents lack of process control, poor design of the mounting system with high strain, and excess contamination through improper solder sealing of the enclosure. The lower value represents what can be done with good design and careful control even in solder-sealed metal

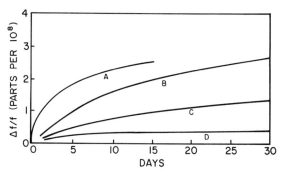

FIG. 6-3 Aging of metal and glass enclosed 5-MHz crystal units. A, Metal-enclosed units; B, glass-enclosed units; C, gettered-glass-enclosed units; D, high-temperature bonded-metal-enclosed units.

holders. The use of glass holders brought improvements through the necessity of using a high-temperature process during sealing. Also, glass is easier to clean by conventional techniques, which results in aging rates often comparable with the best for metal holders. Cold-welded metal holders, together with mouting systems that will allow high-temperature bakeout prior to sealing, have yielded additional improvement in aging. Figure 6-3 illustrates what has been done in recent years (including the addition of getters) for a particular case (Byrne and Reynolds, 1964). Most recently, hermetic enclosures fashioned of alumina ceramic have been used with excellent success (Wilcox *et al.*, 1975); in contrast with glass, these are impervious to He, withstand high-temperature processing, and have the form of IC-compatible flat packs.

6.2.3 Progress through Mounting and Crystal Plate Design

Frequency changes that are caused by either adsorption and desorption of gases or by strains set up between the crystal and its electrodes are illustrated in Fig. 6-4, which shows the effect of shutting off the oven and stopping

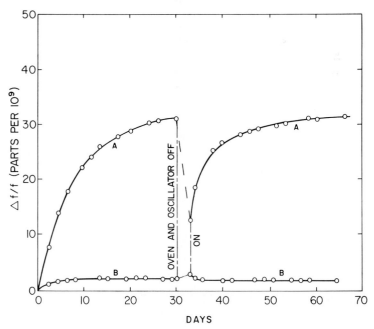

FIG. 6-4 Frequency change in precision quartz crystal units due to stopping quartz plate vibrations and to interruption of oven control. A, solder bonding, glass encapsulation; B, thermocompression bonding, high-temperature processing.

the quartz plate vibration for three days (Armstrong et al., 1966). As can be seen, recovery from temperature control or power supply interruptions can be greatly improved by using a mounting system for the quartz plate that may be vacuum-baked and by using the cold-welded metal enclosures mentioned previously. The mounting system to support the quartz resonator for the metal enclosures makes use of higher-temperature-bonding alloys than the glass units, so the complete unit may be vacuum-baked in an oil-free system and then cold-welded while under vacuum. This results in less contamination and strain in the mounting, with a consequent shortening of the initial stabilization time. Figure 6-5 shows the effect of small abrupt temperature changes for perpendicular and for lateral field resonators (Warner, 1963). It can be seen that the transient frequency excursion due to the 1°C thermal shock is decreased by more than an order of magnitude in the lateral field resonator, the reason probably being that no strain can be set up between the active part of the crystal and the electrode since the center of the vibrator is free of any metal plating. Another method of minimizing the effects of electrode stress relaxation on the aging of high-precision thickness-shear resonators is the use of doubly rotated quartz plates known as SC-cuts (refer to Chapter 2, Section 2 and Chapter 4). They simultaneously permit control of thermal gradient effects, and their introduction (EerNisse, 1975; Ballato, 1977; Ballato et al., 1978) promises to eliminate the major part of the aging components due to electrode and mounting stress relaxation.

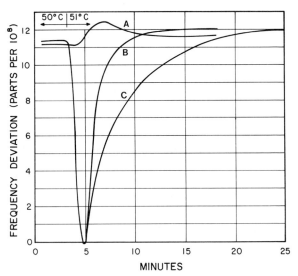

FIG. 6-5 Change in frequency of perpendicular- and lateral-field crystal vibrators due to a 1°C change in temperature. A, 5 MHz, $n = 1$, lateral field; B, 5 MHz, $n = 1$, perpendicular field; C, 5 MHz, $n = 5$, perpendicular field.

The effects of edge forces applied to the plate by the mounting supports can also be largely eliminated by using crystals of rhomboid geometry, with the orientation of the rhomboid sides determined by the particular cut being used (Lukaszek and Ballato, 1979). All detrimental effects of the coating on stability can be avoided by reviving (with the so-called B.V.A[§] design) uncoated resonators mounted between airgap electrodes (Bechmann, 1942; Besson, 1976). Preliminary stability measurements look promising.

6.2.4 Isolation of Aging Causes

During the past few years several authors have tried to isolate the various causes for the aging mentioned above, learn to control them, and thus come up with precision vibrators with a very low and reproducible aging rate. An indication of the long-range stability of a quartz resonator can be obtained by observing (Belser and Hicklin, 1969; Byrne and Hokanson, 1968; Dick and Silver, 1970; Dybwad, 1977; Hafner and Blewer, 1968a,b)

(1) the length of the initial stabilization period,
(2) the magnitude of transient effects due to power interruption or temperature cycling, and
(3) the fit of the aging curve to mathematical models describing various rate processes.

The best results were obtained when all fabrication processes were performed under the same very high, uninterrupted vacuum. In this case the aging of aluminum-plated 62-MHz fifth-overtone quartz crystals followed the equation for a one-rate process (Hafner and Blewer, 1968a,b):

$$\Delta f/f = k \ln(1 + t/T), \qquad (6\text{-}1)$$

where t is time and k and T are constants. It appeared likely, since the free energy of formation for aluminum oxide is lower than that for silicon dioxide, that the quartz lattice could serve as a source of oxygen for the atomic aluminum film and that this process would be related to the observed aging. In contrast to aluminum, copper electrodes yielded consistently much better results with aging rate reduced to $\leq 5 \times 10^{-10}$ per week (see Fig. 6-6). This was to be expected since copper attracts oxygen less strongly than does the silicon of quartz. It could also be possible that the main aging-reducing factor of copper electrodes is that compared with other metals they possess a small intrinsic stress (Hoffman, 1974). Copper also has less tendency to diffuse into quartz than gold, silver, or aluminum, thus reducing this reason

---

[§] B. V. A. stands for "inside a box with lower aging," *en boîtier à vieillissement amélioré*.

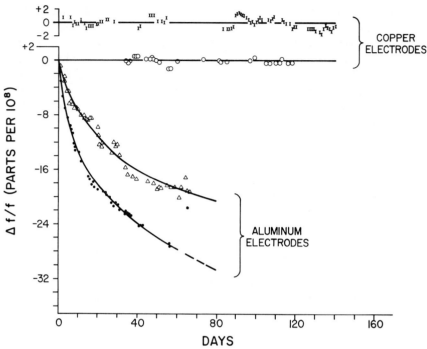

FIG. 6-6 Comparison of aging behavior of aluminum- and copper-plated crystal units.

for instability (Belser and Hicklin, 1969). Obviously, 5-MHz fundamental-mode cystal units fabricated in the same way are expected to be significantly more stable since strain and mass transfer effects influence the resonance frequency inversely proportionally to crystal thickness. This is borne out by measurements on 2.5- and 5-MHz fifth-overtone planoconvex gold-plated quartz crystals that were also fabricated by similar very careful methods (Byrne and Hokanson 1968). In this case, the length of the initial stabilization period and the recovery time after power interruption was less than one week and the aging rate $1-3 \times 10^{-10}$/month. It is assumed that due to the applied hydrogen anneal followed by a vacuum bake, a stable mass loading on the quartz plate is achieved, which in this case is apparently the principal reason for the low aging rate.

A study of the effects of impurities in quartz along with an examination of the role of sorption phenomena and thermally induced strains made use of three crystal vibrators in a common vacuum system to differentiate thermal and mass effects (Warner et al., 1965). Evidence of the role of impurities, particularly alkali ions, was noted. The presence of carbon monoxide also decreases frequency stability. This effect may be diminished by using the lateral-field vibrator.

It has been shown experimentally that strain, mass transfer, and other effects may by chance mutually compensate each other to a certain extent to produce the low aging rates observed by several authors (Sykes et al., 1963). However, as the fabrication processes are better and better controlled and the various causes for aging better recognized and isolated, the likelihood for compensation by chance becomes rather remote. On the contrary, one can expect by extrapolating the progress made to date that 2.5- and 5-MHz quartz crystal standards will become competitive with Rubidium and Cesium gas-cell standards as far as stability is concerned and will additionally have much reduced weight, size, and cost.

6.2.5 Influence of Temperature

Aging is influenced to a large extent by temperature. The dependence of the aging rate on temperature is affected by the processes used to fabricate the resonator. Nearly all aging rates due to the various aging mechanisms have an exponential dependence on temperature, albeit with very different values of the constants in the rate equations. Therefore, care must be taken in interpreting the results of accelerated aging tests conducted at high temperatures. On the other hand, aging can be substantially reduced if the resonator is kept at very low temperatures. An instability of only 4×10^{-14} over a period of 100 sec was observed (Smagin, 1975) when a 1-MHz quartz resonator was kept at liquid He temperature. However, crystals held at this low temperature proved to be very sensitive to shock (Simpson and Morgan, 1959).

6.2.6 Influence of Radiation

The exposure to combined neutron and gamma radiation (Bloch and Denman, 1974) or to ultraviolet light and ozone (Vig et al., 1975) also appears to reduce the aging rate of crystal plates. The UV–ozone diminishes aging because it oxidizes hydrocarbons from the surface, producing CO_2 and a stable carbonaceous ash on the surface.

6.3 SURFACE-WAVE DEVICES

Aging of surface-acoustic-wave (SAW) devices used for frequency control purposes is in many ways similar to aging of bulk-wave resonators. However, there are some important differences. First, the physical size of the crystal does not affect the frequency of a SAW device. Therefore, any foreign material adsorbed on the surface does not decrease the frequency due to an increase in thickness but may change the frequency either upward or downward due to a modification of the acoustic properties of the material. Second, SAW devices operate at much higher frequencies than bulk resonators and are therefore more sensitive against all surface-disturbing effects. Also, compared with standard quartz bulk-wave resonators, the fabrication methos for SAW devices are substantially different and, consequently, the processes causing aging may affect them in a different way. Since SAW frequency control devices have come into being rather recently, only a limited amount of information on the stability of these devices is now available. However, progress has been made recently to explain and control the various processes that lead to aging of SAW devices. The hope is justified that, following the lead of bulk devices (Lukaszek and Ballato, 1980), a final solution of this complicated and vexing problem may be forthcoming in the near future.

6.3.1 SAW Resonators

The aging of 184-MHz plasma-etched SAW resonators was measured for 48 days at 250°C (Bell and Miller, 1976; Bell, 1977). Thus, aging was accelerated to minimize the time required to determine the effect of the most important processes and their influencing on room-temperature aging. Surface preparation, cleaning and lithographic processes, storage, and package were the variables chosen. Substantial aging occurred in devices on substrates polished without postpolish etch. The best results were obtained with unsealed packages, which allowed outgassing products to escape (see Fig. 6-7). This result looks unusual when compared with the experience on

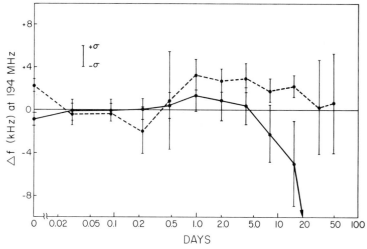

FIG. 6-7 Isothermal aging of 184-MHz SAW resonators at 250°C. Solid curve, post-polish etch; dashed curve, unsealed can.

precision bulk-wave resonators. It shows clearly that more effort is necessary to control and eliminate the sources of contaminations during packaging. Rates of less than 2 ppm per doubling of time were obtained over the entire range of times to 48 days at 250°C for devices aged in air. But the correct acceleration factor is not yet known. There is little or no difference between aging at room temperature, 50, 100, or 150°C. The aging rate increases, however, at 200 and 250°C. This is concluded from extrapolated results on plasma-etched two-port quartz resonators with a frequency of 194 MHz (Shreve, 1977).

Aging results at lower temperatures were obtained on 160-MHz one-port resonators made with etched groove arrays and aluminum transducers on rotated Y-cut quartz substrates (Shreve et al., 1978; Adams and Kusters, 1978). They were fabricated and packaged using proven bulk-wave resonator techniques, such as chemical and ultrasonic cleaning, vacuum baking, and mounting in cold-weld containers. The long-term aging rates approached 1 ppb/day. This is still 2 orders of magnitude above the aging rate of the best 2.5- and 5-MHz bulk-wave quartz resonators, but taking into account the much higher frequency and the relatively new SAW technology, it must be considered a very significant result. There is not much difference between aging rates of devices in a copper–ceramic and a KOVAR TO-style header; the best rates measured over a period of more than one year are -0.064 ppm/year and -0.31 ppm/year, respectively. These tests were also run with 160-MHz one-port SAW resonators made on quartz substrates with etched groove reflector arrays and aluminum transducers (Shreve, 1980).

One reason for measuring aging rate of resonators is to be able to predict future performance with some degree of confidence. For this purpose, extrapolation of shorter tests is required. The accuracy of this extrapolation is affected by both the quality of the model and the amount of time beyond the end of the aging test over which one extrapolates. The best fit for the data measured on 160-MHz SAW resonators was obtained with two logarithmic processes (Shreve et al., 1978), in contrast to only one process needed for 62-MHz bulk devices discussed above and to more than two processes needed in other cases. Table 6-1 shows the results of an extrapolation test by comparing the measured aging over different periods with the calculated aging. Extrapolation from 29 days of data does not give good predictions for the aging at 130 days or beyond. The data from 57 days, however, resulted in a prediction for 130 days only 4% from the measured value for 130 days. If a similar accuracy can be expected for the 1 year prediction based on 160 days of testing, then the expected frequency change after 1 year should be 1.04 ± 0.04 ppm. If only one aging process is involved, then one can extrapolate even farther out compared with the results for two processes (Shreve, 1982).

TABLE 6-1

Extrapolation Accuracy for Different Aging Periods[a]

Days aged	Frequency change (ppm), measured	Frequency change (ppm), calculated		
		130 days	1 year	5 years
29	1.98	−2.67	−18.6	−73.5
57	2.31	2.07	1.26	−0.6
130	2.01	1.99	0.97	−1.4
162	1.80	2.04	1.04	−1.4

[a] From Shreve et al. (1978). © 1978 IEEE.

Aging measurements on 300 MHz two-port SAW resonators, also with plasma etched grooves, point to similar aging rates (Latham, Saunders 1978). The units were also properly cleaned, baked, and vacuum-sealed. The results confirm the large influence of the various mounting methods and adhesives on stability (see, for example, Fig. 6-8). If polyimide adhesives or gold–tin solders are used for mounting the resonators, then aging rates of 1 to 2 ppm/year can be expected.

The aging of SAW resonators can change as the operating power is varied (Shreve et al., 1981). Increasing the drive level in resonators with pure Al

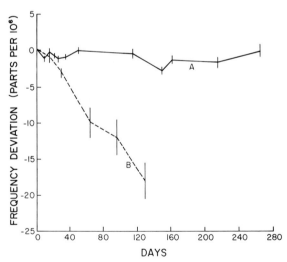

FIG. 6-8 Isothermal aging of 300-MHz SAW resonators at room temperature, mounted with room temperature vulcanizing (RTV) adhesive and without adhesive. A, vacuum sealed, no adhesive; B, hermetic sealed, RTV 6-1104. [From Latham and Saunders (1978). © 1978 IEEE.]

6 LONG-TERM STABILITY AND AGING OF RESONATORS

metalization can cause the aging to vary from a few parts per million per year to more than a part per thousand per year. The major cause of this effect is the acoustically induced migration of Al into the coupling transducers. This process can be largely eliminated by adding a small amount of Cu to the metalization. Copper-alloyed Al plating also improves the low-drive aging compared with Cr–Al electrodes (Schoenwald et al., 1981).

6.3.2 SAW Delay Lines

In addition to one- and two-port SAW resonators, SAW delay lines are also being investigated as frequency controlling elements in oscillators. Since their Q increases linearly with frequency, they may be advantageous in the frequency range up to and above 1 GHz. As an example, delay lines with a frequency of 1.4 GHz sealed in all-quartz packages show aging rates of several parts per million over more than 52 weeks (Gilden et al., 1980). As is to be expected, the aging rate of delay-line oscillators is also influenced by the

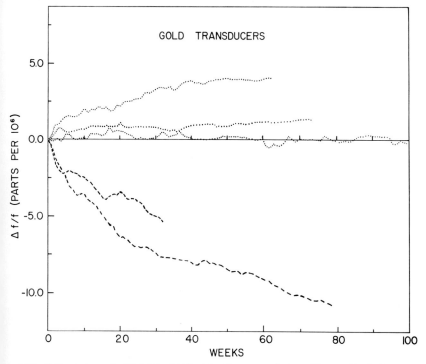

FIG. 6-9 Aging of 401-MHz SAW delay-line-controlled oscillators. Dashed curves, flatpacks (brazed); dotted curves, HC 36/U (cold-weld). Delay lines use gold interdigital transducers.

methods of fabrication and processing. This is shown in Fig. 6-9, which illustrates the aging behavior of five devices with gold transducers operating at 401 MHz and fabricated in different enclosures. As can be seen, the best aging rate is only a few tenths of a part per million for a period of 100 weeks (Parker, 1980). The units were made from 40°-rotated Y-cut quartz plates, and the best ones were packaged in HC 36/U enclosures. Twenty-five units with aluminum transducers show somewhat different, but not significantly worse, aging behavior. It follows from these measurements that the investigated delay lines show a drift of less than 2 ppm for a significant fraction of the devices. Also, Parker's data strongly suggest that the transducer metallization is very likely the source of relaxation of a mechanism that causes a modification of acoustic properties of the material and, thus, frequency drift. Recent measurements on 400-MHz devices cold-weld-sealed in TO-8 packages (Parker, 1982a,b) confirm the results obtained in HC-36/U enclosures. It seems possible that long-term aging rates well under 1 ppm in the first year can be obtained under production conditions. However, it was found (Parker, 1983a,b) that random frequency fluctuations with periods up to months and years are superimposed on the systematic drift and may thus decrease the overall stability. This type of noise is less significant for resonators than for delay lines; its source is not yet known.

Bibliography

Eduard A. Gerber
Arthur Ballato

U.S. Army Electronics Technology and Devices Laboratory
Fort Monmouth, New Jersey

Introduction	285
General Bibliography	287
Books	287
Conference and Symposia Proceedings; Special Issues	288
General Papers; Reviews; Book Chapters	289
Chapter Bibliographies	293
Chapter 1	293
Sections 2.1 and 2.2	310
Section 2.3	345
Chapter 3	357
Chapter 4	360
Sections 5.1 and 5.2	369
Sections 5.3 and 5.4	389
Chapter 6	413

INTRODUCTION

The editors have assembled a reasonably complete bibliography covering the years 1968–1982. More than 5000 references were found for this period. The pre-1968 period is covered by inclusion of reviews and a selection of references to seminal articles. Science Abstracts A (Physics) and B (Electrical and Electronics) and Chemical Abstracts were particularly helpful in leading us to the pertinent literature. The organization of the main part of the bibliography follows that of the rest of the book, that is, the references are arranged in alphabetical order under each chapter heading. Thus, it will be easy for the reader to find the references quoted in the text. In the case of Chapters 2, 5, and 8, however, the arrangement is made according to sections, due to different authorships.

The references for the various chapters include papers on applications of specific devices, such as application of crystal units and filters. References describing more than one topic are listed under the chapter where that subject is predominant. Multi-subject papers without emphasis on a specific topic, items of a more general nature, survey papers, books, and chapters of books are listed separately in the general bibliography section.

Obviously, some overlap of topics is unavoidable. It is, therefore, recommended that the reader review related subjects in search for a paper of a specific nature. For instance, looking for a paper on the properties of a crystal unit, it would be well to search the bibliographies for Chapters 2 and 4.

In the area of quantum electronic devices and standards, we have included only references on gaseous masers. On the other hand, a large number of references may be found in the area of laser frequency standards. It is thought that in this rapidly developing field, valuable information is to be found in papers on gas lasers even though not directly concerned with frequency standards.

Papers on tuning forks, particularly for wristwatches, are considered to be a special case and are listed under Chapter 14; paper regarding their applications are listed under Chapter 15.

Where an article or book exists in English translation, normally only the English version is cited. This is particularly germane to the Russian literature, but since this book is primarily intended for the English-literate reader and since the original citation will appear in the translated journal, no essential information is omitted.

Certain documents, particularly U. S. government reports, are cited as obtainable from NTIS; these are usually catalogued according to a six digit number prefixed by the letters AD or ADA. The address is National Technical Information Service, U. S. Department of Commerce, 5285 Port Royal Road, Springfield, Virginia 22161, U.S.A.

References cited as *Proc. Annu. Freq. Control Symp.* refer to proceedings of the Annual Frequency Control Symposia. sponsored by the U. S. Army Electronics Research and Development Command, Fort Monmouth, New Jersey 07703, U.S.A. Proceedings from the 10th (1956) through the 31st (1977) may be obtained from NTIS. The 32nd through 35th are available through Electronic Industries Association, 2001 Eye Street, N.W., Washington, D.C. 20006, U.S.A. The 36th through 38th may be obtained from Systematics General Corporation, 2711 Jefferson Davis Highway, Arlington, Virginia 22202, U.S.A.

GENERAL BIBLIOGRAPHY

Books

Androsova, V. G., Bankov, V. N., Dikidzhi, A. N., Il'ichev, V. A., Karaul'nik, A. Ye., Pozdnyakova, P. G. Rakhmaninov, S. V., Fedotov, I. M., and Khristoforov, V. N. (1978). "Quartz Resonators Reference Book" (P. G. Pozdnyakova, ed.) Moscow:Svyaz', 288 pp. In Russian.

Auld, B. A. (1973). "Acoustic Fields and Waves in Solids," vols. 1 and 2. Wiley, New York.

Bennett, R. E. (1960). "Quartz Resonator Handbook: Manufacturing Guide for "AT" Type Units." Union Thermoelectric Division, Niles, Illinois. U. S. Army Contract DA36-039-SC-71061, 225 pp.

Blair, B. E. (1971). "Time and Frequency: a Bibliography of NBS Literature Published July 1955 to December 1970." Report Spec. Publ. 350, National Bureau of Standards, Washington, D.C., 50 pp.

Blair, B. E. (ed.) (1974). "Time and Frequency: Theory and Fundamentals." NBS Monograph 140. U. S. Government Printing Office, Washington, D.C.

Bottom, V. E. (1982). "Introduction to Quartz Crystal Unit Design." Van Nostrand-Reinhold, New York.

Buchanan, J. P. (1956). "Handbook of Piezo-Electric Crystals for Radio Equipment Designers." Wright Air Development Center, Wright-Patterson AFB, Ohio, Number WADC TR-56-156, 692 pp.

Cady, W. G. (1964). "Piezoelectricity," vols. 1 and 2, Dover Publications, New York.

Edson, W. A. (1953). "Vacuum Tube Oscillators." Wiley, New York.

Firth, D. (1965). "Quartz Crystal Oscillator Circuits Design Handbook," Magnavox Serial No. TP64-1072, Magnavox Company, Fort Wayne, Indiana. U. S. Army Contract No. DA-36-039-AMC-00043 (E).

Frerking, M. E. (1978). "Crystal Oscillator Design and Temperature Compensation." Van Nostrand-Reinhold, New York.

Groszkowski, J. (1964). "Frequency of Self-Oscillations." Macmillan, New York.

Heising, R. A. (1946). "Quartz Crystals for Electrical Circuits." Van Nostrand, New York.

Herrmann, G. (ed.) (1974). "R. D. Mindlin and Applied Mechanics." Pergamon, New York.

Herzog, W. (1949). "Siebschaltungen mit Schwingkristallen." Dieterich'sche Verlagsbuchhandlung, Wiesbaden, Germany. In German.

Holland, R., and EerNisse, E. P. (1969). "Resonant Piezoelectric Devices, Design of." MIT Press, Cambridge, Massachusetts.

Johnson, R. A. (1983). "Mechanical Filters in Electronics." Wiley, New York.

Kartaschoff, P. (1978). "Frequency and Time." Academic Press, London.

Mason, W. P. (1942). "Electromechanical Transducers and Wave Filters." Van Nostrand, New York.

Mason, W. P. (1950). "Piezoelectric Crystals and Their Application to Ultrasonics." Van Nostrand, New York.

Mason, W. P. (1966). "Crystal Physics of Interaction Processes." Academic Press, New York.

Mason, W. P., and Thurston, R. N. (eds.) (1964ff). "Physical Acoustics: Principles and Methods." Vol. 1 (1964)–Vol. 16 (1983). Academic Press, New York.

Matthews, H. (ed.) (1977). "Surface Wave Filters." Wiley, New York.

Matthys, R. J. (1983). "Crystal Oscillator Circuits." Wiley, New York.

Mindlin, R. D. (1955). "An Introduction to the Mathematical Theory of Vibrations of Elastic Plates." U. S. Army Signal Corps Engineering Laboratories, Fort Monmouth, New Jersey. Contract No. DA-36-039 SC-56772.

Parzen, B. (1983). "Design of Crystal and Other Harmonic Oscillators." Wiley, New York.

Post Office Research Station Staff (R. Bechmann, et al.) (1957). "Piezoelectricity." Her Majesty's Stationery Office, London.
Ramsey, N. F. (1969) "Molecular Beams." Oxford Univ. Press (Clarendon), London and New York.
Ristic, V. M. (1983). "Principles of Acoustic Devices." Wiley, New York.
Sanders, J. H., and Wapstra, A. H. (1976). "Atomic Masses and Fundamental Constants." Plenum, New York.
Scheibe, A. (1938). "Piezoelektrizitaet des Quarzes." Steinkopff, Dresden and Leipzig. In German.
Sheahan, D. F., and Johnson, R. A. (eds.) (1977). "Modern Crystal and Mechanical Filters." IEEE Press, New York.
Siegmann, A. E. (1971). "An Introduction to Lasers and Masers." McGraw-Hill, New York.
Smagin, A. G. (1964). "Precision Quartz Resonators." Izdatyelestvo Standarfov, Moscow. In Russian.
Tiersten, H. F. (1969). "Linear Piezoelectric Plate Vibrations." Plenum, New York.
Townes, C. H., and Schawlow, A. L. (1955). "Microwave Spectroscopy." McGraw-Hill, New York.
Vanier, J. (1971). "Basic Theory of Lasers and Masers." Gordon and Breach Science Publishers, London and New York.
Vigoureux, P. (1939). "Quartz Oscillators and Their Applications." His Majesty's Stationary Office, London.
Vigoureux, P., and Booth, C. (1950). "Quartz Vibrators and Their Applications." His Majesty's Stationery Office, London.
Voigt, W. (1928). "Lehrbuch der Kristallphysik." B. G. Teubner, Leipzig und Berlin. In German.
Vuylsteke, A. A. (1965). "Elements of Maser Theory." D. Van Nostrand Co., Inc., New York.
Zverev, A. I. (1967). "Handbook of Filter Synthesis." Wiley, New York.

Conferences and Symposia Proceedings; Special Issues

Cooper, M. (1967). The quartz crystal comes of age. *18th Annual Conference of the IEEE Vehicular Technology Group (New York: IEEE)*, pp. 72–77.
First open symposium on Time and Frequency of URSI Commission A. *Radio Sci.* **14**, (4), July–August 1979, 519–731.
Gagnepain, J.-J., Meeker, T. R., Nakamura, T., and Shuvalov, L. A. (eds.) (1982). Special Issue on Piezoelectricity. *Ferroelectrics*; **40** (3/4), 123–256 (Part I); **41** (1/2/3/4), 1–254 (Part II); **42** (1/2/3/4), 1–246 (Part III); **43** (1/2), 1–96 (Part IV).
IEEE (1965). Special issue on ultrasonics. *Proc. IEEE* **53**, 1292–1680.
IEEE (1966). Special issue on frequency stability. *Proc. IEEE* **54**, 103–338.
IEEE (1967). Special issue on radio measurement methods and standards. *Proc. IEEE* **55**, 741–1126.
IEEE (1969). Special issue on microwave acoustics. *IEEE Trans. Microwave Theory Tech.* **MTT-17**, 800–1046.
IEEE (1976). Special issue on surface acoustic wave devices and applications. *Proc. IEEE* **64**, 581–807.
IEEE Ultrasonics Symposium Proceedings (1972ff). IEEE Group on Sonics and Ultrasonics, IEEE, 345 East 47th Street, New York, New York 10017.
Proceedings of the Annual Frequency Control Symposium (1956ff). U. S. Army Electronics Research and Development Command, Fort Monmouth, New Jersey 07703.
Proceedings of the Annual Precise Time and Time Interval Applications and Planning Meeting (1969ff). NASA/Goddard Space Flight Center, Greenbelt, Maryland 20715.

Proceedings of the IEEE-NASA Symposium on Short-Term Frequency Stability (1964). Report NASA SP-80. Goddard Space Flight Center, Greenbelt, Maryland 20715.

General Papers; Reviews; Book Chapters

Allan, D. W., Gray, J. E., and Machlan, H. E. (1974). The National Bureau of Standards atomic time scale: Generation, stability, accuracy and accessibility. In "Time and Frequency: Theory and Fundamentals" (B. E. Blair, ed.), pp. 205-231. Natl. Bur. Stand. Monogr. 140.
Allan, D. W., Shoaf, J. H., and Halford, D. (1974). Statistics of time and frequency data analysis. In "Time and Frequency: Theory and Fundamentals (B. E. Blair, ed.), pp.153-204. Natl. Bur. Stand. Monogr. 140.
Alley, C. (1979). Relativity and clocks. *Proc. 33rd Annu. Freq. Control Symp.*, pp. 4-39.
Anonymous (1969). Applications of mechanical oscillations in communication technology. *Technica* **18**, 1451-1454. In German.
Anonymous (1973). Piezoelectronics. *Pr. Inst. Tele- Radiotech.* **17**, 72-83. In Polish.
Anonymous (1976). The technology of surface acoustic waves and their applications. *Eletton. Oggi (Italy)* (**10**), 1269-1277. In Italian.
Anonymous (1979). The SAW [surface acoustic wave] technology. [Function theory]. *Mikrowellen Mag.* (4), 235-238. In German.
Anonymous (1980). Quartz crystals for accuracy. *Telonde* (1), 38-41.
Balek, R., and Bunc, V. (1975). The applicability of surface elastic waves. *Sdelovaci Tech.* **23**, 130-132. In Czech.
Ballato, A. (1977). Doubly rotated thickness mode plate vibrators. In "Physical Acoustics," vol. XIII, (W. P. Mason and R. N. Thurston, eds.), pp. 115-181. Academic Press, New York.
Ballato, A., and Gerber, E. A. (1979). Piezoelectricity. In "Digest of Literature on Dielectrics" (M. R. Wertheimer and A. Yelon, eds.), vol. 41, pp. 393-436. National Academy of Sciences, National Research Council, Washington, D.C. 20418.
Ballato, A., Staudte, J. H., Horton. W. H., Takeuchi, T., and Fischer, R. (1981). The future of the quartz crystal industry—worldwide. *Proc. 35th Annu. Freq. Control Symp.*, pp. 576-595.
Barnes, J. A. (1974). Basic concepts of precise time and frequency. In "Time and Frequency: Theory and Fundamentals" (B. E. Blair, ed.), pp. 3-14. Natl. Bur. Stand. Monogr. 140.
Barnes, J. A., and Winkler, G. M. R. (1974). The standards of time and frequency in the U.S.A. In "Time and Frequency: Theory and Fundamentals" (B. E. Blair, ed.), pp. 317-327. Natl. Bur. Stand. Monogr. 140.
Beehler, R. E. (1974). A historical review of atomic frequency standards. In "Time and Frequency: Theory and Fundamentals" (B. E. Blair, ed.), pp. 85-100. Natl. Bur. Stand. Monogr. 140.
Beehler, R. E. (1974). Recent progress on atomic frequency standards. In "Time and Frequency: Theory and Fundamentals" (B. E. Blair, ed.), pp. 101-109. Natl. But. Stand. Monogr. 140.
Bidart, L. (1982). New design of very high frequency sources. *Ferroelectrics* **40**, 231-236.
Blair, B. E. (1974). Time and frequency dissemination: An overview of principles and techniques. In "Time and Frequency: Theory and Fundamentals" (B. E. Blair, ed.), pp. 233-313. Natl. Bur. Stand. Monogr. 140.
Bowers, K. D. (1971). Ultrasonics in communications. *Bell Lab. Rec.* **49**, 139-145.
Butler, M. B. N. (1980). Surface acoustic wave devices. I. [Analogue signal processing]. *Electron. Eng.* **52**, 37-39, 41, 43-44.

Butler, M. B. N. (1980). Surface acoustic wave devices. II. *Electron. Eng.* **52**, 57, 59, 61, 63, 65.
Butler, M. B. N. (1981). Surface acoustic wave devices. III. Applications. *Electron. Eng.* **53**, 54–63.
Chao, G. (1977). Surface acoustic wave devices move into high-volume markets. *EDN* **22**, 96–99.
Cometta, M. (1974). Piezoelectricity, (devices) and its applications. *Elettrificazione* (2), 71–75. In Italian.
Contoni, A. (1981). Surface acoustic wave devices. I. *Antenna* **53**, 435–438. In Italian.
Cook, W. R., and Jaffe, H. (1979). Piezoelectric, electrostrictive and dielectric constants, and electromechanical coupling factors of piezoelectric crystals. *In* "Landolt–Boernstein, New Series" (K.-H. Hellwege, ed.), pp. 287–470. Springer-Verlag, Berlin, Heidelberg, New York.
De Klerk, J. (1975). Elastic surface wave devices. *In* "Physical Acoustics," vol. XI (W. P. Mason and R. N. Thurston, eds.), pp. 213–244. Academic Press, New York.
Dressner, J. M., and Del Vecchio, J. P. (1969). Frequency control requirements for the Mallard communication system. *Proc. 23rd Annu. Freq. Control Symp.*, pp. 1–7.
Essen, L. (1971). Electrical standards in terms of frequency and physical constants. *Alta Freq.* **40**, 693–696.
Eykholt, A., and Lipetz, N. (1970). Surface acoustic wave devices—an emerging technology. *Proc. Natl. Electron. Conf.*, p. 991.
Fisk, J. R. (1971). New acoustic surface-wave devices. *Electron. Aust.* **33**, 24–27.
Folts, H. (1972). Precise time and frequency in a communication system. *Proc. 26th Annu. Freq. Control Symp.*, pp. 4–7.
Forsbergh, P. W., Jr. (1956). Piezoelectricity, electrostriction and ferroelectricity. *In* "Handbuch der Physik," vol. XVII (S. Flugge, ed.). Springer-Verlag, Berlin, Gottingen, Heidelberg.
Franx, C. (1970). A report on IEC technical committee TC-49. *Proc. 24th Annu. Freq. Control Symp.*, pp. 172–176.
Gerber, E. A. (1948). Piezoelektrische Eigenschaften (piezoelectric properties). *In* "Physik des festen Korpers (Physics of Solids)" (G. Joos, ed.), Chapter 52. Dieterich'sche Verlagsbuchhandlung, Wiesbaden, Germany.
Gerber, E. A., and Sykes, R. A. (1966). State of the art—Quartz crystal units and oscillators. *Proc. IEEE* **54**, 103–116.
Gerber, E. A., and Sykes, R. A. (1967). Quartz frequency standards. *Proc. IEEE* **55**, 783–791.
Gerber, E. A., and Sykes, R. A. (1971). A quarter century of progress in the theory and development of crystals for frequency control and selection. *Proc. 25th Annu. Freq. Control Symp.*, pp. 1–45.
Gerber, E. A., and Sykes, R. A. (1974). Progress in the development of quartz crystal units and oscillators since 1966. *In* "Time and Frequency: Theory and Fundamentals" (B. E. Blair, ed.), pp. 59–66. Natl. Bur. Stand. Monogr. 140.
Glaze, D. J. (1974). Improvements in atomic cesium beam frequency standards at the National Bureau of Standards. *In* "Time and Frequency: Theory and Fundamentals" (B. E. Blair, ed.), pp. 111–117. Natl. Bur. Stand. Monogr. 140.
Glaze, D. J., Hellwig, H., Jarvis, S., Wainwright, A. E., and Allan, D. W. (1974). The new primary cesium beam frequency standard. *In* "Time and Frequency: Theory and Fundamentals" (B. E. Blair, ed.), pp. 121–124. Natl. Bur. Stand. Monogr. 140.
Greebe, C. A. A. J. (1974). Electromagnetic, elastic and electro-elastic waves. *Philips Tech. Rev.* **33**, 311–349.
Hafner, E. (1972). Frequency control aspects in Army communications and surveillance. *Proc. 26th Annu. Freq. Control Symp.*, pp. 15–19.
Hawkins, D. T. (1973). Absolute configuration of piezoelectrics—A bibliography. *Ferroelectrics* **5**, 111–123.

GENERAL PAPERS; REVIEWS; BOOK CHAPTERS 291

Hearman, R. F. S. (1979). The elastic constants of crystals and other anisotropic materials. *In* "Landolt-Boernstein, New Series" (K.-H. Hellwege, ed.), pp. 1-244. Springer-Verlag, Berlin, Heidelberg, New York.
Hearman, R. F. S. (1979). The third- and higher-order elastic constants. In "Landolt-Boernstein, New Series" (K.-H. Hellwege, ed.), pp. 245-286. Springer-Verlag, Berlin, Heidelberg, New York.
Hellwig, H. (1974). Areas of promise for the development of future primary frequency standards. *In* "Time and Frequency: Theory and Fundamentals" (B. E. Blair, ed.), pp. 125-136. Natl. Bur. Stand. Monogr. 140.
Hellwig, H., and Halford, D. (1974). Accurate frequency measurements: Relevance to some other areas of metrology. *In* "Time and Frequency: Theory and Fundamentals" (B. E. Blair, ed.), pp. 137-149. Natl. Bur. Stand. Monogr. 140.
Henaff, J. (1980). Applications of surface-acoustic-wave devices in satellite communication systems. *Proc. 10th European Microwave Conf.*, pp. 202-207.
Holland, M. G., and Claiborne, L. T. (1974). Practical surface acoustic wave devices. *Proc. IEEE* **62**, 582-611.
Inamura, T., and Yoshikawa, S. (1976). Applications of surface acoustic wave devices. *Oyo Buturi*, **45**, 717-728. In Japanese.
Ivanek, F. (1979). New developments in frequency control for microwave communications. *3rd World Telecommunication Forum, Pt. II (Switzerland: Telecommunication Union)*, p. 1.3.3/1-8.
Joshi, N. K., and Singh, A. (1978). Surface acoustic wave devices in electronic warfare. *Def. Sci. J.* **27**, 157-162.
Kartaschoff, P. (1977). Frequency control and timing requirements for communication. *Proc. 31st Annu. Freq. Control Symp.*, pp. 478-483.
Kentley, E. (1972). The practical aspects of international standardization in the frequency control field. *Proc. 26th Annu. Freq. Control Symp.*, pp. 159-163.
Kino, Y., Nakayama, Y., Furuya, N., Watanabe, Y., Aihara, Y., and Inaba, R. (1976). Application for surface acoustic waves. *Natl. Tech. Rep. (Japan)* **22**, 644-650. In Japanese.
Kinsman, R., and Gunn, D. (1981). Frequency control requirements for 800 MHz mobile communication. *Proc. 35th Annu. Freq. Control Symp.*, pp. 501-510.
Kurokawa, S. (1978). Japan's quartz crystal industry sees international cooperation as its panacea. *JEE J. Electron. Eng.* (144), 28-30.
Lagasse, P. E. (1980), State of the art in design and technology of SAW devices. *Electrocompon. Sci. Technol.* **6**, 199-204.
Langer, E. (1979). The piezoelectric effect and its applications. A simple tutorial on the physics and techniques for piezoelectric devices. IV. *Nachr. Elektron.* **33**, 237-238. In German.
Langer, E. (1979). The piezoelectric effect and its applications. V. *Nachr. Elektron.* **33**, 273-274. In German.
Lawrence, M. W. (1975). Surface acoustic waves and their applications. *Aust. Electron. Eng.* **8**, 19-22.
Lawrence, W. D. (1972). Frequency control requirements for remote sensor systems. *Proc. 26th Annu. Freq. Control Symp.*, pp. 113-119.
Lemanov, V. V., and Ilisavsky, Yu. V. (1982). Piezoelectricity and acoustoelectronics. *Ferroelectrics* **42**, 77-101.
Lubimov, L. A., and Zhoukina, L. I. (1975). Piezoelectric devices. *Eltek. Aktuell Elektron.* **18**, 59-61.
Lubimov, L. A., and Zhourkina, L. I. (1975). Piezoelectric devices. *IEC Bull.* **9**, 1-2.
Lundbom, P.-O. (1974). On the activities of URSI Commission I [for radio measurements and standards] during the last 20 years and plans for the future. *CPEM Digest: Conference on Precision Electromagnetic Measurements (London: IEE)*, p. 330.

Mains, J. D. (1977). Surface wave devices for radar equipment. *In* "Surface Wave Filters" (H. Matthews, ed.), pp. 443–476. Wiley, Chichester, England.
Maines, J. D., and Paige, E. G. S. (1976). Surface acoustic-wave devices for signal processing applications. *Proc. IEEE* **64**, 639–652.
Mason, W. P. (1964). Use of piezoelectric crystals and mechanical resonators in filters and oscillators. *In* "Physical Acoustics," vol. 1, Part A (W. P. Mason, ed.), pp. 336–417. Academic Press, New York.
Mason, W. P. (1975). Professor Walter G. Cady's contributions to piezoelectricity and what followed from them. *J. Acoust. Soc. Am.* **58**, 301–309.
Mason, W. P. (1981). Piezoelectricity, its history and applications. *J. Acoust. Soc. Am.* **70**, 1561–1566.
McDermott, J. (1973). Focus on piezoelectric crystals and devices. *Electron. Des.* **21**, 44–54.
McQuiddy, D. N., Jr. (1981). Frequency stability requirements for a 95 GHz instrumentation radar system. *Proc. 35th Annu. Freq. Control Symp.*, pp. 516–524.
Mockler, R. C. (1961). Atomic beam frequency standards. *In* "Advances in Electronics and Electron Physics," vol. 15, pp. 1–71. Academic Press, New York.
Nekrasov, M. M., Lavrinenko, V. V., Bozhko, A. A., Kartashev, I. A., Koval, V. S., Miroshinchenko, A. P., and Yakimenko, Yu. I. (1971). Elements of piezo-electronics and prospects of application in electrical engineering. *Elec. Technol. USSR* **4**, 134–135.
Nunamaker, R. J. (1971). Frequency control devices for mobile communications. *Proc. 25th Annu. Freq. Control Symp.*, p. 74.
Nyffeler, F. (1982). High-performance signal processing using SAW components. *Bull. Assoc. Suisse Electr.* **73**, 883–888. In German.
Ohnuki, A. (1977). Quartz crystal units for CB transceivers. *JEE J. Electron. Eng.* (121), 45–49.
Paige, E. G. S. (1971). Acoustic surface waves and their applications in electronics. *Proc. 7th International Congress on Acoustics* (*Hungary: Akademiai Kiado*), pp. 141–151.
Pajewski, W. (1971). Development in research and technology of piezoelectric elements. *Elektronika* (1), 25–34. In Polish.
Pajewski, W. (1973). Piezoelectronics and acoustoelectronics. *Elektronika* (1–2), 50–54. In Polish.
Redwood, M. (1980). Piezoelectric devices in electronics. *Phys. Educ.* **15**, 9–14.
Risley, A. S. (1974). The physical basis of atomic frequency standards. *In* "Time and Frequency: Theory and Fundamentals" (B. E. Blair, ed.), pp. 65–84. Natl. Bur. Stand. Monogr. 140.
Robrock, R. B. (1971). Quartz crystal applications in digital transmission. *Proc. 25th Annu. Freq. Control Symp.*, pp. 70–73.
Sabine, H., and Cole, P. H. (1971). Acoustic surface wave devices: a survey. *Proc. Inst. Radio Electron. Eng. Aust.* **32**, A12.
Sabine, H., and Cole, P. H. (1971). Acoustic surface wave devices: a survey. *Proc. Inst. Radio Electron. Eng. Aust.* **32**, 445–458.
Schmidt, L. (1970). The piezoelectric effect and its application to modern technology. *VOE Fachber.* **26**, 139–143.
Schmidt, L. (1970). The piezoelectric effect and its application in the modern technology. *Messen U. Prufen* **6**, 811–816. In German.
Shaw, H. J. (1971). An overview of surface-wave methods and devices. *81st Meeting of the Acoustical Society of America*, p. 29.
Shibayama, K. (1982). SAW materials and devices in Japan. *Ferroelectrics* **42**, 153–159.
Stevenson, S. A. (1969). Developments in the application of synthetic quartz. *Electronic Compon.* **10**, 1465–1466.
Stone, R. R., Jr., Phillips, D. H., and Berg, W. B., Jr. (1971). Technology forecasts: frequency and time. *Digest of Technical Papers of IEEE Region 2 Conference on Technology Forecasting and Assessment of Electrotechnology* (*New York: IEEE*), pp. 54–55.

Szlavy, A. (1980). Quartz crystals in electronics. II. *Finommech. Mikrotech.* **19**, 172–180. In Hungarian.
Tanaka, T. (1973). Recent developments and promising applications of piezo-resonators. *JEE J. Electron. Eng.* (80), 58–60, 62.
Tanaka, T. (1982). Piezoelectric devices in Japan. *Ferroelectrics* **40**, 167–187.
Tiesler, G. (1979). Quartz crystals for microprocessors. *Components Rep.* **14**, 91–93.
Warashina, N. (1978). Crystal oscillators and units improve frequencies for measuring instruments. *JEE* (144), 44–47.
White, R. M. (1970). Surface elastic waves. *Proc. IEEE* **58**, 1238–1276.
Wiefelsputz, F. G. (1978). Frequency control in radiocommunications. *Telecommun. J. (Engl. Ed.)* **45**, 395–402.
Williamson, R. C. (1978). Case studies of successful surface-acoustic-wave devices. *Ultrason. Symp. Proc.*, pp. 460–468.
Yakovkin, I., and Kohler, E. (1975). Components based on elastic surface waves. *Nachrichtentech. Elektron.* **25**, 186–189. In German.
Zelenka, J. (1980). The centennial of discovering piezoelectricity. *Slaboproudy Obz.* **41**, 237–244. In Czech.

CHAPTER BIBLIOGRAPHIES

Chapter 1

Abrahams, S. C., Reddy, J. M., and Bernstein, J. L. (1966a). Ferroelectric lithium niobate. 3. Single crystal x-ray diffraction study at 24°C. *J. Phys. Chem. Solids* **27**, 997–1012.
Abrahams, S. C., and Bernstein, J. L. (1967). Ferroelectric lithium tantalate-1. Single crystal x-ray diffraction study at 24°C. *J. Phys. Chem. Solids* **28**, 1685–1692.
Abrahams, S. C., Hamilton, W. C., and Reddy, J. M. (1966b). Ferroelectric lithium niobate. 4. Single crystal neutron diffraction study at 24°C. *J. Phys. Chem. Solids* **27**, 1013–1018.
Abrahams, S. C., Levinstein, H. J., and Reddy, J. M. (1966c). Ferroelectric lithium niobate. 5. Polycrystal x-ray diffraction study between 24° and 1200°C. *J. Phys. Chem. Solid* **27**, 1019–1026.
Abrahams, S. C., Jamieson, P. B., and Bernstein, J. L. (1967). Crystal structure of piezoelectric bismuth germanium oxide $Bi_{12}GeO_{20}$. *J. Chem. Phys.* **47**, 4034–4041.
Adachi, M., and Kawabata, A. (1972). Piezoelectric properties of potassium tantalate-niobate single crystal. *Jpn. J. Appl. Phys.* **11**, 1855.
Adachi, M., and Kawabata, A. (1978). Elastic and piezoelectric properties of potassium lithium niobate (KLN) crystals. *Jpn. J. Appl. Phys.* **17**, 1969–1974.
Adams, C. A., Enslow, G. M., Kusters, J. A., and Ward, R. W. (1970). Selected topics in quartz crystal research. *Proc. 24th Annu. Freq. Control Symp.*, pp. 55–63.
Adhav, R. S. (1975). Piezoelectric effect in tetragonal crystals. *J. Appl. Phys.* **46**, 2808.
Aleksandrov, K. S., Sorokin, B. P., Kokorin, Y. I., Chetvergov, N. A., and Grekhova, T. I. (1982). Non-linear piezoelectricity in sillenite structure crystals. *Ferroelectrics* **41**, 27–33.
Anderson, T. L., Newnham, R. E., Cross, L. E., and Laughner, J. W. (1976). Laser-induced twinning in quartz. *Phys. Status Solidi A* **37**, 235–245.
Anderson, T. L., Newnham, R. E., and Cross, L. E. (1977). Coercive stress for ferrobielastic twinning in quartz. *Proc. 31st Annu. Freq. Control. Symp.*, pp. 171–177.
Annaka, S., and Nemoto, A. (1977). Piezoelectric constants of α-quartz determined from dynamical X-ray diffraction curves. *J. Appl. Crystallogr.* **10**, 354–355.
Arlt, G., and Quadflieg, P. (1968). Piezoelectricity in III-V compounds with a phenomenological analysis of the piezoelectric effect. *Phys. Status Solidi A* **25**, 323–330.

Armington, A. F., Bruce, J. A., Halliburton, L. E., and Markes, M. (1980). Defects induced by seed orientation during quartz growth. *J. Cryst. Growth* **49**, 739–742.
Armington, A. F., Larkin, J. J., O'Connor, J. J., and Horrigan, J. A. (1981). Initial results with the Air Force hydrothermal research facility. *Proc. 35th Annu. Freq. Control Symp.*, pp. 297–303.
Armington, A. F., Larkin, J. J., O'Connor, J. J., and Horrigan, J. A. (1982). Recent results with the Air Force hydrothermal facility. *Proc. 36th Annu. Freq. Control Symp.*, pp. 55–61.
Armington, A. F., Larkin, J. J., O'Connor, J. J., and Cormier, E. (1983). Effect of seed treatment on quartz dislocations. *Proc. 37th Annu. Freq. Control Symp.*, pp. 177–180.
Arnold, G. W. (1957). Defects in quartz crystals. *Proc. 11th Annu. Freq. Control Symp.*, pp. 112–129.
Arnold, G. W. (1965). Defect structure of crystalline quartz. I. Radiation-induced optical absorption. *Phys. Rev. A* **139**, 1234–1239.
Asahara, J. (1978). Superior large autoclaves produce synthetic quartz crystals. *JEE J. Electron. Eng.* (144), 36–39.
Asahara, J., and Taki, S. (1972). Physical properties of synthetic quartz and its electrical characteristics. *Proc. 26th Annu. Freq. Control Symp.*, pp. 93–103.
Asahara, J., Takazawa, K., Yazaki, E., Okuda, J., and Asanuma, N. (1974). Defects in synthetic quartz crystals and their influence on the electrical characteristics of quartz resonators. *Proc. 28th Annu. Freq. Control Symp.*, pp. 117–124.
Aucoin, T. R., Savage, R. O., Wade, M. J., Gualtieri, J. G., and Schwartz. A. (1980). Large high quality single crystal aluminum phosphate for acoustic wave devices. *Proc. Army Sci. Conf.* (West Point, NY) **1**, 121–133.
Avdienko, K. I., Kidyarov, B. I., and Sheloput, D. V. (1977a). Growth of α-LiIO$_3$ crystals of high piezoelectric and optical quality. *J. Cryst. Growth* **42**, 228–233.
Avdienko, K. I., Bogdanov, S. V., Kidyarov, B. I., Semenov, V. I., and Sheloput, D. V. (1977b). Optical, acoustic, and piezoelectric properties of α-LiIO$_3$ crystals. *Bull. Acad. Sci. USSR, Phys. Ser. (Engl. Transl.)* **41**, 40–45.
Bailey, D. S., Soluch, W., Lee, D. L., Vetelino, J. F., Andle, J. C., and Chai, B. H. T. (1982). The elastic, piezoelectric and dielectric constants of berlinite. *Proc. 36th Annu. Freq. Control Symp.*, pp. 124–132.
Balascio, J. F., and Lias, N. C. (1980). Factors affecting the quality and perfection of hydrothermally grown quartz. *Proc. 34th Annu. Freq. Control Symp.*, pp. 65–71.
Balascio, J. F., and Lias, N. C. (1983). Standard characterization methods for the determination of the quality of hydrothermally grown quartz. *Proc. 37th Annu. Freq. Control Symp.*, pp. 157–163.
Ballman, A. A. (1965). Growth of piezoelectric and ferroelectric materials by the Czochralski technique. *J. Am. Ceram. Soc.* **48**, 112–113.
Ballman, A. A. (1967). The growth and properties of piezoelectric bismuth germanium oxide Bi$_{12}$GeO$_{20}$. *J. Cryst. Growth* **1**, 37–40.
Ballman, A. A., and Rudd, D. W. (1972). Czochralski growth of ferroelectric lithium tantalate for piezoelectric filter devices. *2nd National Conf. on Crystal Growth (abstracts only)*, p. 110.
Ballman, A. A., Laudise, R. A., and Rudd, D. W. (1966). Synthetic quartz with a mechanical Q equivalent to natural quartz. *App. Phys. Lett.* **8**, 53–54.
Baranovskii, S. N., and Panov, V. I. (1974). Effect of uniaxial compression on the piezoelectric modulus of quartz. *Sov. Phys. Acoust. (Engl. Transl.)* **20**, 71–72.
Barns, R. L., Kolb, E. D., Key, P. L., Laudise, R. A., Simpson, E. E., and Kroupa, K. M. (1975). Production and perfection of 'r-face' quartz. *Proc. 29th Annu. Freq. Control Symp.*, pp. 98–104.
Barns, R. L., Kolb, E. D., Laudise, R. A., Simpson, E. E., and Kroupa, K. M. (1976). Production and perfection of 'z-face' quartz. *J. Cryst. Growth* **34**, 189–197.

Barns, R. L., Freeland, P. E., Kolb, E. D., Laudise, R. A., and Patel, J. R. (1978). Dislocation-free and low-dislocation quartz prepared by hydrothermal crystallization. *J. Cryst. Growth* **43**, 676-686.
Barsch, G. R. (1976). X-ray determination of piezoelectric constants. *Acta Crystallogr. Sect. A* **A32**, 575-586.
Barybin, A. A. (1967). A theorem for the kinetic power of the charge carrier motion in piezoelectric crystals. *Radiotekh. Elektron.* **12**, 1842-1844. In Russian.
Battaglia, A., and Croce, U. D. (1972). Apparatus for determining the piezoelectricity of crystalline powders. *Alta Freq.* **41**, 375-378.
Baughman, R. J. (1982). A comparison of quartz crystals grown from fused silica and from crystalline nutrient. *Proc. 36th Annu. Freq. Control Symp.*, pp. 82-89.
Bechmann, R. (1951). Contour modes of square plates excited piezoelectrically and determination of elastic and piezoelectric coefficients. *Proc. Phys. Soc. London B* **64**, 323-337.
Bechmann, R. (1958). Elastic and piezoelectric constants of alpha-quartz. *Phys. Rev.* **110**, 1060-1061.
Bechmann, R. (1966). The elastic, piezoelectric, and dielectric constants of piezoelectric crystals. *In* "Landolt-Boernstein Series, Group III," vol. 1 (K. H. Hellwege, ed.), p. 83. Springer, Berlin.
Bechmann, R., Ballato, A. D., and Lukaszek, T. J. (1962). Higher-order temperature coefficients of the elastic stiffnesses and compliances of alpha-quartz. *Proc. IRE* **50**, 1812-1822, 2451.
Bekisz, J., Pacho, Z., and Wlosinski, W. (1981). Some materials for electronic components manufactured in the Polish Electronic Materials Research and Production Centre (CNPME), *Elektronika* **22**, 3-8. In Polish.
Belikova, G. S., Pisarevskii, Y. V., and Silvestrova, I. M. (1975). Piezoelectric and elastic properties of crystals of rubidium biphthalate. *Sov. Phys. Crystallogr. (Engl. Transl.)* **19**, 545-546.
Belyaev, A. D., Olikh, Y. M., Miselyuk, E. G., and Taborov, V. F. (1979). Acoustic properties of Li_2GeO_3. *Ukr. Fiz. Zh. (Russ. Ed.)* **24**, 1896-1898. In Russian.
Bernstein, J. L. (1967). The unit cell and space group of piezoelectric bismuth germanium oxide ($Bi_{12}GeO_{20}$). *J. Crystal Growth* **1**, 45-46.
Berry, B. S., and Nowick, A. S. (1966). Anelasticity and internal friction due to point defects in crystals. *In* "Physical Acoustics" (W. P. Mason, ed.), vol. III, Part A, pp. 1-42. Academic Press, New York.
Besson, R. (1974). Measurement of nonlinear elastic, piezoelectric, [and] dielectric coefficients of quartz crystal. Applications. *Proc. 28th Annu. Freq. Control Symp.*, pp. 8-13.
Besson, R. (1974). Nonlinearity of the direct piezoelectric effect of quartz. *C.R. Hebd. Seances Acad. Sci. Ser. B* **279**, 325-328. In French.
Besson, R., and Gagnepain, J.-J. (1972). Determination of the nonlinear electric polarization coefficients of quartz. *C.R. Acad. Sci. Ser. B* **274**, 835-838. In French.
Besson, R., and Mesnage, P. (1970). Electrostriction of quartz. New high sensitivity static method for the measurement of electrostriction and piezoelectric coefficients of an α-quartz crystal layer. *C.R. Acad. Sci. Ser. B* **270**, 994-996. In French.
Bond, W. L. (1976). "Crystal Technology." Wiley, New York.
Booyens, H., Vermaak, J. S., and Proto, G. R. (1977). Dislocations and the piezoelectric effect in III-V crystals. *J. Appl. Phys.* **48**, 3008-3013.
Bottom, V. E. (1969). Measurement of the piezoelectric coefficient of quartz using the Fabry-Perot dilatometer. *Proc. 23rd Annu. Freq. Control Symp.*, pp. 21-25.
Bottom, V. E. (1970). Measurement of the piezoelectric coefficient of quartz using the Fabry-Perot dilatometer. *J. Appl. Phys.* **41**, 3941-3944.
Bottom, V. E. (1972). Dielectric constants of quartz. *J. Appl. Phys.* **43**, 1493-1495.

Brandle, C. D., and Miller, D. C. (1974). Czochralski growth of large diameter $LiTaO_3$ crystals. *J. Cryst. Growth* **2425**, 432–436.
Breazeale, M. A., and Philip, J. (1981). Relation of the third-order elastic constants to other nonlinear quantities. *Ultrason. Symp. Proc.*, pp. 425–431.
Brice, J. C., and Cole, A. M. (1978). The characterization of synthetic quartz by using infra-red absorption. *Proc. 32nd Annu. Freq. Control Symp.*, pp. 1–10.
Brice, J. C., and Cole, A. M. (1979). Infrared absorption in α-quartz. *J. Phys. D.* **12**, 459–463.
Brown, R. N., and Kahan, A. (1975). Optical absorption of irradiated quartz in the near IR. *J. Phys. Chem. Solids* **36**, 467–476.
Brown, R. N., O'Connor, J. J., and Armington, A. F. (1980). Sweeping and Q measurements at elevated temperatures in quartz. *Mater. Res. Bull.* **15**, 1063–1067.
Brownlow, D. I. (1976). Fracture resistance of synthetic α-quartz seed plates. *Proc. 30th Annu. Freq. Control Symp.*, pp. 23–31.
Buisson, M. X. (1975). Industrial production of synthetic quartz crystals. *International Symposium on Materials for Electronic Components*, pp. 106–112. In French.
Bye, K. L., and Cosier, R. S. (1979). An x-ray double crystal topographic assessment of defects in quartz resonators. *J. Mater. Sci.* **14**, 800–810.
Byer, R. L., Young, J. F., and Feigelson, R. S. (1970). Growth of high-quality $LiNbO_3$ crystals from the congruent melt. *J. Appl. Phys.* **41**, 2320–2325.
Cady, W. G. (1964). "Piezoelectricity." 2 vols. Dover Publications, New York. (Revised and enlarged version; published originally by McGraw-Hill, 1946).
Capone, B. R., Kahan, A., and Sawyer, B. (1971). Evaluation of quartz for high Q resonators. *Proc. 25th Annu. Freq. Control Symp.*, pp. 109–112.
Carruthers, J. R., Peterson, G. E., Grasso, M., and Bridenbaugh, P. M. (1971). Non-stoichiometry and crystal growth of lithium niobate. *J. Appl. Phys.* **42**. 1846–1851.
Cecchi, L., Vacher, R., and Danyach, L. (1970). Brillouin scattering measurement of the elastic constants of quartz. *J. Phys.* **31**, 501–506. In French.
Chai, B. H.-T., Shand, M. L., Buehler, E., and Gilleo, M. A. (1979). Experimental data on the piezoelectric properties of berlinite. *Ultrason. Symp. Proc.*, pp. 577–583.
Chakraborty, D. (1978). Suitability of coloured quartz crystals for resonator plates. *J. Mater. Sci.* **13**, 2529–2530.
Chang, Z.-P., and Barsch, G. R. (1976). Elastic constants and thermal expansion of berlinite. *IEEE Trans. Sonics Ultrason.* **SU-23**, 127–135.
Chao, H. L., and Parker, T. E. (1983). Tensile fracture strength of ST cut quartz. *Proc. 37th Annu. Freq. Control Symp.*, pp. 116–124.
Chen, P. J., and Montgomery, S. T. (1978). Boundary effects on the normal-mode response of linear transversely isotropic piezoelectric materials. *J. Appl. Phys.* **49**, 900–904.
Ching, H. C., Yang, K. C., Wang, C. M., and Chang, C. C. (1965). Studies on synthetic quartz. *Acta Electron. Sin.* (1), 127–136. In Chinese.
Chizhikov, S. I., Sorokin, N. G., and Petrakov, V. S. (1982). The elastoelectric effect in the non-centrosymmetric crystals. *Ferroelectrics* **41**, 9–25.
Chow, K., McKnight, H. G., and Rothrock, L. R. (1974). The congruently melting composition of $LiNbO_3$. *Mater. Res. Bull.* **9**, 1067–1072.
Chowdhury, K. L., and Glockner, P. G. (1976). Constitutive equations for elastic dielectrics. *Int. J. Non-Linear Mech.* **11**, 315–324.
Chuvyrov, A. N., Balitskii, V. S., Kuznetsov, A. F., Chernyi, L. N., and Khairetdinov, I. A. (1978). Growth features and crystal-physical properties of quartz grown in fluoride environment. *Sov. Phys.-Dokl. (Engl. Transl.)* **23**, 872–873.
Cline, T. W., Laughner, J. W., Newnham, R. E., and Cross, L. E. (1978). Electrical and acoustic emission during ferrobielastic twinning in quartz. *Proc. 32nd Annu. Freq. Control Symp.*, pp. 43–49.

Cohen, A. J., and Sumner, G. G. (1958). Relationships among impurity contents, color centers and lattice constants in quartz. *Am. Mineral.* **43**, 58–68.

Conlee, L., and Reifel, D. (1974). Analysis of synthetic quartz for stringent frequency versus temperature applications. *Proc. 28th Annu. Freq. Control Symp.*, pp. 125–128.

Cook, R. K., and Weissler, P. G. (1950). Piezoelectric constants of alpha- and beta-quartz at various temperatures. *Phys. Rev.* **80**, 712–716.

Cook, W. R., and Jaffe, H. (1979). Piezoelectric, electrostrictive, and dielectric constants, and electromechanical coupling factors of piezoelectric crystals. *In* "Landolt-Boernstein New Series," Chapter 3 (K.-H. Hellwege and A. M. Hellwege, eds.). Springer, Berlin, pp. 287–470.

Corruccini, R. J., and Gniewek, J. J. (1961). Thermal expansion of technical solids at low temperature—A compilation from the literature. *Nat. Bur. Stand. Monogr.* **29**, 13.

Cousins, C. S. G. (1972). Standard choice of axes for crystal class $\bar{6}m2$, with reference to elastic and piezoelectric properties. *J. Phys. C* **5**, L201–203.

Crisler, D. F., Cupal, J. J., and Moore, A. R. (1968). Dielectric, piezoelectric, and electromechanical coupling constants of zinc oxide crystals. *Proc. IEEE* **56**, 225–226.

Cross, L. E., and Newnham, R. E. (1970). Ferroelectric, piezoelectric, and electrooptic materials. *In* "Digest of Literature on Dielectrics," **34**, 374–432.

Croxall, D. F., Christie, I. R. A., Holt, J. M., Isherwood, B. J., and Todd, A. G. (1982). Growth and characterisation of high purity quartz. *Proc. 36th Annu. Freq. Control Symp.*, pp. 62–65.

Dan'kov, I. A., Pado, G. S., Kobyakov, I. B., and Berdnik, V. V. (1979). Elastic, piezoelectric, and dielectric properties of cadmium sulfide in the temperature range 4.2–300 K. *Sov. Phys.-Solid State (Engl. Transl.)* **23**, 1481–1483.

Danielsson, S., Grenthe, I., and Oskarsson, Å. (1976). A low-temperature apparatus for single-crystal diffractometry. The unit-cell dimensions of α-quartz in the temperature range 86–298 K. *J. Appl. Crystalogr.* **9**, 14–17.

De Jong, M. (1971). Materials with stronger piezoelectric properties. *Ingenieur (The Hague)*, **83**, E65–74. In Dutch.

DeVries, A. J., Everett, P., Gilchrist, D. F., Hansen, K., and Wojcik, T. J. (1979). Acoustic effects of filamentary defects in Y-Z $LiNbO_3$. *Proc. Ultrason. Symp. Proc.*, pp. 584–588.

Deshmukh, K. G., and Ingle, S. G. (1972). Optical and etching studies on single crystals of potassium niobate. *Indian J. Pure Appl. Phys.* **10**, 881–884.

Détaint, J., Poignant, H., and Toudic, Y. (1980). Experimental thermal behavior of berlinite resonators. *Proc. 34th Annu. Freq. Control Symp.*, pp. 93–101.

Dodd, D. M., and Fraser, D. B. (1965). The 3000–3900 cm^{-1} absorption bands and anelasticity in crystalline α-quartz. *J. Phys. Chem. Solids* **26**, 673–686.

Doherty, S. P., Martin, J. J., Armington, A. F., and Brown, R. N. (1980). The effects of irradiation and electrodiffusion on the sodium acoustic loss peak in synthetic quartz. *J. Appl. Phys.* **51**, 4164–4168.

Dokmeci, M. C. (1973). Variational principles in piezoelectricity. *Lett. Nuovo Cimento* **7**, 449–454.

Dolino, G., and Bachheimer, J. P. (1982). Effect of the α–β transition on mechanical properties of quartz. *Ferroelectrics* **43**, 77–86.

Donnay, J. D. H., and Le Page, Y. (1978). The vicissitudes of the low-quartz crystal setting or the pitfalls of enantiomorphism. *Acta Crystallogr.* **A34**, 584–594.

Duffy, M. T. (1978). The preparation and properties of heteroepitaxial III-V and II-VI compounds for surface acoustic wave and electrooptic devices. *In* "Heteroepitaxial Semiconductors for Electronic Devices," (G. W. Cullen and C. C. Wang, eds.), pp. 150–181. Springer-Verlag, Berlin.

Ecolivet, C., and Poignant, H. (1981). Berlinite and quartz $\alpha \leftrightarrow \beta$ phase transition analogy as seen by Brillouin scattering. *Phys. Status Solidi A* **63**, K107–K109.

Euler, F., Lipson, H. G., and Ligor, P. A. (1980). Radiation effects in quartz oscillators, resonators and materials, *Proc. 34th Annu. Freq. Control Symp.*, pp. 72–80.
Euler, F., Lipson, H. G., Kahan, A., and Armington, A. F. (1982). Characterization of alkali impurities in quartz. *Proc. 36th Annu. Freq. Control Symp.*, pp. 115–123.
Fedulov, S. A., Shapiro, Z. I., and Ladyzhinskii, P. B. (1965). The growth of crystals of $LiNbO_3$, $LiTaO_3$, and $NaNbO_3$ by the Czochralski method. *Sov. Phys.-Crystallogr. (Engl. Transl.)* **10**, 218–220.
Feigl, F. J., Fowler, W. B., and Yip, K. L. (1974). Oxygen vacancy model for the E'_1 center in SiO_2. *Solid State Commun.* **14**, 225–229.
Firsova, M. M. (1974). Piezoelectric and elastic properties of hexagonal α-ZnS. *Sov. Phys.-Solid State* **16**, 350–351.
Fontanella, J., Andeen, C., and Schuele, D. (1974). Low-frequency dielectric constants of α-quartz, sapphire, MgF_2, and MgO. *J. Appl. Phys.* **45**, 2852–2854.
Fowles, R. (1967). Dynamic compression of quartz. *J. Geophys. Res.* **72**, 5729–5742.
Fraser, D. B. (1964). Anelastic effects of alkali ions in crystalline quartz. *J. Appl. Phys.* **35**, 2913–2918.
Fraser, D. B. (1968). Impurities and anelasticity in crystalline quartz. *In* "Physical Acoustics" (W. P. Mason, ed.), V, 59–110. Academic Press, New York.
Fraser, D. B., Dodd, D. M., Rudd, D. W., and Carroll, W. J. (1966). Using infrared to find the mechanical Q of α-quartz. *Frequency* **4**, 18–21.
Fukuda, T., and Hirano, H. (1975). Growth and characteristics of $LiNbO_3$ plate crystals. *Mater. Res. Bull.* **10**, 801–806.
Fukuda, T., Matsumura, S., Hirano, H., and Ito, T. (1979). Growth of $LiTaO_3$ single crystal for SAW device applications. *J. Cryst. Growth* **46**, 179–184.
Fukuyo, H., Oura, N., and Shishido, F. (1976). The quality evaluation of quartz by measuring the Q-value of the second overtone of a Y-bar. *Bull. Tokyo Inst. Technol.* (131), 1–5.
Fukuyo, H., Oura, N., and Shishido, F. (1977). A new quality evaluation method of raw quartz by measuring the Q-value of Y-bar resonator. *Proc. 31st Annu. Freq. Control Symp.*, pp. 117–121.
Gautschi, G. (1981). Discovery of a new electrical effect (Piezo-electricity). *Rev. Polytech. Suisse* (5), 627, 629. In French.
Gavrilyachenko, V. G., and Fesenko, E. G. (1971). Piezoelectric effect in lead titanate single-crystals. *Kristallografiya* **16**, 640–641. In Russian.
Glass, A. M., Peterson, G. E., and Negran, T. J. (1972). Optical index damage in electrooptic crystal. *In* "Laser Induced Damage in Optic Materials" (A. Glass, ed.), pp. 15–26. National Bureau of Standards Special Publication No. 372, Washington, D.C.
Golan'sk, R., and Kosecki, T. (1971). Synthetic quartz in the modern manufacturing of quartz resonators. *Elekronika* (1), 18–24. In Polish.
Graham, R. A. (1972a). Strain dependence of longitudinal piezoelectric, elastic, and dielectric constants of X-cut quartz. *Phys. Rev. B.* **6**, 4779–4792.
Graham, R. A. (1972b). Strain dependence of the longitudinal piezoelectric, elastic, and dielectric constants of x-cut quartz as determined by shock compression techniques. *IEEE Trans. Sonics Ultrason.*, **SU-19**, 408.
Graham, R. A. (1973). Strain dependence of the piezoelectric polarization of z-cut lithium niobate. *Solid State Commun.* **12**, 503–506.
Graham, R. A. (1974). Linear and nonlinear piezoelectric constants of quartz, lithium niobate and lithium tantalate. *Satellite Symp. of the 8th International Congress on Acoustics on Microwave Acoustics*, pp. 124–129.
Graham, R. A. (1977). Second- and third-order piezoelectric stress constants of lithium niobate as determined by the impact-loading technique. *J. Appl. Phys.* **48**, 2153–2163.

Graham, R. A., and Halpin, W. J. (1968). Dielectric breakdown and recovery of X-cut quartz under shock-wave compression. *J. Appl. Phys.* **39**, 5077–5082.
Graham, G. M., and Pereira, F. N. D. D. (1971). Temperature variation of the piezoelectric effect in quartz. *J. Appl. Phys.* **42**, 3011.
Graham, R. A., and Chen, P. J. (1975). A new electrical to mechanical coupling effect for nonlinear piezoelectric solids. *Solid State Commun.* **17**, 469–471.
Greaves, R. W. (1976). The piezoelectric effect, transducers and their applications. *Ferroelectrics* **14**, 691.
Griscom, D. L. (1979). Point defects in α-quartz. *Proc. 33rd Annu. Freq. Control Symp.*, pp. 98–109.
Gross, U., and Kersten, R. (1972). Automatic crystal pulling with optical diameter control using a laser beam. *J. Cryst. Growth* **15**, 85–88.
Guseva, I. N. (1968). Optical investigation of inhomogeneities of crystals grown by various methods. *J. Cryst. Growth* **3–4**, 723–727.
Hair, M. L. (1973). The molecular nature of absorption on silica surfaces. *Proc. 27th Annu. Freq. Control Symp.*, pp. 73–78.
Halliburton, L. E., Markes, M. E., and Martin, J. J. (1980a). Point defects in synthetic quartz: A survey of spectroscopic results. *Proc. 34th Annu. Freq. Control Symp.*, pp. 1–8.
Halliburton, L. E., Kappers, L. A., Armington, A. F., and Larkin, J. (1980b). Radiation effects in synthetic berlinite ($AlPO_4$). *J. Appl. Phys.* **51**, 2193–2198.
Halliburton, L. E., Koumvakalis, N., Markes, M. E., and Martin, J. J. (1981). Radiation effects in crystalline SiO_2: The role of aluminum. *J. Appl. Phys.* **52**, 3565–3574.
Hamid, S. A. (1977). Piezoelectric properties of $Li_{1-x}H_xIO_3$. *Phys. Status Solidi A* **43** K29–30.
Hanson, W. P. (1983). Computer controlled quartz electrodiffusion (sweeping) with real time data collection. *Proc. 37th Annu.Freq. Control Symp.*, pp. 261–264.
Hanyu, T. (1964). Dislocation etch tunnels in quartz crystals. *J. Phys. Soc. Japan* **19**, 1489.
Hartemann, P. (1976). Characteristics of non-crystalline films obtained by implanting ions in piezoelectric materials. *Rev. Phys. Appl.* **12**, 843–848. In French.
Hartemann, P. (1982). Effects of ion implantation on crystalline quartz. *Ferroelectrics* **42**, 197–201.
Haussuhl, S. (1968). Piezoelectric and electrical behavior of lithium iodate. *Phys. Status Solidi* **29**, 159–162. In German.
Haussuhl, S. (1972). Piezoelectric, elastic and optical properties of $Al(IO_3)_3 2HIO_3 \cdot 6H_2O$. *Z. Kristallogr.* **135**, 287–293. In German.
Hellwege, K. H., and Hellwege, A. M. (eds.) (1979). "Elastic, Piezoelectric, Pyroelectric Piezooptic, Electrooptic Constants and Nonlinear Dielectric Susceptibilities of Crystals," group III, vol. II. Landolf–Boernstein, New Series, Springer-Verlag, Berlin.
Henaff, J., Feldmann, M., and Kirov, M. A. (1982). Piezoelectric crystals for surface acoustic waves (Quartz, $LiNbO_3$, $LiTaO_3$, Tl_3VS_4, Tl_3TaSe_4, $AlPO_4$, GaAs). *Ferroelectrics* **42**, 161–185.
Henry, N. F. M., and Lonsdal, K. (1952). International tables for X-ray crystallography, Vol. 1. Kynoch Press, Birmingham, England.
Hofacker, M., and v Löhneysen, H. (1981). Low temperature thermal properties of crystalline quartz after electron irradiation. *Z. Phys. B* **42**, 291–296.
Holland, R., and EerNisse, E. P. (1969). Accurate method for measuring piezoelectric coefficients. *IEEE Trans. Sonics Ultrason.* **SU-16**, 21.
Homma, S., and Iwata, M. (1973). X-ray topography and EPMA studies of synthetic quartz. *J. Cryst. Growth* **19**, 125–132.

Hosaka, M., Taki, S., Nagai, K., and Asahara, J. (1981). Synthetic quartz crystals grown in NaCl, KCl solution and pure water, and their low temperature infrared absorption. *Proc. 35th Annu. Freq. Control Symp.*, pp. 304–311.
Hruska, K. (1969). Verification of the interpretation of the polarizing effect with piezoelectric cuts. *Czech. J. Phys., Sect. B*, **19**, 1092.
Hruska, K., and Janik, L. (1968). Change in elastic coefficients and moduli of α-quartz in an electric field. *Czech. J. Phys. Sect. B*. **18**, 112–116.
Hruska, K., and Kazda, V. (1968). The polarizing tensor of the elastic coefficients and moduli for α-quartz. *Czech. J. Phys. Sect. B* **18**, 500–503.
Hughes, R. C. (1975). Electronic and ionic charge carriers in irradiated single crystal and fused quartz. *Radia. Eff.* **26**, 225–235.
IEEE Standard on Piezoelectricity (1978). ANSI/IEEE Std 1976–1978. IEEE, New York.
IRE Standards on Piezoelectricity (1949). *Proc. IRE* **37**, 1378–1395.
Isoya, J., Weil, J. A., and Halliburton, L. E. (1981). EPR and ab initio SCF-MO studies of the Si·H-Si system in the E'_4 center of α-quartz. *J. Chem. Phys.* **74**, 5436–5448.
Iwasaki, F. (1977). Line defects and etch tunnels in synthetic quartz. *J. Cryst. Growth* **39**, 291–298.
Iwasaki, F. (1980). Hydrogen bonded OH in synthetic quartz. *Jpn. J. Appl. Phys.* **19**, 1247–1256.
Iwasaki, F., and Kurashige, M. (1978). Lattice distortions and optical inhomogeneities in synthetic quartz. *Jpn. J. Appl. Phys.* **17**, 817–824.
Iwasaki, F., and Kurashige, M. (1982). Defects in synthetic quartz and their effects on the vibrational characteristics. *Ferroelectrics* **43**, 43–50.
Iwasaki, F., Shinada, T., and Matsuki, G. (1970). Lattice defects in artificial quartz. *Nippon Kessho Gakkaishi* **12**, 25–34.
Iwasaki, H., Miyazawa, S., Yamada, T., Uchida, N., and Niizeki, N. (1972). Single crystal growth and physical properties of $LiTaO_3$. *Rev. Electr. Commun. Lab.* **20**, 129–137.
Jain, H., and Nowick, A. S. (1982a). Electrical conductivity of synthetic and natural quartz crystals. *J. Appl. Phys.* **53**, 477–484.
Jain, H., and Nowick, A. S. (1982b). Radiation-induced conductivity in quartz crystals. *J. Appl. Phys.* **53**, 485–489.
Jalilian-Nosraty, M., and Martin, J. J. (1981). The effects of irradiation and electrolysis on the thermal conductivity of synthetic quartz. *J. Appl. Phys.* **52**, 785–788.
Jarzebski, Z. M. (1974). Review of proposed defect structures in $LiNbO_3$. *Mater. Res. Bull.* **9**, 233–240.
Jemna, I., and Ioan, C. (1979). Preparation and properties of piezoelectric materials such as $Sr\,Pb_{1-x}(Zr_{0.545}Ti_{0.455})O_3$. *Bul. Inst. Polytech. Jassy Sect. I (Rumania)*, 25, 117–123. In Rumanian.
Jhunjhunwala, A., Vetelino, J. F., and Field, J. C. (1976). Berlinite, a temperature compensated material for surface acoustic wave applications. *Ultrason. Symp. Proc.*, pp. 523–527.
Jhunjhunwala, A., Field, J. C., and Vetelino, J. F. (1977). Berlinite, a temperature-compensated material for surface acoustic wave applications. *J. Appl. Phys.* **48**, 887–892.
Jipson, V. B., Vetelino, J. F., Jhunjhunwala, A., and Field, J. C. (1976). Lithium iodate—a new material for surface acoustic wave applications. *Proc. IEEE* **64**, 568–569.
Jones, G. R., Young, I. M., Burgess, J. W., O'Hara, C., Whatmore, R. W. (1980). The growth and piezoelectric properties of $Te_2V_2O_9$ [for surface acoustic wave device application]. *J. Phys. D* **13**, 2143–2149.
Julian, C. L., and Lane, F. O., Jr. (1968a). Calculation of the elastic constants of alpha quartz from a model. *J. Appl. Phys.* **39**, 3931–3932.
Julian, C. L., and Lane, F. O., Jr. (1968b). Calculation of the piezoelectric constants of quartz on Born's theory. *J. Appl. Phys.* **39**, 2316–2324.

Kahan, A. (1977). Radiation effects in quartz. *Proc. Natl. Bur. Stand.* Seminar on Time and Frequency: Standards Measurements, Usage. Boulder, Colorado, 10.1–10.32.
Kahan, A. (1982). Elastic constants of quartz and their temperature coefficients. *Proc. 36th Annu. Freq. Control Symp.*, pp. 159–169.
Kaminow, I. P. (1974). "An Introduction to Electrooptic Devices." Academic Press, New York.
Katz, S., Halperin, A., and Schieber, M. (1982). Growth tunnels in quartz crystals. *Proc. 36th Annu. Freq. Control Symp.*, pp. 193–196.
Katz, S., Halperin, A., and Ronen, M. (1983a). Thermoluminescence from different growth sectors in synthetic quartz crystals. *Proc. 37th Annu. Freq. Control Symp.*, pp. 181–184.
Katz, S., Halperin, A., and Schieber, M. (1983b). Characterization of quartz crystals by cathodoluminescence. *Proc. 37th Annu. Freq. Control Symp.*, pp. 185–186.
Khromova, N. N. (1975). Elastic, piezoelectric, and dielectric constants of polydomain and nonstoichiometric crystals of lithium niobate. *Inorg. Mater.* **11**, 1233–1236.
Kim, C. K., and Kim, Y. I. (1971). Edge defect of synthetic quartz and its optimum growth conditions. *Suhak Kwa Mulli* **15**, 33–36. In Korean.
Kim, Y. S., and Smith, R. T. (1969). Thermal expansion of lithium tantalate and lithium niobate single crystals. *J. Appl. Phys.* **40**, 4637–4641.
Kimura, M., Doi, K., Nanamatsu, S., and Kawamura, T. (1973). A new piezoelectric crystal $Ba_2Ge_2TiO_8$. *Appl. Phys. Lett.* **23**, 531–532.
King, J. C. (1959). The anelasticity of natural and synthetic quartz at low temperatures. *Bell Syst. Tech. J.* **38**, 573–602.
King, J. C., and Sander, H. H. (1972). Rapid annealing of frequency change in crystal resonators following pulsed x-irradiation. *IEEE Trans. Nuc. Sci.* **NS-19**, 23–32.
King, J. C., and Sander, H. H. (1975). Transient changes in quartz resonators following exposure to pulse ionization. *Radiat. Eff.* **26**, 203–212.
King, J. C., Ballman, A. A., and Laudise, R. A. (1962). Improvement of the mechanical Q of quartz by the addition of impurities to the growth solution. *J. Phys. Chem. Solids* **23**, 1019–1021.
Kiselev, D. F., and Firsova, M. M. (1973). Measurements of the piezoelectric coefficient of lithium niobate by the interference dilatometer method. (Apparatus description.) *Sov. Phys.-Solid State* **15**, 198–199.
Klemens, P. G. (1965). Effect of thermal and phonon processes on ultrasonic attenuation. *In* "Physical Acoustics" (W. P. Mason, ed.), vol. III, Part B, pp. 201–234. Academic Press, New York.
Knolmayer, E. (1980). Exact netplane data for quartz, $LiTaO_3$ and $LiNbO_3$. *Proc. 34th Annu. Freq. Control Symp.*, pp. 102–111.
Koehler, D. R. (1981). Radiation-induced conductivity and high temperature Q changes in quartz resonators. *Proc. 35th Annu. Freq. Control Symp.*, pp. 322–328.
Kolb, E. D. (1979). Solubility, crystal growth and perfection of aluminum orthophosphate. *Proc. 33rd Annu. Freq. Control Symp.*, pp. 88–97.
Kolb, E. D., and Laudise, R. A. (1976). The phase diagram, $LiOH-Ta_2O_5-H_2O$ and the hydrothermal synthesis of $LiTaO_3$ and $LiNbO_3$. *J. Cryst. Growth* **33**, 145–149.
Kolb, E. D., Laudise, R. A. (1977). Hydrothermal synthesis of aluminum orthophosphate. *Proc. 31st Annu. Freq. Control Symp.*, pp. 178–181.
Kolb, E. D., and Laudise, R. A. (1978). Hydrothermal synthesis of aluminum orthosphosphate. *J. Cryst Growth* **43**, 313–319.
Kolb, E. D., and Laudise, R. A. (1981). Recent progress on aluminum phosphate crystal growth. *Proc. 35th Annu. Freq. Control Symp.*, pp. 291–296.
Kolb, E. D., Nassau, K., Laudise, R. A., Simpson, E. E., and Kroupa, K. M. (1976). New sources of quartz nutrient for the hydrothermal growth of quartz. *J. Cryst. Growth* **36**, 93–100.

Kolb, E. K., Barns, R. L., Laudise, R. A., and Rosenberg, R. L. (1979). Solubility, crystal growth and perfection of aluminum orthophosphate. *Proc. 33rd Annu. Freq. Control Symp.*, pp. 88–97.
Kolb, E. D., Glass, A. M., Rosenberg, R. L., Grenier, J. C., and Laudise, R. A. (1981a). Dielectric and piezoelectric properties of aluminum Phosphate. *Ultrason. Symp. Proc.*, pp. 332–336.
Kolb, E. D., Grenier, J. C., and Laudise, R. A. (1981b). Solubility and growth of $AlPO_4$ in a hydrothermal solvent: HCl. *J. Cryst. Growth* **51**, 178–182.
Kolb, E. D., Key, P. L., and Laudise, R. A. (1983). Pressure–volume–temperature behavior in the system H_2O-NaOH-SiO_2 and its relationship to the hydrothermal growth of quartz. *Proc. 37th Annu. Freq. Control Symp.*, pp. 153–156.
Korobov, A. I., and Lyamov, V. E. (1975). Nonlinear piezoelectric coefficients of $LiNbO_3$. *Sov. Phys.-Solid State* **17**, 932–933.
Korolyuk, A. P., Matsakov, L. Y., and Vasil'chenko, V. V. (1970). Determination of the elastic and piezoelectric constants of lithium niobate single crystals. *Kristallografiya* **15**, 1028–1032. In Russian.
Kosmodamianskii, A. S., and Lozhkin, V. N. (1977). Generalized plane stress state of a thin piezoelectric plate. *Prikl. Mekh.* **13**, 75–79. In Russian.
Koumvakalis, N. (1980). Defects in crystalline SiO_2: Optical absorption of the aluminum-associated hole center. *J. Appl. Phys.* **51**, 5528–5532.
Kraut, E. A., Tittmann, B. R., Graham, L. J., and Lim, T. C. (1970). Acoustic surface waves on metallized and unmetallized $Bi_{12}GeO_{20}$. *Appl. Phys. Lett.* **17**, 271–272.
Krefft, G. B. (1976). Effects of high-temperature electrolysis on the coloration characteristics and OH-absorption bands in alpha-quartz. *Radiat. Eff.* **26**, 249–259.
Krishna, M. M. (1979). Rotated elastic constants, coupling factors and conversion efficiencies of AC and BC quartz. *Indian J. Pure Appl. Phys.* **17**, 735–738.
Kurashige, M. (1979). Q value evaluation by IR absorption facilitates high Q quartz production. *JEE J. Electron, Eng.*, **16**, 62–65.
Kusters, J. A. (1970). The effects of static electric fields on the elastic constants of α-quartz. *Proc. 24th Annu. Freq. Control Symp.*, pp. 46–54.
Kusters, J. A., and Adams, C. A. (1980). Production statistics of SC (or TTC) crystals. *Proc. 34th Annu. Freq. Control Symp.*, pp. 167–174.
Laermans, C., De Goer, A. M., and Locatelli, M. (1980). Evidence for the presence of two level systems related to point defects in electron irradiated crystalline quartz. *Phys. Lett. A* **80**, 331–336.
Landolt-Börnstein, New Series (1979). "Elastic, Piezoelectric, Pyroelectric, Piezooptic, Electrooptic Constants and Nonlinear Dielectric Susceptibilities of Crystals." Group III, vol. 2 (K.-H. Hellwege and A. M. Hellwege, eds.) Springer-Verlag, Berlin.
Lang, A. R., and Miuscov, V. F. (1967). Dislocations and fault surfaces in synthetic quartz. *J. Appl. Phys.*, **38**, 2477–2483.
Lang, R., Calvo, C., and Datars, W. R. (1977). Phase transformation in $AlPO_4$ and quartz studied by electron paramagnetic resonance of Fe^{3+}. *Can. J. Phys.*, **55**, 1613–1620.
Laudise, R. A., and Sullivan, R. A. (1959). Pilot plant production, synthetic quartz. *Chem. Eng. Prog.* **55**, 55–59.
Laudise, R. A., and Nielsen, J. W. (1961). Hydrothermal crystal growth. *In* "Solid State Physics." (F. Seitz and D. Turnbull, eds.) **12**, 149–222. Academic Press, New York.
Laudise, R. A., Ballman, A. A., and King, J. C. (1965). Impurity content of synthetic quartz and its effect upon mechanical Q. *J. Phys. Chem. Solids* **26**, 1305–1308.
Le Page, Y., and Donnay, G. (1976). Refinement of the crystal structure of low-quartz. *Acta. Crystallogr. Sect. B* **32**, 2456–2459.

Le Page. Y., Calvert, L. D., and Gabe, E. J. (1980). Parameter variation in low-quartz between 94 and 298 K. *J. Phys. Chem. Solids* **41**, 721–725.
Lee, R., and Soluch, W. (1977). The elastic, piezoelectric, dielectric and acoustic properties of $LiIO_3$ crystals. *Ultrason. Symp. Proc.*, pp. 389–392.
Lerner, P., Legras, C., and Dumas, J. P. (1968). Stoichiometry of lithium metaniobate single crystal. *J. Cryst. Growth* **3/4**, 231–235.
Levinstein, H. J., Ballman, A. A., and Capio, C. D. (1966). Domain structure and Curie temperature of single-crystal lithium tantalate. *J. Appl. Phys.* **37**, 4585–4586.
Lias, N. C., Grudenski, E. E., Kolb, E. D., and Laudise, R. A. (1973). Growth of high acoustic Q quartz at high growth rates. *J. Cryst. Growth*, **18**, 1–6.
Lipson, H. G., Euler, F., and Armington, A. F. (1978). Low temperature infrared absorption of impurities in high grade quartz. *Proc. 32nd Annu. Freq. Control Symp.*, pp. 11–23.
Lipson, H. G., Euler, F., and Ligor, P. A. (1979). Radiation effects in swept Premium-Q quartz material, resonators and oscillators. *Proc. 33rd Annu. Freq. Control Symp.*, pp. 122–133.
Lipson, H. G., Kahan, A., Brown, R. N., and Euler, F. K. (1981). High temperature resonance loss and infrared characterization of quartz. *Proc. 35th Annu. Freq. Control Symp.*, pp. 329–334.
Lipson, H. G., Kahan, A., and O'Connor, J. J. (1983). Aluminum and hydroxide defect centers in vacuum swept quartz. *Proc. 37th Annu. Freq. Control Symp.*, pp. 169–176.
Lissalde, F., and Peuzin, J. C. (1976). X-ray determination of piezoelectric coefficient in lithium niobate. *Ferroelectrics* **14**, 579–582.
Ljamov, V. E. (1972). Nonlinear acoustical parameters of piezoelectric crystals. *J. Acoust. Soc. Am.* **52**, 199–202.
Lynch, W. T. (1972). Calculation of electric field breakdown in quartz as determined by dielectric dispersion analysis. *J. Appl. Phys.* **43**, 3274–3278.
Marculescu, L., and Hauret, G. (1973). Study of the Brillouin effect at room temperature in $LiNbO_3$. *C.R. Hebd. Seances Acad. Sci. Ser. B* **276**, 555–558. In French.
Markes, M. E., and Halliburton, L. E. (1979). Defects in synthetic quartz: Radiation-induced mobility of interstitial ions. *J. Appl. Phys.* **50**, 8172–8180.
Martin, J. J., and Doherty, S. P. (1980). The acoustic loss spectrum of 5 MHz 5th overtone AT-cut deuterated quartz resonators. *Proc. 34th Annu. Freq. Control Symp.*, pp. 81–84.
Martin, J. J., Halliburton, L. E., and Bossoli, R. B. (1981). Point defects in cultured quartz: Recent acoustic loss, infrared, and magnetic resonance results. *Proc. 35th Annu. Freq. Control Symp.*, pp. 317–321.
Martin, J. J., Halliburton, L. E., Bossoli, R. B., and Armington, A. F. (1982). The influence of crystal growth rate and electrodiffusion (sweeping) on point defects in α-quartz. *Proc. 36th Annu. Freq. Control Symp.*, pp. 77–81.
Martin, J. J., Bossoli, R. B., Halliburton, L. E., Subramaniam, B., and West, J. D. (1983). Electrodiffusion of charge-compensating ions in alpha-quartz. *Proc. 37th Annu. Freq. Control Symp.*, pp. 164–168.
Mason, W. P. (1950). Piezoelectric crystals and their applications to ultrasonics. Van Nostrand, Princeton, New Jersey.
Mason. W. P. (1965). Effect of impurities and phonon processes on the ultrasonic attenuation of germanium, crystal quartz, and silicon. *In* "Physical Acoustics" (W. P. Mason, ed.), vol. III, Part B, pp. 235–236. Academic Press, New York.
Mathur, S. S., and Gupta, P. N. (1970). Higher order elastic constants in piezoelectric crystals in the presence of ultrasonic waves. *Acustica* **23**, 160–164.
Mathur, S. S., Gupta, P. N., and Sharma, Y. P. (1971). Effect of piezoelectricity on the second order elastic constants of piezoelectric crystals. *Acustica* **24**, 108–110.
Matsumura, S. (1981). Growth conditions for large diameter X-axis $LiTaO_3$ crystals. *J. Cryst. Growth* **51**, 41–46.

Matthias, B. T., and Remeika, J. P. (1949). Ferroelectricity in the ilmenite structure. *Phys. Rev.* **76**, 1886–1887.

McLaren, A. C., Osborne, C. F., and Saunders, L. A. (1971). X-ray topographic study of dislocations in synthetic quartz. *Phys. Status Solidi A* **4**, 235–247.

McMahon, D. H., (1968). Acoustic second-harmonic generation in piezoelectric crystals. *J. Acoust. Soc. Am.* **44**, 1007–1013.

McSkimin, H. J. (1962). Measurement of the 25°C zero-field elastic moduli of quartz by high frequency plane-wave propagation. *J. Acoust. Soc. Am.* **34**, 1271–1274.

Megaw, H. D. (1973). "Crystal structures: A working approach." W. B. Saunders Company, Philadelphia.

Mérigoux, H., Darces, J. F., Zecchini, P., Lamboley, J., and Alcatel, Q. E. (1983). Nondestructive observation of random electrical twinning in cultured quartz. *Proc. 37th Annu. Freq. Control Symp.*, pp. 111–115.

Miller, R. L., Arai, G., Cocco, A., Knowles, T., and Wallner, J. (1979). Filamentary domain-reversal defects in Y-Z LiNbO$_3$: structure, composition, and visualization techniques. *Ultrason. Symp. Proc.*, pp. 589–594.

Mindlin, R. D., and Toupin, R. A. (1971). Acoustical and optical activity in alpha quartz. *Proc. 25th Annu. Freq. Control Symp.*, pp. 58–62.

Miyazawa, S., and Iwasaki, H. (1971). Congruent melting composition of lithium metatantalate. *J. Cryst. Growth* **10**, 276–278.

Morency, D. G., Soluch, W., Vetelino, J. F., Mittelman, S. D., Harmon, D., Surek, S., Field, J. C., and Lehmann, G. (1978). Experimental measurement of the SAW properties of berlinite. *Appl. Phys. Lett.* **33**, 117–119.

Moriya, K., and Ogawa, T. (1978). Observation of growth defects in synthetic quartz crystals by light-scattering tomography. *J. Cryst. Growth* **44**, 53–60.

Moriya, K., and Ogawa, T. (1980). Observation of dislocations in a synthetic quartz crystal by light scattering tomography. *Philos. Mag. A* **41**, 191–200.

Mullen, A. J. (1969). Temperature variation of the piezoelectric constant of quartz. *J. Appl. Phys.* **40**, 1693–1696.

Muller, L. (1980). The reciprocal theorem for piezoelectrics as interpreted by R. D. Mindlin. *Bull. Acad. Pol. Sci. Ser. Sci. Tech.* **28**, 27–32.

Nadratowska, B., and Pacewicz, J. (1972). Structural investigation of quartz for resonators. *Pr. Inst. Tele. Radiotech.* **16**, 57–68. In Polish.

Nagai, K. (1980). Present state of synthetic quartz. *JEE J. Electron. Eng.* **17**, 44–46.

Nakazawa, M. (1974). Generalized Hooke's law in piezoelectric and elastic body. *J. Fac. Eng. Shinshu Univ.* (34), 147–153. In Japanese.

Nakagawa, Y. (1979). Measurements of third-order piezoelectric constants and third-order dielectric constants in piezoelectric thin film using surface acoustic waves. *Oyo Butsuri* **48**, 1065–1072. In Japanese.

Nakagawa, Y., Yamanouchi, K., and Shibayama, K. (1973). Third-order elastic constants of lithium niobate. *J. Appl. Phys.* **44**, 3969–3974.

Nanamatsu, S., Doi, K., and Takahashi, M. (1973). Piezoelectric, elastic and dielectric properties of LiGaO$_2$ single crystal. *NEC Res. Dev.* no. 28, 72–79.

Naparidze, N. (1970). Piezoelectric properties of quartz. *Tr. Metrol. Inst. SSSR*, no. 106, 98–105. In Russian.

Nasalsi, W., and Zorski, H. (1975). On the equations of classical and quantum piezoelectricity. *Bull Acad. Pol. Sci. Ser. Sci. Tech.* **23**, 479–487. In Polish.

Nassau, K., and Lines, M. E. (1970). Stacking-fault model for stoichiometry deviations in LiNbO$_3$ and LiTaO$_3$ and the effect on the Curie temperature. *J. Appl. Phys.* **41**, 533–537.

Nassau, K., Levinstein, H. J., and Loiacono, G. M. (1965). The domain structure and etching of ferroelectric lithium niobate. *Appl. Phys. Lett.* **6**, 228–229.
Nassau, K., Levinstein, H. J., and Loiacono, G. M. (1966a). Ferroelectric lithium niobate. 1. Growth, domain structure, dislocations and etching. *J. Phys. Chem. Solids* **27**, 983–988.
Nassau, K., Levinstein, H. J., and Loiacono, G. M. (1966b). Ferroelectric lithium niobate. 2. Preparation of single domain crystals. *J. Phys. Chem. Solids* **27**, 989–996.
Nelson, C. M., and Weeks, R. A. (1960). Trapped electrons in irradiated quartz and silica: I. Optical absorption. *J. Am. Ceram. Soc.* **43**, 396–399.
Newnham, R. E., and Cross, L. E. (1976). Tailored domains in quartz and other piezoelectrics. *Proc. 30th Annu. Freq. Control Symp.*, pp. 71–77.
Newnham, R. E., Miller, C. S., Cross, L. E., and Cline, T. W. (1975). Tailored domain patterns in piezoelectric crystals. *Phys. Status Solidi A* **32**, 69–78.
Nicola, J. H., Scott, J. F., and Ng, H. N. (1978). Raman study of the cristobalite phase transition in $AlPO_4$. *Phys. Rev. B.* **18**, 1972–1976.
Nielsen, J. W., and Foster, F. G. (1960). Unusual etch pits in quartz crystals. *Am. Mineral.* **45**, 299–310.
Niizeki, N., Yamada, T., and Toyoda, H. (1967). Growth ridges, etched hillocks, and crystal structure of lithium niobate. *Jpn. J. Appl. Phys.* **6**, 318–327.
Nikoforov, L. G. (1977). Directed search for new piezoelectric materials. *Inorg. Mater.* **13**, 872–874.
Nowick, A. S., and Jain, H. (1980). Electrical conductivity and dielectric loss of quartz crystals before and after irradiation. *Proc. 34th Annu. Freq. Control Symp.*, pp. 9–13.
Nowick, A. S., and Stanley, M. W. (1969). Dielectric relaxation due to the Al-Na defect in α-quartz. *J. Appl. Phys.* **40**, 4995–4997.
Nuttall, R. H., and Weil, J. A. (1981). The magnetic properties of the oxygen-hole aluminum centers in crystalline SiO_2. I. $[AlO_4]°$. *Can. J. Phys.* **59**, 1696–1708.
Nye, J. F. (1957). "Physical Properties of Crystals." Oxford Univ. Press, London.
O'Connell, R. M., and Carr, P. H. (1977a). High piezoelectric coupling temperature-compensated cuts of berlinite ($AlPO_4$) for SAW applications. *IEEE Trans. Sonics Ultrason.* **SU-24**, 376–384.
O'Connell, R. M., and Carr, P. H. (1977b). New materials for surface acoustic wave (SAW) devices. *Opt. Eng.* **16**, 440–445.
O'Connell, R. M., and Carr, P. H. (1978). New materials for surface acoustic wave devices. *Ultrason. Symp. Proc.*, pp. 590–593.
Ohnishi, N., and Iizuka, T. (1975). Etching study of microdomains in $LiNbO_3$ single crystals. *J. Appl. Phys.* **46**, 1063–1067.
Onoe, M., Warner, A. W., and Ballman, A. A. (1967). Elastic and piezoelectric characteristics of bismuth germanium oxide $Bi_{12}GeO_{20}$. *IEEE Trans. Sonics Ultrason.* **SU-14**, 165–167.
Otomo, J. (1968). Hydrothermal crystal growth with special reference to quartz crystals. *Kobutsugaku Zasshi* **8**, 383–396. In Japanese.
Ozimek, E. J., and Chai, B. H.-T. (1979). Piezoelectric properties of single crystal berlinite. *Proc. 33rd Annu. Freq. Control Symp.*, pp. 80–87.
Ozkan, H. (1979). Elastic constants of tourmaline. *J. Appl. Phys.* **50**, 6006–6008.
Pajewski, W. (1969). New piezoelectric materials. *Przegl. Elektron.* **10**, 53–65. In Polish.
Palmer, A. W., and Lynch, A. C. (1979). The measurement of piezoelectric coefficients and permittivity with small specimens. *Proc. 3rd International Conference on Dielectric Materials*, pp. 385–387.
Park, D. S., and Nowick, A. S. (1974). Dielectric relaxation of point defects in α-quartz. *Phys. Status Solidi A* **26**, 617–626.

Parpia, D. Y. (1977). Growth sector boundaries and their influence on quartz resonator performance. *J. Mater. Sci.* **12**, 844–848.

Parshad, R., and Singh, V. R. (1972). Observations on the mechanical strain in quartz crystals under electric field using strain-gauge instrumentation and their application for determining the goodness of raw quartz crystals. *Proc. 26th Annu. Freq. Control Symp.*, pp. 106–107.

Pellegrini, P., Euler, F., Kahan, A., Flanagan, T. M., and Wrobel, T. F. (1978). Steady-state and transient radiation effects in precision quartz oscillators. *IEEE Trans. Nucl. Sci.* **NS-25**, 1267–1273.

Peterson, G. E., Ballman, A. A., Lenzo, P. V., and Bridenbaugh, P. M. (1964). Electro-optic properties of $LiNbO_3$. *Appl. Phys. Lett.* **5**, 62–64.

Picot, C. (1974). Piezoelectricity. I. *Toute Electron.* (385), 27–32. In French.

Poll, R. A., and Ridgway, S. L. (1966). Effects of pulsed ionizing radiation on some selected quartz oscillator crystals. *IEEE Trans. Nuc. Sci.* **NS-13**, 130–140.

Pulvari, C. F. (1982). Piezoelectricity and the emergence of the polar molecule. *Ferroelectrics* **40**, 131–132.

Rabbets, R. W. T. (1967). Man-made quartz crystal. *Electron. Commun.* **42**, 73–83.

Raeuber, A. (1978). Chemistry and physics of lithium niobate. *In* "Current Topics in Materials Science," vol. 1 (E. Kaldis, ed.). North-Holland, New York, pp. 481–601.

Raymond, B., and Gagnepain, J.-J. (1972). Determination of the nonlinear elastic polarization coefficients of quartz. *C.R. Acad. Sci. Ser. B* **274**, 835–838. In French.

Regreny, A. (1973). Hydrothermal recrystallisation of quartz and characterisation with a view to radioelectrical applications. *Ann. Telecommun.* **28**, 111–222.

Regreny, A., and Autmont, R. (1970). Hydrothermal recrystallisation of quartz. *Ann. Telecommun.* **25**, 294–306. In French.

Rudd, D., and Ballman, A. (1972). The growth of lithium tantalate, a wideband, low impedance and zero temperature coefficient of frequency piezoelectric. *Proc. 26th Annu. Freq. Control Symp.*, p. 92.

Rudd, D. W., and Ballman, A. A. (1973). Growth of lithium tantalate crystals for transmission resonator and filter devices. *West. Electr. Eng.*, **17**, 14–18.

Rudd, D. W., Fiore, A. R., and Lias, N. C. (1969). Computerized process control for synthetic quartz growth. *Proc. 23rd Annu. Freq. Control Symp.*, pp. 171–177.

Rudd, D. W., Fiore, A. R., and Lias, N. C. (1970). Computerized process control for hydrothermal growth of quartz. *J. Cryst. Growth* **7**, 29–36.

Rusu, E. (1975). A reciprocity theorem and a variational theorem for coupled mechanical and thermoelectric fields in piezoelectric crystals. *Bul. Inst. Politeh. Iasi Sect. 1* (Rumania), **21**, 85–88.

Sawyer, B. (1972). Q capability indications from infrared absorption measurements for Na_2CO_3 process cultured quartz. *IEEE Trans. Sonics Ultrason.* **SU-19**, 41–44.

Sawyer, B. (1983). Recalibration of Q capability indications from infrared measurements on cultured quartz. *Proc. 37th Annu. Freq. Control Symp.*, pp. 151–152.

Scheiding, C., and Schmidt, G. (1972). Piezoelectric d_{36} coefficient of gadolinium molybdate. *Phys. Status Solidi* **53**, K95–98.

Schichl, H. (1976). Piezoelectric materials for electromechanical filters. *Nachrichtentech. Z.* **29**, 299–301. In German.

Schnadt, R., and Schneider, J. (1970). The electronic structure of the trapped-hole center in smoky quartz. *Phys. Kondens. Materie* **11**, 19–42.

Schwarzenbach, D. (1966). Refinement of the structure of the low-quartz modification of $AlPO_4$. *Z. Kristallogr.* **123**, 161–185.

Shand, M. L., and Chai, B. H.-T. (1980). H_2O in berlinite detected by Raman scattering. *J. Appl. Phys.* **51**, 1489–1490.

Sherman, J. H. (1980). Intrinsic Q of quartz. *IEEE Trans. Sonics Ultrason.* **SU-27**, 45–46.
Sherman, J. H., Jr. (1974). Characterization of cultured quartz for use in precision AT-cut quartz resonators. *Proc. 28th Annu. Freq. Control Symp.*, pp. 129–142.
Sherman, A. B., and Lemanov, V. V. (1971). Absorption of elastic waves in reduced $LiNbO_3$. *Sov. Phys.-Solid State* **13**, 1413–1415.
Shevel'ko, M. M., and Yakovlev, L. A. (1977). Precision measurements of the elastic characteristics of synthetic piezoelectric quartz. *Sov. Phys.-Acoust* **23**, 187–188.
Shimura, F., Fujino, Y. (1977). Crystal growth and fundamental properties of $LiNb_{1-y}Ta_yO_3$. *J. Cryst. Growth* **38**, 293–302.
Shiro, Y. (1968a). The force fields and the elastic constants of crystals. I. A general treatment of the calculation of the elastic constants from force constants. *J. Sci. Hiroshima Univ. Ser. A Div. 2* **32**, 59–67.
Shiro, Y. (1968b). The force fields and the elastic constants of crystals. II. The force field and the elastic constants of quartz. *J. Sci. Hiroshima Univ. Ser. A Div. 2* **32**, 69–76.
Shmin, Y. I. (1970). Theoretical and experimental investigation of the elasticity constants of quartz. *Sov. Phys. J*, **22**, 956–961.
Shorrocks, N. M., Whatmore, R. W., Ainger, F. W., and Young, I. M. (1981). Lithium tetraborate—A new temperature compensated piezoelectric substrate material for surface acoustic wave devices. *Ultrason. Symp. Proc.*, pp. 337–340.
Sibley, W. A., Martin, J. J., Wintersgill, M. C., and Brown, J. D. (1979). The effect of radiation on the OH^- infrared absorption of quartz crystals. *J. Appl. Phys.* **50**, 5449–5452.
Silsbee, R. H. (1961). Electron spin resonance in neutron-irradiated quartz. *J. Appl. Phys.* **32**, 1459–1462.
Singh, K., and Deshmukh, K. G. (1973). Surface dendrite in lithium niobate crystals. *Mater. Res. Bull* **8**, 1139–1141.
Slobodnik, A. J., and Budreau, A. J. (1972). Acoustic surface wave loss mechanisms on $Bi_{12}GeO_{20}$ at microwave frequencies. *J. Appl. Phys.* **43**, 3278–3283.
Slobodnik, A. J., and Sethares, J. C. (1972). Elastic, piezoelectric, and dielectric constants of $Bi_{12}GeO_{20}$. *J. Appl. Phys.* **43**, 247–248.
Smith, R. T. (1967). Elastic, piezoelectric and dielectric properties of lithium tantalate. *Appl. Phys. Lett.* **11**, 146–148.
Smith, R. T., and Welsh, F. S. (1971). Temperature dependence of the elastic, piezoelectric, and dielectric constants of lithium tantalate and lithium niobate. *J. Appl. Phys.* **42**, 2219–2230.
Snow, E. H., and Gibbs, P. (1964). Dielectric loss due to impurity cation migration in α-quartz. *J. Appl. Phys.* **35**, 2368–2374.
Soluch, W., Ksiezopolski, R., Vetelino, J. F., and Jhunjhunwala, A. (1979). The piezoelectric and surface acoustic wave properties of $Bi_{12}SiO_{20}$ crystal. *Ultrason. Symp. Proc.* pp. 602–605.
Sorge, G., and Beige, H. (1975). Determination of the piezocoefficients d_{mi} from the frequency dependence of the dielectric permittivity. *Exp. Tech. Phys.* **23**, 489–493.
Soroka, V. V. (1969). Effect of impurities on the temperature dependence of the elastic constants of quartz. *Sov. Phys.-Solid State* **10**, 2241.
Sosin, A. (1973). Hydrogen diffusion in quartz: The kinetics of a one-dimensional process. *Proc. 27th Annu. Freq. Control Symp.*, pp. 136–138.
Spencer, W. J., and Haruta, K. (1966). Defects in synthetic quartz. *J. Appl. Phys.* **37**, 549–553.
Srinivasan, T. P. (1970). Invariant piezoelectric coefficients for crystals. *Phys. Status Solidi* **41**, 615–620.
Stanley, J. M. (1954). Hydrothermal synthesis of large aluminum phosphate crystals. *Ind. Eng. Chem.* **46**, 1684–1689.

Stern, R., and Smith, R. T. (1968). On the third-order elastic moduli of quartz. *J. Acoust. Soc. Am.* **44**, 640–641.
Stevels, J. M., and Volger, J. (1962). Further experimental investigations on the dielectric losses of quartz crystals in relation to their imperfection. *Philips Res. Rep.* **17**, 283–314.
Suchanek, J. (1971). The use of infrared absorption in the comparison of synthetic quartz quality. *Cesk. Cas. Fis. A.* **21**, 393–395. In Czech.
Sugii, K., and Iwasaki, H. (1973). Observation of Pendelloesung fringes in a melt-grown lithium niobate single crystal. *J. Appl. Crystallogr.* **6**, 97–98.
Sugii, K., Iwasaki, H., Miyazawa, S., and Niizeki, N. (1973). An X-ray topographic study on lithium niobate single crystals. *J. Cryst. Growth* **18**, 159–166.
Sugii, L. Koizumi, H., Miyazawa, S., and Kondo, S. (1976). Temperature variations of lattice parameters of $LiNbO_3$, $LiTaO_3$ and $Ki(Nb_{1-y}Ta_y)O_3$ solid solutions. *J. Cryst. Growth* **33**, 199–202.
Suzuki, C. K., Iwasaki, F., and Kohra, K. (1980). Studies of micron order defects in quartz by a high angular resolved x-ray small angle scattering technique. *Proc. 34th Annu. Freq. Control Symp.*, pp. 14–24.
Taki, S., and Asahara, J. (1972). Physical properties of synthetic quartz and its electrical characteristics. *Proc. 26th Annu. Freq. Control Symp.*, pp. 93–105.
Taylor, A. L., and Fernell. G. W. (1964). Spin-lattice interaction experiments on color centers in quartz. *Can. J. Phys.* **42**, 595–607.
Teague, J. R., Rice, R. R., and Gerson, R. (1975). High-frequency dielectric measurements on electro-optic single crystals. *J. Appl. Phys.* **46**, 2864–2866.
Thomas, L. A., and Wooster, W. A. (1951). Piezocrescence—The growth of Dauphiné twinning in quartz under stress. *Proc. R. Soc. London Ser. A* **208**, 43–62.
Tichy, J., Zelenka, J., Chalupa, B., Michalec, R., and Petrzilka, V. (1969). The application of neutron diffraction to the determination of the quality of synthetic quartz crystals. *Brit. J. Appl. Phys. Ser. 2* **2**, 1041–1044.
Tiersten, H. F. (1969). "Linear Piezoelectric Plate Vibrations." Plenum Press, New York.
Touloukian, Y. S., and Buyco, E. H. (1970). "Specific heat, nonmetallic solids." *Thermophysical Properties of Matter Series* **5**, 207–209. IFI/Plenum, New York.
Touloukian, Y. S., Powell, R. W., Ho, C. Y., and Klemens, P. G. (1970). "Thermal Conductivity, Nonmetallic Solids." *Thermophysical Properties of Matter Series* **2**, 174–182. IFI/Plenum, New York.
Touloukian, Y. S., Kirby, R. K., Taylor, R. E., and Lee, T. Y. R. (1977). "Thermal Expansion Nonmetallic Solids." *Thermophysical Properties of Matter Series* **13**, 350–357. IFI/Plenum, New York.
Toulouse, J., Green, E. R., and Nowick, A. S. (1983). Effect of alkali ions on electrical conductivity and dielectric loss of quartz crystals. *Proc. 37th Annu. Freq. Control Symp.*, pp. 125–129.
Uchino, K., and Cross, L. E. (1979). A high-sensitivity AC dilatometer for the direct measurement of piezoelectricity and electrostriction. *Proc. 33rd Annu. Freq. Control Symp.*, pp. 110–117.
Ushakovskii, V. T., Kashkurov, K. F., and Simonov, A. V. (1968). The overgrowth of holes in artificial quartz crystals. *Kristallografiya* **13**, 559–560. In Russian.
Uspyenskaya, A. B., and Sobolev, G. A. (1969). The piezoelectric field of a quartz layer. *Izv. Akad. Sci. USSR Phys. Solid Earth* **4**, 234–237.
Vasilchenko, V. V., Golik, O. V., Korolyuk, O. P., and Matsakov, L. Y. (1982). Investigation of elastic and electric characteristics of synthetic piezoquartz at temperature of 4.2 K. *Dopov. Akad. Nauk Ukr. RSR. Ser. A* (2), 74–76. In Ukrainian.

Vig, J. R., LeBus, J. W., and Filler, R. L. (1977). Chemically polished quartz. *Proc. 31st Annu. Freq. Control Symp.*, pp. 131–143.
Vig, J. R., Brandmayr, R. J., and Filler, R. L. (1979). Etching studies on singly and doubly rotated quartz plates. *Proc. 33rd Annu. Freq. Control Symp.*, pp. 351–358.
Voigt, W. (1910). "Lehrbuch der Kristallphysik." B. G. Teubner, Leipzig.
Warner, A. W. (1960). Design and performance of ultraprecise 2.5-mc quartz crystal units. *Bell Syst. Tech. J.* **39**, 1193–1217.
Warner, A. W., Onoe, M., and Coquin, G. A. (1967). Determination of elastic and piezoelectric constants for crystals in class (3m). *J. Acoust. Soc. Am.* **42**, 1223–1231.
Wasim, S., and Nava, R. (1979). Thermal conductivity of γ-irradiated α-quartz below 10 K. *Phys. Status Solid: A* **51**, 359–366.
Weeks, R. A. (1956). Paramagnetic resonance of lattice defects in irradiated quartz. *J. Appl. Phys.* **27**, 1376–1381.
Weeks, R. A. (1963). Paramagnetic spectra of E'_2 centers in crystalline quartz. *Phys. Rev.* **130**, 570–576.
Weeks, R. A., and Nelson, C. M. (1960). Trapped electrons in irradiated quartz and silica: II, Electron spin resonance. *J. Am. Ceram. Soc.* **43**, 399–404.
Weil, J. A. (1973). The aluminum centers in α-quartz. *Proc. 27th Annu. Freq. Control Symp.*, pp. 153–156.
Weil, J. A. (1975). The aluminum centers in α-quartz. *Radiat. Eff.* **26**, 261–265.
Weinert, R. W., and Isaacs, T. J. (1975). New piezoelectric materials which exhibit temperature stability for surface waves. *Proc. 29th Annu. Freq. Control Symp.*, pp. 139–142.
Wert, C. (1966). Determination of the diffusion coefficient of impurities by anelastic methods. *In* "Physical Acoustics" (W. P. Mason, ed.), vol. III, Part A, pp. 43–75. Academic Press, New York.
Westphal, W. B. (1961). Personal communication referred to in Landolt–Boernstein New Series, Group III, vol. 2, 453, Springer-Verlag, Berlin.
Whatmore, R. W. (1980). New polar materials: their application to SAW and other devices. *J. Cryst. Growth* **48**, 530–547.
Whatmore, R. W., O'Hara, C., Cockayne, B., Jones, G. R., and Lent, B. (1979). $Ca_{12}Al_{14}O_{33}$: a new piezoelectric material. *Mater. Res. Bull.* **14**, 967–972.
Whatmore, R. W., Shorrocks, N. M., O'Hara, C., Ainger, F. W., and Young, I. M. (1981). Lithium tetraborate: a new temperature-compensated SAW substrate material. *Electron. Lett.* **17**, 11–12.
White, G. K. (1964). Thermal expansion of silica at low temperatures. *Cryogenics* **4**, 2–7.
Willard, G. W. (1946). Use of the etch technique for determining orientation and twinning in quartz crystals. *In* "Quartz Crystals for Electrical Circuits" (R. A. Heising, ed.), pp. 164–204. Von Nostrand, New York.
Willibald, E., Born, E., and Knauer, U. (1982). Electrical twinning in quartz SAW devices *Ultrason. Symp. Proc.*, pp. 372–375.
Wyckoff, R. W. G. (1966). "Crystal Structures," 2nd ed. pp. 1–4. Wiley, New York.
Yakovlev, L. A., and Kirov, E. A. (1971). Ultrasonic method of determining the piezoelectric and elastic constants of piezoelectric materials. *Ind. Lab.* **37**, 1881–1884.
Yamada, T. (1973). Single-crystal growth and piezoelectric properties of lead potassium niobate. *Appl. Phys. Lett.* **23**, 213–214.
Yamada, T. (1975). Elastic and piezoelectric properties of lead potassium niobate. *J. Appl. Phys.* **46**, 2894–2898.
Yamada, T., Iwasaki, H., and Niizeki, N. (1969). Piezoelectric and elastic properties of $LiTaO_3$: Temperature characteristics. *Jap. J. Appl. Phys.* **8**, 1127–1132.

Yamaguchi, S., and Wada, H. (1971). Electron diffraction measuring technique for the investigation of the pyro- or piezoelectricity of crystals. *Messtechnik* **79**, S135–136. In German.
Yip, K. L., and Fowler, W. B. (1975). Electronic structure of E'_1 centers in SiO_2. *Phys. Rev. B* **11**, 2327–2338.
Yoda, H. (1972). Synthetic crystals of Japan. *JEE Jpn. Electron. Eng.* (67), 41–48.
Yoda, H. (1973). Technical aspects of crystals: from material cultivation to finished product. *JEE Jpn. Electron. Eng.* (79), 39–45.
Yoda, H. (1978). Japan achieves new synthetic quartz crystal technology. *JEE Jpn. Electron. Eng.* (144), 31–35.
Yoda, H., Taki, S., Asahara, J., and Okano, S. (1968). Quality and cost of synthetic quartz. *Proc. 22nd Annu. Freq. Control Symp.*, pp. 15–34.
Yoshimudra, J., and Kohra, K. (1976). Studies on growth defects in synthetic quartz by x-ray topography. *J. Cryst. Growth* **33**, 311–323.
Young, R. A., and Post, B. (1962). Electron density and thermal effects in alpha quartz. *Acta. Crystallogr.* **15**, 337–346.
Zachariasen, W. H., and Plettinger, H. A. (1965). Extinction in quartz. *Acta. Crystallogr.* **18**, 710–716.
Zakirova, A. D., Balitskii, V. S., and Chuvyrov, A. N. (1977). Investigation of optical and piezoelectric properties of quartz with germanium additions. *Pis'ma V. Zh. Tekh. Fiz.* **3**, 1206–1208. In Russian.
Zelenka, J. (1978). Electromechanical properties of bismuth germanium oxide ($Bi_{12}GeO_{20}$). *Czech. J. Phys. Sect. B* **B28**, 165–169.
Zelenka, J., and Lee, P. C. Y. (1971). Temperature coefficients of the elastic stiffness and compliances of alpha quartz. *IEEE Trans. Sonics Ultrason.* **SU-18**, 79–80.
Zhdanova, V. V., Klyuev, V. P., Lemanov, V. V., Smirnov, I. A., and Tikhonov, V. V. (1968). Thermal properties of lithium niobate crystals. *Sov. Phys.-Solid State* **10**, 1360–1362.
Zubov, V. G., and Firsova, M. M. (1974). Use of an interference dilatometer for measurement of the piezoelectric properties of crystals. *Instrum. Exp. Tech. (Engl. Transl.)* **17**, 1166–1167.
Zwikker, C. (1954). "Physical Properties of Solid Materials." Interscience (Pergamon Press), New York.

Sections 2.1 and 2.2

Adachi, M., and Kawabata, A. (1972). Piezoelectric properties of potassium tantalate-niobate single crystal. *Jpn. J. Appl. Phys.* **11**, 1855.
Adachi, M., and Kawabata, A. (1978). Elastic and piezoelectric properties of potassium lithium niobate (KLN) crystals. *Jpn. J. Appl. Phys.* **17**, 1969–1974.
Adachi, T., Tsuzuki, Y., and Takeuchi, C. (1981). Investigation of spurious modes of convex DT-cut quartz crystal resonators. *Proc. 35th Annu. Freq. Control Symp.*, pp. 149–156.
Adachi, T. Okazaki, M., and Tsuzuki, Y. (1983). Improvements of laser interferometric measure system of vibration displacements. *Proc. 37th Annu. Freq. Control Symp.*, pp. 187–193.
Adams, C. A., and Kusters, J. A. (1981). The SC cut, a brief summary. *Hewlett-Packard J.* **32**, 22–23.
Adams, C. A., Enslow, C. M., Kusters, J. A., and Ward, R. W. (1970). Selected topics in quartz crystal research. *Proc. 24th Annu. Freq. Control Symp.*, pp. 55–63.
Aleksandrov, K. S., Zaitseva, M. P., Sysoev, A. M., and Kokorin, Y. I. (1982). The piezoelectric resonator in a DC electric field. *Ferroelectrics* **41**, 3–8.

Aleksandsov, K. S., Sorokin, B. P., Kokorin, Y. I., Chetvergov, N. A., and Grekhova, T. I. (1982). Non-linear piezoelectricity in sillenite structure crystals. *Ferroelectrics* **41**, 27–33.
Alekseev, A. N., and Bondarenko, V. S. (1976). Electroelastic coefficients of a $Bi_{12}GeO_{20}$ crystal. *Sov. Phys.-Solid State* (*Engl. Transl.*) **18**, 1595–1597.
Alippi, A. (1982). Nonlinear acoustic propagation in piezoelectric crystals. *Ferroelectrics* **42**, 109–116.
Allik, H., and Hughes, T. J. R. (1970). Finite element method for piezoelectric vibration. *Int. J. Numerical Methods Eng.* **2**, 151–157.
Allington, R. W. (1975). A length–thickness flexure mode quartz resonator. *Proc. 29th Annu. Freq. Control Symp.*, pp. 195–201.
Androsova, V. G., Pozdnyakov, P. G., and Biryukov, V. I. (1975). Temperature dependence of shear oscillation frequency in quartz bars. *Sov. Phys.-Dokl.* (*Engl. Transl.*) **19**, 723–724.
Anonymous (1976). Quartz crystals, mechanical resonators for electronics. *Radio Mentor Electron.* **42**, 296. In German.
Anonymous. (1979). Surface-skimming bulk wave devices. *Radio Electron. Eng.* **49**, 288.
Anonymous (1980). New cut for quartz oscillator plates. *Radio Elektron.* **28**, 29. In Dutch.
Anonymous (1980). Crystals with TTC-cut. *Elektronik* (*Denmark*) (3), 14–15. In Danish.
Anonymous (1981). Techniques of quartz crystals. *Toute Electron.* (*France*) (468), 67–68. In French.
Anonymous (1981). Unconventional cuts for quartz resonators. *Electron. Components Appl.* (*Netherlands*) **2**, 251–252.
Apostolov, A. V., and Slavov, S. H. (1982). Frequency spectrum and modes of vibration in circular, convex AT-cut bevelled-design quartz resonators. *Appl. Phys. A* **29**, 33–37.
Apostolov, A. V., Slavov, S. H., and Krustev, V. P. (1979). Acoustic vibrations in quartz plates as visualized by the Berg–Barrett method. *C.R. Acad. Bulg. Sci.* **32**, 1061–1063.
Ariga, M., and Sato, M. (1974). A high electromechanical-coupling resonator and its application to filter-synthesis. *Mem. Fac. Technol. Tokyo Metrop. Univ.* (23), 2033–2042.
Asahara, J., Yazaki, E., Takazawa, K., and Kita, K. (1975). Influences of the inclusions in synthetic quartz crystal on the electrical characteristic of quartz crystal resonator. *Proc. 29th Annu. Freq. Control Symp.*, pp. 211–219.
Ashida, T., Sawamoto, K., and Niizeki, N. (1970). Temperature dependence of X-cut $LiTaO_3$ crystal resonator vibrating in thickness shear mode of motion. *Rev. Electr. Commun. Lab.* **18**, 854–861.
Auld, B. A. (1981). Wave propagation and resonance in piezoelectric materials. *J. Acoust. Soc. Am.* **70**, 1577–1585.
Avdienko, K. I., Bogdanov, S. V., Kidyarov, B. I., Semenov, V. I., and Sheloput, D. V. (1977). Optical, acoustic, piezoelectric properties of α-$LiIO_3$ crystals. *Bull. Acad. Sci. USSR Phys. Ser.* (*Engl. Transl.*) **41**, 40–45.
Bahadur, H., and Parshad, R. (1975). Effect of magnetic field on oscillating properties of quartz crystals. *Indian J. Pure Appl. Phys.* **13**, 696–698.
Bahadur, H., and Parshad, R. (1975). Operation of quartz crystals in their overtones: new methods. *Indian J. Pure Appl. Phys.* **13**, 862–865.
Bahadur, H., and Parshad, R. (1976). Effect of magnetic field on oscillating properties of quartz crystals vibrating in the fundamental and overtone modes. *Indian J. Pure Appl. Phys.* **14**, 855–856.
Bahadur, H., and Parshad, R. (1980). Scanning electron microscopy of vibrating quartz crystals—a review. *Scanning Electron Microsc.* pt. 1, 509–522.
Bahadur, H., and Parshad, R. (1980). Simple experimental method for demonstrating transient frequency shifts in X- and gamma-irradiated quartz crystals. *Rev. Sci. Instrum.* **51**, 1420–1421.

Bahadur, H., Parshad, R., Hepworth, A., and Lall, V. K. (1978). Electron contrast effects from oscillating quartz crystals seen by the scanning electron microscope. *IEEE Trans. Sonics Ultrason.* **SU-25**, 309–313.
Bahadur, H., Parshad, R., Hepworth, A., and Lall, V. K. (1978). Some observations using scanning electron microscope for studying the ultrasonic vibrations of quartz crystals. *Proc. 32nd Annu. Freq. Control Symp.*, pp. 207–219.
Bahadur, H., Parshad, R., Hepworth, A., and Lall, V. K. (1979). Studies on SEM surface patterns of oscillating quartz crystals and applications to energy trapping. *Scanning Electron Microsc.* pt. 1, 333–338.
Balakirev, M. K., and Gorchakov, A. V. (1977). Leakage of an elastic wave across a gap between piezoelectrics. *Sov. Phys.-Solid State* (*Engl. Transl.*) **19**, 327–328.
Balbi, J. H., and Dulmet, M. I. (1981). Simple model for an AT cut rectangular quartz plate. *Proc. 35th Annu. Freq. Control Symp.*, pp. 187–192.
Balbi, J. H., Duffaud, J. A., and Besson, R. J. (1978). A new nonlinear analysis method and its application to quartz crystal resonator problems. *Proc. 32nd Annu. Freq. Control Symp.*, pp. 162–168.
Balbi, J. H., Dulmet, M., and Thirard, A. (1982). Non-rational frequency division in a quartz resonator. *Rev. Phys. Appl.* **17**, 1–7. In French.
Ballato, A. (1966). Resonance phenomena in piezoelectric vibrators. Report ECOM 3181 (AD-697114), U. S. Army Electronics Command, Fort Monmouth, New Jersey, 18 pp.
Ballato, A. (1972a). Transmission-line analogs for stacked piezoelectric crystal devices. *Proc. 26th Annu. Freq. Control Symp.*, pp. 86–91.
Ballato, A. (1972b). Transmission line analogs for piezoelectric layered structures. Dissertion, Polytechnic Institute of Brooklyn, 245 pp.
Ballato, A. (1974). Bulk and surface acoustic wave excitation and network representation. *Proc. 28th Annu. Freq. Control Symp.*, pp. 279–289.
Ballato, A. (1976). Apparent orientation shifts of mass-loaded plate vibrators. *Proc. IEEE* **64**, 1449–1450.
Ballato, A. (1977). Doubly rotated thickness mode plate vibrators. *In* "Physical Acoustics: Principles and Methods" (W. P. Mason and R. N. Thurston, eds.), Academic Press, New York, pp. 115–181.
Ballato, A. (1978). Force–frequency compensation applied to four-point mounting of AT-cut resonators. *IEEE Trans. Sonics Ultrason.* **SU-25**, 223–236.
Ballato, A. (1978). The fluency matrix of quartz. *IEEE Trans. Sonics Ultrason.* **SU-25**, 107–108.
Ballato, A. (1979). Static and dynamic behavior of quartz resonators. *IEEE Trans. Sonics Ultrason.* **SU-26**, 299–306.
Ballato, A. (1979). Resonators compensated for acceleration fields. *Proc. 33rd Annu. Freq. Control Symp.*, pp. 322–336.
Ballato, A. (1980). Crystal resonators with increased immunity to acceleration fields. *IEEE Trans. Sonics Ultrason.* **SU-27**, 195–201.
Ballato, A., and Iafrate, G. J. (1976). The angular dependence of piezoelectric plate frequencies and their temperature coefficients. *Proc. 30th Annu. Freq. Control Symp.*, pp. 141–156.
Ballato, A., and Lukaszek, T. (1973). Mass effects on crystal resonators with arbitrary piezo-coupling. *Proc. 27th Annu. Freq. Control Symp.*, pp. 20–29.
Ballato, A., and Lukaszek, T. (1974). Mass-loading effects on crystal resonators excited by thickness electric fields. Report ECOM-4270, U. S. Army Electronics Command, Fort Monmouth, New Jersey, 80 pp.
Ballato, A., and Lukaszek, T. J. (1974). Mass-loading of thickness-excited crystal resonators having arbitrary piezo-coupling. *IEEE Trans. Sonics Ultrason.* **SU-21**, 269–274.
Ballato, A., and Lukaszek, T. (1975). Distributed network modelling of bulk acoustic waves in

crystal plates and stacks. Report ECOM-4311, U. S. Army Electronics Command, Fort Monmouth, New Jersey, 19 pp.
Ballato, A., and Lukaszek, T. (1975). Higher-order temperature coefficients of frequency of mass-loaded piezoelectric crystal plates. *Proc. 29th Annu. Freq. Control Symp.*, pp. 10–25.
Ballato, A., and Lukaszek, T. (1979). Shallow bulk acoustic wave devices. *IEEE MTT-S Int. Microwave Symp. Dig.*, pp. 162–164.
Ballato, A., and Lukaszek, T. J. (1979). Shallow bulk acoustic wave progress and prospects. *IEEE Trans. Microwave Theory Tech.* **MTT-27**, 1004–1012.
Ballato, A., Lukaszek, T., and Iafrate, G. J. (1980). Subtle effects in high stability vibrators. *Proc. 34th Annu. Freq. Control Symp.*, pp. 431–444.
Ballato, A., and Lukaszek, T. (1980). Waves in piezoelectric crystals for frequency control and signal processing. *Proc. Soc. Photo-Opt. Instrum. Eng.* **239**, 162–169.
Ballato, A., and Mizan, M. (1984). Simplified expressions for the stress-frequency coefficients of quartz plates. *IEEE Trans. Sonics Ultrason.* **SU-31**, 11–17.
Ballato, A., and Vig, J. R. (1978). Static and dynamic frequency–temperature behavior of singly and doubly rotated, oven-controlled quartz resonators. *Proc. 32nd Annu. Freq. Control Symp.*, pp. 180–188.
Ballato, A., Bertoni, H. L., and Tamir, T. (1972). Transmission-line analogs for stacked crystals with piezoelectric excitation. *Program 83rd Meeting of the Acoustical Society of America. Abstracts only* (*New York: Acoust. Soc. America*), p. 85.
Ballato, A., Bertoni, H. L., and Tamir, T. (1974). Systematic design of stacked-crystal filters by microwave network methods. *IEEE Trans. Microwave Theory Tech.* **MTT-22**, 14–25.
Ballato, A., EerNisse, E. P., and Lukaszek, T. (1977). The force–frequency effect in doubly rotated quartz resonators. *Proc. 31st Annu. Freq. Control Symp.*, pp. 8–16.
Ballato, A., EerNisse, E. P., and Lukaszek, T. J. (1978). Experimental verification of stress compensation in the SC cut. *Ultrason. Symp. Proc.*, pp. 144–147.
Ballato, A., Lukaszek, T. J., and Iafrate, G. J. (1982). Subtle effects in high-stability quartz resonators. *Ferroelectrics* **43**, 25–41.
Baranovskii, S. N., and Shestopal, V. O. (1975). Variation of the frequency of a piezoelectric resonator under the influence of friction forces as a function of the pressure on the contact surfaces. *Sov. Phys.-Acoust.* **21**, 7–9.
Barcus, L. C. (1975). Nonlinear effects in the AT cut quartz resonator. *IEEE Trans. Sonics Ultrason.* **SU-22**, 245–250.
Barcus, L. C. (1978). Holographic displacement amplitude measurements of four anharmonic AT modes. *Proc. 32nd Annu. Freq. Control Symp.*, pp. 202–206.
Bardati, F., Barzilai, G., and Gerosa, G. (1968). Elastic wave excitation in piezoelectric slabs. *IEEE Trans. Sonics Ultrason.* **SU-15**, 193–202.
Barsch, G. R., and Newnham, R. E. (1975). Piezoelectric materials with positive elastic constant temperature coefficients. AFCRL-TR-75-0163, Final Report, Contract No. F19628-73-C-0108.
Baryshnikova, L. F., and Lyamov, V. E. (1978). Excitation of an acoustic harmonic and its energy flux in piezoelectric crystals. *Sov. Phys.-Solid State* (*Engl. Transl.*) **20**, 640–643.
Baryshnikova, L. F., and Lyamov, V. E. (1978). Harmonic generation in piezoelectric crystals under the action of homogeneous external fields. *Sov. Phys.-Acoust.* (*Engl. Transl.*) **24**, 167–170.
Baryshnikova, L. F., and Lyamov, V. E. (1980). Elliptical polarization of acoustic waves in piezoelectric crystals under the action of an electric field. *Sov. Phys.-Acoust.* (*Engl. Transl.*) **26**, 465–467.
Bateman, T. B., Mason, W. P., and McSkimin, H. J. (1961). Third-order elastic moduli of gemanium. *J. Appl. Phys.* **32**, 928–936.

Baumhauer, J. C., and Tiersten, H. F. (1973). Nonlinear electroelastic equations for small fields superposed on a bias. *J. Acoust. Soc. Am.* **54**, 1017–1034.
Bayle, D. (1974). On the resonance vibrations of quartz and the disturbance of the normal frequencies following two preferred directions. *Acustica* **30**, 336–341. In French.
Beaver, W. D. (1968). Analysis of elastically coupled piezoelectric resonators. *J. Acoust. Soc. Amer.* **43**, 972–981.
Beaver, W. D. (1973). Design equations for bi- and plano-convex AT-cut resonators. *Proc. 27th Annu. Freq. Control Symp.*, pp. 11–19.
Beaver, W. D. (1983). Thickness modes in circular AT-cut quartz plates with circular electrodes. *Proc. 37th Annu. Freq. Control Symp.*, pp. 226–231.
Bechmann, R. (1934). The temperature coefficients of the natural frequencies of piezoelectric quartz plates and bars. *Hochfrequenztech. Electroakust.* **44**, 145–160.
Bechmann, R. (1939). Piezoelectric Plate. U. S. Patent No. 2,249,933, issued 22 July 1941.
Bechmann, R. (1954). Contour modes of plates excited piezoelectrically and determination of elastic and piezoelectric coefficients. *IRE Nat. Conv. Rec.* **2** (pt. 6), 77–85.
Bechmann, R. (1955a). Some applications of the linear piezoelectric equations of state. *IRE Trans. Ultrason. Eng.* **3**, 43–62.
Bechmann, R. (1955b). Influence of the order of overtone on the temperature coefficient of frequency of AT-cut quartz resonators. *Proc. IRE* **43**, 1667–1668.
Bechmann, R. (1956). Elastic, piezoelectric, and dielectric constants of polarized barium titanate ceramics and some applications of the piezoelectric equations. *J. Acoust. Soc. Amer.* **28**, 347–350.
Bechmann, R. (1958). Filter crystals. *Proc. 12th Annu. Freq. Control Symp.*, pp. 437–473.
Bechmann, R. (1961). Thickness-shear mode quartz cut with small second and third order temperature coefficients of frequency (RT-cut). *Proc. IRE* **49**, 1454.
Bechmann, R. (1962). The higher order temperature coefficients of the elastic moduli and the elastic coefficients of alpha quartz. *Arch. Elektr. Uebertragung* **16**, 307–313; 534.
Bechmann, R., and Ayers, S. (1957). The theory of dynamical determination of elastic and piezoelectric constants. *In* "Piezoelectricity," Eng. Rept. No. 4, Her Majesty's Stationery Office, London, pp. 73–92.
Bechmann, R., and Ayers, S. (1951). The shear elastic constants of quartz and their behavior with temperature. Post Office Research Station Engineering Dept., Dollis Hill, England, Research Report No. 13524.
Bechmann, R., Ballato, A. D., and Lukaszek, T. J. (1962). Higher order temperature coefficients of the elastic stiffnesses and compliances of alpha quartz. *Proc. IRE* **50**, 1812–1822; 2451.
Bechmann, R., Ballato, A. D., and Lukaszek, T. J. (1963). Higher order temperature coefficients of the elastic stiffnesses and compliances of alpha quartz. USAELRDL Technical Report 2261, U. S. Army, Fort Monmouth, New Jersey, 79 pp.
Bedowski, S., and Chelmonski, J. (1978). The electrodes effect on inharmonic thickness-shear vibration in AT-cut quartz crystal units. *Pr. Inst. Tele- Radiotech.* (79), 41–55. In Polish.
Belikova, G. S., Pisarevskii, Yu. V., and Sil'vestrova, I. M. (1975). Piezoelectric and elastic properties of crystals of rubidium biphthalate. *Sov. Phys.-Crystallogr. (Engl. Transl.)* **19**, 545–546.
Bellucci, A. G., and Stacchiotti, G. V. (1981). A new type of AT-cut quartz resonator with Q-factor and frequency of the unwanted modes controlled. *IEEE Trans. Sonics Ultrason.* **SU-28**, 460–467.
Berlincourt, D. A., Curran, D. R., and Jaffe, H. (1964). Piezoelectric and piezomagnetic materials and their function in transducers. *In* "Physical Acoustics: Principles and

Methods" (W. P. Mason and R. N. Thurston, eds.), Academic Press, New York, pp. 169–270.
Berté, M. (1977). Acoustic-bulk-wave resonators and filters operating in the fundamental mode at frequencies greater than 100 MHz. *Electron. Lett.* **13**, 248–250.
Berté, M., and Hartemann, P. (1978). Quartz resonators at fundamental frequencies greater than 100 MHz. *Ultrason. Symp. Proc.*, pp. 148–151.
Berthaut, A., and Besson, R. (1979). Bulk resonators at very high excitation level. *C.R. Hebd. Seances Acad. Sci. Ser. B* **289**, 223–234. In French.
Besson, R. (1974). Measurement of nonlinear elastic, piezoelectric, and dielectric coefficients of quartz crystals. Applications. *Proc. 28th Annu. Freq. Control Symp.*, pp. 8–13.
Besson, R. (1979). A new type of quartz resonator design. *C.R. Hebd. Seances Acad. Sci. Ser. B* **288**, 245–248. In French.
Besson, R. J., and Peier, U. R. (1980). Further advances on B.V.A. quartz resonators. *Proc. 34th Annu. Freq. Control Symp.*, pp. 175–182.
Besson, R. J., Dulmet, B. M., and Gillet, D. P. (1978). Analysis of the influence of some parameters in quartz bulk resonators. *Ultrason. Symp. Proc.* pp. 152–156.
Besson, R., Gagnepain, J.-J., Janiaud, D., and Valdois, M. (1979). Design of a bulk wave quartz resonator insensitive to acceleration. *Proc. 33rd Annu. Freq. Control Symp.*, pp. 337–345.
Besson, R. J., Groslambert, J. M., and Walls, F. L. (1982). Quartz crystal resonators and oscillators, recent developments and future trends. *Ferroelectrics* **43**, 57–65.
Bhattacharya, A. (1976). On disturbances in a piezoelectric slab with dissipative characteristics. *Indian J. Pure Appl. Math* **7**, 1096–1103.
Birch, J., and Marriott, S. P. (1979). Appraisal of the inverted-mesa AT-cut quartz resonator for achieving low-inductance high-Q single-response crystal units. *Electron. Lett.* **15**, 641–643.
Birch, J., and Weston, D. A. (1976). Frequency/temperature, activity/temperature anamalies in high frequency quartz crystal units. *Proc. 30th Annu. Freq. Control Symp.*, pp. 32–39.
Bleustein, J. L., and Tiersten, H. F. (1968). Forced thickness-shear vibrations of discontinuously plated piezoelectric plates. *J. Acoust. Soc. Amer.* **43**, 1311–1318.
Blinick, J. S., and Maris, H. J. (1970). Velocities of first and zero sound in quartz. *Phys. Rev. B* **3**, 2139–2146.
Bonczar, L. J., and Barsch, G. R. (1975). Elastic and thermoelastic constants of nepheline $((KAlSiO_4)(NaAlSiO_4))$. *J. Appl. Phys.* **46**, 4339–4340.
Bond, W. L. (1943). The mathematics of the physical properties of crystals. *Bell Syst. Tech. J.* **22**, 1–72.
Borisov, V. Y., Zyuryukin, Y. A., and Shevchik, V. N. (1977). Features of the propagation of elastic waves through hexagonal piezoelectric crystals. *Izv. Vyssh. Uchebn. Zaved. Fiz.* (5), 101–105. In Russian.
Borissov, M., Stoytchev, K. T., and Kovachev, M. I. (1976). Holographic study of vibrations of piezoelectric AT-cut quartz resonators. *C. R. Acad. Bulg. Sci.* **29**, 477–480.
Bottom, V. E. (1982). Private Communication.
Bourgeois, C. (1980). Decoupled families of contour modes of planar thin plates. *Proc. 34th Annu. Freq. Control Symp.*, pp. 419–425.
Bourquin, R., Nassour, D., and Hauden, D. (1982). Amplitude frequency effect of SC-cut quartz trapped energy resonators. *Proc. 36th Annu. Freq. Control Symp.*, pp. 200–207.
Boy, J.-J., and Gillet, D. (1978). Electro-acoustic effects on the phase velocities of the acoustic and electromagnetic waves in a piezoelectric crystal. *C.R. Hebd. Seances Acad. Sci. Ser. B* **287**, 203–203. In French.

Braginskii, L. S., and Gilinskii, I. A. (1979). Generalized shear surface waves in piezoelectric crystals. *Sov. Phys.-Solid State (Engl. Transl.)* **21**, 2035–2037.

Brendel, R.. and Gagnepain. J.-J. (1982). Electroelastic effects and impurity relaxation in quartz resonators. *Proc. 36th Annu. Freq. Control Symp.*, pp. 97–107.

Brice, J. C. (1981). Quartz resonators. *Phys. Educ.* **16**, 162–166.

Brice, J. C., and Fletcher, E. D. (1981). The influence of the quality factor of quartz on some device properties. *Proc. 35th Annu. Freq. Control Symp.*, pp. 312–316.

Brice, J. C., and Metcalf, W. S. (1982). Quartz-crystal resonators using an unconventional cut. *Philips Tech. Rev.* **40**, 1–11.

Browning, T. I., and Lewis, M. F. (1977). New family of bulk-acoustic-wave devices employing interdigital transducers. *Electron. Lett.* **13**, 128–130.

Brugger, K. (1964). Thermodynamic definition of higher order elastic coefficients. *Phys. Rev. A.* **133**, A1611–A1612.

Brugger, K. (1965a). Pure modes for elastic waves in crystals. *J. Appl. Phys.* **36** (pt. 1), 759–768.

Brugger, K. (1965b). Determination of third-order elastic constants in crystals. *J. Appl. Phys.* **36** (pt. 1), 768–773.

Buist, D. G. (1961). Measurement of the constants of piezoelectric vibrators. *A.M.A. Tech. Rev. (Amalgamated Wireless, Australia)* **11**, 195–202.

Burgess, J. W., and Hales, M. C. (1976). Temperature coefficients of frequency in $LiNbO_3$ and $LiTaO_3$ plate resonators. *Proc. Inst. Electr. Eng.* **123**, 499–504.

Burgess, J. W., and Muir, A. J. L. (1975). Quartz-resonator inductance. *Electron. Lett.* **11**, 502–503.

Burgess, J. W., and Porter, R. J. (1973). Single mode resonance in lithium niobate/lithium tantalate for monolithic crystal filters. *Proc. 27th Annu. Freq. Control Symp.*, pp. 246–252.

Burgess, J. W., Hales, M. C., and Porter, R. J. (1975). Equivalent-circuit parameters for $LiTaO_3$ plate resonators. *Electron. Lett.* **11**, 449–450.

Burlii, P. V., Gololobov, Y. P., Kucherov, I. Y., and Rozhko, S. K. (1979). Simultaneous excitation and investigation of inverse and direct elastic waves. *Ukr. Fiz. Zh. (Russ. Ed.)*, **24**, 1905–1907. In Russian.

Burlii, P. V., Il'in, P. P., and Kucherov, I. Ya. (1979). Control of elastic wave velocity in piezoelectric plates. *Ukr. Fiz. Zh. (Russ. Ed.)* **24**, 561–562. In Russian.

Burov, J., and Anastasova, N. (1978). Synchronous amplification and generation of flexural waves in resonator Platelets of cadium sulphide. *Bulg. J. Phys.* **5**, 164–170.

Butterworth, S. (1914). On a null method of testing vibration galvanometers. *Proc. Phys. Soc. London* **26**, 264–273.

Bye, K. L., and Cosier, R. S. (1979). An x-ray double crystal topographic assessment of defects in quartz resonators. *J. Mater. Sci.* **14**, 800–810.

Byrne, R. J., Lloyd, P. and Spencer, W. J. (1968). Thickness-shear vibration in rectangular AT-cut quartz plates with partial electrodes. *J. Acoust. Soc. Amer.* **43**, 232–238.

Cady, W. G. (1922). The piezoelectric resonator. *Proc. IRE* **10**, 83–114.

Cady, W. G. (1946). "Piezoelectricity," McGraw-Hill, New York. Rev. edition, vols. 1 and 2, (1964). Dover Publications, Inc. New York.

Camarero, E. G. (1976). Acousto-electromagnetic propagation modes in hexagonal piezoelectric crystals. *An. Fis.* **72**, 138–144. In Spanish.

Capparelli, F., and Liberatore, A. (1970). On the possibility of increasing the separation between the series and parallel frequencies of a piezoelectric resonator. *RC Riun. Ass. Elettrotec. Ital.* **45**, 6 pp. In Italian.

Carr, P. H., and O'Connell, R. M. (1976). New temperature compensated materials with high piezoelectric coupling. *Proc. 30th Annu. Freq. Control Symp.*, pp. 129–131.

Cernick, K. (1967). The problems of evaluating piezoelectric units and of the accuracy of calculating the elements of their equivalent circuits. *XII International Scientific Colloquium, Ilmenau, Germany*, pp. 25–32. In German.
Chai, B. H.-T., and Ozimet, E. J. (1979). Experimental data on the piezoelectric properties of berlinite. *Ultrason. Symp. Proc.*, pp. 577–583.
Chang, Z.-P., and Barsch, G. R. (1976). Elastic constants and thermal expansion of berlinite. *IEEE Trans. Sonics Ultrason.* **SU-23**, 127–135.
Chen, P. J., and Montgomery, S. T. (1977). Normal mode responses of linear piezoelectric materials with hexagonal symmetry. *Int. J. Solids Struct.* **13**, 947–955.
Cheng, N. C., and Sun, C. T. (1975). Wave propagation in two-layered piezoelectric plates. *J. Acoust. Soc. Am.* **57**, 632–638.
Chernykh, G. G., Bankov, V. N., and Pozdnyakov, P. G. (1971). The calculation of the characteristic frequencies of torsional vibrations of quartz piezoelements. *Kristallografiya* **16**, 792–795. In Russian.
Chernykh, G. G., Bogush, M. E., and Fedorkov, A. P. (1975). Spectral and temperature–frequency characteristics of regular X-cut quartz piezoelements. *Sov. Phys.-Dokl.* (*Eng. Transl.*) **19**, 818–819.
Chernykh, G. G., Yaroslavskii, M. I., and Gruzinenko, V. B. (1975). Temperature–frequency characteristics of quartz piezoelements developing displacement oscillations at the boundary. *Sov. Phys.-Crystallogr.* (*Engl. Transl.*) **19**, 512–513.
Chin, T. H., and Grossman, P. L. (1970). Flexural vibrations of circular plates including those with elastic support at the surface. *Trans. ASME Ser. E.* **37**, 535–537.
Choi, S. K., Chung, K. H., and Lee, M. Y. (1976). Utilization of domestic natural quartz as electric resonators; feasibility study. *J. Korean Inst. Electron. Eng.* **13**, 84–88. In Korean.
Chowdhury, B. R. (1978). A note on disturbances in a piezoelectric resonator. *Indian J. Theor. Phys.* **26**, 89–95.
Christoffel, E. B. (1877). On the propagation of percussions through elastic solid bodies. *Ann. Mat. Pura Appl. Milano* [2] **8**, 193–243. In German.
Comparini, A. A., and Honnon, J. J. (1974). Flexural, width-shear, and width-twist vibrations of thin rectangular crystal plates. *IEEE Trans. Sonics Ultrason.* **SU-21**, 130–135.
Cooper, M. (1968). The quartz crystal come of age. *18th Annu. Conf. IEEE Vehic. Tech. Grp.*, pp. 72–77.
Coquin, G. A. (1964). A footnote on page 93 of Tiersten (1964) ascribes the derivation of the tan x/x sum form of the impedance of the plate to a private communication.
Cowdrey, D. R., and Willis, J. R. (1973). Finite element calculations relevant to AT-cut quartz resonators. *Proc. 27th Annu. Freq. Control Symp.*, pp. 7–10.
Crisler, D. F., Cupal, J. J., and Moore, A. R. (1968). Dielectric, piezoelectric, and electromechanical coupling constants of zinc oxide crystals. *Proc. IEEE* **56**, 225–226.
Curran, D. R., and Koneval, D. J. (1964). Energy trapping and the design of single and multi-electrode filter crystals. *Proc. 18th Annu. Freq. Control Symp.*, pp. 93–119.
Curran, D. R., and Koneval, D. J. (1965). Factors in the design of VHF filter crystals. *Proc. 19th Annu. Freq. Control Symp.*, pp. 213–268.
Datta, S., and Hunsinger, B. J. (1977). Analysis of line acoustical waves in general piezoelectric crystals. *Phys. Rev. B* **16**, 4224–4229.
Dauwalter, C. R. (1972). The temperature dependence of the force sensitivity of AT-cut quartz crystals. *Proc. 26th Annu. Freq. Control Symp.*, pp. 108–112.
De Jong, G. (1974). Generation of thickness-twist modes in a piezoceramic plate. *J. Appl. Phys.* **45**, 996–1000.
De Jong, G. (1974). Piezoelectric trapped energy resonators vibrating in thickness-twist modes. *J. Acoust. Soc. Am.* **56**, 1158–1164.

Demidenko, A. A., Piskovoi, V. N., and Nguen, C. T. (1975). Calculation of the electromechanical parameters of substances by the resonance–antiresonance method in the presence of losses. *Ukr. Fiz. Zh. (Russ. Ed.)* **20**, 1028–1031. In Russian.

Détaint, J. (1977). Zero temperature coefficients in overtone lithium tantalate thickness-mode resonators. *Electron. Lett.* **13**, 20–21.

Détaint, J., and Lançon, R. (1976). Temperature characteristics of high frequency lithium tantalate plates. *Proc. 30th Annu. Freq. Control Symp.*, pp. 132–140.

Détaint, J., Poignant, H., and Toudic, Y. (1980). Experimental thermal behaviour of berlinite resonators. *Proc. 34th Annu. Freq. Control Symp.*, pp. 93–101.

Dieulesaint, E., Royer, D., and Rakouth, H. (1982). Optical excitation of quartz resonators. *Electron. Lett.* **18**, 381–382.

Dinger, R. (1981). A miniature quartz resonator vibrating at 1 MHz. *Proc. 35th Annu. Freq. Control Symp.*, pp. 144–148.

Dinger, R. J. (1982). The torsional tuning fork as a temperature sensor. *Proc. 36th Annu. Freq. Control Symp.*, pp. 265–269.

Doerffler, H. (1930). Bent and transverse oscillations of piezoelectrically excited quartz plates. *Z. Phys.* **63**, 30–53.

Dokmeci, M. C. (1974). A theory of high frequency vibrations of piezoelectric crystal bars. *Int. J. Solids Struct.* **10**, 401–409.

Dokmeci, M. C. (1974). On the higher order theories of piezoelectric crystal surfaces. *J. Math. Phys.* **15**, 2248–2252.

Dokmeci, M. C. (1980). Vibrations of piezoelectric crystals. *Int. J. Eng. Sci.* **18**, 431–448.

Drapal, S. (1980). Properties of materials used in electromechanical filter resonators. *Slaboproudy Obz.* **41**, 224–229. In Czech.

Driscoll, M. M. (1973). Q-multiplied quartz crystal resonator for improved HF and VHF source stabilization. *Proc. 27th Annu. Freq. Control Symp.*, pp. 157–169.

Drukker, Y. M., Gruzinenko, V. B., and Yaroslavskii, M. I. (1968). Mechanically coupled bending and contour shear vibrations of piezoelectric plates. *Kristallografiya* **13**, 714–717. In Russian.

Dulmet, B., Gillet, D., and Maitre, P. (1979). Definition of a Q-factor for electrodeless resonators operating at high frequencies. *C.R. Hebd. Seances Acad. Sci. Ser. B* **287**, 293–296. In French.

Dworsky, L. N. (1978). Discrete element modelling of AT-quartz devices. *Proc. 32nd Annu. Freq. Control Symp.*, pp. 142–149.

Dworsky, L. (1983). Properties of AT quartz resonators on wedgy plates. *Proc. 37th Annu. Freq. Control Symp.*, pp. 232–238.

Dworsky, L., and Kennedy, G. (1981). Air-gap probe evaluation of thin quartz plates. *Proc. 35th Annu. Freq. Control Symp.*, pp. 237–243.

Dybwad, G. L. (1980). Analysis of quartz resonator electrodes using the Rutherford Backscattering Technique. *Proc. 34th Annu. Freq. Control Symp.*, pp. 46–51.

Dye, D. W. (1926). The piezoelectric quartz resonator and its equivalent electrical circuit. *Proc. Phys. Soc. London* **35**, 399–458.

Edson, W. A. (1953). "Vacuum Tube Oscillators." Wiley, New York, pp. 111–120.

EerNisse, E. P. (1968). Parametric and other nonlinear elastic effects in piezoelectric resonators. *Proc. 22nd Annu. Freq. Control Symp.*, pp. 2–14.

EerNisse, E. P. (1975). Quartz resonator frequency shifts arising from electrode stress. *Proc. 29th Annu. Freq. Control Symp.*, pp. 1–4.

EerNisse, E. P. (1976). Calculation on the stress compensated (SC-cut) quartz resonator. *Proc. 30th Annu. Freq. Control Symp.*, pp. 8–11.

EerNisse, E. P. (1978). Rotated X-cut quartz resonators for high temperature applications. *Proc. 32nd Annu. Freq. Control Symp.*, pp. 255–259.
EerNisse, E. P. (1979). Temperature dependence of the force frequency effect for the rotated X-cut. *Proc. 33rd Annu. Freq. Control Symp.*, pp. 300–305.
EerNisse, E. P. (1980). Temperature dependence of the force frequency effect for the SC-cut and FC-cut. *Proc. 34th Annu. Freq. Control Symp.*, pp. 426–430.
Eer Nisse, E. P., and Paros, J. M. (1983). Practical considerations for miniature quartz resonator force transducers. *Proc. 37th Annu. Freq. Control Symp.*, pp. 255–260.
EerNisse, E. P., Lukaszek, T. J., and Ballato, A. (1978). Variational calculation of force-frequency constants of doubly rotated quartz resonators. *IEEE Trans. Sonics Ultrason.* **SU-25**, 132–138.
Emin, C. J. D., and Werner, J. F. (1983). The bulk acoustic wave properties of lithium tetraborate. *Proc. 37th Annu. Freq. Control Symp.*, pp. 136–143.
Epstein, S. (1973). Elastic-wave formulation for electroelastic waves in unbounded piezoelectric crystals. *Phys. Rev. B.* **7**, 1636–1644.
Epstein, S., and Benson, J. (1974). Rotation of characteristic vectors with piezoelectric coupling. *IEEE Trans. Sonics Ultrason.* **SU-21**, 214–216.
Fano, R. M., and Lawson, A. W. (1965). The theory of microwave filters. *In* "Microwave Transmission Circuits" (G. L. Ragan, ed.). Dover Publications, New York.
Feldmann, M. (1973). Piezoelectric waves in insulators. *Onde Electr.* **53**, 189–193. In French.
Filler, R. L., and Vig, J. R. (1980). Fundamental mode SC-cut resonators. *Proc. 34th Annu. Freq. Control Symp.*, pp. 187–193.
Filler, R. L., and Vig, J. R. (1981). The acceleration and warmup characteristics of four-point-mounted SC and AT-cut resonators. *Proc. 35th Annu. Freq. Control Symp.*, pp. 110–116.
Filler, R. L., Kosinski, J. A., and Vig, J. R. (1982). The effect of blank geometry on the acceleration sensitivity of AT & SC-cut quartz resonators. *Proc. 36th Annu. Freq. Control Symp.*, pp. 215–219.
Filler, R. L., Kosinski, J. A., and Vig, J. R. (1983). Further studies on the acceleration sensivitity of quartz resonators. *Proc. 37th Annu. Freq. Control Symp.*, pp. 265–271.
Fink, G. (1969). Diffraction of X-rays by piezo-electric vibration of quartz-crystals. *Acta Phys. Austriaca* **29**, 100–104. In German.
Firsova, M. M. (1974). Piezoelectric and elastic properties of hexagonal α-ZnS. *Sov. Phys.-Solid State (Engl. Transl.)* **16**, 350–351.
Fischer, H. (1977). Some results of frequency response measurements of 1 MHz precision quartzes. *Nachrichtentech. Elektron.* **27**, 154–155. In German.
Fischler, C. (1971). Propagation and amplification of shear-horizontal waves in piezoelectric plates. *J. Appl. Phys.* **42**, 919–924.
Flanagan, T. M. (1974). Hardness assurance in quartz crystal resonators. *IEEE Trans. Nucl. Sci.* **NS-21**, 390–392.
Fletcher, E. D., and Douglas, A. J. (1979). A comparison of the effects of bending moments on the vibrations of AT and SC (or TTC) cuts of quartz *Proc. 33rd Annu. Freq. Control Symp.*, pp. 346–350.
Foerster, H.-J. (1982). Thermal hysteresis of AT and SC- cut quartz crystal resonators automated measurement method and results. *Proc. 36th Annu. Freq. Control Symp.*, pp. 140–158.
Forrer, M. P. (1969). A flexure-mode quartz for an electronic wrist watch. *Proc. 23rd Annu. Freq. Control Symp.*, pp. 157–162.
Fowles, A. (1967). Dynamic compression of quartz. *Geophys. Res.* **72**, 5729–5742.
Frensch, T. (1980). Doubly-rotated quartz crystal cut improves stability and performance of oscillators. *Nachrichtentech. Elektron.* **34**, 201–203. In German.

Fujiwara, Y., Yamada, S., and Wakatsuki, N. (1983). Miniaturized $LiTaO_3$ strip resonator. *Proc. 37th Annu. Freq. Control Symp.*, pp. 343–348.
Fukumoto, A., and Watanabe, A. (1968). Temperature dependence of resonant frequencies of $LiNbO_3$ plate resonators. *Proc. IEEE* **56**, 1751–1753.
Fukuyo, H., Oura, N., Yokoyama, A., and Nonaka, S. (1974). Thickness-shear vibration of circular biconvex AT-cut plates. *Electr. Eng. Jap.* **94**, 36–42.
Gagnepain, J.-J. (1968). Study of non-linearities of quartz and demonstration of the isochronism defect. *C.R. Acad. Sci. B* **266**, 711–713. In French.
Gagnepain, J.-J. (1973). Influence of a continuous electric field on the resonance frequency of a quartz crystal. *C.R. Hebd. Seances Acad. Sci. B* **276**, 491–494. In French.
Gagnepain, J.-J. (1973). Nonlinear evaluation of the moving elements of the equivalent scheme of quartz. *C.R. Hebd. Seances Acad. Sci. B* **276**, 231–233. In French.
Gagnepain, J.-J. (1976). Fundamental noise studies of quartz crystal resonators. *Proc. 30th Annu. Freq. Control Symp.*, pp. 84–91.
Gagnepain, J.-J. (1981). Nonlinear properties of quartz crystals and quartz resonators: a review. *Proc. 35th Annu. Freq. Control Symp.*, pp. 14–30.
Gagnepain, J.-J., and Besson, R. (1972). A study of quartz crystal nonlinearities: application to X-cut resonators. *Phys. Lett. A*. **41A**, 443–444.
Gagnepain, J.-J., and Besson, R. (1975). Nonlinear effects in piezoelectric quartz crystals. *In* "Physical Acoustics: Principles and Methods" (W. P. Mason and R. N. Thurston, eds.), vol. 11, Academic Press, New York, pp. 245–288.
Gagnepain, J.-J., and Theobald, J. G. (1981). Influence of a magnetic field on the frequency of a quartz resonator. *C.R. Seances Acad. Sci. Ser. II* **292**, 283–285. In French.
Gagnepain, J.-J., Ponçot, J. C., and Pegeot, C. (1977). Amplitude-frequency behavior of doubly rotated quartz resonators. *Proc. 31st Annu. Freq. Control Symp.*, pp. 17–22.
Gagnepain, J.-J., Theobald, G., and Uebersfeld, J. (1981). Analysis of $1/f^2$ and $1/f$ frequency noises in quartz resonators. *J. Phys. Colloq.* **42**, 201–209.
Gagnepain, J.-J., Uebersfeld, J., Goujon, G., and Handel, P. (1981). Relation between $1/f$ noise and Q factor in quartz resonators at room and low temperatures. First theoretical interpretation. *Proc. 35th Annu. Freq. Control Symp.* pp. 476–483.
Gagnepain, J.-J., Olivier, M., and Walls, F. L. (1983). Excess noise in quartz crystal resonators. *Proc. 37th Annu. Freq. Control Symp.*, pp. 218–225.
Ganguly, S. N. (1975). A note on vibration of a piezo-electric crystal with dissipative mechanical parameters. *Indian J. Theor. Phys.* **23**, 63–68.
Garcia-Camarero, E. (1974). Equations of motion of acousto-electromagnetic waves. *An. Fis.* **70**, 289–291. In Spanish.
Gerber, E. A. (1966). Quartz crystal units for very high frequencies. U. S. Army Electronics Command, Fort Monmouth, New Jersey, Tech. Rept. ECOM-2697, 14 pp.
Gerber, E. A. (1966). Unwanted responses in VHF quartz crystal units. *Proc. IEEE* **54**, 1613–1635.
Gerdes, R. J., and Wagner, C. E. (1970). Scanning electron microscopy of oscillating quartz crystals. *Proc. 3rd Annu. Scanning Electron Microsc. Symp.*, pp. 441–448.
Gerdes, R. J., and Wagner, C. E. (1971). Scanning electron microscopy of resonating quartz crystals. *Appl. Phys. Lett.* **18**, 39–41.
Gerdes, R. J., and Wagner, C. E. (1971). Study of frequency control devices in the scanning electron microscope. *Proc. 25th Annu. Freq. Control Symp.*, pp. 118–124.
Ghosh, N. C. (1979). Heating of a finite piezo-electric rod due to some time-dependent distributed heat sources. *Indian J. Theor. Phys.* **27**, 283–294.
Ghosh, N. C., and Pal, A. K. (1978). Wave propagation in an infinite piezoelectric crystal under a thermal field. *Czech. J. Phys. Sect. B* **B28**, 807–812.

Giebe, E., and Blechschmidt, E. (1940). On torsional vibrations of quartz rods and their use as standards of frequency. *Hochfrequenztech. Electroakust.* **56**, 65–87.
Gilinskiy, I. A., and Popov, V. V. (1978). Theory of wave excitation in piezoelectric crystals by narrow metallic electrodes. *Radio Eng. Electron. Phys.* **23**, 102–110.
Glowinski, A. (1972). Thickness vibrations of piezoelectric slabs. *Ann. Telecommun.* **27**, 147–158. In French.
Glowinski, A., and Lançon, R. (1969). Resonance frequencies of monolithic quartz structures. *Proc. 23rd Annu. Freq. Control Symp.*, pp. 39–55.
Glowinski, A., Lançon, R., and Lefevre, R. (1973). Effects of asymmetry in trapped energy piezoelectric resonators. *Proc. 27th Annu. Freq. Control Symp.*, pp. 233–242.
Gniewinska, B. (1977). Effect of oscillator circuit and ambient conditions on parameters of quartz crystal unit. *Elektronika* **18**, 309–312. In Polish.
Gniewinska, B., and Kalinowska, B. (1980). The dynamic thermal properties of quartz crystal resonators. *Pr. Inst. Tele- Radiotech.* (84), 5–17. In Polish.
Gniewinska, B., and Smolarski, A. (1970). Influence of the non-linearity of a quartz resonator upon the parameters of frequency standards. *Pomiary Autom. Kontrola* **16**, 247–248. In Polish.
Godenhielm, B. (1972). The piezoelectric characteristics of the oscillator crystal. *ERT Elektron. Radio Telev.* **25**, 17–19, In Finnish.
Godenhielm, B. (1979). Influence of tolerances on acoustic bulk wave resonators made of thin AT-cut quartz crystals. *Helsinki Univ. Technol., Espoo, Finland.* Thesis.
Godenhielm, B. (1979). Influence of tolerances on acoustic bulk wave resonators made of thin AT-cut quartz crystals. *Tech. Res. Cent. Finland Electr. Nucl. Technol. Publ.* (25), 5–45.
Goldfrank, B., and Warner, A. (1980). Further developments on 'SC' cut crystals. *Proc. 34th Annu. Freq. Control Symp.*, pp. 183–186.
Goldfrank, B., Ho, J., and Warner, A. W. (1981). Update of SC cut crystal resonator technology. *Proc. 35th Annu. Freq. Control Symp.*, pp. 92–98.
Goodall, F. N., and Wallace, C. A. (1975). The analysis of unwanted-mode vibration patterns in AT-cut quartz oscillator crystals, revealed by X-ray diffraction topography. II. A partial theoretical description of the unwanted mode. *J. Phys. D* **8**, 1843–1850.
Goodell, J., and Sundelin, R. (1979). Frequency characteristics of bulk microwave acoustic resonators. *Ultrason. Symp. Proc.*, pp. 123–127.
Goruk, W. S., and Stegeman, G. I. (1978). Surface to bulk mode conversion at interfaces on Y-Z $LiNbO_3$. *Appl. Phys. Lett.* **32**, 265–266.
Graf, E. P., and Peier, U. R. (1983). BVA quartz crystal resonators and oscillator production — A statistical review. *Proc. 37th Annu. Freq. Control Symp.*, pp. 492–500.
Graham, R. A. (1972). Strain dependence of longitudinal piezoelectric, elastic, and dielectric constants of X-cut quartz. *Phys. Rev. B* **6**, 4779–4792.
Graham, R. A. (1973). Strain dependence of the piezoelectric polarization of z-cut lithium niobate. *Solid State Commun.* **12**, 503–506.
Graham, R. A. (1974). Shock-wave compression of X-cut quartz as determined by electrical response measurements. *J. Phys. Chem. Solids* **35**, 355–372.
Gray, D. E. (1963). "American Institute of Physics Handbook," 2nd ed., McGraw-Hill, New York, pp. 2–40.
Gregoire, J. P. Nedelec, J. C., and Planchard, J. (1973). Calculating the eigenfrequencies of an acoustic resonator. *Bull. Dir. Etud. Rech. Ser. C* (2), 5–22. In French.
Gridnev, S. A., Postnikov, V. S., Prasolov, B. N., and Turkov, S. K. (1978). Damping of low-frequency elastic vibrations in $LiNbO_3$. *Sov. Phys.-Solid State* (*Engl. Transl.*) **20**, 747–750.

Grinchenko, V. T., Karlash, V. L., Meleshko, V. V., and Ulitko, A. F. (1976). Planar vibrations of rectangular piezoelectric plates. *Prikl. Mekh.* **12**, 71–78. In Russian.
Grishchenko, E. K. (1975). Acoustic and electric fields in a piezoelectric plate. *Sov. Phys.-Acoust.* (*Engl. Transl.*) **20**, 328–331.
Grishchenko, E. K. (1980). Longitudinal and shear mode conversion by a controlled lithium niobate piezoelectric element. *Sov. Phys.-Acoust.* (*Engl. Transl.*) **26**, 476–480.
Grudkowski, T. W., Black, J. F., Reeder, T. M., Cullen, D. E., and Wagner, R. A. (1980). Fundamental mode VHF/UHF bulk acoustic wave resonators and filters on silicon. *Ultrason. Symp. Proc.* **2**, 829–833.
Grudkowski, T. W., Black, J. F., Reeder, T. M., Cullen, D. E., and Wagner, R. A. (1980). Fundamental-mode VHF/UHF miniature acoustic resonators and filtes on silicon [IC applications]. *Appl. Phys. Lett.* **37**, 993–995.
Gulyaev, Y. V., and Plessky, V. P. (1979). Shear surface acoustic waves at the periodically nonuniform boundary between two solids. *Electron. Lett.* **15**, 63–64.
Hadley, G. (1961). "Linear Algebra," Addison-Wesley, Reading, Massachusetts.
Hafner, E. (1968). The role of crystal parameters in circuit design. *Proc. 22nd Annu. Freq. Control Symp.*, pp. 269–281.
Hafner, E. (1974). Filters and resonators—a review. I. Crystal resonators. *IEEE Trans. Sonics Ultrason.* **SU-21**, 220–237.
Haines, D. W. (1969). Bechmann's number for harmonic overtones of thickness-shear vibrations of rotated-Y-cut quartz-plates. *Int. J. Solids Struc.* **5**, 1–16.
Hales, M. C., Burgess, J. W., and Porter, R. J. (1974). Energy trapped vibrations in lithium tantalate and lithium niobate resonators. *Proc. 28th Annu. Freq. Control Symp.*, p. 43.
Halperin, A., Katz, Z., Ronen, M., and Schieber, M. (1982). Photoluminescence from worked surface layers and frequency instability in quartz resonators. *Proc. 36th Annu. Freq. Control Symp.*, pp. 187–192.
Hammond, D. L., and Benjaminson, A. (1969). The crystal resonator—a digital transducer. *IEEE Spectrum* **6**, 53–58.
Hammond, D., Adams, C., and Cutler, L. (1963). Precision crystal units. *Proc. 17th Annu. Freq. Control Symp.*, pp. 215–232.
Hammond, D. L., Adams, C. A., and Benjaminson, A. (1968). Hysteresis effects in quartz resonators. *Proc. 22nd Annu. Freq. Control Symp.*, pp. 55–66.
Handel, P. H. (1979). Nature of $1/f$ frequency fluctuations in quartz crystal resonators. *Solid-State Electron.* **22**, 875–876.
Hannon, J. J., Lloyd, P., and Smith, R. T. (1970). Lithium tantalate and lithium niobate piezoelectric resonators in the medium frequency range with low ratios of capacitance and low temperature coefficients of frequency. *IEEE Trans. Sonics Ultrason.* **SU-17**, 239–246.
Harmon, D. L., Josse, F., and Vetelino, J. F. (1979). Surface skimming bulk waves in Y-rotated quartz-experimental characterization and filter device implementation. *Ultrason. Symp. Proc.*, pp. 791–796.
Hatch, E. R., and Ballato, A. (1983). Lateral-field excitation of quartz plates. *Ultrason. Symp. Proc.*, pp. 512–515.
Hearmon, R. F. S. (1961). "An Introduction to Applied Anisotropic Elasticity," Oxford University Press, London, pp. 48–52.
Hearn, E. W., and Schwuttke, C. H. (1970). Defects and frequency mode patterns in quartz plates. *Proc. 24th Annu. Freq. Control Symp.*, pp. 64–73.
Heising, R. A. (Ed.) (1946). "Quartz Crystals for Electrical Circuits—Their Design and Manufacture." Van Nostrand, New York.
Hermann, J. (1975). Determination of the electromechanical coupling factor of quartz bars vibrating in flexure or length-extension. *Proc. 29th Annu. Freq. Control Symp.*, pp. 26–34.

Hermann, J. (1977). DT-cut torsional resonators. *Proc. 31st Annu. Freq. Control Symp.*, pp. 55–61.
Hermann, J., and Bourgeois, C. (1979). A new quartz crystal cut for contour mode resonators. *Proc. 33rd Annu. Freq. Control Symp.*, pp. 255–262.
Hermann, J. (1977). Quartz torsion resonators. *Bull. Annu. Soc. Siosse Chronom. Lab. Suisse Rech. Horlog.* **7**, 299–303. In French.
Heyman, J. S., and Miller, J. G. (1973). Verification of sensitivity enchancement factors for cw ultrasonic resonators. *J. Appl. Phys.* **44**, 3398–3400.
Hirose, Y., Tsuziki, Y., and Iijima, K. (1970). Measurement of contour vibrations of quartz plates by holographic technique. *Electron. Commun. Jap.* **53**, 49–54.
Holbeche, R. J., and Morley, P. E. (1981). Investigation into temperature variation of equivalent-circuit parameters of AT-cut quartz crystal resonators. *IEE Proc. A* **128**, 507–510.
Holland, R. (1968). Contour extensional resonant properties of rectangular piezoelectric plates. *IEEE Trans. Sonics Ultrason.* **SU-15**, 97–105.
Holland, R. (1968). Resonant properties of piezoelectric ceramic rectangular parallelepipeds. *J. Acoust. Soc. Am.* **43**, 988–997.
Holland, R. (1974a). Nonuniformly heated anisotropic plates. I. Mechanical distortion and relaxation [quartz resonator]. *IEEE Trans. Sonics Ultrason.* **SU-21**, 171–178.
Holland, R. (1974b). Nonuniformly heated anisotropic plates: II. Frequency transients in AT and BT quartz plates. *Ultrason. Symp. Proc.*, pp. 592–598.
Holland, R., and EerNisse, E. P. (1968). Variational evaluation of admittances of multi-electroded three-dimensional piezoelectric structures. *IEEE Trans. Sonics Ultrason.* **SU-15**, 119–132.
Holland, R., and EerNisse, E. P. (1969). "Resonant piezoelectric devices, Design of." M.I.T. Press, Cambridge, Massachusetts, 258 pp.
Hoppmann, W. H., II, and Kunukkasseril, V. X. (1967). Normal-mode vibrations of model of centrosymmetric cubic crystal. *J. Acoust. Soc. Am.*, **42**, 42–49.
Horton, W. H., and Smythe, R. C. (1967). The work of Mortley and the energy-trapping theory for thickness-shear piezoelectric. *Proc. IEEE* **55**, 222.
Horton, W. H., and Smythe, R. C. (1973). Experimental investigation of intermodulation in monolithic crystal filters. *Proc. 27th Annu. Freq. Control Symp.*, pp. 243–245.
Hruska, K. (1968). Relation between the general and the simplified condition for the velocity of propagation of ultrasonic waves in a piezoelectric medium. *Czech. J. Phys. B.* **18**, 214–221.
Hruska, K. (1969). Vibration of piezoelectric plates with a small non-zero potential across their electrodes. *Czech. J. Phys. B.* **19**, 1315–1321. In Polish.
Hruska, K. (1970). Frequency–temperature dependence of thickness vibrations of piezoelectric plates. *Proc. 24th Annu. Freq. Control Symp.*, pp. 33–45.
Hruska, K. (1971). Polarizing effect with piezoelectric plates and second-order effects. *IEEE Trans. Sonics Ultrason.* **SU-18**, 1–7.
Hruska, C. K. (1977). The electroelastic tensor and other second-order phenomena in quasi-linear interpretation of the polarizing effect with thickness vibration of α-quartz plates. *Proc. 31st Annu. Freq. Control Symp.*, pp. 159–170.
Hruska, C. K. (1978). Second-order phenomena in α-quartz and the polarizing effect with plates of orientation $(xzl)\psi$. *IEEE Trans. Sonics Ultrason.* **SU-25**, 198–203.
Hruska, C. K. (1980). Zero polarizing effect with doubly rotated quartz plates vibrating in thickness. *IEEE Trans. Sonics Ultrason.* **SU-27**, 87–89.
Hruska, C. K. (1981). On the linear polarizing effect with α-quartz AT plates. *IEEE Trans. Sonics Ultrason.* **SU-28**, 108–110.

Hruska, K., and Kazda, V. (1968). The polarizing tensor of the elastic coefficients and moduli for α-quartz. *Czech. J. Phys. B.* **18**, 500–503.

Hruska, K., and Khogali, A. (1971). Polarizing effect with alpha-quartz rods and the electroelastic tensor. *IEEE Trans. Sonics Ultrason.* **18**, 169–174.

Hutson, A. R. (1960). Piezoelectricity and conductivity in ZnO and CdS. *Phys. Rev. Lett.* **4**, 505–507.

Hutson, A. R., and White, D. L. (1962). Elastic wave propagation in piezoelectric semiconductors. *J. Appl. Phys.* **33**, 40–47.

Hutson, A. R., McFee, J. H., and White, D. L. (1961). Ultrasonic amplification in CdS. *Phys. Rev. Lett.* **7**, 237–239.

IEEE (1966). Standard definitions and methods of measurement for piezoelectric vibrators. IEEE Standard No. 177, IEEE, New York, 19 pp.

IEEE (1978). IEEE standard on piezoelectricity. IEEE STD 176-1978, IEEE, New York, 55 pp.

Ikeda, T. (1972). On the relations between electromechanical coupling coefficients and elastic constants in a piezoelectric crystal. *Jap. J. Appl. Phys.* **11**, 463–471.

Ikeda, T. (1982). Depolarizing-field effect in piezoelectric resonators. *Ferroelectrics* **43**, 3–15.

Ikeda, and Inawashiro, S. (1978). Depolarizing- or demagnetizing-field in electromechanical extensional vibrators. I. Depolarizing-field effect in a piezoelectric rod under the longitudinal-effect extensional vibration. *J. Acoust. Soc. Jpn.* **34**, 223–233.

Ikegami, S., Ueda, I., Kobayashi, S. (1974). Frequency spectra of resonance vibration in disk plates of $PbTiO_3$ piezoelectric ceramic. *J. Acoust. Soc. Am.* **55**, 339–344.

IRE (1957). IRE standards on piezoelectric crystals—The piezoelectric vibrator: Definitions and methods of measurement (Standard 57 IRE 14.S1). *Proc. IRE* **45**, 353–358.

IRE (1958). IRE standards on piezoelectric crystals: Determination of the elastic, piezoelectric, and dielectric constants—The electromechanical coupling factor (Standard 58 IRE 14.S1). *Proc. IRE* **46**, 763–778.

IRE (1961). IRE standard on piezoelectric ceramics: Measurements of piezoelectric ceramics (Standard 61 IRE 14.S1). *Proc. IRE* **49**, 1161–1169.

Isherwood, B. J., and Wallace, C. A. (1975). The analysis of unwanted-mode vibration patterns in AT-cut quartz oscillator crystals, revealed by X-ray diffraction topography. I. Interpretation of the X-ray diffraction topographs. *J. Phys. D* **8**, 1827–1842.

Jamiolkowski, J. (1973). High stability quartz resonators for extended frequency range 4.5 to 5.5 MHz. *Pr. Inst. Tele- Radiotech.* **12**, 55–57. In Polish.

Jamiolkowski, J. (1977). High-stability quartz crystal unit of short frequency stabilization time. *Pr. Inst. Tele- Radiotech.* (75), 5–16. In Polish.

Jamiolkowski, J. (1977). Modernized quartz crystal unit of linear temperature characteristic. *Pr. Inst. Tele- Radiotech.* (72), 57–61. In Polish.

Janiaud, D. (1977). Influence of supports on the accelerometric sensitivity of quartz resonators. *C.R. Hebd. Seances Acad. Sci. Ser. B.* **285**, 69–72. In French.

Janiaud, D., Nissim, L., and Gagnepain, J.-J. (1978). Analytical calculation of initial stress effects on aniosotropic crystals: application to quartz resonators. *Proc. 32nd Annu. Freq. Control Symp.*, pp. 169–179.

Janiaud, D., Besson, R., Gagnepain, J.-J., and Valdois, M. (1981). Quartz resonator thermal transient due to crystal support. *Proc. 35th Annu. Freq. Control Symp.*, pp. 340–344.

Jhunjhunwala, A., Vetelino, J. F., and Field, J. C. (1976a). Berlinite, a temperature compensated material for surface acoustic wave applications. *Ultrason. Symp. Proc.*, pp. 523–527; (1977). *J. Appl. Phys.* **48**, 887–892.

Jhunjhunwala, A., Vetelino, J. F., and Field, J. C. (1976b). Temperature compensated cuts with zero power flow in Tl_3VS_4 and $Tl_3 TaSe_4$. *Electron. Lett.* **12**, 683–684.

Jipson, V. B., Vetelino, J. F., and Jhunjhunwala, A., and Field, J. C. (1976). Lithium iodate—A new material for surface wave applications. *Proc. IEEE* **64**, 568–569.

Jones, G. L., and Henneke, E. G. II. (1972). Reflection of an elastic wave from a free surface in quartz. *Program 83rd Meeting of the Acoustical Society of America*, abstracts, (New York: Acoust. Soc. America). pp. 83–84.

Josse, F., and Lee, D. L. (1982). Analysis of the excitation, interaction, and detection of bulk and surface acoustic waves on piezoelectric substrates. *IEEE Trans. Sonics Ultrason.* **SU-29**, 261–273.

Jumonji, H. (1970). Analysis of characteristics of multiple-mode resonators vibrating in longitudinal and flexural modes. *Trans. Inst. Electron. Commun. Eng. Jpn.* Sect. A, B, C, **53**, 3–5.

Kabakov, M. F., and Plonskii, A. F. (1980). Optimization of temperature-sensitive quartz resonators. *Radioelectron. Commun. Syst.* **23**, 21–26.

Kabanovich, I. V., Smagin, A. G., Simonov, A. V., and Kleshchev, G. V. (1970). X-ray diffraction topograms of excited quartz resonators. *Izv. Vyssh. Uchelon. Zaved. Fiz.* **1**, 157–158. In Russian.

Kagawa, Y., and Gladwell, G. M. L. (1970). Finite element analysis of flexure-type vibrators with electrostrictive transducers. *IEEE Trans. Sonics Ultrason.* **SU-17**, 41–49.

Kagawa, Y., and Yamabuchi, T. (1976a). A finite element approach to electromechanical problems with an application to energy-trapped and surface-wave devices. *IEEE Trans. Sonics Ultrason.* **SU-23**, 263–272.

Kagawa, Y., and Yamabuchi, T. (1976b). Finite element approach for a piezoelectric circular rod. *IEEE Trans. Sonics Ultrason.* **SU-23**, 379–385.

Kagawa, Y., and Yamabuchi, T. (1977). Finite element approach for a piezoelectric circular rod. *Trans. Inst. Electron. Commun. Eng. Jpn.* E. **59**, 21–22.

Kagawa, Y., Arai, H., Yakuwa, K., Okuda, S., and Shirai, K. (1975). Finite element simulation of energy-trapped electromechanical resonators. *J. Sound Vib.* **39**, 317–335.

Kagawa, Y., Maeda, H., and Nakazawa, M. (1981). Finite element approach to frequency-temperature characteristic prediction of rotated quartz crystal plate resonators. *IEEE Trans. Sonics Ultrason.* **SU-28**, 257–264.

Kaliski, S. (1967). A perfect self-excited piezoelectric resonator. *Proc. Vib. Probl.* **8**, 3–20.

Kaliski, S. (1968). Self-excited perfect piezoquartz resonator with an external electron stream. *Proc. Vib. Probl.* **9**, 429–436.

Kaliski, S. (1971). Some properties of the equations of piezoelectric vibration of lithium niobate and tantalate crystals. *Proc. Vib. Probl.* **12**, 63–70.

Kantor, V. M. (1975). Calculation of the parameters of energy-trapping piezoelectric resonators. *Sov. Phys.-Acoust.* (*Engl. Transl.*) **20**, 516–518.

Karaul'nik, A. E., and Shin, V. (1969). Spectral characteristics of vibrating crystals as affected by the direction of the exciting electrical field. *Kristallografiya* **14**, 84–89. In Russian.

Kasatkin, B. A., and Lebedev, V. G. (1979). Spectrum of natural frequencies of a loaded piezoelectric plate with a transition layer. *Sov. Phys.-Acoust.* (*Engl. Transl.*) **25**, 224–227.

Katoh, T., and Ueda, H. (1979). Frequency temperature characteristics of rectangular AT-cut quartz plates. *Proc. 33rd Annu. Freq. Control Symp.*, pp. 271–276.

Kaul, R. K. (1968). Low frequency extensional vibrations of thin anisotropic discs. Report ECOM-01731-2 (AD 681348), Columbia Univ., New York, 70 pp. Contract DA-28-043-AMC-0173(E). Available from NTIS.

Kawashima, H., Sato, H., and Ochiai, O. (1980). New frequency temperature characteristics of miniaturised GT-cut quartz resonators. *Proc. 34th Annu. Freq. Control Symp.*, pp. 131–139.

Kazelle, I., and Zelenka, J. (1978). Temperature dependence of the oscillation frequency of AT-cut quartz resonators between resonance and antiresonance frequencies. *Sov. Phys.-Crystallogr.* (*Engl. Transl.*) **23**, 363–364.

Kessenikh, G. G., and Shuvalov, L. A. (1976). Energy flow and group velocity of sound waves in piezoelectric crystals. *Sov. Phys.-Crystallogr. (Engl. Transl.)* **21**, 587–588.

Keuning, D. H. (1972). Exact resonant frequencies for the thickness-twist trapped energy mode in a piezoceramic plate. *J. Eng. Math.* **6**, 143–154.

Keyes, R. W., and Blair, F. W. (1967). Stress dependence of the frequency of quartz plates. *Proc. IEEE* **55**, 565–566.

Kimura, M., Doi, K., Nanamatsu, S., and Kawamura, T. (1973). A new piezoelectric crystal: $Ba_2Ge_2TiO_8$. *Appl. Phys. Lett* **23**, 531–532.

King, P. J., and Luukkala, M. (1973). Subharmonic generation in quartz plates. *J. Phys. D* **6**, 1047–1051.

Kittinger, E., Seil, K., and Tichy, J. (1979). Electroelastic effect in tourmaline. *Z. Naturforsch. A* **34A**, 1352–1354.

Klimenko, B. I., Perelomova, N. V., Blistanov, A. A., and Bondarenko, V. S. (1978). Anisotropy and phase velocities of propagation of elastic waves in lithium niobate. *Sov. Phys.-Crystallogr. (Engl. Transl.)* **23**, 114–117.

Koga, I. (1932). Thickness vibrations of piezoelectric oscillating crystals. *Physics* **3**, 70–80. *Rep. Radio Res. Work Jpn.* **2**, 157–173; *J. Inst. Telegr. Teleph. Eng. Jpn.* **115**, 1223–1253.

Koga, I. (1934). Thermal characteristics of piezoelectric oscillating quartz plates. *Rep. Radio Res. Work Jpn.* **4**, 61–76.

Koga, I. (1969). Anomalous vibrations in AT-cut plates. *Proc. 23rd Annu. Freq. Control Symp.*, pp. 128–131.

Koga, I. (1969). Anomalous vibration (activity dips) in AT-cut (R_1) plates. *Electron. Commun. Jpn.* **52A** (6), 1–12.

Koga, I., and Fukuyo, H. (1953). Vibration of piezoelectric quartz plate. *J. Inst. Electr. Commun. Eng. Jpn.* **36**, 59–67.

Koga, I., Aruga, M., and Yoshinaka, Y. (1958). Theory of plane elastic waves in a piezoelectric crystalline medium and the determination of the elastic and piezoelectric constants of quartz. *Phys. Rev.* **109**, 1467–1473.

Kogure, S., and Momosaki, E. (1980). New type twin mode resonator. *Proc. 34th Annu. Freq. Control Symp.*, pp. 160–166.

Kojima, H., Kinoshita, Y., Hikita, M., and Tabuchi, T. (1980). Velocity, electromechanical coupling factor and acoustic loss of surface shear waves propagating along x-axis on rotated Y-cut plates of $LiNbO_3$. *Electron. Lett.* **16**, 445–446.

Kokorin, Y. I., Zaitseva, M. P., Sysoev, A. M. (1974). Mechanical nonlinearity accompanying the oscillations of ferroelectric crystal resonators. *Izv. Akad. Nauk SSSR Ser. Fiz.* **39**, 816–821. In Russian.

Kokorin, Y. I., Zaitseva, M. P., Kidyarov, B. I., Sysoev, A. M., and Malyshevskii, N. G. (1978). Polarization effect in piezoelectric resonators executing longitudinal vibrations. *Sov. Phys.-Crystallogr. (Engl. Trans.)* **23**, 118–119.

Komiyama, B. (1981). Quartz crystal oscillator at cryogenic temperature. *Proc. 35th Annu. Freq. Control Symp.*, pp. 335–339.

Koneval, D. J., and Sliker, T. R. (1969). Shear mode CdS–quartz composite resonators. *IEEE Trans. Sonics Ultrason.* **SU-16**, 21.

Kosek, M. (1972). The influence of electrode thickness on the frequency response of a piezoelectrical transducer. *Slaboproudy Obzor* **33**, 311–315. In Czech.

Kossof, G. (1966). The effects of backing and matching on the performance of piezoelectric ceramic transducers. *IEEE Trans. Sonics Ultrason.* **SU-13**, 20–30.

Kraut, E. A. (1969). New methematical formulation for piezoelectric wave propagation. *Phys. Rev.* **188**, 1450–1455.

Krishna, M. M. (1978). Rotated elastic constants, coupling factors and conversion efficiencies of AC and BC quartz. *Indian J. Pure Appl. Phys.* **16**, 735–738.

Kubata, V. G., Prokimov, A. A., Sarafanov, V. F., and Sviridov, A. V. (1978). Freedom from spurious oscillations in UHF quartz resonators. *Radiotekhnika Kharkov* (45), 126–128. In Russian.

Kulikov, E. L., Gubenkov, A. N., and Kazankin, L. K. (1979). A variational method for design of piezoelectric devices. *Radio Eng. Electron. Phys.* **24**, 123.

Kunath, P., and Magerl, R. (1975). Analysis of a circular piezoelectric plate. *Z. Elektr. Inf. Energietech.* **5**, 435–442. In German.

Kusakabe, C. (1978). Suppression of spurious responses of a flexural piezoelectric vibrator with split-electrode. *J. Acoust. Soc. Jpn.* **34**, 73–78. In Japanese.

Kusakabe, C. (1979). An analysis of a longitudinal resonator for suppression of spurious mode. *Bull. Yamagata Univ.* **15**, 155–165. In Japanese.

Kusters, J. A. (1976). Transient thermal compensation for quartz resonators. *IEEE Trans. Sonics Ultrason.* **SU-23**, 273–276.

Kusters, J. A. (1981). The SC cut crystal—an overview. *Ultrason. Symp. Proc.*, pp. 402–409.

Kusters, J. A., and Leach, J. G. (1977). Further experimental data on stress and thermal gradient compensated crystals. *Proc. IEEE* **65**, 282–284.

Kusters, J. A., and Leach, J. G. (1978). Dual mode operation of temperature and stress compensated crystals. *Proc. 32nd Annu. Freq. Control Symp.*, pp. 389–397.

Kyame, J. J. (1951). Wave propagation in piezoelectric crystals. *J. Acoust. Soc. Am.* **21**, 159–167.

Kyame, J. J. (1954). Conductivity and viscosity effects on wave propagation in piezoelectric crystals. *J. Acoust. Soc. Am.* **26**, 990–993.

Lafitte, F., Pouliquen, J., and Segard, N. (1969). Photoelastic properties of quartz vibrating at resonance. Synchronization of a pulsed laser. *Ann. Soc. Sci. Bruxelles 1* **83**, 145–169. In French.

Lagasse, G., Ho, J., and Bloch, M. (1972). Research and development of a new type of crystal—the FC cut. *Proc. 26th Annu. Freq. Control Symp.*, pp. 148–151.

Lakes, R. S. (1979). Prediction of anelastic loss in piezoelectric solids: Effect of geometry. *Appl. Phys. Lett* **34**, 729–730.

Lakes, R. (1980). Shape-dependent damping in piezoelectric solids. *IEEE Trans. Sonics Ultrason.* **SU-27**, 208–213.

Lakin, K. M. (1983). Analysis of composite resonator geometries. *Proc. 37th Annu. Freq. Control Symp.*, pp. 320–324.

Lakin, K. M., and Wang, J. S. (1980). UHF composite bulk wave resonators. *Ultrason. Symp. Proc.* **2**, 834–837.

Lakin, K. M., Wang, J. S., and Landin, A. R. (1982). Aluminum nitride thin film and composite bulk wave resonators. *Proc. 36th Annu. Freq. Control Symp.*, pp. 517–524.

Lançon, R. (1973). Mechanical resonances of asymmetric monolithic structures. *Ann. Telecommun.* **28**, 406–412. In French.

Landolt-Boernstein (1966). "Numerical Data and Functional Relationships in Science and Technology," New Series, Group III, vol. 1, Springer-Verlag, Berlin and New York, pp. 40–123. (Crystal material constants.)

Landolt-Boernstein (1969). "Numerical Data and Functional Relationships in Science and Technology," New Ser., Group III, vol. 2, Springer-Verlag, Berlin and New York, 838–839. (Crystal material constants.)

Lau, K. F., Yen, K. H., Wilcox, J. Z., and Kagiwada, R. S. (1979). Analysis of shallow bulk acoustic wave excitation by interdigital transducers. *Proc. 33rd Annu. Freq. Control Symp.*, pp. 388–395.

Lau, K. F., Stokes, R. B., and Yen, K. H. (1981). Investigation of temperature stable SBAW in berlinite. *Ultrason. Symp. Proc.*, pp. 286–290.
Lau, K. F., Yen, K. H., Stokes, R. B., and Kagiwada, R. S. (1981). Shallow bulk acoustic waves in berlinite. *Proc. 35th Annu. Freq. Control Symp.*, pp. 401–405.
Lawson, A. W. (1941). Comment on the elastic constants of alpha quartz. *Phys. Rev.* **59**, 838–839.
Lawson, A. W. (1942). The vibration of piezoelectric plates. *Phys. Rev.* **62**, 71–76.
Lazutkin, V. N., and Mikhailov, A. I. (1972). Equivalent circuit of a radially vibrating disk. *Sov. Phys. Acoust.* (*Engl. Transl.*) **18**, 45–48.
Lee, D. L. (1981). Analysis of energy trapping effects for SH-type waves on rotated Y-cut quartz. *IEEE Trans. Sonics Ultrason.* **SU-28**, 330–341.
Lee, P. C. Y. (1971a). Extensional, flexural, and width-shear vibrations of thin rectangular crystal plates. *J. Appl. Phys.* **42**, 4139–4144.
Lee, P. C. Y. (1971b). Extensional, flexural and width-shear vibrations of thin rectangular quartz resonator plates. *Proc. 25th Annu. Freq. Control Symp.*, pp. 63–69.
Lee, P. C. Y. (1972). An approximate theory for high-frequency vibrations of elastic plates. *Proc. 26th Annu. Freq. Control Symp.*, p. 85.
Lee, P. C. Y., and Chen, S. (1969). Vibrations of contoured and partially plated, contoured, rectangular, AT-cut quartz plates. *J. Acoust. Soc. Am.* **46**, 1193–1202.
Lee, P. C. Y., and Lam, C. S. (1981). Stresses in rectangular cantilever crystal plates under transverse loading. *Proc. 35th Annu. Freq. Control Symp.*, pp. 193–204.
Lee, P. C. Y., and Lam, C. S. (1982). Effect of transverse force on the thickness-shear resonance frequencies in rectangular, doubly rotated crystal plates. *Proc. 36th Annu. Freq. Control Symp.*, pp. 29–36.
Lee, P. C. Y., and Spencer, W. J. (1969). Shear-flexure-twist vibrations in rectangular AT-cut quartz plates with partial electrodes. *J. Acoust. Soc. Am.* **45**, 637–645.
Lee, P. C. Y., and Syngellakis, S. (1975). Waves and vibrations in an infinite piezoelectric plate. *Proc. 29th Annu. Freq. Control Symp.*, pp. 65–70.
Lee, P. C. Y., and Wu, K.-M. (1976). Effects of acceleration on the resonance frequencies of crystal plates. *Proc. 30th Annu. Freq. Control Symp.*, pp. 1–7.
Lee, P. C. Y., and Wu, K.-M. (1978). The influence of support-configuration on the acceleration sensitivity of quartz resonator plates. *IEEE Trans. Sonics Ultrason.* **SU-25**, 220–223.
Lee, P. C. Y., and Wu, K.-M. (1980). Nonlinear effect of initial stresses in doubly-rotated crystal resonator plates. *Proc. 34th Annu. Freq. Control Symp.*, pp. 403–411.
Lee, P. C. Y., and Yong, Y. K. (1983). Temperature derivatives of elastic stiffnesses derived from the frequency-temperature behavior of quartz plates. *Proc. 37th Annu. Freq. Control Symp.*, pp. 200–207.
Lee, P. C. Y., and Zelenka, J. (1971). The resonance frequency temperature dependence of coupled vibrations of X-cut quartz plates. *Proc. 7th Int. Congr. Acoust.* (Budapest: Academiai Kiado), pp. 201–204.
Lee, P. C. Y., and Zelenka, J. (1972). The frequency temperature dependence of coupled extensional, flexural, and width-shear vibrations of rotated X-cut quartz plates. *J. Appl. Phys.* **43**, 3325–3328.
Lee, P. C. Y., Wang, Y. S., and Markenscoff, X. (1973). Elastic waves and vibrations in deformed crystal plates. *Proc. 27th Annu. Freq. Control Symp.*, pp. 1–6.
Lee, P. C. Y., Wang, Y. S., and Markenscoff, X. (1974). Effects of initial bending on the resonance frequencies of crystal plates. *Proc. 28th Annu. Freq. Control Symp.*, pp. 14–18.
Lee, P. C. Y., Tameroglu, S., and Cakmak, A. S. (1975). Extensional, flexural, width-shear and width-stretch vibrations of thin rectangular resonators. *Ultrasonics International* (London, England: IPC Sci. & Technol. Press), pp. 162–165.

Lee, P. C. Y., Zee, C., and Brebbia, C. A. (1978). Thickness-shear, thickness-twist, and flexural vibrations of rectangular AT-cut quartz plates with patch electrodes. *Proc. 32nd Annu. Freq. Control Symp.*, pp. 108–119.
Lee, P. C. Y., Nakazawa, M., and Hou, J. P. (1981). Extensional vibrations of rectangular crystal plates. *Proc. 35th Annu. Freq. Control Symp.*, pp. 222–229.
Lemanov, V. V., and Yushin, N. K. (1973). Generation of harmonics of transverse elastic waves in crystals. *Sov. Phys.-Solid State (Engl. Transl.)* **14**, 2053–2056.
Lemanov, V. V., and Yushin, N. K. (1974). Nonlinear effects in the propagation of elastic waves in piezoelectric crystals. *Sov. Phys.-Solid State (Engl. Transl.)* **15**, 2140–2142.
Lemanov, V. V., Sherman, A. B., Smolenskii, G. A., Angert, N. B., and Klyuev, V. P. (1968). Excitation and propagation of longitudinal elastic waves of 200-2000 MHz frequency in lithium niobate crystals. *Fiz. Tverdogo Tela* **10**, 1720–1724. In Russian.
Lewis, O., and Lu, C.-S. (1975). Relationship of resonant frequency of quartz crystal to mass loading, *Proc. 29th Annu. Freq. Control Symp.*, pp. 5–9.
Liu, H. J., Zhao, Z. Y., and Shi, Z. J. (1981). Measurement of all components of the electroelastic tensor for α-LiIO$_3$ crystal. *Acta Phys. Sin.* **30**, 297–305. In Chinese.
Ljamov, V. E. (1972). Nonlinear acoustical parameters of piezoelectric crystals. *J. Acoust. Soc. Am.* **52**, 199–202.
Lloyd, P. (1967). Equations governing the electrical behaviour of an arbitrary piezoelectric resonator having N electrodes. *Bell Syst. Tech. J.* **46**, 1881–1900.
Lohmann, H. D., and Tholl, H. (1969). Holographic investigations of vibrating piezoelectric quartz plates. *Z. Naturforsch.* **24A**, 1806–1809. In German.
Love, A. E. H. (1944). "A Treatise on the Mathematical Theory of Elasticity." Cambridge University Press, London, 4th Edition, and Dover Publications, Inc., New York.
Loza, Y. K., Molchanov, A. A., and Yalovega, G. I. (1974). Analysis of triggering phenomena in a quartz-crystal resonator, controlled by p-n junction capacitance. *Izv. Vyssh. Uchebn. Zaved. Radioelektron.* **17**, 90–92. In Russian.
Lozhkin, V. N. (1976). Flexure of a thin piezoelectric plate with an elliptic opening. *Mat. Fiz.* **20**, 90–94. In Russian.
Lukaszek, T. J. (1965). Improvements of quartz filter crystals. *Proc. 19th Annu. Freq. Control Symp.*, pp. 269–296.
Lukaszek, T. J. (1970). Mode control and related studies of VHF quartz filter crystals. *Proc. 24th Annu. Freq. Control Symp.*, pp. 126–140.
Lukaszek, T. J. (1971). Mode control and related studies of VHF quartz filter crystals. *IEEE Trans. Sonics Ultrason.* **SU-18**, 238–246.
Lukaszek, T. J., and Ballato, A. (1979). Resonators for severe environments. *Proc. 33rd Annu. Freq. Control Symp.*, pp. 311–321.
Lyubimov, V. N. (1970). Elastic waves in crystals in the presence of the piezoelectric effect. *Fiz. Tverdogo Tela* **12**, 947–949. In Russian.
Madorskii, V. V., and Ustinov, Y. A. (1976). Construction of a system of homogeneous solutions and analysis of the roots of the dispersion equation of antisymmetric vibrations of a piezoelectric plate. *J. Appl. Mech. Tech. Phys.* **17**, 867–873.
Malyshev, V. A. (1976). On the equivalency of resonators and tank circuits. *Radio Eng. Electron. Phys.* **21**, 37–43.
Mamatova, T. A., Fokina, L. A., and Lyamov, V. E. (1973). Effect of domain structure on the propagation of elastic waves in lithium niobate. *Sov. Phys.-Solid State (Engl. Transl.)* **15**, 393–394.
Markenscoff, X. (1974). Effects of initial bending on resonance frequencies of crystal plates. Princeton Univ., New Jersey, Thesis, 122 pp. Available from Univ. Microfilms, Ann Arbor, Michigan, Order No. 75-20648.

Markenscoff, X. (1977). Higher-order effects of initial deformation on the vibrations of crystal plates. *J. Acoust. Soc. Am.* **61**, 436–438.
Martin, J. J., and Doherty, S. P. (1980). The acoustic loss spectrum of 5 MHz 5th overtone AT-cut deuterated quartz resonators. *Proc. 34th Annu. Freq. Control Symp.*, pp. 81–84.
Mashonis, A. (1975). Pulse response of radial vibration of piezoelectric disk. *Period. Polytech. Electr. Eng.* **19**, 3–11.
Mason, W. P. (1940). A new crystal plate designated the GT, which produces a very constant frequency. *Proc. IRE* **28**, 220–223.
Mason, W. P. (1948). "Electromechanical Transducers and Wave Filters." 2nd ed. (1st ed. published in 1942), Van Nostrand, Princeton, New Jersey.
Mason, W. P. (1950). "Piezoelectric Crystals and Their Application to Ultrasonics." Van Nostrand, Princeton, New Jersey.
Mason, W. P. (1964). Piezoelectric Crystals and Mechanical Resonators. *In* "Physical Acoustics: Principles and Methods" (W. P. Mason, ed.), vol. IA, pp. 335–416. Academic Press, New York.
Matistic, A. S. (1976). Quartz-crystal timing accuracy is hard to beat. *Electron. Des.* **24**, 74–79.
Matthaei, G. L. (1978). Variational solutions for acoustic resonator problems from the 'reaction' point of view. *Ultrason. Symp. Proc.*, pp. 157–162.
Mavrov, N. P. (1977). Calculation of parameters of circular quartz resonators with energy-trapping. *Bulg. J. Phys.* **4**, 167–178. In Russian.
McAvoy, B. R., and DeKlerk, J. (1976). High Q microwave bulk mode resonators. *Ultrason. Symp. Proc.*, pp. 590–592.
McMahon, D. H. (1968). Acoustic second-harmonic generation in piezoelectric crystals. *J. Acoust. Soc. Am.* **44**, 1007–1013.
McSkimin, H. J., Andreatch, P., Jr., and Thurston, R. N. (1965). Elastic moduli of quartz versus hydrostatic pressure at 25 degrees and -195.5 degrees C. *J. Appl. Phys.* **36**, 1624–1632.
Meeker, T. R. (1972). Thickness mode piezoelectric transducers. *Ultrasonics* **10**, 26–36.
Meeker, T. R. (1975). Plate constants and dispersion relations for width–length effects in rotated Y-cut quartz plates. *Proc. 29th Annu. Freq. Control Symp.*, pp. 54–64.
Meeker, T. R. (1977). X_1 and X_3 flexure, face-shear, extension, thickness-shear, and thickness-twist modes in rectangular rotated Y-cut quartz plates. *Proc. 31st Annu. Freq. Control Symp.*, pp. 35–43.
Meeker, T. R. (1979a). Extension, flexure, and shear modes in rotated X-cut quartz rectangular bars. *Proc. 33rd Annu. Freq. Control Symp.*, pp. 286–292.
Meeker, T. R. (1979b). A review of the new standard of piezoelectricity. *Proc. 33rd Annu. Freq. Control Symp.*, pp. 176–180.
Meeker, T. R. (1982). Development and technology of piezoelectric bulk wave resonators and transducers. *Proc. 36th Annu. Freq. Control Symp.*, pp. 549–561.
Meeker, T. R., and Miller, A. J. (1980). The temperature coefficient of frequency of AT-cut resonators made from cultured r-face quartz. *Proc. 34th Annu. Freq. Control Symp.*, pp. 85–92.
Mesnage, P. (1976). Bulk waves quartz resonator. *Onde. Electr.* **56**, 289–292. In French.
Mesnage, P. (1980). One hundred years of piezoelectricity studies. *Bull. Annu. Soc. Suisse Chronom. Lab. Suisse Rech. Horlog.* **9**, 165–168. In French.
Miller, A. J. (1979). Private communication based on Ballato (1976).
Milsom, R. F. (1979). Three-dimensional variational analysis of small crystal resonators. *Proc. 33rd Annu. Freq. Control Symp.*, pp. 263–270.
Milsom, R. F., Elliott, D., and Redwood, D. (1981). Three-dimensional mode-matching theory of rectangular bar resonators using complex wavenumbers. *Proc. 35th Annu. Freq. Control Symp.*, pp. 174–186.

Mindlin, R. D. (1955). An introduction to the mathematical theory of vibrations of elastic plates. Monograph, Signal Corps Contract DA-36-039 SC-56772; Fort Monmouth, New Jersey.
Mindlin, R. D. (1956). Mathematical theory of vibration of elastic plates. *Proc. 10th Annu. Freq. Control Symp.*, pp. 10–44.
Mindlin, R. D. (1957). Mathematical theory of vibration of elastic plates. *Proc. 11th Annu. Freq. Control Symp.*, pp. 1–40.
Mindlin, R. D. (1961). High frequency vibrations of crystal plates. *Quart. Appl. Math.* **19**, 51–61.
Mindlin, R. D. (1965). Studies in the mathematical theory of vibrations of crystal plates. (Abstract only.) *Proc. 19th Annu. Freq. Control Symp.*, p. 212.
Mindlin, R. D. (1967). Bechmann's number for harmonic overtones of thickness-twist vibrations of rotated Y-cut quartz plates. *J. Acoust. Soc. Am.* **41**, Pt. 2, 969–973.
Mindlin, R. D. (1968). Electro-mechanical vibrations in centrosymmetric crystals. *Proc. 22nd Annu. Freq. Control Symp.*, p. 1.
Mindlin, R. D. (1968). Lattice and continuum theories of simple modes of vibration in cubic crystal plates and bars. Report TR-11 (AD-677041), 17 pp. Contract Nonr-4259(13). Available from NTIS.
Mindlin, R. D. (1968). Optimum sizes and shapes of electrodes for quartz resonators. *J. Acoust. Soc. Am.* **43**, 1329–1331.
Mindlin, R. D. (1970). Thickness-twist vibrations of a quartz strip. *Proc. 24th Annu. Freq. Control Symp.*, pp. 17–20.
Mindlin, R. D. (1970). Thickness-twist vibrations of quartz strip. Report TR-3 (AD-708057), Columbia Univ., New York, 13 pp. Contract DA-28-043-AMC-01731(E). Available from NTIS.
Mindlin, R. D. (1972a). High frequency vibrations of piezoelectric crystal plates. Report TR-8 (AD-735956), Columbia Univ., New York, 27 pp. Contract DAAB07-72-Q-0166. Available from NTIS.
Mindlin, R. D. (1972b). High frequency vibrations of piezoelectric crystal plates. *Int. J. Solids Struct.* **8**, 895–906.
Mindlin, R. D. (1972c). Electromagnetic radiation from a vibrating quartz plate. Report TR-23. (AD-744312), Columbia Univ., New York, 18 pp. Contract N00014-67-A-0027. Available from NTIS.
Mindlin, R. D. (1972d). Theory of vibrations of plates. *Proc. 26th Annu. Freq. Control Symp.*, p. 84.
Mindlin, R. D. (1973). Electromagnetic radiation from a vibrating quartz plate. *Int. J. Solids Struct.* **9**, 697–702.
Mindlin, R. D. (1974). Coupled piezoelectric vibrations of quartz plates. *Int. J. Solids Struct.* **10**, 453–459.
Mindlin, R. D. (1979). The sesquicentennial of the first crystal plate equations. *Proc. 33rd Annu. Freq. Control Symp.*, pp. 1–3.
Mindlin, R. D. (1982). Third overtone quartz resonator. *Int. J. Solids Struct.* **18**, 809–817.
Mindlin, R. D. (1982). Third overtone quartz resonator. *Proc. 36th Annu. Freq. Control Symp.*, pp. 3–21.
Mindlin, R. D., and Spencer, W. J. (1967). Anharmonic, thickness-twist overtones of thickness-shear and flexural vibrations of rectangular AT-cut quartz plates. *J. Acoust. Soc. Am.* **42**, 1268–1277.
Mitchell, W. S., and Muster, D. (1969). The vibrational response of a thick, circular, piezoelectric plate. *Acustica* **22**, 321–328.
Mitra, G. B., and Kar, L. (1979). A preliminary study of the changes in X-rays reflected by a piezoelectric crystal when subjected to an oscillatory electric field. *Indian J. Phys. A* **53A**, 158–167.

Mizan, M., and Ballato, A. (1983). The stress coefficient of frequency of quartz plate resonators. *Proc. 37th Annu. Freq. Control Symp.*, pp. 194–199.

Mochizuki, Y. (1975). Simple exact solutions for thickness shear mode of vibration of a crystal strip. *Proc. 29th Annu. Freq. Control Symp.*, pp. 35–41.

Mochizuki, Y. (1978). A new method to analyse vibrations by the combination of plate equations and finite element method. *Proc. 32nd Annu. Freq. Control Symp.*, pp. 120–133.

Moore, R. A., Hopwood, F. W., Haynes, T., and McAvoy, B. R. (1979). Bulk acoustic resonator for microwave frequencies. *Proc. 33rd Annu. Freq. Control Symp.*, pp. 444–448.

Moore, R. A., Goodell, J., Hopwood, W., Sundelin, R. A., Zahorchak, A., Murphy, J., and McAvoy, B. R. (1980). High Q bulk acoustic resonators for direct microwave oscillator stabilization. *Proc. 34th Annu. Freq. Control Symp.*, pp. 243–251.

Moore, R. A., Haynes, J. T., and McAvoy, B. R. (1981). High overtone bulk resonator stabilized microwave sources. *Ultrason. Symp. Proc.*, pp. 414–424.

Morrison, J. A., Seery, J. B., and Wilson, L. O. (1977). Propagation of high-frequency elastic surface waves along cylinders with various cross-sectional shapes. *Bell Syst. Tech. J.* **56**, 77–114.

Mortley, W. S. (1957). Frequency modulated quartz oscillators for broadcasting equipment. *Proc. Inst. Elect. Eng.* **104B**, 239–253.

Mortley, W. S. (1966). Priority in energy trapping. *Phys. Today* **19**, 11–12.

Mowbray, D. F. (1970). Vibrations of an infinite plate based on the dynamic theory of thermoelasticity. Rensselaer Polytech. Inst., Troy, New York, thesis, 185 pp.

Mukherjee, S. (1973). A note on vibration of a piezo-electric flexural vibrator. *Indian J. Theor. Phys.* **21**, 89–93.

Muller, F. (1981). Oscillating quartz crystals and their application. *Funk-Tech.* **36**, 246–251. In German.

Munk, S. (1977). Facts about the quartz crystal. *Elektronik* (4), 10, 12, 14. In Danish.

Munn, R. W. (1981). External and internal strains in piezoelectric crystals. *Phys. Status Solidi* B **104**, 325–330.

Murphy, J., and Gad, M. M. (1978). A versatile program for computing and displaying the bulk acoustic wave properties of anisotropic crystals. *Ultrason. Symp. Proc.*, pp. 172–181.

Musha, T. (1975). 1/f resonant frequency fluctuation of a quartz crystal. *Proc. 29th Annu. Freq. Control Symp.*, pp. 308–310.

Nagata, T., Nakajima, Y., and Sasaki. R. (1972). $PbTiO_3$ ceramic resonators operating in the VHF band. *Electron. Commun. Jap.* **55**, 93–98.

Nagata, T., Nakajima, Y., and Sasaki, R. (1974). A new trapped-energy mode of the thickness-dilational wave. *Electron. Commun. Jap.* **57**, 19–24.

Nakagawa, Y., Dobashi, T., and Saigusa, Y. (1978). Bulk-wave generation due to the nonlinear interaction of surface acoustic waves. *J. Appl. Phys.* **49**, 5294–5927.

Nakajima, A., Honma, T., Yamagata, S., and Fukai, I. (1980). The distribution of stress T in a rectangular AT-cut quartz plate. *Trans. Inst. Electron. Commun. Eng. Jpn. Sect.* E **63**, 826.

Nakamura, K., and Shimizu, H. (1976). Analysis of two-dimensional energy trapping in piezoelectric plates with rectangular electrodes. *Ultrason. Symp. Proc.* pp. 606–609.

Nakamura, K., Watanabe, H., and Shimizu, H. (1976). Analysis of trapped-energy resonators and monolithic filters by the equivalent distributed-constant circuits. *Trans. Inst. Electron. Commun. Eng. Jpn. Sect.* E **59**, 26–27.

Nakazawa, M. (1973). Pure- and quasi-plane elastic waves in thickness quartz crystal plate. *J. Fac. Eng. Shinshu Univ.* (34), 155–156. In Japanese.

Nakazawa, M. (1979). Specific directions of plane elastic waves in thin $LiTaO_3$ and $LiNbO_3$ crystal plates. *Int. J. Electron.* **46**, 289–298.

Nakazawa, M. (1980). Improving frequency–temperature characteristics of grooved AT-cut plates. *Proc. 34th Annu. Freq. Control Symp.*, pp. 152–159.
Nakazawa, M. (1980). Specific directions of plane elastic waves in thin quartz crystal plates. *Int. J. Electron.* **48**, 275–282.
Nakazawa, M., and Kozima, S. (1982). A study of GT-type quartz crystal plates. *IEEE Trans. Sonics Ultrason.* **SU-29**, 121–127.
Nakazawa, M., Ito, H., Lukaszek, T., and Ballato, A. (1982). Improved ring-supported resonators. *Proc. 36th Annu. Freq. Control Symp.*, pp. 513–516.
Nakazawa, M., Ito, H., Usui, A., Ballato, A., and Lukaszek, T. (1982). New quartz resonators with precision frequency-linearity over a wide temperature range. *Proc. 36th Annu. Freq. Control Symp.*, pp. 290–296.
Nakazawa, M., Lukaszek, T., and Ballato, A. (1981). Force- and acceleration-frequency effects in grooved and ring-supported resonators. *Proc. 35th Annu. Freq. Control Symp.*, pp. 71–91.
Nakazawa, M., Lukaszek, T., and Ballato, A. (1981). Resonators with reduced frequency sensitivity to static and dynamic stresses. *Ultrason. Symp. Proc.*, pp. 410–413.
Nanamatsu, S., Doi, K., and Takahaski, M. (1973). Piezoelectric, elastic, and dielectric properties of $LiGaO_2$ single crystal. *NEC Res. Dev.* **28**, 72–79.
Nelson, D. F. (1978). Nonlinear electroacoustics of crystals. *Ultrason. Symp. Proc.*, pp. 357–362.
Nelson, D. F. (1978). Three-field electroacoustic parametric interactions in piezoelectric crystals. *J. Acoust. Soc. Am.* **64**, 891–895.
Nelson, R. A., Jr., and Royster, L. H. (1969). On the vibration of a thin piezoelectric disk with an arbitrary impedance on the boundary. *J. Acoust. Soc. Am.* **46**, 828–830.
Nelson, R. A., Jr., and Royster, L. H. (1969). On the electromechanical coupling and electrical impedance of a thin piezoelectric disk with an arbitrary impedance on the boundary. Report TR-4 CAS-39(AD689206), North Carolina State Univ., Raleigh, 13 pp. Contract NONR-486(11). Available from NTIS.
Niizeki, N., and Sawamoto, K. (1970). Zero temperature coefficient of resonant frequency in $LiTaO_3$ length expander bars. *Proc. IEEE* **58**, 1289–1290.
Nonaka, S., Yuuki, T., and Hara, K. (1971). The current dependency of crystal unit resistance at low drive level. *Proc. 25th Annu. Freq. Control Symp.*, pp. 139–147.
Nye, J. F. (1960). "Physical Properties of Crystals." Oxford Univ. Press (Clarendon), London. (First published in 1957).
O'Connell, R. M., and Carr, P. H. (1977). Temperature compensated cuts of berlinite and β-eucryptite for SAW devices. *Proc. 31st Annu. Freq. Control Symp.*, pp. 182–186.
Ochiai, O., Kudo, A., Nakajima, A., and Kawashima, H. (1982). A method of adjusting resonant frequency and frequency-temperature coefficients of miniaturized GT-cut quartz resonator. *Proc. 36th Annu. Freq. Control Symp.*, pp. 90–96.
Okano, S., Kudama, T., Yamazaki, K., and Kotake, H. (1981). 4.19 MHz cylindrical AT-cut miniature resonator. *Proc. 35th Annu. Freq. Control Symp.*, pp. 166–173.
Okazaki, M., and Watanabe, S. (1983). Miniature $LiTaO_3$ X-cut strip resonator. *Proc. 37th Annu. Freq. Control Symp.*, pp. 337–342.
Onoe, M. (1969). Relationship between temperature behavior of resonant and antiresonant frequencies and the electromechanical coupling factors of piezoelectric resonators. *Proc. IEEE* **57**, 702–703.
Onoe, M. (1969). Thickness-twist vibrations of a piezoelectric plate. *J. Acoust. Soc. Am.* **46**, 1087–1089.
Onoe, M. (1969). Thickness-twist vibrations of a piezoelectric plate. *Electron. Commun. Jpn.* **52**, 38–43.
Onoe, M. (1972). General equivalent circuit of a piezoelectric transducer vibrating in thickness modes. *Trans. IECE Jpn.* **55A**, 239–244. In Japanese.

Onoe, M. (1977). Piezoelectric application of lithium tantalate. *Bull. Acad. Sci. USSR, Phys. Ser* **41**, 52–56.
Onoe, M., and Araki, F. (1974). Simplified equivalent circuit for a piezoelectric resonator with high electromechanical coupling, *J. Acoust. Soc. Jap.* **30**, 492–497.
Onoe, M., and Araki, F. (1974). Simplified equivalent circuit for a piezoelectric resonator with high electromechanical coupling. *Proc. IEEE* **62**, 1392–1393.
Onoe, M., and Jumonji, H. (1965). Analysis of piezoelectric resonators vibrating in trapped energy modes. *Electron. Commun. Jap.* **43**, 84–93.
Onoe, M., and Kasai, Y. (1971). Quarter or half wavelength mounting of low frequency flexural vibrators. *J. Acoust. Soc. Jap.* **27**, 515–523. In Japanese.
Onoe, M., and Okada, K. (1969). Analysis of contoured piezoelectric resonators vibrating in thickness-twist modes. *Proc. 23rd Annu. Freq. Control Symp.*, pp. 26–38.
Onoe, M., and Okazaki, M. (1975). Miniature AT-cut strip resonators with tilted edges. *Proc. 29th Annu. Freq. Control Symp.*, pp. 42–48.
Onoe, M., and Yamagishi, I. (1975). Thickness-twist vibration of a plate with tilted edges. *Electron. Commun. Jap.* **58**, 9–18.
Onoe, M., and Yano, T. (1968). Analysis of flexural vibrations of a circular disk. *IEEE Trans. Sonics Ultrason.* **SU-15**, 182–185.
Onoe, M., Ashida, T., and Sawamoto, K. (1969). Zero temperature coefficient of resonant frequency in an X-cut lithium tantalate at room temperature. *Proc. IEEE* **57**, 1446–1447.
Onoe, M., Shinada, T., Itoh, K., and Miyazaki, S. (1973). Low frequency resonators of lithium tantalate. *Proc. 27th Annu. Freq. Control Symp.*, pp. 42–49.
Oomura, Y. (1976). Miniaturized circular disk R_1-cut (AT) crystal vibrator by means of side wire mounting. *Trans. Inst. Electron. Commun. Eng. Jpn. Sect. E* **E59**, 17–18.
Oomura, Y. (1976). Miniaturized circular disk AT-cut crystal vibrator. *Proc. 30th Annu. Freq. Control Symp.*, pp. 202–208.
Oomura, Y. (1978). The frequency temperature behavior of miniaturized circular disk AT-cut vibrator. *Trans. Inst. Electron. Commun. Eng. Jpn. Sec. E* **E16**, 93–94.
Oomura, Y. (1980). Frequency–temperature behavior of miniaturized circular disk AT-cut crystal resonator. *Proc. 34th Annu. Freq. Control Symp.*, pp. 140–151.
Ostrovskii, I. V., and Polovina, A. I. (1978). Acoustic correlation in a piezoelectric plate resonator. *Sov. Phys.-Solid State (Engl. Transl.)* **20**, 1979–1980.
Oura, N., Fukuyo, H., and Yokoyama, A. (1976). The vibration of a biconvex circular AT-cut plate. *Proc. 30th Annu. Freq. Control Symp.*, pp. 191–195.
Oura, N., Kuramochi, N., Matsuda, I., Kono, H., and Fukuyo, H. (1981). Effect of the worked layer in quartz-crystal plates on their frequency stabilities. *IEEE Trans. Instrum. Meas.* **IM-30**, 139–143.
Ozimet, E. J., and Chai, B. H.-T. (1977) Piezoelectric properties of single crystal berlinite. *Proc. 31st Annu. Freq. Control Symp.*, pp. 80–87.
Paige, E. G. S., and Whittle, P. (1979). Coupling of surface skimming bulk waves with a multistrip coupler. *Ultrason. Symp. Proc.*, pp. 802–805.
Paige, E. G. S., and Whittle, P. (1979). Multistrip coupler performance for surface-skimming bulk waves on lithium tantalate. *Electron. Lett.* **15**, 374–376.
Paldas, M. M. (1971). Propagation of torsional impulse in semi-infinite piezoelectric quartz crystal. *Acta Phys. Pol. A* **A39**, 91–101.
Paradise, E. D. (1980). Switchable 'Q' network. *IBM Tech. Disc. Bull.* **23**, 2910–2911.
Paul, H. S. (1968). Vibrational waves in a thick infinite plate of piezoelectric crystal. *J. Acoust. Soc. Am.* **44**, 478–482.
Paul, H. S., and Sarma, K. V. (1978). Forced torsional vibrations of a semi-infinite piezoelectric medium of (622) class. *Proc. Indian Natl. Sci. Acad. Part A* **44**, 362–368.

Pauley, K. E., and Dong, S. B. (1976). Analysis of plane waves in laminated piezoelectric plates. *Wave Electron.* **1**, 265-285.
Peach, R. C. (1982). The design of partially contoured quartz crystal resonators. *Proc. 36th Annu. Freq. Control Symp.*, pp. 22-28.
Pearman, G. T. (1969). Thickness-twist vibrations in beveled AT-cut quartz plates. *J. Acoust. Soc. Am.* **45**, 928-934.
Pegeot, C. (1979). Comparative evaluation on crystal resonators between AT cut and SC cut; (single and double rotated cuts). *Onde Electr.* **59**, 65-69. In French.
Peschl, H. (1979). The quartz crystal. Structure, mode of operation and characteristics. I. *Funkschau* **51**, 369-372. In German.
Petrakov, V. S., Sorokin, N. G., Chizhikov, S. I., Blistanov, A. A., and Shaskol'skaya, M. P. (1979). Influence of electric fields on hypersonic wave propagation in $LiNbO_3$ and $LiTaO_3$ crystals. *Sov. Phys.-Crystallogr.* (*Engl. Transl.*) **24**, 495-497.
Pietrzak, J., Ostafin, M., and Blaszak, K. (1976). Piezoelectric radio-frequency resonance in triglycine sulphate crystals. *Acta Phys. Pol. A* **A50**, 713-717.
Pine, A. S. (1970). Direct observation of acoustical activity in α-quartz. *Phys. Rev. B.* **3**, 2049-2054.
Planat, M., and Hauden, D. (1982). Nonlinear properties of bulk and surface acoustic waves in piezoelectric crystals. *Ferroelectrics* **42**, 117-136.
Planat, M., Theobald, G., Gagnepain, J.-J., and Siffert, P. (1980). Intermodulation in X-cut lithium tantalate resonators [used in high-power filters]. *Electron. Lett.* **16**, 174-175.
Pol'skii, A. I., and Rizunenko, V. I. (1976). Bending resonance vibrations of a free two-layer disc. *Fiz. Met. Metalloved.* **42**, 1095-1098. In Russian.
Pozdnyakov, P. G. (1970). Quartz resonators for torsional vibrations. *Kristallografiya* **15**, 78-85. In Russian.
Pozdnyakov, P. G. (1971). Piezoelectric excitation of torsional vibrations in crystalline rods. *Kristallografiya* **16**, 944-946. In Russian.
Pozdnyakov, P. G., and Vasin, I. G. (1969). Quartz torsional-vibration resonators. *Dokl. Adad. Nauk. SSSR* **185**, 809-812. In Russian.
Pozdnyakov, P. G., and Fedotov, I. M. (1973). Thermal probing of oscillating piezoelectric plates. *Sov. Phys.-Dokl.* (*Engl. Transl.*) **17**, 807-809.
Pozdnyakov, P. G., Bankov, V. N., and Vasin, I. G. (1970). The overtones of torsional vibrations of quartz rods. *Kristallografiya* **15**, 1033-1037. In Russian.
Preobrazhenskii, N. S., and Il'in, V. K. (1976). Method of determining quality of piezoelectrics. *Ind. Lab.* **42**, 1421-1423.
Rao, B. S. (1978). Wave propagation in piezoelectric medium of hexagonal symmetry. *Proc. Indian Acad. Sci. Sect. A.* **87**, 125-136.
Ratajski, J. M. (1968). Force-frequency coefficient of singly rotated vibrating quartz crystals. *IBM J. Res. Dev.* **12**, 92-99.
Ristic, V. M., and Gehrels, J. F. (1973). Real time visualization of thickness modes in crystal resonators. *Int. J. Electron.* **35**, 281-283.
Roberts, G. E. (1975). The design of coupled-resonator AT-cut quartz crystals for operation on the third thickness-shear overtone. *Proc. IEEE* **63**, 1527-1529.
Rohde, R. W., and Jones, O. E. (1968). Mechanical and piezoelectric properties of shock-loaded X-cut quartz at 573 K. *Rev. Sci. Instrum.* **39**, 313-316.
Rosati, V., and Filler, R. L. (1981). Reduction of the effects of vibration on SC cut quartz crystal oscillators. *Proc. 35th Annu. Freq. Control Symp.*, pp. 117-121.
Rossman, H., and Haynes, J. T. (1983). Determination of acceleration sensivitity of bulk mode resonator plates. *Proc. 37th Annu. Freq. Control Symp.*, pp. 272-274.
Roy, P. (1969). Note on responses in a piezoelectric crystal with divided electrodes. *Proc. Nat. Inst. Sci. India A* **35**, 612-618.

Royer, J. J. (1981). Unwanted modes in 5° X-cut crystal units. *Proc. 35th Annu. Freq. Control Symp.*, pp. 250–256.
Sang, K. C. (1974). Temperature characteristics of elastic layer mode propagating on piezoelectric crystal. *J. Korean Inst. Electron. Eng.* **10**, 56–61. In Korean.
Sannomiya, T., and Chubachi, N. (1977). An overtone composite resonator for UHF range. *Rec. Electr. Commun. Eng. Conversazione Tohoku Univ.* **46**, 53–59. In Japanese.
Sasaki, E., and Jumonji, H. (1973). Effects of electrode dimensions on resonant Q of energy-trapped AT-cut quartz resonators. *Electron. Commun. Jap.* **56**, 69–77.
Sato, J., Kawabuchi, M., and Fukumoto, A. (1980). A calculating method of electromechanical coupling coefficient of piezoelectric vibrators. *Trans. Inst. Electron. Commun. Eng. Jpn. Sect. E* **63**, 161.
Sauerbrey, G., and Jung, G. (1968). Vibration modes of plano convex quartz plates. *Z. Angew. Phys.* **24**, 100–108. In German.
Sawamoto, K. (1971). Energy trapping in a lithium tantalate X-cut resonator. *Proc. 25th Annu. Freq. Control Symp.*, pp. 246–250.
Schmidt, G. H. (1972). Extensional vibrations of piezoelectric plates. *J. Eng. Math.* **6**, 133–142.
Schmidt, G. H. (1977). On anti-symmetric waves in an unbounded piezoelectric plate with axisymmetric electrodes. *Int. J. Solids Struct.* **13**, 179–195.
Schmidt, G. H. (1977). Resonances of an unbounded piezoelectric plate with circular electrodes. *Int. J. Eng. Sci.* **15**, 495–510.
Schmitt, P. K. (1974). High-Q BT-cut resonators in flat configuration. *Proc. 28th Annu. Freq. Control Symp.*, pp. 67–72.
Schmitt, P. (1979). On the Q value calculation for planar quartz resonators. *Wiss. Ber. AEG-Telefunken* **52**, 218–227. In German.
Schnabel, P. (1969). Frequency equations for n mechanically coupled piezoelectric resonators. *Acustica* **21**, 351–357.
Schulz, M. B., Holland, M. G., Fukumoto, A., and Watanabe, A. (1970). Comments on 'temperature dependence of resonant frequencies of $LiNbO_3$ plate resonators.' *Proc. IEEE* **58**, 477.
Schwarz, R. (1977). Digital computer simulation of a piezoelectric thickness vibrator. *J. Acoust. Soc. Am.* **62**, 463–467.
Schwarz, R. (1980). The impulse response of piezoelectric thickness vibrators. *Arch. Elektron. Uebertragungstech.* **34**, 413–420. In German.
Schweppe, H. (1971). High-coupling piezoelectric flexural resonator. *Proc. 7th Int. Congr. Acoust.* (Budapest: Academiai Kiado), pp. 389–392.
Sedlacek, J. (1977). Torsional thickness oscillations in a $Bi_{12}GeO_{20}$ plate featuring a (110) cut. *Slaboproudy Obz.* **38**, 106–108. In Czech.
Seed, A. (1962). Ph.D. Thesis, London; *Proc. 4th Int. Congr. Acoust.* (Copenhagen).
Sekiguchi, Y., and Funakubo, H. (1980). Strained surface layers of quartz plates produced by lapping and polishing and their influence on quartz resonator performance. *J. Mater. Sci.* **15**, 3066–3070.
Sekimoto, H., and Ariga, M. (1978). Dispersion characteristics near cutoff frequencies for thickness waves in high coupling piezoelectric plates. *Trans. Inst. Electron. Commun. Eng. Jpn. Sect. E* **61**, 830–831.
Sekimoto, H., and Ariga, M. (1979). Analysis of trapped-energy resonators with circular electrodes. *Trans. Inst. Electron. Commun. Eng. Jpn. Sect. E* **62**, 554–555.
Sharma, R. S. (1977). Effect of magnetic field on oscillating properties of a quartz crystal. *Indian J. Pure Appl. Phys.* **15**, 293–295.
Sheahan, D. F. (1970). An improved resonance equation for AT-cut quartz crystals. *Proc. IEEE* **58**, 260–261.

Sheiko, Y. A. (1978). Equivalent circuit of a flexurally vibrating multielectrode piezoelectric bar. *Sov. Phys.-Acoust. (Eng. Transl.)* **24**, 154–156.
Sherman, J. H., Jr. (1969). A novel algorithm for the design of the electrodes of single-mode AT-cut resonators. *Proc. 23rd Annu. Freq. Control Symp.*, pp. 143–156.
Sherman, J. H., Jr. (1974). Characterization of cultured quartz for use in medium precision AT-cut quartz resonators. *Proc. 28th Annu. Freq. Control Symp.*, pp. 129–142.
Sherman, J. H., Jr. (1976). Dimensioning rectangular electrodes and arrays of electrodes on AT-cut quartz bodies. *Proc. 30th Annu. Freq. Control Symp.*, pp. 54–64.
Sherman, J. H., Jr. (1981). Some properties of electroded piezoelectrics realized and illustrated in AT-cut quartz. *Ultrason. Symp. Proc.*, pp. 448–451.
Shick, D. V., and Tiersten, H. F. (1982). An analysis of thickness-extensional trapped energy mode transducers. *Ultrason. Symp. Proc.*, pp. 509–514.
Shick, D. V., Tiersten, H. F., and Sinha, B. K. (1981). Forced thickness-extensional trapped energy vibrations of piezoelectric plates. *Ultrason. Symp. Proc.*, pp. 452–457.
Shimizu, H., and Yamada, K. (1979). Energy-trapping for backward-wave-mode thickness-vibrations by controlling piezoelectric reaction. *Trans. Inst. Electron. Commun. Eng. Jpn. Sect. E* **62**, 21–22.
Shimizu, H., and Yamada, K. (1980). Energy-trapping for backward-wave-mode thickness vibrations by using multistrip electrodes. *Trans. Inst. Electron. Commun. Eng. Jpn. Sect. E.* **63**, 678.
Shockley, W., Curran, D. R., and Koneval, D. J. (1963). Energy trapping and related studies of multiple electrode filter crystals. *Proc. 17th Annu. Freq. Control Symp.*, pp. 88–126.
Shockley, W., Curran, D. R., and Koneval, D. J. (1967). Trapped energy modes in quartz filter crystals. *J. Acoust. Soc. Am.* **41**, 981–993.
Singh, A. (1968). Transverse wave interaction in piezo-electric material. *Int. J. Electron.* **25**, 495–496.
Sinha, B. K. (1980). Stress induced frequency shifts in thickness-mode quartz resonators. *Ultrason. Symp. Proc.* **2**, 813–818.
Sinha, B. K. (1981). Stress compensated orientations for thickness-shear quartz resonators. *Proc. 35th Annu. Freq. Control Symp.*, pp. 213–221.
Sinha, B. K. (1982). Elastic waves in crystals under a bias. *Ferroelectrics* **41**, 61–73.
Sinha, B. K., and Tiersten, H. F. (1978). Temperature derivatives of the fundamental elastic constants of quartz. *Proc. 32nd Annu. Freq. Control Symp.*, pp. 150–154.
Sinha, B. K., and Tiersten, H. F. (1979a). Temperature induced frequency changes in electroded contoured quartz crystal resonators. *Proc. 33rd Annu. Freq. Control Symp.*, pp. 228–234.
Sinha, B. K., and Tiersten, H. F. (1979b). First temperature derivatives of the fundamental elastic constants of quartz. *J. Appl. Phys.* **50**, 2732–2739.
Sinha, B. K. (1969). Mechanical response of a free piezoelectric plate. *Indian J. Phys.* **43**, 516–518.
Sinha, B. K., and Tiersten, H. F. (1980). Transient thermally induced frequency excursions in doubly-rotated quartz thickness-mode resonators. *Proc. 34th Annu. Freq. Control Symp.*, pp. 393–402.
Sinkevich, E. L., and Fedorchenko, A. M. (1978). Thermal fluctuations in a piezoelectric plate. *Izv. Vyssh. Uchebn. Zaved. Fiz.* (6), 95–101. In Russian.
Sliker, T. R., and Koneval, D. J. (1968). Frequency–temperature behavior of X-cut lithium tantalate resonators. *Proc. IEEE* **56**, 1402.
Sliker, T. R.. Koneval. D. J., and Hora, C. J., Jr. (1969). Frequency-temperature behavior of CdS thickness mode resonators. *IEEE Trans. Sonics Ultrason.* **SU-16**, 15–18.
Slobodnik, A. J., Jr. (1970). Microwave acoustic resonances in thin piezoelectric disks. *IEEE Trans. Sonics Ultrason.* **SU-17**, 196–199.

Slobodnik, A. J., Jr., and O'Brien, J. V. (1971). Complete theory of acoustic bulk wave propagation in anisotropic piezoelectric media. Report AFCRL-71-0601 (AD-739162), Air Force Cambridge Res. Labs., Bedford, Massachusetts, 89 pp. Available from NTIS.

Smagin, A. G. (1975). A 1-MHz quartz resonator with a Q factor of 4.2×10^9 at a temperature of 2 K. *Instrum. Exp. Tech.* **17**, 1721–1723.

Smagin, A. G. (1975). A quartz resonator for a frequency of 1 MHz with a Q-value of 4.2×10^9 at a temperature of 2 K. *Cryogenics* **15**, 483–485.

Smagin, A. G. (1977). Amplitude–frequency effect as an evidence of anharmonicity of oscillations. *Radio Eng. Electron. Phys.* **22**, 143–145.

Smagin, A. G. (1978). Fluctuations of the macroscopic quantities of a piezoelectric crystal. *Radio Eng. Electron. Phys.* **23**, 154–156.

Smagin, A. G., and Nikol'skaya, V. I. (1967). Quartz resonator with Q close to 120×10^6 at temperature 2 K. *JETP Lett.* **6**, 72–74.

Smith, A. B., Kestigian, M., Kedzie, R. W., and Grace, M. I. (1967). Shear-wave attenuation in lithium niobate. *J. Appl. Phys.* **38**, 4928–4929.

Smith, R. T. (1967). Elastic, piezoelectric, and dielectric properties of lithium tantalate. *Appl. Phys. Lett.* **11**, 146–148.

Smith, R. T., and Welsh, F. S. (1971). Temperature dependence of the elastic, piezoelectric. and dielectric constants of lithium tantalate and lithium niobate. *J. Appl. Phys.* **42**, 2219–2230.

Smith, W. L. (1968). The application of piezoelectric coupled-resonator devices to communication systems. *Proc. 22nd Annu. Freq. Control Symp.*, pp. 206–225.

Smolarski, A. (1968). Nonlinear modification of the quartz crystal equivalent circuit. *Przeglad Elektron.* **9**, 475–489. In Polish.

Smythe, R. C. (1974). Intermodulation in thickness-shear resonators. *Proc. 28th Annu. Freq. Control Symp.*, pp. 5–7.

Solymar, L., and Lashmore-Davies, C. N. (1967). Transverse wave interactions in piezoelectric materials in the presence of applied electric and magnetic fields. *Int. J. Electron.* **22**, 549–556.

Spencer, W. J. (1972). Monolithic crystals filters. *In* "Physical Acoustics: Principles and Methods" (W. P. Mason and R. N. Thurston, eds.), Vol. IX, Academic Press, New York, pp. 167–220.

Spencer, W. J., and Pearman, G. T. (1970). X-ray diffraction from vibrating quartz plates. *Adv. X-Ray Anal. (Denver, Colorado)*, pp. 507–525.

Spivak, G. V., Antoshin, M. K., Luk'yanov, A. E., Pau, E. I., Ushakov, O. A., Akishin, A. I., Tokarev, G. A., Lyamov, V. E., and Bartel', I. (1972). Observations of the propagation of elastic waves in piezoelectrics using scanning and mirror electron microscopes. *Izv. Akad. Nauk SSSR Ser. Fiz.* **36**, 1954–1956. In Russian.

Stanley, J. M. (1954). Hydrothermal synthesis of large aluminum phosphate crystals. *Ind. Eng. Chem.* **46**, 1684–1689.

Stefan, O. (1969). The coupling between the radial and thickness vibration of the circular resonator. *Czech. J. Phys. B.* **19**, 1425–1428.

Stevens, D. S., and Tiersten, H. F. (1980). Temperature induced frequency changes in electroded AT-cut quartz trapped energy resonators. *Proc. 34th Annu. Freq. Control Symp.*, pp. 384–392.

Stevens, D. S. and Tiersten, H. F. (1981). An analysis of SC-cut quartz trapped energy resonators with rectangular electrodes. *Proc. 35th Annu. Freq. Control Symp.*, pp. 205–212.

Stevens, D. S., and Tiersten, H. F. (1982). Temperature induced frequency changes in electroded contoured SC-cut quartz crystal resonators. *Proc. 36th Annu. Freq. Control Symp.*, pp. 46–54.

Stevens, D. S., and Tiersten, H. F. (1983). Transient thermally induced frequency excursions in AT- and SC-cut quartz trapped energy resonator. *Proc. 37th Annu. Freq. Control Symp.*, pp. 208–217.
Stevenson, J. K., and Redwood, M. (1969). The motional reactance of a piezoelectric resonator —A more accurate and simpler representation for use in filter design. *IEEE Trans. Circuit Theory* CT-16, 568–572.
Stoddard, W. G. (1963). Design equations for plano-convex AT filter resonators. *Proc. 17th Annu. Freq. Control Symp.*, pp. 272–282.
Strashilov, V. L., Borisov, M. I., Branzalov, K. P., and Mirtcheva, D. S. (1979). An application of thickness-shear approximation to the analysis of multielectroded monolithic [piezoelectric] structures. *Wave Electron.* **4**, 67–80.
Suchanek, J. (1982). The influence of electrodes on frequency of piezoelectric crystal resonators. *Ferroelectrics* **43**, 17–23.
Suk, J. (1969). A contribution to the question of the frequencies of piezoelectric plates. *Czech. J. Phys. B.* **19**, 1271–1280.
Suk, J. (1970). An influence of the electrodes on the frequency of piezoelectric bars. *Czech. J. Phys. B.* **20**, 441–446.
Suk, J. (1977). The dependence of inter-resonator coupling coefficient on the orientation of electrodes [crystal resonators]. *Slaboproudy Obz.* **38**, 159–162. In Czech.
Suk, J. (1980). The influence of the density of the environment on the frequency and temperature dependence of the frequency of piezoelectric resonators vibrating in the flexure mode. *Cesk. Cas. Fyz. Sekce A* **30**, 492–495. In Czech.
Sykes, R. A., Smith, W. L., and Spencer, W. J. (1967). Monolithic crystal filters. *IEEE Conv. Rec.* **15**, 78–93.
Syngellakis, S., and Lee, P. C. Y. (1976). An approximate theory for the high-frequency vibrations of piezoelectric crystal plates. *Proc. 30th Annu. Freq. Control Symp.*, pp. 184–190.
Tabarrok, B., and Sakaguchi, R. L. (1969). Complementary formulation for plate vibration problems. *Z. Angew. Math. Phys.* **20**, 423–428.
Takahashi, S., Oyama, S., and Konno, M. (1977). Vibration nodal point movement of free–free flexural bar vibrator. *Trans. Inst. Electron. Commun. Eng. Jpn. Sect. E* **59**, 20–21.
Tanaka, H., and Shimizu, H. (1972). Flexural modes of wave propagation in a piezoelectric plate poled in its plane. *Rec. Elec. Commun. Eng. Conversazione Tohoku Univ.* **41**, 146–151. In Japanese.
Tanaka, H., Shimizu, H., and Yamada, K. (1979). Methods for energy trapping of thickness extensional mode and thickness shear mode in piezoelectric ceramic plate. *Trans. Inst. Electron. Commun. Eng. Jpn. Sec. E* **62**, 549–550.
Theobald, G., and Gagnepain, J.-J. (1979). Frequency variations in quartz crystal resonators due to internal dissipation. *J. Appl. Phys.* **50**, 6309–6315.
Theobald, G., Marianneau, G., Pretot, R., and Gagnepain, J.-J. (1979). Dynamic thermal behavior of quartz resonators. *Proc. 33rd Annu. Freq. Control Symp.*, pp. 239–246.
Thery, P., Bridoux, E., and Moriamez, M. (1970). Interactions between high-frequency acoustical waves in lithium niobate. *J. Acoust. Soc. Am.* **48**, 772–773.
Thurston, R. N. (1964). Wave propagation in fluids and normal solids. *In* "Physical Acoustics: Principles and Methods" (W. P. Mason, ed.), vol. I, pp. 1–110. Academic Press, New York.
Thurston, R. N. (1965). Effective elastic coefficients for wave propagation in crystals under stress. *J. Acoust. Soc. Am.* **37**, 348–356.
Thurston, R. N., and Brugger, K. (1964). Third-order elastic constants and the velocity of small amplitude elastic waves in homogeneously stressed media. *Phys. Rev.* **133**, A1604–A1610.

Thurston, R. N., McSkimin, H. J., and Andreatch, P., Jr. (1966). Third order elastic constants of quartz. *J. Appl. Phys.* **37**, 267–275.

Tichy, J., Tolman, J., and Zelenka, J. (1967). The dependence of the resonance frequency of quartz oscillator crystals on the electric field. *Conf. Freq. Gener. Control Radio Syst.* (London: IEE), pp. 72–77.

Tiersten, H. F. (1963). Thickness vibrations of piezoelectric plates. *J. Acoust. Soc. Am.* **35**, 53–58.

Tiersten, H. F. (1969a). "Linear Piezoelectric Plate Vibrations." Plenum Press, New York.

Tiersten, H. F. (1969b). The relation of electromechanical coupling factors to the fundamental material constants for thickness vibrating piezoelectric plates. *IEEE Trans. Sonics Ultrason.* **SU-16**, 30.

Tiersten, H. F. (1970). Electromechanical coupling factors and fundamental material constants of thickness vibrating piezoelectric plates. *Ultrasonics* **8**, 19–23.

Tiersten, H. F. (1971). On the nonlinear equations of thermo-electroelasticity. *Int. J. Eng. Sci.* **9**, 587–604.

Tiersten, H. F. (1974a). Analysis of intermodulation in rotated Y-cut quartz thickness-shear resonators. *Proc. 28th Annu. Freq. Control Symp.*, pp. 1–4.

Tiersten, H. F. (1974b). Analysis of trapped energy resonators operating in overtones of thickness-shear. *Proc. 28th Annu. Freq. Control Symp.*, pp. 44–48.

Tiersten, H. F. (1975a). Nonlinear electroelastic equations cubic in the small field variables. *J. Acoust. Soc. Am.* **57**, 660–666.

Tiersten, H. F. (1975b). Analysis of intermodulation in thickness-shear and trapped energy resonators. *J. Acoust. Soc. Am.* **57**, 667–681.

Tiersten, H. F. (1975c). Analysis of nonlinear resonance in rotated Y-cut quartz thickness-shear resonators. *Proc. 29th Annu. Freq. Control Symp.*, pp. 49–53.

Tiersten, H. F. (1975d). Analysis of trapped energy resonators operating in overtones of coupled thickness-shear and thickness-twist. *Proc. 29th Annu. Freq. Control Symp.*, pp. 71–75.

Tiersten, H. F. (1976a). Analysis of nonlinear resonance in thickness-shear and trapped-energy resonators. *J. Acoust. Soc. Am.* **59**, 866–878.

Tiersten, H. F. (1976b). Analysis of trapped-energy resonators operating in overtones of coupled thickness shear and thickness twist. *J. Acoust. Soc. Am.* **59**, 879–888.

Tiersten, H. F. (1978). Perturbation theory for linear electroelastic equations for small fields superposed on a bias. *J. Acoust. Soc. Am.* **64**, 832–837.

Tiersten, H. F., and Ballato, A. (1979). Nonlinear vibrations of quartz rods. *Proc. 33rd Annu. Freq. Control Symp.*, pp. 293–299.

Tiersten, H. F., and Ballato, A. (1983). Nonlinear extensional vibrations of quartz rods. *J. Acoust. Soc. Am.* **73**, 2022–2033.

Tiersten, H. F., and Mindlin, R. D. (1961). Forced vibrations of piezoelectric plates. Office of Naval Research, Contract Nonr-266(09), Technical Report No. 43, CU-56-61-ONR-266(09)-CE; Department of the Army Contract DA36-039 SC-87414, Interim Report, CU-6-61 CS-87414; July 24 pp.

Tiersten, H. F. and Mindlin, R. D. (1962). Forced vibrations of piezoelectric plates. *Q. Appl. Math.* **20**, 107–119.

Tiersten, H. F., and Sinha, B. K. (1977). Temperature induced frequency changes in electroded AT-cut quartz thickness shear resonators. *Proc. 31st Annu. Freq. Control Symp.*, pp. 23–28.

Tiersten, H. F., and Sinha, B. K. (1978a). Temperature induced frequency changes in electroded doubly-rotated quartz thickness-mode resonators. *Proc. 32nd Annu. Freq. Control Symp.*, pp. 155–161.

Tiersten, H. F., and Sinha, B. K. (1978b). An analysis of extensional modes in high coupling trapped energy resonators. *Ultrason. Symp. Proc.*, pp. 167–171.
Tiersten, H. F., and Sinha, B. K. (1979). Temperature dependence of the resonant frequency of electroded doubly-rotated quartz thickness-mode resonators. *J. Appl. Phys.* **50**, 8038–8051.
Tiersten, H. F., and Smythe, R. C. (1977). An analysis of overtone modes in contoured crystal resonators. *Proc. 31st Annu. Freq. Control Symp.*, pp. 44–47.
Tiersten, H. F., and Smythe, R. C. (1979). An analysis of contoured crystal resonators operating in overtones of coupled thickness shear and thickness twist. *J. Acoust. Soc. Am.* **65**, 1455–1460.
Tiersten, H. F., and Smythe, R. C. (1981). Coupled thickness-shear and thickness-twist resonances in unelectroded circular AT-cut quartz plates. *Proc. 35th Annu. Freq. Control Symp.*, pp. 230–236.
Tiersten, H. F., and Stevens, D. S. (1982). An analysis of contoured SC-cut quartz crystal resonators. *Proc. 36th Annu. Freq. Control Symp.*, pp. 37–45.
Tiersten, H. F., and Stevens, D. S. (1983). An analysis of thickness-extensional trapped energy resonators with rectangular electrodes in the zinc-oxide thin film on silicon configuration. *Proc. 37th Annu. Freq. Control Symp.*, pp. 325–336.
Tiersten, H. F., Sinha, B. K., McDonald, J. F., and Das, P. K. (1978). On the influence of a tuning inductor on the bandwidth of extensional trapped energy mode transducers. *Ultrason. Symp. Proc.*, pp. 163–166.
Tiersten, H. F., Sinha, B. K., and Meeker, T. R. (1981). Intrinsic stress in thin films deposited on anisotropic substrates and its influence on the natural frequencies of piezoelectric resonators. *J. Appl. Phys.* **52**, 5614–5624.
Togami, Y., and Chiba, T. (1978). Observation of bulk waves and surface waves through side planes of several cuts of $LiNbO_3$ SAW devices. *J. Appl. Phys.* **49**, 3587–3589.
Toki, M., and Tsuzuki, Y. (1979). Motional inductance of plano-convex AT-cut quartz crystal resonators. *Trans. Inst. Electron. Commun. Eng. Jpn. Sect. E* **62**, 154–155.
Toki, M., Tsuzuki, Y., and Mikami, T. (1978). Precise design of motional inductance of longitudinal mode X-cut quartz crystal resonator. *Trans. Inst. Electron. Commun. Eng. Jpn. Sect. E* **61**, 315–316.
Tomikawa, Y. (1978). Analysis of electrical equivalent circuit elements of piezo-tuning forks by the finite element method. *IEEE Trans. Sonics Ultrason.* **SU-25**, 206–212.
Tomikawa, Y., Hirose, S., and Konno, M. (1972). Equivalent constants of a disk vibrator with double resonance. *Bull. Yamagata Univ. Eng.* **12**, 67–91. In Japanese.
Tomikawa, Y., Onodera, T., Sugawara, S., and Konno, M. (1980). Torsional resonator and transducer with modified circular cross section. *Trans. Inst. Electron. Commun. Eng. Jpn. Sect. E* **63**, 680.
Topa, V. I., Mateescu, I., and Velicescu, B. (1973). The design and the realization of quartz low-frequency resonators. *Posta Telecommunicatii* **3**, 304–307. In Rumanian.
Toplevsky, R. B., and Redwood, M. (1975). A general perturbation theory for elastic resonators and its application to the monolithic crystal filter. *IEEE Trans. Sonics Ultrason.* **SU-22**, 152–161.
Tsok, O. E. (1980). Influence of the dimensions of piezoelectric plates on the nature of their vibrational modes. *Sov. Phys.-Acoust.* (*Engl. Transl.*) **26**, 524–525.
Tsuzki, Y. (1968). Parametric interaction between two contour modes of vibration in X-cut quartz bars. *Proc. IEEE* **56**, 98.
Tsuzuki, Y., Hirose, Y., and Iijima, K. (1968). Holographic observation of the parametrically excited vibrational mode of an X-cut quartz plate. *Proc. IEEE* **56**, 1229–1230.
Tsuzuki, Y., Hirose, Y., and Iijima, K. (1971). Measurement of vibrational modes of piezoelectric resonators by means of holography. *Proc. 25th Annu. Freq. Control Symp.*, pp. 113–117.

Tyutekin, V. V., and Shkvarnikov, A. P. (1968). Calculation of the resonance frequencies of flexurally vibrating nonuniform rods by the impedance method. *Sov. Phys.-Acoust.* (*Engl. Transl.*) **14**, 257–258.
Ueda, I., and Ikegami, S. (1968). Piezoelectric properties of modified $PbTiO_3$ ceramics. *Jpn. J. Appl. Phys.* **7**, 236–242.
Uno, T. (1979). Mode dependency of temperature characteristics of trapped energy mode $LiTaO_3$ resonators excited by parallel electric field. *Trans. Inst. Electron. Commun. Eng. Jpn. Sect. E* **62**, 258.
Valdois, M., and Janiaud, D. (1978). The quartz resonator: existence of a direction of zero acceleration effect. *Mes. Regul. Autom.* **43**, 57–58. In French.
Valdois, M., Besson, J., and Gagnepain, J.-J. (1974). Influence of environment conditions on a quartz resonator. *Proc. 28th Annu. Freq. Control Symp.*, pp. 19–32.
Valentin, J.-P., and Besson, R. (1979). Quartz resonators with internal thermal regulations. *C.R. Hebd. Seances Acad. Sci. Ser. B* **289**, 119–122. In French.
Valentin, J.-P., Michel, J. P., and Besson, R. (1979). Ultra thin crystals for "electrodeless" VHF quartz resonators. *C.R. Hebd. Seances Acad. Sci. Ser. B* **289**, 155–158. In French.
Valentin, J.-P., Guerin, C. P., and Besson, R. J. (1981). Indirect amplitude frequency effect in resonators working on two frequencies. *Proc. 35th Annu. Freq. Control Symp.*, pp. 122–129.
Van Ballegooyen, E. C., Boersma, F., and Van der Steen, C. (1977). Influence of the thickness of tabs on the resonating properties of a quartz crystal. *J. Acoust. Soc. Am.* **62**, 1189–1195.
Van Dalen, P. A. (1972). Propagation of transverse electroacoustic waves in a piezoelectric plate of symmetry C_{6v} or $C_{\infty v}$. *Philips Res. Rep.* **27**, 323–339.
Van Dyke, K. S. (1925). The electric network equivalent of a piezoelectric resonator. (Abstract.) *Phys. Rev.* **25**, 895.
Van Dyke, K. S. (1928). The piezoelectric resonator and its equivalent network. *Proc. IRE* **16**, 724–764.
Van der Pauw, L. J. (1973). Impedance matrix of coupled piezoelectric resonators. *Philips Res. Rep.* **28**, 158–178.
Van der Steen, C., Boersma, F., and Van Ballegooyen, E. C. (1977). The influence of mass loading outside the electrode area on the resonant frequencies of a quartz-crystal. *J. Appl. Phys.* **48**, 3201–3205.
Vangheluwe, D. C. L. (1978). Finite element analysis of AT-cut crystals. *Proc. 32nd Annu. Freq. Control Symp.*, pp. 134–141.
Vangheluwe, D. C. L. (1980). The frequency and motional capacitance of partly contoured crystals. *Proc. 34th Annu. Freq. Control Symp.*, pp. 412–418.
Vangheluwe, D. C. L., and Fletcher, E. D. (1981). The edge mode resonator. *Proc. 35th Annu. Freq. Control Symp.*, pp. 157–165.
Vekovishcheva, I. A. (1975). Plane electroelasticity theory for a piezoelectric plate. *Prikl. Mekh.* **11**, 85–89. In Russian.
Vekovishcheva, I. A. (1977). Simple cases of flexure of thin piezoelectric plates. *Prikl. Mekh.* **13**, 127–130.
Viehmann, L. (1973). A monolithic quartz for portable and mobile communications equipment. *Funk-Tech.* (23), 891–893. In German.
Vig, J. R., Filler, R. L., and Kosinski, J. (1982). SC-cut resonators for temperature compensated oscillators. *Proc. 36th Annu. Freq. Control Symp.*, pp. 181–186.
Volluet, G. (1978). Computed characteristics of Tl_3VS_4 bulk wave resonators. *Ultrason. Symp. Proc.*, pp. 182–187.
Vrzal, J., Petrzilka, V., and Holan, P. (1977). Electrical properties of an electromechanical bar-resonator excited to flexural vibrations. *Acoustica* **37**, 167–174. In German.

Waldron, R. A. (1971). Perturbation formulas for elastic resonators and waveguides. *IEEE Trans. Sonics Ultrason.* **SU-18**, 16–20.
Wallace, C. A., and Isherwood, B. J. (1968). Ultrasonic vibrations in quartz crystals. *GEC J. Sci. Technol.* **35**, 87–90.
Wang, C. H., and Zhao, Z. Y. (1981). Thickness vibration of electrically loaded piezoelectric plate. *Acta Acust.* (4), 263–267.
Wang, C. H., Zhao, Z. Y., and Ma, Y. L. (1981). The effect of an electric load on the characteristics of a piezoelectric vibrational system. *Acta Acust.* (2), 92–102.
Wang, J. S., and Lakin, K. M. (1982). Low-temperature coefficient bulk acoustic wave composite resonators. *Appl. Phys. Lett.* **40**, 308–310.
Wang, J. S., Lakin, K. M., and Landin, A. R. (1983). Sputtered c-axis inclined piezoelectric films and shear wave resonators. *Proc. 37th Annu. Freq. Control Symp.*, pp. 144–150.
Ward, R. W. (1981). Design of high performance SC resonators. *Proc. 35th Annu. Freq. Control Symp.*, pp. 99–103.
Warner, A. W. (1960). Design and performance of ultra-precise 2.5 Mc quartz crystal units. *Bell Syst. Tech. J.* **39**, 1193–1218.
Warner, A. W. (1963). Use of parallel-field excitation in the design of quartz crystal units. *Proc. 17th Annu. Freq. Control Symp.*, pp. 248–255.
Warner, A. W. (1981). Private communication based on Ballato (1976).
Warner, A. W., Onoe, M., and Coquin, G. A. (1967). Determination of elastic and piezoelectric constants for crystals in class ($3m$). *J. Acoust. Soc. Am.* **42**, 1223–1231.
Warner, A., Goldfrank, B., Meirs, M., and Rosenfeld, M. (1979). Low "g" sensitivity crystal units and their testing. *Proc. 33rd Annu. Freq. Control Symp.*, pp. 306–310.
Warner, A. W., Goldfrank, B., and Tsaclas, J. (1982). Further development in SC cut crystal resonator technology. *Proc. 36th Annu. Freq. Control Symp.*, pp. 208–214.
Watanabe, H., and Shimizu, H. (1981). Energy trapping of width-shear vibrations in thin piezoelectric strips. *Jpn. J. Appl. Phys.* **20**, 105–108.
Watanabe, A., and Yano, T. (1978). Direction of displacement in $LiNbO_3$ X-cut plate shear transducer, *IEEE Trans. Sonics Ultrason.* **25**, 159–160.
Watanabe, H., Nakamura, K., and Shimizu, H. (1978). Energy trapping of width-extensional vibrations caused by a contribution of complex branches of dispersion curves. *Trans. Inst. Electron Commun. Eng. Jpn. Sect. E.* **61**, 980–981.
Watanabe, H., Nakamura, K., and Shimizu, H. (1980). A new type of energy trapping of width-extensional vibrations of piezoelectric plates caused by contributions from the complex branches of dispersion curves. *Rec. Electron. Commun. Eng. Conversazione Tohoku Univ.* **49**, 137–143. In Japanese.
Watanabe, H., Nakamura, K., and Shimizu, H. (1980). A new-type of energy trapping caused by contributions from the complex branches of dispersion curves. *Ultrason. Symp. Proc.* **2**, 825–828.
Watanabe, H., Nakamura, K., and Shimizu, H. (1981). A new type of energy trapping caused by contributions from the complex branches of dispersion curves. *IEEE Trans. Sonics Ultrason.* **SU-28**, 265–270.
Weinert, R. W., and Isaacs, T. J. (1975). New piezoelectric materials which exhibit temperature stability for surface waves. *Proc. 29th Annu. Freq. Control Symp.*, pp. 139–142.
Werner, J. F., Edwards, H. W., and Smith, M. (1982). Unwanted responses in quartz crystal low frequency X-cut bars. *Proc. 36th Annu. Freq. Control Symp.*, pp. 529–536.
Whatmore, R. W., O'Hara, C., Cohayne, B., Jones, G. R., and Lent, B. (1979). $Ca_{12}Al_{14}O_{33}$: A new piezoelectric material. *Mater. Res. Bull.* **14**, 967–972.
White, D. L. (1962). Amplification of ultrasonic waves in piezoelectric semiconductors. *J. Appl. Phys.* **33**, 2547–2554.

Wilfinger, R. J., Bardell, P. H., and Chhabra, D. S. (1968). The resonistor: a frequency selective device utilizing the mechanical resonance of a silicon substrate. *IBM J. Res. Develop.* **12**, 113–118.

Wilson, C. J. (1973). X-ray topography of quartz crystal resonators: an investigation of band breaks. *Proc. 27th Annu. Freq. Control Symp.*, pp. 35–38.

Wilson, C. J. (1974). Vibration modes of AT-cut convex quartz resonators. *J. Phys. D.* **7**, 2449–2454.

Wilson, L. O. (1977). Vibrations of a lithium niobate fiber. *Bell Syst. Tech. J.* **56**, 1387–1404.

Yamabuchi, T., and Kagawa, Y. (1976). Finite element approach for a piezoelectric circuit rod. *IEEE Trans. Sonics Ultrason.* **SU-23**, 379–385.

Yamabuchi, T., and Kagawa, Y. (1976). Finite element method for electromechanical problems with application to energy-trapped and surface-wave devices. *J. Acoust. Soc. Jpn.* **32**, 65–75. In Japanese.

Yamabuchi, T., and Kagawa, Y. (1977). Finite element approach for a piezoelectric circular rod. *Trans. Inst. Electron. Commun. Eng. Jpn. Sect. E.* **59**, 21–22.

Yamada, K., and Shimizu, H. (1980). Equivalent-network analyses of energy-trapping of backward-wave thickness vibrations. *Trans. Inst. Electron. Commun. Eng. Jpn. Sect. E.* **63**, 449–450.

Yamada, T. (1973). Single crystal growth and piezoelectric properties of lead potassium niobate. *Appl. Phys. Lett.* **23**, 213–214.

Yamada, T. (1975). Elastic and piezoelectric properties of lead potassium niobate. *J. Appl. Phys.* **46**, 2894–2898.

Yamada, T., and Niizeki, N. (1970a). Admittance of piezoelectric plates vibrating under the perpendicular field excitation. *Proc. IEEE* **58**, 941–942.

Yamada, T., and Niizeki, N. (1970b). Formulation of admittance for parallel field excitation of piezoelectric plates. *J. Appl. Phys.* **41**, 3604–3609.

Yamada, T., and Niizeki, N. (1971). A new formulation of piezoelectric plate thickness vibration. *Rev. Elec. Commun. Lab.* **19**, 705–713.

Yamagata, S., Fukai, I., and Yasuda, I. (1978). Analysis of vibrations in AT-cut quartz crystal plates using a light modulated oscillator. *IEEE Trans. Sonics Ultrason.* **SU-25**, 192–198.

Yamagata, S., Nara, S., Fukai, I., and Yasuda, I. (1979). Analysis of vibration modes for planoconvex AT-cut quartz crystal resonator by Schrödinger equation and finite element method. *Trans. Inst. Electron. Commun. Eng. Jpn. Sect. E.* **E62**, 494–495.

Yamagata, S., Yamamoto, K., Fulkai, I., and Yasuda, I. (1978). Thickness-shear stress distributions of the oscillating planoconvex AT-cut quartz-crystal resonator. *Electron. Lett.* **14**, 450–451.

Yamagata, S., Nara, S., and Fukai, I. (1979). Measurement of thickness-shear-stress distribution of the oscillating plano-convex AT-cut quartz-crystal resonator. *IEE J. Microwave Opt. Acoust.* **3**, 265–271.

Yamashita, S., Echigo, N., Kawamura, Y., Watanabe, A., and Kubota, K. (1978). A 4.19 MHz beveled miniature rectangular AT-cut quartz resonator. *Proc. 32nd Annu. Freq. Control Symp.*, pp. 267–276.

Yamashita, S., Motte, S., Takahashi, K., Echigo, N., Watanabe, A., and Kubota, K. (1979). New frequency-temperature characteristics of 4.19 MHz beveled rectangular AT-cut quartz resonator. *Proc. 33rd Annu. Freq. Control Symp.*, pp. 277–285.

Yao, S. K., and Young, E. H. (1976). Properties and applications of composite bulk acoustic resonators. *Ultrason. Symp. Proc.*, pp. 593–596.

Yaroslavskii, M. I., and Fedorkov, A. P. (1978). Low-frequency heat-sensitive quartz resonators. *Meas. Tech.* **21**, 110–111.

Yen, K. H., Wang, K. L., and Kagiwada, R. S. (1977). Efficient bulk wave excitation on ST quartz. *Electron. Lett.* **13**, 37–38.
Yen, K. H., Lau, K. F., and Kagiwada, R. S. (1978). Temperature stable shallow bulk acoustic wave devices. *Proc. 32nd Annu. Freq. Control Symp.*, pp. 95–101.
Yen, K. H., Lau, K. F., and Kagiwada, R. S. (1979). Recent advances in shallow bulk acoustic wave devices. *Ultrason. Symp. Proc.*, pp. 776–785.
Yoo, K. B., Ueberall, H., and Williams, W., Jr. (1982). Spurious resonances in bulk acoustic wave resonators. *Ultrason. Symp. Proc.*, pp. 490–493.
Yoo, K. B., Ueberall, H., Ashrafi, D., and Ashrafi, S. (1983). Spurious resonances and modelling of composite resonators. *Proc. 37th Annu. Freq. Control Symp.*, pp. 317–319.
Zelenka, J. (1969). Temperature dependence of resonant frequency in accurate AT- and BT-cut quartz resonators and its dependence on excitation level. *Tesla Electron.* **2**, 67–74.
Zelenka, J. (1970). Type DY quartz resonators, having the form of rectangular plates. *Slaboproudy Obz.* **31**, 437–442. In Czech.
Zelenka, J. (1971). Thickness vibrations of rectangular piezoceramic plates. *4th Conference on Ceramics for Electronics* (Hradec Kralove, Czechoslovakia: Vzykumny Ustav Elektrotechnicke Keramiky), 8 pp. In Czech.
Zelenka, J. (1973). The frequency spectrum of NT-cut quartz plates. *Czech. J. Phys. B* **B23**, 696–702.
Zelenka, J. (1975). The influence of electrodes on the resonance frequency of AT-cut quartz plates. *Int. J. Solids Struct.* **11**, 871–876.
Zelenka, J. (1976). A calculation of the resonant frequencies of metallized AT-cut square quartz plates. *Sov. Phys.-Crystallogr.* (*Engl. Transl.*) **20**, 590–593.
Zelenka, J. (1978). Electromechanical properties of bismuth germanium oxide ($Bi_{12}GeO_{20}$). *Czech. J. Phys. B* **28**, 165–169.
Zelenka, J. (1982). The influence of the polarizing field on piezoelectric resonators with high electromechanical coupling. *Ferroelectrics* **41**, 35–38.
Zelenka, J., and Lee, P. C. Y. (1971). On the temperature coefficients of the elastic stiffness and compliances of alpha-quartz. *IEEE Trans. Sonics Ultrason.* **SU-18**, 79–80.
Zelenka, J., Petrzilka, V., Vrzal, J., Michalec, R., Chalupa, B., and Sedlakova, L. (1972). Equivalent electrical circuit of a bar excited in longitudinal vibrations as an electromechanical resonator. *Acustica* **27**, 159–165. In German.
Zelenka, J., Petrzilka, V., Michalec, R., and Mikula, P. (1975). The comparison of the thickness vibrations of circular and square quartz plates. *Cesk. Cas. Fis. A* **25**, 492–494. In Czech.
Zonkhiev, M. A. (1970). Amplitude-phase holograms of vibration patterns. *Soviet Phys.-Acoust.* (*Engl. Transl.*) **16**, 209–212.
Zumsteg, A. E., and Suda, P. (1976). Properties of a 4 MHz miniature flat rectangular quartz resonator vibrating in a coupled mode. *Proc. 30th Annu. Freq. Control Symp.*, pp. 196–201.
Zumsteg, A. E., and Suda, P. (1978). Energy trapping of coupled modes in rectangular AT-cut resonators. *Proc. 32nd Annu. Freq. Control Symp.*, pp. 260–266.

Section 2.3

Adams, C. A., and Kusters, J. A. (1977). Deeply etched SAW resonators. *Proc. 31st Annu. Freq. Control Symp.*, pp. 246–250.
Adams, C. A., and Kusters, J. A. (1978). Improved long-term aging in deeply etched SAW resonators. *Proc. 32nd Annu. Freq. Control Symp.*, pp. 74–76.
Albuquerque, E. L., and Chao, N. C. (1981). On the propagation of elastic surface waves in piezoelectric crystals. *Phys. Status Solidi B* **104**, K11–K14.

Alippi, A., Palma, A., Palmieri, L., and Socine, G. (1977). Phase and amplitude relations between fundamental and second harmonic acoustic surface waves on SiO_2 and $LiNbO_3$. *J. Appl. Phys.* **48**, 2182–2190.
Anonymous. (1973). Progress in micro-electronics; acoustic-electrical surface waves-building elements. *VDI Z.* **115**, 639–641. In German.
Anonymous. (1977). Surface acoustic wave technology. *New Electron.* **10**, 64, 67, 71.
Ash, E. A. (1970). Surface wave grating reflectors and resonators. *IEEE Int. Microwave Symp., Dig. Tech. Papers*, pp. 385–386.
Auld, B. A. (1973). "Acoustic Fields and Waves in Solids," vol. II. Wiley, New York.
Auld, B. A., and Yeh, B. H. (1977). Piezoelectric shear surface wave grating resonators. *Proc. 31st Annu. Freq. Control Symp.*, pp. 251–257.
Bailey, D., Lee, D. L., Andle, J., Vetelino, J. F., and Chai, B. H.-T. (1981). An experimental study of the SAW properties of several berlinite samples. *Ultrason. Symp. Proc.*, pp. 341–345.
Bailey, D. S., Josse, F., Lee, D. L., Andle, J., Soluch, W., and Vetelino, J. F. (1982). Study of SAW properties of temperature-compensated orientations in berlinite. *Electron. Lett.* **18**, 168–170.
Balakirev, M. K., and Gorchakov, A. V. (1977). Coupled surface waves in piezoelectrics. *Fiz. Tverdogo Tela* **19**, 613–614. In Russian.
Bale, R., and Lewis, M. F. (1974). Improvements to the SAW oscillator. *Ultrason. Symp., Proc.*, pp. 272–275.
Ballato, A., Lukaszek, T. J., Yen, K. H., and Kagiwada, R. S. (1979). SAW and SBAW on doubly rotated cut quartz. *Ultrason. Symp. Proc.*, pp. 797–801.
Ballato, A., Lukaszek, T., Williams, D. F., and Cho, F. Y. (1981). Power flow angle and pressure dependence of SAW propagation characteristics in quartz. *Ultrason. Symp. Proc.*, pp. 346–349.
Bell, D. T., Jr. (1972). Phase errors in long surface wave devices. *Ultrason. Symp., Proc.*, pp. 420–423.
Bell, D. T., Jr. (1977). Aging processes in SAW resonators. *Ultrason. Symp. Proc.*, pp. 851–856.
Bell, D. T., Jr., and Li, R. C. M. (1976). Surface acoustic wave resonators. *Proc. IEEE* **64**, 711–721.
Bergmann, H. (1975). Acoustic surface waves-a review. *Elektron Int.* (10), 346–347. In German.
Berté, M., and Hartemann, P. (1979). Quartz resonators operating at frequencies higher than 100 MHz. *Onde Electr.* **59**, 75–80. In French.
Boroson, D. M., and Oates, D. E. (1981). Experimental and theoretical analysis of temperature dependence of wideband SAW RAC devices on quartz. *Ultrason. Symp. Proc.*, pp. 38–43.
Browning, I., and Lewis, M. (1978). A new cut of quartz giving improved temperature stability to SAW oscillators. *Proc. 32nd Annu. Freq. Control Symp.*, pp. 87–94.
Browning, T. I., Lewis, M. F., and Milsom, R. F. (1978). Surface acoustic waves on rotated Y-cut $TiTaO_3$. *Ultrason. Symp. Proc.*, pp. 586–589.
Budreau, A. J., and Carr, P. H. (1971). Temperature dependence of the attenuation of microwave frequency elastic surface waves in quartz. *Appl. Phys. Lett.* **18**, 239–241.
Buff, W. (1980). Acoustoelectronic building elements. II. *Radio Fernsehen Elektron.* **29**, 563–565. In German.
Cambiaggio, E., Azan, F., and Lantz, A. (1978). S.A.W. reflection from metallic gratings of periodicity λ and applications to resonators. *Ultrason. Symp. Proc.*, pp. 643–646.
Carr, P. H., Silva, J. H., and Chai, B. H.-T. (1981). Second-order temperature coefficients of singly and doubly rotated cuts of berlinite. *Ultrason. Symp. Proc.*, pp. 328–331.
Chang, D., Shui, Y.-a., and Jiang, W. H. (1982). Surface quasi-transverse wave resonators on rotated Y-cut lithium niobate. *Ultrason. Symp. Proc.*, pp. 53–56.

Coldren, L. A. (1977). Characteristics of surface acoustic wave resonators obtained from cavity analysis. *IEEE Trans. Sonics Ultrason.* **SU-24**, 212–217.

Coldren, L. A. (1977). Effects of anisotropy of SAW resonator transverse modes. *Appl. Phys. Lett.* **31**, 409–412.

Coldren, L. A., and Lemons, R. A. (1978). Variable frequency SAW resonators on ferroelectric-ferroelastics. *Appl. Phys. Lett.* **32**, 129–131.

Coldren, L. A., and Rosenberg, R. L. (1976). Scattering matrix approach to SAW resonators. *Ultrason. Symp. Proc.*, pp. 266–271.

Coldren, L. A., Haus, H. A., and Wang, K. L. (1977). Experimental verification of mode shape in s.a.w. grating resonators. *Electron. Lett.* **13**, 642–644.

Coussot, G., and Bridoux, E. (1972). Study of the harmonic generation of a surface wave in piezoelectric crystals. *J. Phys.* (France), 33, C6/276-280. In French.

Cross, P. S. (1976). Properties of reflective arrays for surface acoustic resonators. *IEEE Trans. Sonics Ultrason.* **SU-23**, 255–262.

Cross, P. S. (1978). Surface acoustic wave resonator filters using tapered gratings. *IEEE Trans. Sonics Ultrason.* **SU-25**, 313–319.

Cross, P. S., and Elliott, S. S. (1981). Surface-acoustic-wave resonators. *Hewlett-Packard J.* **32**, 9–11, 13–14, 16–17.

Cross, P. S., and Schmidt, R. V. (1977). Coupled surface-acoustic-wave resonators. *Bell Syst. Tech. J.* **56**, 1447–1482.

Cross, P. S., and Shreve, W. R. (1981). Frequency trimming of surface acoustic wave devices. U.S. Patent 4,278,492.

Cross, P. S., Smith, R. S., and Haydl, W. H. (1975). Electrically cascaded surface-acoustic-wave resonator filters. *Electron. Lett.* **11**, 244–245.

Cross, P. S., Haydl, W. H., and Smith, R. S. (1976). Design and applications of two-port SAW resonators on YZ-lithium niobate. *Proc. IEEE* **64**, 682–685.

Cross, P. S., Haydl, W. H., and Smith, R. S. (1976). Electronically variable surface-acoustic-wave velocity and tunable SAW resonators. *Appl. Phys. Lett* **28**, 1–3.

Cross, P. S., Shreve, W. R., and Tan, T. S. (1979). Synchronous IDT SAW resonators with Q above 10,000. *Ultrason. Symp. Proc.*, pp. 824–829.

Cross, P. S., Rissman, P., and Shreve, W. R. (1980). Microwave SAW resonators fabricated with direct-writing electron-beam lithography. *Ultrason. Symp. Proc.*, pp. 158–163.

Cross, P. S., Shreve, W. R., Elliott, S., and Bray, R. C. (1982). Very low loss SAW resonators using parallel coupled cavities. *Ultrason. Symp. Proc.*, pp. 284–289.

Czechowska, Z., and Weiss, K. (1974). Elastic surface waves in piezoelectric materials. *Elektronika* **15**, 383–387.

Danicki, E. (1978). Acoustic surface waves in electronic devices. *Elektronika* **19**, 52–60. In Polish.

Daniel, M. R., de Klerk, J., Jones, C. K., and Patterson, A. (1969). Surface wave propagation in crystalline quartz. *IEEE Trans. Sonics Ultrason.* **SU-16**, 23.

Deka, M., and Claus, R. O. (1979). Measurements of surface waves on an optical flat by reflective interferometry. *Proc. Southeastcon*, pp. 193–196. (IEEE, New York).

Détaint, J., Feldmann, M., Henaff, J., Poignant, H., and Toudic, Y. (1979). Bulk and surface acoustic wave propagation in berlinite. *Proc. 33rd Annu. Freq. Control Symp.*, pp. 70–79.

Dias, J. F., Karrer, H. E., Kusters, J. A., Matsinger, J. H., and Schulz, M. B. (1975). The temperature coefficient of delay-time for x-propagating acoustic surface waves on rotated y-cuts of alpha quartz. *IEEE Trans. Sonics Ultrason.* **SU-22**, 46–50.

Dias, J. F., Karrer, H. E., Kusters, J. A., and Adams, C. A. (1976). Frequency/stress stability of S.A.W. resonators. *Electron. Lett.* **12**, 580–582.

Dunnrowicz, C., Sandy, F., and Parker, T. (1976). Reflection of surface waves from periodic discontinuities. *Ultrason. Symp. Proc.*, pp. 386–390.
Ebata, Y., Sato, K., Morishita, S. (1981). A $LiTaO_3$ SAW resonator and its application to video cassette recorder. *Ultrason. Symp. Proc.*, pp. 111–116.
Elliott, S., Mierzwinski, M., and Planting, P. (1981). The production of surface acoustic wave resonators. *Ultrason. Symp. Proc.*, pp. 89–93.
Engan, H., Hanebrekke, H., Ingebrigsten, K. A., and Jergan, E. (1969). Numerical calculations on surface waves in piezoelectrics. *Appl. Phys. Lett.* **15**, 239–241.
Feldmann, M., and Henaff, J. (1977). Acoustic surface wave propagation; atlas of calculated configurations for quartz, lithium tantalate, lithium niobate and thallium vanadosulphide in relation to temperature coefficients and piezoelectric coupling. *Rev. Phys. Appl.* **12**, 1775–1788. In French.
Field, M. E., and Chen, C. L. (1978). Bistable surface acoustic wave resonators. *Ultrason. Symp. Proc.*, pp. 469–473.
Fildes, R. D., and Hunsinger, B. J. (1976). Application of unidirectional transducers to resonator cavities. *Ultrason. Symp. Proc.*, pp. 303–305.
Gilden, M., and Grudkowski, T. W. (1981). GaAs SAW resonator oscillators with electronic tuning. *Proc. 35th Annu. Freq. Control Symp.*, pp. 395–400.
Gilden, M., Montress, G. K., and Wagner, R. A. (1980). Long-term aging and mechanical stability of 1.4 GHz SAW oscillators. *Ultrason. Symp. Proc.*, pp. 184–187.
Grudkowski, T. W., Montress, G. K., Gilden, M., and Black, J. F. (1980). GaAs monolithic SAW devices for signal processing and frequency control. *Ultrason. Symp. Proc.*, pp. 88–97.
Gurevich, G. L., Sandler, M. S., and Sveshnikov, B. V. (1979). Radiation losses in an acoustic surface wave resonator formed by interdigital type reflectors. *Radio Eng. Electron. Phys.* **24**, 23–36.
Harmon, D., Morency, D., Soluch, W., Vetelino, J. F., and Mittleman, S. D. (1978). Experimental determination of the SAW properties of X-axis boule cuts in berlinite. *Ultrason. Symp. Proc.*, pp. 594–597.
Hartemann, P. (1975). Acoustic surface wave resonator using ion-implanted gratings. *Ultrason. Symp. Proc.*, pp. 303–306.
Hartemann, P. (1976). Ion-implanted acoustic-surface-wave resonator. *Appl. Phys. Lett.* **28**, 73–75.
Hartemann, P. (1978). Influence of annealing on the surface-acoustic-wave velocity increase induced by ion implantation in quartz. *J. Appl. Phys.* **49**, 5334–5335.
Hauden, D., Michel, M., and Gagnepain, J.-J. (1978). Higher order temperature coefficients of quartz SAW oscillators. *Proc. 32nd Annu. Freq. Control Symp.*, pp. 77–86.
Hauden, D., Rousseau, S., and Gagnepain, J.-J. (1980). Sensitivities of SAW oscillators to temperature, forces and pressure: application to sensors. *Proc. 34th Annu. Freq. Control Symp.*, pp. 312–319.
Hauden, D., Planat, M., and Gagnepain, J.-J. (1981). Nonlinear properties of surface acoustic waves: applications to oscillators and sensors. *IEEE Trans. Sonics Ultrason.* **SU-28**, 342–348.
Haus, H. A. (1976). Scattering loss of S.A.W. resonators. *Electron. Lett.* **12**, 214–215.
Haus, H. A. (1977a). Modes in SAW grating resonators. *Electron. Lett.* **13**, 12–13.
Haus, H. A. (1977b). Modes in SAW grating resonators. *J. Appl. Phys.* **48**, 4955–4961.
Haydl, W. H. (1976). Surface acoustic wave resonators. *Microwave J.* **19**, 43–46.
Haydl, W. H., Dischler, B., and Hiesinger, P. (1976). Multimode SAW resonators—a method to study the optimum resonator design. *Ultrason. Symp. Proc.*, pp. 287–296.

Haydl, W. H., Hiesinger, P., Smith, R. S., Dischler, B., and Heber, K. (1976). Design of quartz and lithium niobate SAW resonators using aluminium metallization. *Proc. 30th Annu. Freq. Control Symp.*, pp. 346–357.

Haydl, W. H., Hiesinger, P., Kohlbacher, G., and Schmitt, P. (1977). Design and performance of SAW-resonators and resonator-filters. *Proc. AGARD Conf. No. 230 (Neuilly-sur-Seine, France: AGARD)*, 3.3/1–7.

Heighway, J. (1976). Surface acoustic wave devices and appliances. *Syst. Technol.* (23), 2–6.

Helmick, C. N., Jr., and White, D. J. (1980). Frequency and temperature effects induced by dielectric coatings on SSBW and SAW devices. *Proc. 34th Annu. Freq. Control Symp.*, pp. 307–311.

Henaff, J., and Feldmann, M. (1977). Computed coupling and temperature coefficients of ASW in Tl_3VS_4 and Tl_3TaSe_4. *Ultrason. Symp. Proc.*, pp. 696–700.

Hiesinger, P., and Schmidt, K. H. (1978). Scanning electron microscopy of resonating surface acoustic wave devices. *Ultrason. Symp. Proc.*, pp. 611–616.

Hirano, H. (1982). SAW devices utilizing $LiTaO_3$ single crystal. *Ferroelectrics* **42**, 203–214.

Inaba, R., and Wasa, K. (1981). Temperature-stable $SiO_2/LiTaO_3$ structure for SAW devices. *Jpn. J. Appl. Phys.* **20**, 153–155.

Ingebrigtsen, K. A. (1969). Surface waves in piezoelectrics. *J. Appl. Phys.* **40**, 2681–2686.

Ishihara, F., and Yoshikawa, S. (1974). Elastic surface wave devices, their application to communication and processing. *J. Acoust. Soc. Jpn.* **30**, 557–562. In Japanese.

Janus, A. R. (1976). Progress report on surface acoustic wave device MMT. *Proc. 30th Annu. Freq. Control Symp.*, pp. 157–166.

Jhunjhunwala, A., Vetelino, J. F., and Field, J. C. (1976). Temperature compensated cuts with zero power flow in Tl_3VS_4 and Tl_3TaSe_4. *Electron. Lett.* **12**, 683–684.

Joseph, T. R., and Lakin, K. M. (1975). Equivalent circuit and properties of surface wave planar resonators. *Proc. 29th Annu. Freq. Control Symp.*, pp. 158–166.

Joshi, S. G. (1980). Effect of a static electric field on the propagation of surface acoustic waves in piezoelectric media. *Ultrason. Symp. Proc.*, pp. 438–441.

Josse, F., and Vetelino, J. F. (1980). Acoustic wave properties and device characterization of 55°-rotated y-cut quartz. *Appl. Phys. Lett.* **37**, 1062–1064.

Kim, C. S. (1973). Temperature characteristics of elastic surface wave. *J. Korean Inst. Electron. Eng.* **10**, 141–148. In Korean.

Kim, C. S., Yamanouchi, K., Karasawa, S., and Shibayama, K. (1974). Temperature dependence of the elastic surface wave velocity on $LiNbO_3$ and $LiTaO_3$. *Jpn. J. Appl. Phys.* **13**, 24–27.

Kinoshita, Y. (1980). Review on the research of surface acoustic devices. *J. Inst. Electron. Commun. Eng. Jpn.* **63**, 639–641. In Japanese.

Kinoshita, Y., Kojima, H., and Tabuchi, T. (1978). Two-port SAW resonator utilizing piezoelectric surface shear wave mode. *IEEE MTT-S Int. Microwave Symp. Dig.*, pp. 472–474.

Kohlbacher, G., and Schmitt, P. (1978). Components with surface acoustic waves. *Nachr. Elektron.* **32**, 181–187. In German.

Kojima, T., and Tominaga, J. (1981). Split open metal strip arrays and their application to resonators. *Ultrasonics Symp. Proc.*, pp. 117–122.

Kotecki, C. (1982). Effects of RIE tuning on the electrical and temperature characteristics of quartz SAW resonators. *Proc. 36th Annu. Freq. Control Symp.*, pp. 459–469.

Koyamada, Y., and Yoshikawa, S. (1977). A two-port SAW resonator using long IDTs. *J. Acoust. Soc. Jpn.* **33**, 557–564. In Japanese.

Koyamada, Y., and Yoshikawa, S. (1978). Long IDT coupled mode analysis. *Rev. Electr. Comm. Lab.* **27**, 1557–1572. In Japanese.

Koyamada, Y., and Yoshikawa, S. (1979). Coupled mode analysis for a long IDT. *Rev. Electr. Commun. Lab.* **27**, 432–444.

Koyamada, Y., Yoshikawa, S., and Ishihara, F. (1977). Analysis of SAW resonators using long IDTs and their applications. *Trans. Inst. Electron. Commun. Eng. Jpn. Sect. E.* **E60**, 481–482.

Koyamada, Y., Yoshikawa, S., and Ishihara, F. (1978). One-port SAW resonators using long IDTs and their application to narrow band filters. *Electr. Commun. Lab. Tech. J.* **27**, 1663–1677. In Japanese.

Koyamada, Y., Yoshikawa, S., and Ishihara, F. (1979). One-port SAW resonators using long IDTs and their application to narrow band filters. *Rev. Electr. Commun. Lab.* **27**, 445–458.

Laker, K. R., Szabo, T. L., and Kearns, W. J. (1977). High-Q-factor SAW resonators at 780 MHz. *Electron. Lett.* **13**, 97–99.

Lakin, K. M. (1974). Electrode resistance effects in interdigital transducers. *IEEE Trans. Microwave Theory Tech.* **MTT-22**, 418–424.

Lakin, K. M., and Joseph, T. R. (1975). Equivalent circuit and properties of surface wave planar resonators. *Proc. 29th Annu. Freq. Control Symp.*, pp. 158–166.

Lakin, K. M., Joseph, T., and Penunuri, D. (1974). A surface acoustic wave planar resonator employing an interdigital electrode transducer. *Appl. Phys. Lett.* **25**, 363–365.

Lardat, C. (1976). Experimental performance of grooved reflective array compressors and resonators. *Ultrason. Symp., Proc.*, pp. 272–276.

Larosa, R., Vasile, C. F., and Zagardo, D. V. (1973). Comparison of surface-wave reflection coefficients for different metals on quartz. *Electron. Lett.* **9**, 495–496.

Lau, K. F., Stokes, R. B., Yen, K. H., and Chai, B. H.-T. (1981). Investigation of temperature stable SBAW in berlinite. *Ultrason. Symp., Proc.*, pp. 286–290.

Lau, K. F., Yen, K. H., Stokes, R. B., Kagiwada, R. S., and Chai, B. H.-T. (1981). Shallow bulk acoustic waves in berlinite. *Proc. 35th Annu. Freq. Control Symp.*, pp. 401–405.

Lean, E. G. H., Powell, C. G., and Kuhn, L. (1969). Acoustic surface wave mixing on α-quartz. *Appl. Phys. Lett.* **15**, 10–12.

Lec, R., Vetelino, J. F., Josse, F., Bailey, D. S., and Ehsasi, M. (1980). High temperature stable overlay configurations on X-rotated quartz. *Ultrason. Symp., Proc.*, pp. 424–428.

Lee, D. L. (1979). Design considerations for electronically compensated SAW delay line oscillators. *Ultrason. Symp., Proc.*, pp. 849–854.

Levesque, P., Valdois, M., Hauden, D., Gagnepain, J.-J., Hartemann, P., and Uebersfeld, J. (1979). Theoretical and experimental analysis of SAW quartz oscillator acceleration sensitivity. *Ultrason. Symp., Proc.*, pp. 896–899.

Lewis, M. F. (1973). Some aspects of SAW oscillators. *Ultrason. Symp., Proc.*, 344–347.

Lewis, M. F. (1979). Temperature compensation techniques for SAW devices. *Ultrason. Symp. Proc.*, pp. 612–622.

Lewis, M. F., and Patterson, E. (1971). Some properties of acoustic surface waves on X-cut quartz. *J. Acoust. Soc. Am.* **49**, 1667–1668.

Li, R. C. M. (1977). 310-MHz SAW resonator with Q at the material limit. *Appl. Phys. Lett.* **31**, 407–409.

Li, R. C. M., and Melngailis, J. (1975). The influence of stored energy at step discontinuities on the behavior of surface-wave gratings. *IEEE Trans. Sonics Ultrason.* **SU-22**, 189–198.

Li, R. C. M., Alusow, J. A., and Williamson, R. C. (1975a). Experimental exploration of the limits of achievable Q of grooved surface-wave resonators. *Ultrason. Symp. Proc.*, pp. 279–283.

Li, R. C. M., Alusow, J. A., and Williamson, R. C. (1975b). Surface-wave resonators using grooved reflectors. *Proc. 29th Annu. Freq. Control Symp.*, pp. 167–176.

SECTION 2.3

Lukaszek, T., and Ballato, A. (1980). What SAW can learn from BAW: implications for future frequency control, selection, and signal processing. *Ultrason. Symp. Proc.*, pp. 173–183.

Martin, S. J., Gunshor, R. L., and Pierret, R. F. (1980). High Q, temperature stable ZnO-on-silicon SAW resonators. *Ultrason. Symp. Proc.*, pp. 113–117.

Martin, S., Datta, S., Gunshor, R. L., Melloch, M. R., Pierret, R. F., and Staples, E. J. (1982). SAW resonators on silicon. *Ultrason. Symp. Proc.*, pp. 290–294.

Martin, S. J., Gunshor, R. L., Miller, T. J., Datta, S., Pierret, R. F., and Melloch, M. R. (1983). Surface wave resonators on silicon. *Proc. 37th Annu. Freq. Control Symp.*, pp. 423–427.

Mason, I. M., Chambers, J., and Lagasse, P. E. (1975). Laser-probe analysis of field distributions within acoustic-surface-wave planar resonators. *Electron. Lett.* **11**, 288–290.

Matthaei, G. L., and Barman, F. (1976). SAW resonators using low-loss 'waffle-iron' reflectors. *Ultrason. Symp. Proc.*, pp. 415–418.

Matthaei, G. L., and Barman, F. (1978). A study of the Q and modes of SAW resonators using metal 'waffle-iron' and strip arrays. *IEEE Trans. Sonics Ultrason.* **SU-25**, 138–146.

Matthaei, G. L., Barman, F., and Savage, E. B. (1976). Acoustic-surface-wave resonators for band-pass filter applications. *IEEE MTT-S Int. Microwave Symp.*, pp. 283–285.

Matthaei, G. L., O'Shaughnessy, B. P., and Barman, F. (1976). Relations for analysis and design of surface-wave resonators. *IEEE Trans. Sonics Ultrason.* **SU-23**, 99–107.

McAvoy, B. R., Murphy, J., and de Klerk, J. (1977). Temperature compensation in bulk mode microwave resonators. *Ultrason. Symp. Proc.*, pp. 403–407.

Mierzwinski, M. E., and Terrien, M. E. (1981). 280-MHz production SAWR *Hewlett-Packard J.* **32**, 15–16.

Miller, S. P., Stigall, R. E., and Shreve, W. R. (1975). Plasma etched quartz SAW resonators. *Ultrason. Symp. Proc.*, pp. 474–477.

Milstein, L. B., and Das, P. K. (1979). Surface acoustic wave devices. *IEEE Commun. Mag.* **17**, 25–33.

Minagawa, S., Okamoto, T., Tsubouchi, K., and Mikoshiba, N. (1978). SAW tunable resonator on monolithic MIS structure. *Ultrason. Symp. Proc.*, pp. 464–468.

Minowa, J. I. (1978). A method for accurately adjusting the center frequency of surface acoustic wave filters. *Rev. Electr. Commun. Lab.* **26**, 797–807.

Mitsuyu, T., Yamazaki, O., Ono, S., Matsuura, S., and Wasa, K. (1979). ZnO thin-film SAW devices for high frequency range. *Natl. Tech. Rep.* **25**, 1053–1064. In Japanese.

Moore, R. A., Newman, B. A., McAvoy, B. R., and Murphy, J. (1979). Temperature characteristics of microwave acoustic resonators. *IEEE MTT-S Int. Microwave Symp. Dig.*, pp. 171–173.

Morency, D., Soluch, W., Harmon, D., Vetelino, J. F., and Mittleman, S. D. (1978). Experimental determination of the SAW properties of X-axis boule cuts in berlinite. *Ultrason. Symp. Proc.*, pp. 594–597.

Morency, D. G., Soluch, W., Vetelino, J. F., Mittleman, S. D., Harmon, D., Surek, S., Field, J. C., and Lehmann, G. (1978). Experimental measurement of the SAW properties of berlinite. *Appl. Phys. Lett.* **33**, 117–119.

Morency, D. G., Soluch, W., Vetelino, J. F., Mittleman, S. D., Harmon, D., Surek, S., and Field, J. C. (1978). Experimental measurement of the SAW properties of berlinite. *Proc. 32nd Annu. Freq. Control Symp.*, pp. 196–201.

Nakamura, K., Kazumi, M., and Shimizu, H. (1977). SH-type and Rayleigh-type surface waves on rotated Y-cut LiTaO$_3$. *Ultrason. Symp. Proc.*, pp. 819–822.

Nevesely, M. (1979). One-port surface wave resonator. *Elektrotech. Cas.* **30**, 744–752.

Nevesely, M., and Szekely, J. (1976). The possibilities of applications of surface (Rayleigh) elastic waves in microwave technology. *Elektrotech. Cas.* **27**, 775–782. In Slovak.

Newton, C. O. (1979). A study of the propagation characteristics of the complete set of SAW paths on quartz with zero temperature coefficient of delay. *Ultrason. Symp. Proc.*, pp. 632–636.

O'Connell, R. M. (1978). Cuts of lead potassium niobate, $Pb_2KNb_5O_{15}$, for surface acoustic wave (SAW) applications. *J. Appl. Phys.* **49**, 3324–3327.

O'Connell, R. M. (1979). A new cut of quartz with orthogonal temperature-compensated propagation directions for surface acoustic wave applications *Proc. 33rd Annu. Freq. Control Symp.*, pp. 402–405.

O'Connell, R. M. (1979). A new surface-acoustic-wave cut of quartz with orthogonal temperature-compensated propagation directions. *Appl. Phys. Lett.* **35**, 217–219.

O'Connell, R. M., and Carr, P. H. (1977). High piezoelectric coupling temperature-compensated cuts of berlinite ($AlPO_4$) for SAW applications. *IEEE Trans. Sonics Ultrason.* **SU-24**, 376–384.

O'Connell, R. M., and Carr, P. H. (1978). Progress in closing the lithium niobate-ST cut quartz piezoelectric coupling gap. *Proc. 32nd Annu. Freq. Control Symp.*, pp. 189–195.

O'Connell, R. M., Slobodnik, A. J., Jr., and Carr, P. H. (1978). Material choice for optimum SAW device performance. *AGARD Conf. Proc. No. 230 (Neuilly-sur-Seine, France: AGARD)*, 2.1/1–19.

Oates, D. E. (1979). A new cut of quartz for temperature-stable SAW dispersive delay lines. *IEEE Trans. Sonics Ultrason.* **SU-26**, 428–430.

Oates, D. E., and Williamson, R. C. (1979). Effects of temperature-dependent anisotropy in RAC devices and a cut of quartz for a temperature-compensated RAC. *Ultrason. Symp. Proc.*, pp. 691–695.

Okano, S., and Mitsuoka, T. (1981). Quartz crystals finding new applications in digital equipment. *JEE J. Electron. Eng.* **18**, 64–67.

Oliner, A. A. (ed.) (1978). "Acoustic Surface Waves." Springer-Verlag, Berlin, Germany.

Ono, S., Yamazaki, O., Ohji, K., Wasa, K., and Hayakawa, S. (1978). SAW resonators using RF-sputtered ZnO films on glass substrates. *Appl. Phys. Lett.* **33**, 217–218.

Onoe, M., and Mochizuki, Y. (1970). Zero temperature coefficient ultrasonic delay lines utilizing synthetic quartz crystals as delay media. *Proc. 24th Annu. Freq. Control Symp.*, pp. 21–32.

Owens, J. M., Smith, C. V., Jr., Adam, J. D., and Patterson, R. (1978). Magnetostatic wave devices: a status report. *Ultrason. Symp. Proc.*, pp. 684–688.

Paige, E. G. S. (1980). Surface acoustic waves and their applications. *Fis. Tecnol.* **3**, 181–189. In Italian.

Pantani, L. (1975). Temperature stabilization and design criteria in some SAW devices. Ultrasonics International, (Guildford, Surrey, England: IPC Sci. & Technol. Press), pp. 138–141.

Parker, T. E. (1977). Aging characteristics of SAW controlled oscillators. *Ultrason. Symp. Proc.* pp. 862–866.

Parker, T. E. (1979). $1/f$ phase noise in quartz delay lines and resonators. *Ultrason. Symp. Proc.*, pp. 878–881.

Parker, T. E. (1980). Analysis of aging data on SAW oscillators. *Proc. 34th Annu. Freq. Control Symp.*, pp. 292–297.

Parker, T. E. (1982). Precision surface acoustic wave (SAW) oscillators. *Ultrason. Symp., Proc.*, pp. 268–274.

Parker, T. E., and Callerame, J. (1981). Sensitivity of SAW delay lines and resonators to vibration. *Ultrason. Symp., Proc.*, pp. 129–134.

Parker, T. E., and Lee, D. L. (1979). Stability of phase shift on quartz SAW devices. *Proc. 33rd Annu. Freq. Control Symp.*, pp. 379–387.

Parker, T. E., and Schulz, M. B. (1975). Stability of SAW controlled oscillators. *Ultrason. Symp. Proc.*, pp. 261–263.
Parker, T. E., and Schulz, M. B. (1975). Temperature stable materials for SAW devices. *Proc. 29th Annu. Freq. Control Symp.*, pp. 143–149.
Parker, T. E., and Wichansky, H. (1979). Temperature-compensated surface-acoustic-wave devices with SiO_2 film overlays. *J. Appl. Phys.* **50**, 1360–1369.
Paul, H. S. (1968). Vibrational waves in a thick infinite plate of piezoelectric crystal. *J. Acoust. Soc. Am.* **44**, 478–482.
Penavaire, L., Seguignes, D., Lardat, C., Bonnier, J. J., and Chevalier, J. Y., and Besson, Y. (1980). A 120 MHz SAW resonator stabilized oscillator with high spectral purity. *Ultrason. Symp. Proc.*, pp. 256–259.
Pirio, F., and Desrousseaux, P., (1980). Surface acoustic wave resonators with Hermito-Gaussian transverse modes. *Proc. 34th Annu. Freq. Control Symp.*, pp. 269–272.
Planat, M., Hauden, D., Groslambert, J., and Gagnepain, J.-J. (1980). Nonlinear propagation of surface acoustic waves on quartz. *Proc. 34th Annu. Freq. Control Symp.*, pp. 255–261.
Planat, M., Marianneau, G., and Gagnepain, J.-J. (1980). Measurement of the amplitude–frequency effect of surface acoustic wave resonators. *C.R. Hebd. Seances Acad. Sci. Ser. B* **290**, 305–307. In French.
Planat, M., and Gagnepain, J.-J., Lardat, C., and Penavaire. L. (1980). Nonlinear characteristics of SAW grooved resonators. *Ultrason. Symp. Proc.*, pp. 153–157.
Renard, A., Henaff, J., and Auld, B. A. (1981). SH surface wave propagation on corrugated surfaces of rotated Y-cut quartz and berlinite crystals. *Ultrason. Symp. Proc.*, pp. 123–128.
Richardson, B. A., and Kino, G. S. (1970). Probing of elastic surface waves in piezoelectric media. *Appl. Phys. Lett.* **16**, 82–85.
Rosenberg, R. L. (1979). Behavior of unidirectional transducers in grating resonators. *IEEE Trans. Sonics Ultrason.* **SU-26**, 377–379.
Rosenberg, R. L., and Coldren, L. A. (1976). Reflection-dependent coupling between grating resonators [passband filter]. *Ultrason. Symp. Proc.*, pp. 281–286.
Rosenberg, R. L., and Coldren, L. A. (1980). Broader-band transducer-coupled SAW resonator filters with a single critical masking step. *Ultrason. Symp. Proc.*, p. 164–168.
Rosenfeld, R. C., O'Shea, T. F., and Arneson, S. H. (1977). Tuning quartz SAW resonators by opening shorted reflectors. *Proc. 31st Annu. Freq. Control Symp.*, pp. 231–239.
Roy, M. K., Dube, N. M., Kurhatti, N. G., and Joshi, S. G. (1979). Rise time of a surface-acoustic-wave resonator. *Ultrason. Symp. Proc.*, pp. 845–848.
Sabine, P. V. H. (1970). Rayleigh wave propagation on a periodically roughened surface. *Electron. Lett.* **6**, 149–151.
Sabine, H., and Cole, P. H. (1971). Surface acoustic waves in communications engineering. *Ultrasonics* **9**, 103–113.
Schmidt, R. V. (1975). Acoustic surface wave velocity perturbations in $LiNbO_3$ by diffusion of metals. *Appl. Phys. Lett.* **27**, 8–10.
Schmitt, P., and Waller, W. (1977). Resonators with surface acoustic waves. *Wiss. Ber. AEG-Telefunken* **50**, 100–106. In German.
Schoenwald, J. S. (1976). Optical waveguide model for SAW resonators. *Proc. 30th Annu. Freq. Control Symp.*, pp. 340–345.
Schoenwald, J. S. (1976). Progress in SAW resonator development. *Proc. IEEE Int. Symp. Circuits System.*, pp. 698–701.
Schoenwald, J. S. (1977). Tunable variable bandwidth/frequency SAW resonators. *Proc. 31st Annu. Freq. Control Symp.*, pp. 240–245.
Schoenwald, J. S., Shreve, W. R., and Rosenfeld, R. C. (1975). Surface acoustic wave resonator development. *Proc. 29th Annu. Freq. Control Symp.*, pp. 150–157.

Schulz, M. B., Matsinger, B. J., and Holland, M. G. (1970). Temperature dependence of surface acoustic wave velocity on quartz. *J. Appl. Phys.* **41**, 2755–2765.
Schwelb, O., and Adler, E. L. (1978). Guided modes and proximity coupling in SAW resonators. *Ultrason. Symp. Proc.*, pp. 651–657.
Schwelb, O., Adler, E. L., and Farnell, G. W. (1977). Effect of anisotropy on waveguide modes in SAW resonators. *Ultrason. Symp. Proc.*, pp. 867–872.
Shibayama, K. (1977). Some aspects on $LiNbO_3$ crystals for surface acoustic wave devices. *Bull. Acad. Sci USSR Phys. Ser.* **41**, 19–23.
Shimizu, H., and Takeuchi, M. (1979). Theoretical studies of the energy storage effects and the second harmonic responses of SAW reflection gratings. *Ultrason. Symp. Proc.*, pp. 667–672.
Shimizu, Y., and Yamamoto, Y. (1980). SAW propagation characteristics on α-quartz with arbitrary cut. *Trans. Inst. Electron. Commun. Eng. Jpn. Sect. E.* **63**, 824–825.
Shimizu, Y., and Yamamoto, Y. (1980). SAW propagation characteristics on α-quartz with arbitrary cut. *Jpn. J. Appl. Phys.* **20**, 127–130.
Shimizu, Y., and Yamamoto, Y. (1980). SAW propagation characteristics of complete cut of quartz and new cuts with zero temperature coefficient of delay. *Ultrason. Symp. Proc.*, pp. 420–423.
Shimizu, Y., Sakuae, T., Terazaki, A., and Kaneaki, T. (1978). Temperature dependence of surface-acoustic-waves on α-quartz. *Trans. Inst. Electron. Commun. Eng. Jpn. Sect. E.* **61**, 94–95.
Shimizu, Y., Terazaki, A., and Sakaue, T. (1976). Temperature dependence of SAW velocity for metal film on α-quartz. *Ultrason. Symp. Proc.*, pp. 519–522.
Shorrocks, N. M., Whatmore, R. W., Ainger, F. W., and Young, I. M. (1981). Lithium tetraborate—a new temperature compensated piezoelectric substrate material for surface acoustic wave devices. *Ultrason. Symp. Proc.*, pp. 337–340.
Shreve, W. R. (1975). Surface-wave two-port resonator equivalent circuit. *Ultrason. Symp. Proc.*, pp. 295–298.
Shreve, W. R. (1976a). Surface wave resonators and their use in narrow-band filters. *Ultrason. Symp. Proc.*, pp. 706–713.
Shreve, W. R. (1976b). Two-port quartz SAW resonators. *Proc. 30th Annu. Freq. Control Symp.*, pp. 328–333.
Shreve, W. R. (1977). Aging in quartz SAW resonators. *Ultrason. Symp. Proc.*, pp. 857–861.
Shreve, W. R., and Stigall, R. E. (1978). Surface acoustic wave devices for use in a high performance television tuner. *IEEE Trans. Consum. Electron.* **CE-24**, 96–104.
Shreve, W. R., Kusters, J. A., and Adams, C. A. (1978). Fabrication of SAW resonators for improved long-term aging. *Ultrason. Symp. Proc.*, pp. 573–579.
Sinha, B. K., and Tiersten, H. F. (1978). On the temperature dependence of the velocity of surface waves in quartz. *Ultrason. Symp. Proc.*, pp. 662–666.
Sinha, B. K., and Tiersten, H. F. (1979). Zero temperature coefficient of delay for surface waves in quartz. *Appl. Phys. Lett.* **34**, 817–819.
Sinha, B. K., Tanski, W. J., Lukaszek, T., and Ballato, A. (1983). Stress induced effects on the propagation of surface waves. *Proc. 37th Annu. Freq. Control Symp.*, pp. 415–422.
Sittig, E. K., and Coquin, G. A. (1968). Filters and dispersive delay lines using repetitively mismatched ultrasonic transmission lines. *IEEE Trans. Sonics Ultrason.* **SU-15**, 111–119.
Sleeckx, F., Naten, T., Van de Capelle, A., and Vandewege, J. (1980). SAW technology for resonator structures. *Electroncompon. Sci. Technol.* **7**, 83–85.
Slobodnik, A. J., Jr. (1974). Surface-quality effects on S.A.W. attenuation on Y-Z $LiNbO_3$. *Electron. Lett.* **10**, 233–234.
Slobodnik, A. J., Jr. (1976). Surface acoustic waves and SAW materials. *Proc. IEEE* **64**, 581–595.
Slobodnik, A. J., Jr., Conway, E. D., and Delmonico, R. T. (1973). Microwave acoustic hand-

book, Volume IA. Surface wave velocities. Tech. Rept. 73-0597, Physical Sciences Research Papers, No. 565, AFCRL, Bedford, Massachusetts, 01731.
Slobodnik, A. A., Jr., Silva, J. H., Kearns, W. J., and Szabo, T. L. (1978). Lithium tantalate SAW substrate minimal diffraction cuts. *IEEE Trans. Sonics Ultrason.* **SU-25**, 92–97.
Smith, W. R., Gerard, H. M., Collins, J. H., Reeder, T. M., and Shaw, H. J. (1969). Analysis of interdigital surface wave transducers by use of an equivalent circuit model. *IEEE Trans. Microwave Theory Tech.* **MTT-17**, 856–864.
Soluch, W., Lec, R., and Latuszek, A. (1972). Properties of elastic surface waves in $LiIO_3$. *Bull. Acad. Pol. Sci. Ser. Sci. Tech.* **20**, 473–476.
Staples, E. J. (1974). UHF surface acoustic wave resonators. *Proc. 28th Annu. Freq. Control Symp.*, pp. 280–285.
Staples, E. J., and Smythe, R. C. (1976). SAW resonators and coupled resonator filters. *Proc. 30th Annu. Freq. Control Symp.*, pp. 322–327.
Staples, E. J., Shoenwald, J. S., Rosenfeld, R. C., and Hartmann, C. S. (1974). UHF surface acoustic wave resonators. *Ultrason. Symp. Proc.*, pp. 245–252.
Stevens, R., White, P. D., Mitchell, R. F., Moore, P., and Redwood, M. (1977). Stopband level of 2-port resonator filters. *Ultrason. Symp., Proc.*, pp. 905–908.
Stigall, R. E., and Shreve, W. R. (1978). SAW devices for use in a high performance television tuner. *Proc. 32nd Annu. Freq. Control Symp.*, pp. 50–57.
Szabo, T. L., and Slobodnik, A. J., Jr. (1973). The effect of diffraction on the design of acoustic surface wave devices. *IEEE Trans. Sonics Ultrason.* **SU-20**, 240–251.
Tancrell, R. H., and Holland, M. G. (1971). Acoustic surface wave filters. *Proc. IEEE* **59**, 393–409.
Tanski, W. J. (1977). Developments in resonators on quartz. *Ultrason. Symp. Proc.*, pp. 900–904.
Tanski, W. J. (1978). A configuration and circuit analysis for one-port SAW resonators. *J. Appl. Phys.* **49**, 2559–2560.
Tanski, W. J. (1978). High Q and GHz SAW resonators. *Ultrason. Symp. Proc.*, pp. 433–437.
Tanski, W. J. (1979a). Surface acoustic wave resonators on quartz. *IEEE Trans. Sonics Ultrason.* **SU-26**, 93–104.
Tanski, W. J. (1979b). SAW resonators utilizing withdrawal weighted reflectors. *IEEE Trans. Sonics Ultrason.* **SU-26**, 404–410.
Tanski, W. J. (1979c). GHz SAW resonators. *Ultrason. Symp. Proc.*, pp. 815–823.
Tanski, W. J. (1980a). SAW resonators approach 1.5 GHz. *Microwave Syst. News.* **10**, 99, 102–104, 106–107.
Tanski, W. L. (1980b). UHF SAW resonators and applications. *Proc. 34th Annu. Freq. Control Symp.*, pp. 278–285.
Tanski, W. J., and van de Vaart, H. (1976). The design of SAW resonators on quartz with emphasis on two ports. *Ultrason. Symp. Proc.*, pp. 260–265.
Tanski, W. J., and Wittels, N. D. (1979). SEM observations of SAW resonator transverse modes. *Appl. Phys. Lett.* **34**, 537–539.
Tanski, W. J., Bloch, M., Vulcan, A. (1980). High performance SAW resonator filters for satellite use. *Ultrason. Symp. Proc.*, pp. 148–152.
Togami, Y. (1979). Observation of surface waves and bulk waves in a surface acoustic wave device by frequency-shift holography. *NHK Tech. J.* **31**, 22–38. In Japanese.
Tominaga, T. K. J., and Suzuki, T. (1981). Reflection characteristics of open metal strip arrays for SAW resonators. *Jpn. J. Appl. Phys.* **20**, 103–106.
Uno, T., Abe, H., Miyamoto, N., and Jumonji, H. (1981). Realization of miniature SAW resonators having high quality factor. *Jpn. J. Appl. Phys.* **20**, 85–88.
Vandewege, J. (1976). Acoustic resonators for VHF and UHF. *Rev. HF* **10**, 111–120. In Dutch.

Vandewege, J. (1979). The surface acoustic wave distributed-feedback resonator. *Rev. E.* **9**, 55–68. In Flemish.

Vandewege, J., Lagasse, P. E. (1981). Analysis of SAW distributed feedback resonators. *IEEE Trans. Sonics Ultrason.* **SU-28**, 42–47.

Vandewege, J., Lagasse, P. E., Tromp, H., Hoffman, G., Naten, T., and Sleeckx, F. (1978). Acoustic surface wave resonators for broadband applications. *Proc. 8th European Microwave Conf.* (Sevenoaks, England: Microwave Exhibitions & Pubs. Ltd.). pp. 663–667.

Vandewege, J., Lagasse, P. E., and Naten, T. (1978). Distributed feedback acoustic surface wave resonators. *Ultrason. Symp. Proc.*, pp. 438–441.

Veilleux, O. (1981). Surface-acoustic-wave devices. *Elektronik* **30**, 35–41. In German.

Vella, P. J., Stegeman, G. I., and Ristic, V. M. (1977). Surface-wave harmonic generation on y-z, x-z, and $41\frac{1}{2} - x$ lithium niobate. *J. Appl. Phys.* **48**, 82–85.

Viktorov, I. A., and Zubova, O. M. (1975). Two types of surface waves in cubic crystals. *Sov. Phys.-Acoust. (Engl. Trans.)* **20**, 556–557.

Wagers, R. S. (1976). Plate mode coupling in acoustic surface wave devices. *IEEE Trans. Sonics Ultrason.* **SU-23**, 113–127.

Wagers, R. S. (1976). Spurious acoustic responses in SAW devices. *Proc. IEEE* **64**, 699–702.

Webb, D. C., Forester, D. W., Ganguly, A. K., and Vittoria, C. (1979). Application of amorphous magnetic-layers in surface-acoustic-wave devices. *IEEE Trans. Magn.* **MAG-15**, 1410–1415.

Weglein, R. D., and Otto, O. W. (1977a). Effect of vibration on S.A.W. oscillator noise spectra. *Electron. Lett.* **13**, 103–104.

Weglein, R. D., and Otto, O. W. (1977b). Microwave SAW oscillators. *Ultrason. Symp. Proc.*, pp. 913–921.

Welsh, F. S. (1971). Surface-wave temperature coefficients on lithium tantalate. *IEEE Trans. Sonics Ultrason.* **SU-18**, 108–109.

Wested, J. (1975). New high frequency component: SAW-crystal resonator. *EC-Nyt* (43), 12–14. In Danish.

White, P. D., Mitchell, R. F., Stevens, R., Moore, P., and Redwood, M. (1978). Synthesis and design of weighted reflector banks for SAW resonators. *Ultrason. Symp. Proc.*, pp. 634–638.

White, P. D., Stevens, R., and Mitchell, R. F. (1978). Surface acoustic wave resonators in communications. *Conf. Commun. Equip. Syst.*, pp. 270–273.

White, R. M. (1970). Surface elastic waves. *Proc. IEEE* **58**, 1238–1276.

Williams, D. F., and Cho, F. Y. (1979). Numerical analysis of doubly-rotated-cut SAW devices. *Ultrason. Symp. Proc.*, pp. 627–631.

Williams, D. F., and Cho, F. Y. (1980). Numerical analysis of doubly rotated cut SAW devices. *Proc. 34th Annu. Freq. Control Symp.*, pp. 302–306.

Williams, D. F., Cho, F. Y., and Sanchez, J. J. (1980). Temperature stable SAW devices using doubly rotated cuts of quartz. *Ultrason. Symp. Proc.*, pp. 429–433.

Williams, D. F., Cho, F. Y., Ballato, A., and Lukaszek, T. (1981). The propagation characteristics of surface acoustic waves on singly and doubly rotated cuts of quartz. *Proc. 35th Annu. Freq. Control Symp.*, pp. 376–382.

Wright, P. V., and Haus, H. A. (1980). Theoretical analysis of second-order effects in surface-wave gratings. *Proc. 34th Annu. Freq. Control Symp.*, pp. 262–268.

Yamanouchi, K. (1979). Acoustic surface wave devices. *J. Inst. Telev. Eng. Jpn.* **33**, 884–892. In Japanese.

Yashiro, K., and Goto, N. (1978). Analysis of bulk waves in surface-acoustic wave devices. *IEE J. Microwave Opt. Acoust.* **2**, 187–193.

Yasuda, T. (1974). Research works on elastic surface waves in United States and European countries. *J. Acoust. Soc. Jpn.* **30**, 578–583. In Japanese.

Yen, K. H., Lau, K. F., and Kagiwada, R. S. (1978). Second harmonic SAW resonators. *Ultrason. Symp. Proc.*, pp. 442–447.

Yen, K. H., Lau, K. F., and Kagiwada, R. S. (1978). Temperature stable shallow bulk acoustic wave devices. *Proc. 32nd Annu. Freq. Control Symp.*, pp. 95–101.

Chapter 3

Akishin, A. I., Vintovkin, S. I., Titov, V. I., and Tokarev, G. A. (1970). Effect of ionizing radiation on the piezoelectric properties of quartz plates. *Radiats. Fiz. Nemet. Krist.*, pp. 220–229. In Russian.

Anderson, T. C., and Merrill, F. G. (1960). Crystal controlled primary frequency standards: latest advances for long-term stability. *IRE Trans. Instrum* **I-9**, 136–140.

Aoki, T., and Wada, K. (1978). Optical and anelastic absorptions and resonator frequency of electron-irradiated quartz. *Jpn. J. Appl. Phys.* **17**, 1015–1022.

Aoki, T., Norisawa, K., and Sakisaka, M. (1975). Frequency change of quartz resonators irradiated by 1 MeV electrons. *Mem. Chubu. Inst. Technol.* **11A**, 113–119.

Aoki, T., Norisawa, K., and Sakisaka, M. (1976a). Frequency change and elastic constants of quartz irradiated by 1 MeV electrons. *Jpn. J. Appl. Phys.* **15**, 749–754.

Aoki, T., Norisawa, K., and Sakisaka, M. (1976b). Optical absorption study for quartz resonators irradiated by electrons. *Jpn. J. Appl. Phys.* **15**, 2131–2135.

Aoki, T., Norisawa, K., and Sakisaka, M. (1976c). Frequency change of quartz resonators irradiated by alpha particles. *Jpn. J. Appl. Phys.* **15**, 2307–2310.

Bahadur, H., and Parshad, R. (1979). Some new findings on the effect of nuclear and n-irradiation on the oscillating characteristics of quartz crystals. *Indian J. Phys. A* **53A**, 239–256.

Bahadur, H., and Parshad, R. (1980). Simple experimental method for demonstrating transient frequency shifts in X and gamma irradiated quartz crystals. *Rev. Sci. Instrum.* **51**, 1420–1421.

Benedikter, H. J., Sherman, J. H., and Gillespie, R. D., III (1974). The effect of gamma irradiation on the temperature-frequency characteristics of AT-cut quartz. *Proc. 28th Annu. Freq. Control Symp.*, pp. 143–149.

Berg, C. A., and Erickson, J. R. (1969). Effects of gamma irradiation on frequency stability of 5th overtone crystal oscillators. *Proc. 23rd Annu. Freq. Control Symp.*, pp. 178–186.

Capone, B. R., Kahan, A., Brown, R. N., and Buckmelter, J. R. (1970). Quartz crystal radiation effects. *IEEE Trans. Nucl. Sci.* **NS-17**, 217–221.

Doherty, S. P., Morris, S. E., Andrews, D. C., and Croxall, D. F. (1982). Radiation effects in synthetic and high purity synthetic quartz: Some recent infrared, electron spin resonance and acoustic loss results. *Proc. 36th Annu. Freq. Control Symp.*, pp. 66–76.

Dowell, M., Lefkowitz, I., and Taylor, G. W. (1969). Radiation damage in $LiNbO_3$. *2nd International Meeting on Ferroelectricity (Japan)* (Phys. Soc. Jpn., Tokyo), pp. 158–159.

Draggoo, V. G., She, C. Y., and Edwards, D. F. (1972). Effects of laser mode structure on damage in quartz. *IEEE J. Quantum Electron.* **QE-8**, 54–57.

EerNisse, E. P. (1971). Permanent radiation effects in swept and unswept optical grade synthetic quartz AT-cut resonators. *IEEE Trans. Nucl. Sci.* **NS-18**, 86–90.

EerNisse, E. P. (1975). Quartz resonator frequency shifts arising from electrode stress. *Proc. 29th Annu. Freq. Control Symp.*, pp. 1–4.

Esquivel, A. L., and Sagara, H. I. (1974). Effects of ionizing radiation on swept and unswept synthetic quartz crystals. *Extended Abstracts, Battery Division, Fall Meeting* (Electrochem. Soc., Princeton, New Jersey, U.S.A.), pp. 395–396.

Euler, F., Ligor, P., Pellegrini, P., Kahan, A., Flanagan, T. M., and Wrobel, T. (1978). Steady state radiation effects in precision quartz resonators. *Proc. 32nd Annu. Freq. Control Symp.*, pp. 24–33.

Euler, F., Lipson, H. G., Kahan, A., and Ligor, P. A. (1980). Radiation effects in quartz oscillators, resonators and materials. *Proc. 34th Annu. Freq. Control Symp.*, pp. 72–80.

Felden, M., Comets, J. C., and Haug, R. (1967). Modification of the resonance frequency of various quartz crystal cuts subjected to various types of particle irradiation. *Ann. Telecommun.* **22**, 284–292. In French.

Flanagan, T. M., and Wrobel, T. F. (1969). Radiation effects in swept-synthetic quartz. *IEEE Trans. Nucl. Sci.* **NS-16**, 130–137.

Flanagan, T. M., and Wrobel, T. F. (1972). Radiation-stable precision quartz oscillators. Available from NTIS as Report No. AD 75284, 59 pp.

Fraser, D. B. (1968). Impurities and anelasticity in crystalline quartz. *In* "Physical Acoustics" (W. P. Mason, ed.), Academic Press, New York, pp. 59–110.

Freymuth, P., and Sauerbrey, G. (1963). Annealing of radiation damage in quartz oscillator crystals. *Phys. Chem. Solids* **24**, 151–155.

Gardner, J. W., and Anderson, A. C. (1981). Low-temperature specific heat and thermal conductivity of neutron-irradiated crystalline quartz. *Phys. Rev. B* **23**, 474–482.

Griscom, D. L. (1979). Point defects and radiation damage processes in α-quartz. *Proc. 33rd Annu. Freq. Control Symp.*, pp. 98–109.

Halliburton, L. E., Kappers, L. A., Armington, A. F., and Larkin, J. (1979). Radiation effects in berlinite. *Proc. 33rd Annu. Freq. Control Symp.*, pp. 62–69.

Halperin, A., Schieber, M., Braner, A. A., Levinson, J., and Zloto, D. (1972). Radiation effects on quartz crystals and oscillators, and growth of doped quartz crystals. Available from NTIS as Report No. AD 750467, 62 pp.

Hartman, E. F., and King, J. C. (1973). Calculation of transient thermal imbalance within crystal units following exposure of pulse irradiation. *Proc. 27th Annu. Freq. Control Symp.*, pp. 124–127.

Hartman, E. F., and King, J. C. (1975). Calculation of transient thermal imbalance within crystal units following exposure to pulse irradiation. *Radiat. Eff.* **26**, 219–223.

Holland, R. (1974). Nonuniformly heated anisotropic plates: 1. Mechanical distortion and relaxation. *IEEE Trans. Sonics Ultrason.* **SU-21**, 171–178.

Hughes, R. C. (1975). Electronic and ionic charge carriers in irradiated single crystal and fused quartz. *Radiat. Eff.* **26**, 225–235.

Jain, H., and Nowick, A. S. (1982a). Electrical conductivity of synthetic and natural quartz crystals. *J. Appl. Phys.* **53**, 477–484.

Jain, H., and Nowick, A. S. (1982b). Radiation-induced conductivity in quartz crystals. *J. Appl. Phys.* **53**, 485–489.

King, J. C. (1958). Anelasticity of synthetic crystalline quartz at low temperatures. *Phys. Rev.* **109**, 1552–1553.

King, J. C. (1959). The anelasticity of natural and synthetic quartz at low temperatures. *Bell Syst. Tech. J.* **38**, 573–602.

King, J. C., and Fraser, D. B. (1962). Effects of reactor irradiation on thickness shear crystal resonators. *Proc. 16th Annu. Freq. Control Symp.*, pp. 7–32.

King, J. C., and Sander, H. H. (1972). Rapid annealing of frequency change in crystal resonators following pulsed X-irradiation. *IEEE Trans. Nucl. Sci.* **NS-19**, 23–32.

King, J. C., and Sander, H. H. (1973a). Rapid annealing of frequency change in high frequency crystal resonators following pulsed X-irradiation at room temperature. *Proc. 27th Annu. Freq. Control Symp.*, pp. 113–119.

King, J. C., and Sander, H. H. (1973b). Transient change in Q and frequency of AT-cut quartz resonators following exposure to pulse X-rays. *IEEE Trans. Nucl. Sci.* **NS-20**, 117–125.
King, J. C., and Sander, H. H. (1975). Transient changes in quartz resonators following exposure to pulse ionization. *Radiat. Eff.* **26**, 203–212.
Koehler, D. R. (1979). Radiation induced frequency transients in AT, BT, and SC cut quartz resonators. *Proc. 33rd Annu. Freq. Control Symp.*, pp. 118–121.
Koehler, D. R. (1981). Radiation induced conductivity and high temperature Q changes in quartz resonators. *Proc. 35th Annu. Freq. Control Symp.*, pp. 322–328.
Koehler, D. R., and Martin, J. J. (1983). Radiation induced transient acoustic loss in quartz crystals. *Proc. 37th Annu. Freq. Control Symp.*, pp. 130–135.
Koehler, D. R., Young, T. J., and Adams, R. A. (1977). Radiation induced transient thermal effects in 5 MHz AT-cut quartz resonators. *Ultrason. Symp. Proc.*, pp. 877–881.
Krebs, M. (1969). The influence of X-rays on the resonance frequency of DKT piezoelectric resonators. *Cesk. Cas. Fis. A* **19**, 415–417. In Czech.
Krebs, M. (1971). Effect of different kinds of radiation on velocity and attenuation of elastic waves in quartz and DKT. *Fiz. Cas. (Czechoslovakia)* **21**, 132–135. In English.
Krebs, M. (1971). The effects of neutrons and X-radiation on the resonance frequency and elastic properties of piezoelectric quartz vibrators. *Cesk. Cas. Fys. A* **21**, 595–602. In Czech.
Krefft, G. B. (1975). Effects of high temperature electrolysis on the coloration characteristics and OH-absorption bands in alpha quartz. *Radiat. Eff.* **26**, 249–259.
Kurin, M. N., Krivobokov, V. P., Koshelev, F. P., Mal'tseva, V. V., and Darymov, V. I. (1978). Radiation corrosion of the silver electrodes of quartz piezoelements. *Izv. Vyssh. Uchebn. Zaved. Radioelektron.* **21**, 116–119. In Russian.
Laermans, C. (1979). Saturation of the 9.4-GHz hypersonic attenuation in fast-neutron-irradiated crystalline quartz. *Phys. Rev. Lett.* **42**, 250–254.
Lipson, H. G., Euler, F., and Ligor, P. A. (1979). Radiation effects in swept premium-Q quartz material. *Proc. 33rd Annu. Freq. Control Symp.*, pp. 122–133.
Lobanov, Ye. M., Chubarov, L. B., and Zverev, B. P. (1968). Effect of radiation in the active region of a nuclear reactor on signal frequency of quartz slabs. *Radiotekh. Elektron.* **13**, 1864–1866.
Ludanov, A. G., Fotchenkov, A. A., and Yakolev, L. A. (1976). Variation of the elastic constants of piezoelectric quartz. *Sov. Phys.-Acoust.* **22**, 343–344.
Markes, M. E., and Halliburton, L. E. (1979). Defects in synthetic quartz: radiation induced mobility of interstitial ions. *J. Appl. Phys.* **50**, 8172–8180.
Martin, J. J., Halliburton, L. E., Markes, M., Koumvakalis, N., Sibley, W. A., Brown, R. N., and Armington, A. (1979). Radiation induced mobility of interstitial ions in synthetic quartz. *Proc. 33rd Annu. Freq. Control Symp.*, pp. 134–147.
Mattern, P. L. (1973). Effects of ^{60}Co gamma ray irradiation on the optical properties of natural and synthetic quartz from 85 to 300 K. *Proc. 27th Annu. Freq. Control Symp.*, pp. 139–152.
Mattern, P. L., Lengweiler, K., and Levy, P. W. (1975). Effects of ^{60}Co gamma-ray irradiation on the optical properties of natural and synthetic quartz from 85 to 300 K. *Radiat. Eff.* **26**, 237–248.
Mitchell, E. W. J., and Paige, E. G. S. (1954). On the formation of colour centers in quartz. *Proc. Phys. Soc. London* **B67**, 262–264.
Nelson, C. M., and Crawford, J. H., Jr. (1958). Optical and spin resonance absorption in irradiated quartz and fused silica. I. Optical absorption. *Bull. Am. Phys. Soc.* **3**, 136.
O'Brien, M. C. M., and Pryce, M. H. L. (1954). Paramagnetic resonance in irradiated diamond and quartz: Interpretation. *Report of Bristol Conference on defects in crystalline solids* (*Phys. Soc., London*), pp. 88–91.

Oura, N., Kuramochi, N., Nakamura, J., and Ogawa, T. (1982). Thermal frequency behavior in contoured quartz crystal plates induced by direct irradiation of laser beam. *Proc. 36th Annu. Freq. Control Symp.*, pp. 133–139.

Paradysz, R. E., and Smith., W. L. (1973). Crystal controlled oscillators for radiation environments. *Proc. 27th Annu. Freq. Control Symp.*, pp. 120–123.

Paradysz, R. E., and Smith, W. L. (1975). Crystal controlled oscillators for radiation environments. *Radiat. Eff.* **26**, 213–218.

Pellegrini, P., Euler, F., Kahan, A., Flanagan, T. M., and Wrobel, T. (1978). Steady state and transient radiation effects in precision quartz oscillators. *IEEE Trans. Nucl. Sci.* **NS-25**, 1267–1273.

Saint-Paul, M., and Lasjaunias, J. C. (1981). Low-temperature specific heat of neutron-irradiated crystalline quartz. *J. Phys. C* **14**, L365–L370.

Smagin, A. G. (1971). Nonstationary acoustic absorption in alpha-quartz crystals, caused by pulsing and continuous radioactive irradiation. *Radiats. Fiz Nemet. Krist.* **3**, 116–120. In Russian.

Smith, R. W. (1971). Gamma radiation effects in lithium niobate. *Proc. IEEE* **59**, 712–713.

Soroka, V. V., and Khromova, N. N. (1977). Elastic and piezoelectric properties of γ-irradiated and annealed lithium niobate crystals. *Sov. Phys.-Solid State* **19**, 371–372.

Soroka, V. V., Suvorova, L. M., and Tazenkov, B. A. (1972). Theory of relaxation of permittivity and dielectric loss tangent in irradiated quartz. *Fiz. Poluprovodn. Elektron., Kratk. Soderzh. Dokl., Gertsenovski Chteniya, 25th, 1972*, pp. 71–75. In Russian.

Sosin, A. (1975). The kinetics of an atom diffusing in one dimension: hydrogen in quartz. *Radiat. Eff.* **26**, 267–271.

Spitsyn, V. I., Pirogova, G. N., Ryabov, A. I., and Kritskaya, V. E. (1978). Pulse radiolysis of fused and crystalline quartz. *Radiat. Eff.* **38**, 29–32.

Vecchi, M. P., and Nava, R. (1977). Propagation of ultrasonic shear waves in γ-ray irradiated quartz crystals. *Appl. Phys.* **13**, 171–173.

Weeks, R. A. (1956). Paramagnetic resonance of lattice defects in irradiated quartz. *J. Appl. Phys.* **27**, 1376–1381.

Weil, J. A. (1973). The aluminum centers in α-quartz. *Proc. 27th Annu. Freq. Control Symp.*, pp. 153–156.

Weil, J. A. (1975). The aluminum centers in α-quartz. *Radiat. Eff.* **26**, 261–265.

Young, T. J., Koehler, D. R., and Adams, R. A. (1978). Radiation induced frequency and resistance changes in electrolyzed high purity quartz. *Proc. 32nd Annu. Freq. Control Symp.*, pp. 34–42.

Chapter 4

Adams, C. A., and Kusters, J. A. (1977). Deeply etched SAW resonators. *Proc. 31st Annu. Freq. Control Symp.*, pp. 246–250.

Adams, C. A., and Kusters, J. A. (1978). Improved long-term aging in deeply etched SAW resonators. *Proc. 32nd Annu. Freq. Control Symp.*, pp. 74–76.

Akhtar, M. S. (1971). Piezoelectricity with special reference to piezoelectric analogous relationships, design and development of testing and measuring methods. *Nucleus (Pakistan)* **8**, 53–58.

Andres, R. P. (1976). Design of a nozzle beam type metal vapor source. *Proc. 30th Annu. Freq. Control Symp.*, pp. 232–236.

Ang, D. (1978). Design and implementation of an etch system for production use. *Proc. 32nd Annu. Freq. Control Symp.*, pp. 282–285.

CHAPTER 4 361

Ang, D. (1979). A microprocessor assisted anodizing apparatus for frequency adjustment. *Proc. 33rd Annu. Freq. Control Symp.*, pp. 364–367.

Ang, D. (1980). A microprocessor assisted baseplating apparatus with improved plateback distribution. *Proc. 34th Annu. Freq. Control Symp.*, pp. 41–45.

Anonymous (1979). The SAW (surface acoustic wave) technology. *Mikrowellen Mag.* (4), 235–238. In German.

Asanuma, N., and Asahara, J. (1980). Highly precise measurement of orientation angle for crystal blanks. *Proc. 34th Annu. Freq. Control Symp.*, pp. 120–130.

Bacigalupi, J. L. (1979). Crystal within everybody's reach [quartz crystals]. *Rev. Telegr. Electron.* (*Argentina*) **67**, 873–874. In Spanish.

Bahadur, H., and Parshad, R. (1975). Operation of quartz crystals in their overtones: new methods. *Indian J. Pure Appl. Phys.* **13**, 862–865.

Ballato, A., EerNisse, E. P., and Lukaszek, T. (1977). The force–frequency effect in doubly rotated quartz resonators. *Proc. 31st Annu. Freq. Control Symp.*, pp. 8–16.

Berenshtein, B. Sh., and Bobrikov, V. S. (1973). The restoration of quartz resonators that are used to measure the thickness and rate of vacuum deposition of thin films. *Instrum. Exp. Tech.* (*Engl. Transl.*) **16**, 1518.

Berenshtein, B. Sh., and Karpova, L. N. (1975). Restoration of quartz resonators. *Instrum. Exp. Tech.* (*Engl. Transl.*) **18**, 626.

Bernstein, M. (1970). Quartz crystal units for high g environments. *Proc. Int. Telemetering Conf.* (Woodland Hills, California, U.S.A.: Int. Found. Telemetering), pp. 452–460.

Bernstein, M. (1971). Quartz crystal units for high g environments. *Proc. 25th Annu. Freq. Control Symp.*, pp. 125–133.

Berté, M. (1977). Acoustic bulk wave resonators and filters operating in the fundamental mode at frequencies greater than 100 MHz. *Proc. 31st Annu. Freq. Control Symp.*, pp. 122–125.

Besson, R. (1976). A new piezoelectric resonator design. *Proc. 30th Annu. Freq. Control Symp.*, pp. 78–83.

Besson, R. J. (1977). A new 'electrodeless' resonator design. *Proc. 31st Annu. Freq. Control Symp.*, pp. 147–152.

Bidart, L. (1982). New design of very high frequency sources. *Ferroelectrics* **40**, 231–236.

Birrell, R. W., Valihura, R. J., Chambers, J. L., Pugh, M. A., and Workman, S. T. (1980). Automated X-ray orientation for quartz crystal resonators. Report DELET-TR-79-0254-1, U. S. Army Electronics Research & Development Command, Fort Monmouth, New Jersey 07703.

Bloch, M. B., Meirs, M. P., and Strauss, A. (1978). Results of temperature slewing quartz crystals for anomalous responses. *Proc. 32nd Annu. Freq. Control Symp.*, pp. 344–353.

Bond, W. L. (1976). "Crystal Technology." Wiley, New York, pp. 42 ff.

Bond, W. L., and Kusters, J. A. (1977). Making doubly rotated quartz plates. *Proc. 31st Annu. Freq. Control Symp.*, pp. 153–158.

Bottom, V. E. (1976). A novel method of adjusting the frequency of aluminium plated quartz crystal resonators. *Proc. 30th Annu. Freq. Control Symp.*, pp. 249–253.

Brandmayr, R., Filler, R., and Vig, J. (1979). Etching studies on quartz. *Proc. 33rd Annu. Freq. Control Symp.*, pp. 351–358.

Bray, R. C., and Chu, Y. C. (1981). SAWR fabrication. *Hewlett-Packard J.* **32**, 11–13.

Bryson, C. E., Sharpen, L. H., and Zajicek, P. L. (1979). An ESCA analysis of several surface cleaning techniques. Unpublished report, Hewlett-Packard Company.

Bulst, W. E., and Willibald, E. (1982). Ultrareproducible SAW resonator production. *Proc. 36th Annu. Freq. Control Symp.*, pp. 442–452.

Byrne, R. J. (1972). Thermocompression bonding to quartz crystals. *Proc. 26th Annu. Freq. Control Symp.*, pp. 71–77.

Caruso, R. D. (1977). Hermetically sealed crystal device having glass lasing window. *Tech. Dig.* (48), 9–10.

Caruso, R. D., and Setter, G. A. (1977). Frequency adjusting a crystal device utilizing the device header as a seal. *Tech. Dig.* (45), 9–10.

Castellano, R. N., and Hokanson, J. L. (1975). A survey of ion beam milling techniques for piezoelectric device fabrication. *Proc. 29th Annu. Freq. Control Symp.*, pp. 128–134.

Castellano, R. N., Meeker, T. R., Sundahl, R. S., and Jacobs, J. C. (1977). The relationship between quartz surface morphology and the Q of high frequency resonators. *Proc. 31st Annu. Freq. Control Symp.*, pp. 126–130.

Chambers, J. L. (1982). An instrument for automated measurement of the angles of cut of doubly rotated quartz crystals. *Proc. 36th Annu. Freq. Control Symp.*, pp. 302–313.

Chambers, J. L. (1983). An instrument for automated measurement of the angles of cut of doubly rotated quartz crystals. *Proc. 37th Annu. Freq. Control Symp.*, pp. 275–283.

Chambers, J. L., Pugh, M. A., Workman, S. T., Birrell, R. W., and Valihura, R. J. (1981). An instrument for automated measurement of the angles of cut of doubly rotated quartz crystals. *Proc. 35th Annu. Freq. Control Symp.*, pp. 60–70.

Cho, F. Y., Chatham, T. B., and Ponce de Leon, R. (1982). Frequency fine tuning of reliable SAW transducers using anodization technique. *Proc. 36th Annu. Freq. Control Symp.*, pp. 470–473.

Clastre, J., Pegeot, C., and Leroy, P. Y. (1978). Goniometric measurements of the angles of cut of doubly rotated quartz plates. *Proc. 32nd Annu. Freq. Control. Symp.*, pp. 310–316.

Costa, P., Piacentini, G. F., and Stacchiotti, G. (1975). LC conventional, electro-mechanical quartz and polylithic filters. *Telettra (Italy)* (27), 3–22.

Coussot, G., and Menager, G. (1975). Experimental investigation of mass-producible acoustic surface wave filter. *Proc. 29th Annu. Freq. Control Symp.*, pp. 181–186.

Cross, P. S., and Shreve, W. R. (1982). Frequency trimming of SAW resonators. *IEEE Trans. Sonics Ultrason.* **SU-29**, 231–234.

Cross, P., Rissman, P., and Shreve, W. (1980). Microwave SAW resonators fabricated with direct-writing electron beam lithography. *Ultrason. Symp. Proc.*, pp. 158–163.

Cutler, L. S., and Hammond, D. L. (1969). Crystal resonators. U. S. Patent Reissue 26,707. Issued November 4, 1969.

Darces, J. F., and Merigoux, H. (1978). Final X-ray control of the orientation of round or rectangular quartz slides for industrial purposes. *Proc. 32nd Annu. Freq. Control Symp.*, pp. 304–309.

Darces, J. F., Lamboley, J., and Merigoux, H. (1982). Dissymmetry used for ϕ, θ angle sign determination of a piezoelectric crystal blank by a non destructive method. *Ferroelectrics* **40**, 245–248.

Deacon, J., and Heighway, J. (1976). SAW technology. *Commun. Int.* **3**, 14, 16–17.

Debély, P. E., and Dinger, R. J. (1980). An on-wafer detection method of the imbalance of quartz tuning fork resonators. *Proc. 34th Annu. Freq. Control Symp.*, pp. 34–40.

Denman, J., Lagasse, G., Bloch, M., and Ho, J. (1968). Kold-seal thermal compression bonded crystals. *Proc. 22nd Annu. Freq. Control Symp.*, pp. 118–135.

Dias, J. F., Karrer, H. E., Kusters, J. A., and Adams, C. A. (1976). Frequency/stress sensitivity of SAW resonators. *Electron. Lett.* **12**, 580–582.

Dolochycki, S. J., Staples, E. J., Wise, J., Schoenwald, J. S., and Lim, T. C. (1979). Hybrid SAW oscillator fabrication and packaging. *Proc. 33rd Annu. Freq. Control Symp.*, pp. 374–378.

Dworsky, L. N. (1978). Discrete element modelling of AT-quartz devices. *Proc. 32nd Annu. Freq. Control Symp.*, pp. 142–149.

Dybwad, G. L. (1978). Simplified fixtures with improved thin film deposition uniformity on quartz crystals. *Proc. 32nd Annu. Freq. Control Symp.*, pp. 286–289.
EerNisse, E. (1975). Quartz resonator frequency shifts arising from electrode stress. *Proc. 29th Annu. Freq. Control Symp.*, pp. 1–4.
Elliott, S., Mierzwinski, M., and Planting, P. (1981). The production of surface acoustic wave resonators. *Ultrason. Symp. Proc.*, pp. 89–93.
Engdahl, J., and Matthey, H. (1975). 32 kHz quartz crystal unit for high precision wrist watch. *Proc. 29th Annu. Freq. Control. Symp.*, pp. 187–194.
Engdahl, J., Zumsteg, A., and Weber, C. (1977). New SSIH miniature 32 kHz quartz resonator, *Bull. Annu. Soc. Suisse Chronom. Lab. Suisse Rech. Horlog.* **7**, 347–349. In French.
Field, M. E., and Chen, C. L. (1976). On the fabrication tolerances of surface acoustic wave resonators, reflectors, and interdigital transducers. *Ultrason. Symp. Proc.*, pp. 510–513.
Filler, R. L., Keres, L. J., Snowden, T. M., and Vig, J. R. (1980). Ceramic flatpack enclosed AT and SC-cut resonators. *Ultrason. Symp. Proc.*, pp. 819–824.
Filler, R. L., and Vig, J. R. (1976a). The effect of bonding on the frequency vs. temperature characteristics of AT-cut resonators. Report ECOM-4433, U. S. Army Electronics Command, Ft. Monmouth, New Jersey, 15pp.
Filler, R. L., and Vig, J. R. (1976b). The effect of bonding on the frequency vs. temperature characteristics of AT-cut resonators. *Proc. 30th Annu. Freq. Control. Symp.*, pp. 264–268.
Filler, R. L., Frank, J. M., Peters, R. D., and Vig, J. R. (1978). Polyimide bonded resonators. *Proc. 32nd Annu. Freq. Control Symp.*, pp. 290–298.
Fischer, H. (1977). Some results of frequency response measurements of 1 MHz precision quartzes. *Nachrichtentech. Elekron.* **27**, 154–155. In German.
Fischer, R., and Schulzke, L. (1976). Direct plating to frequency—a powerful fabrication method for crystals with closely controlled parameters. *Proc. 30th Annu. Freq. Control Symp.*, pp. 209–223.
Frank, J. M. (1981). Vacuum processing for quartz crystal resonators. *Proc. 35th Annu. Freq. Control Symp.*, pp. 40–47.
Frankel, J., Korman, W., and Capsimalis, G. (1980). An in-process thickness determination during electroplating or electropolishing by the ultrasonic pulse-echo technique. *Ultrason. Symp. Proc.*, pp. 887–889.
Fuchs, D. (1978). Basic considerations on metal canned enclosures for the encapsulation of quartz crystal units. *Proc. 32nd Annu. Freq. Control Symp.*, pp. 321–325.
Fuchs, D. (1979). New metal enclosures for resistance welding developed to meet MIL-specifications. *Proc. 33rd Annu. Freq. Control Symp.*, pp. 186–188.
Fukuyo, H., and Oura, N. (1976). Surface layer of a polished crystal plate. *Proc. 30th Annu. Freq. Control Symp.*, pp. 254–258.
Fukuyo, H., Oura, N., and Kuramochi, N. (1979). Some properties of the silver thin film obtained by ion plating on an AT-cut quartz crystal. *Bull. Res. Lab. Precis. Mach. Electron. (Tokyo Inst. Technol.)* (43), 1–7.
Fyfe, W. A. (1972a). Lead attaching machine for quartz crystals. *Tech. Dig.* (25), 25–26.
Fyfe, W. A. (1972b). Quartz crystal turnover mechanism for lead soldering machine. *Tech. Dig.* (25), 27–28.
Giannotto, J. M. (1968). Multiple crystal holder. Report ECOM-2972, U. S. Army Electronics Lab., Ft. Monmouth, New Jersey, 19 pp. Available as No. AD-672064 from NTIS.
Gibert, G. (1968). Improvements in sealing HC-26/U and HC-27/U glass holders. *Proc. 22nd Annu. Freq. Control Symp.*, pp. 155–162.
Gilbert, G., Broussou, S., and Morel, J. (1973). Quartz crystal units for space applications. *Proc. 27th Annu. Freq. Control Symp.*, pp. 39–41.

Goduslawski, K., and Grzyboswki, J. (1971). Improvement of industrial devices for vacuum metallization of quartz resonators. *Przem. Inst. Elektron.* (*Warsaw*) **12**, 77–81. In Polish.

Golan'sk, R., and Kosecki, T. (1971). Synthetic quartz in the modern manufacturing of quartz resonators. *Elektronika* (1), 18–24. In Polish.

Gopinath, A., and Davies, H. (1969). Frequency tuning of CdS platelets. *Electron. Lett.* **5**, 622–623.

Grenier, R. P. (1970). A technique for automatic monolithic crystal filter frequency adjustment. *Proc. 24th Annu. Freq. Control Symp.*, pp. 104–110.

Greuter, A. (1971). Some remarks concerning flexural type quartz vibrators for wristwatches. *In* "Eurocon 71 Digest," IEEE, New York.

Grudkowski, T. W., Black, J. F., Drake, G. W., and Cullen, D. E. (1982). Progress in the development of miniature thin film BAW resonator technology. *Proc. 36th Annu. Freq. Control Symp.*, pp. 537–548.

Grzegorzewicz, J. (1976). High temperature bond to thin films coating quartz crystal plates. *Thin Solid Films* **36**, 409.

Grzegorzewicz, J., and Szulc, W. (1980). Etching of quartz crystal blanks. *Pr. Inst. Tele-Radiotech.* (84), 51–64. In Polish.

Gualtieri, D. M. (1982). Apparatus for determining the polarity of piezoelectric crystal blanks. *Rev. Sci. Instrum.* **53**, 1100–1102.

Guttwein, G. K., Ballato, A. D., and Lukaszek, T. J. (1968). Design considerations for oscillator crystals. *Proc. 22nd Annu. Freq. Control Symp.*, pp. 67–88.

Hafner, E. (1969). The piezoelectric crystal unit-definitions and methods of measurement. *Proc. IEEE* **57**, 179–201.

Hair, M. L. (1973). The molecular nature of adsorption on silica surfaces. *Proc. 27th Annu. Freq. Control Symp.*, pp. 73–78.

Hall, M. (1977). Quartz crystals: familiar component but too much mystique? *Electron* (114), 49–50.

Hammond, D. L. (1961). Precision quartz resonators. *Proc. 15th Annu. Freq. Control. Symp.*, pp. 125–138.

Hanson, W. P. (1983). Chemically polished high frequency resonators. *Proc. 37th Annu. Freq. Control Symp.*, pp. 261–264.

Hara, K. (1978). Surface treatment of quartz oscillator plate by ion implanation. *Oyo Butsuri* **47**, 145–146. In Japanese.

Hart, R. K. (1974). Precision single sideband crystal units. Report ECOM-0172-F on Contract DAAB07-73-C-0172-0004. Availabe from NTIS.

Hart, R. K., and Hicklin, W. H. (1972). Precision and SSB crystal units for temperature compensated crystal oscillators. *Proc. 26th Annu. Freq. Control Symp.*, pp. 152–158.

Hart, R. K., Hicklin, W. H., and Phillips, L. A. (1974). Methods of cleaning contaminants from quartz surfaces during resonator fabrication. *Proc. 28th Annu. Freq. Control Symp.*, pp. 89–95.

Hartemann, P. (1978). Microwave surface-acoustic-wave components. *AGARD* Conference Proceedings No. 230 (Neuilly-sur-Seine, France: AGARD), pp. 2.2/1–9.

Hatschek, R. (1980). Quartz resonators for wristwatches. *Elektroniker* (*Switzerland*) **19**, EL1–EL4. In German.

Hayashi, T. (1980). Quartz crystal units meet stringent quality requirements. *JEE J. Electron. Eng.* **17**, 54–57.

Hayashi, T., and Hara, M. (1979). Crystal units enhance consumer products. *JEE J. Electron. Eng.* **16**, 62–65.

Heighway, J. (1976). Surface acoustic wave devices. *Wireless World* **82**, 39–43.

Heising, R. A., editor. (1946). "Quartz Crystals for Electrical Circuits." Van Nostrand, New York, pp. 95–139.
Hertzog, W. (1971). On the adjustment of the inductance of oscillator crystals in filter circuits. *Bull. Assoc. Suisse Elec.* **62**, 451–454. In French.
Hickernell, F. S., and Bush, H. J. (1978). The monolithic integration of surface acoustic wave and semiconductor circuit elements on silicon for matched filter device development. *AGARD* Conference Proceedings No. 230 (Neuilly-sur-Seine, France: AGARD), 3.5/1–13.
Hicklin, W. H. (1970). Dynamic temperature behavior of quartz crystal units. *Proc. 24th Annu. Freq. Control Symp.*, pp. 148–156.
Hirama, K. (1977). Quartz crystal units in communications equipment. *JEE J. Electron. Eng.* (121), 50–52.
Hirano, H. (1980). LiTaO$_3$ crystals for commercial SAW TV IF filters. *Ferroelectrics* **27**, 151–156.
Hoffman, D. M. (1974). The structure and properties of thin metal films. *Proc. 28th Annu. Freq. Control Symp.*, pp. 85–88.
Hokanson, J. L. (1969). Laser machining thin film electrode arrays on quartz crystal substrates. *Proc. 23rd Annu. Freq. Control Symp.*, pp. 163–170.
Holland, R. (1974). Nonuniformly heated anisotropic plates: II. frequency transients in AT and BT quartz plates. *Ultrason. Symp. Proc.*, pp. 593–598.
Huguenin, R., and Matthey, H. (1974). Manufacture of quartz clock resonators. *Rev. Polytech.* (*Switzerland*) (11), 1517–1519. In French.
Husgen, D., and Calmes, C. C., Jr. (1976). A method of angle correction. *Proc. 30th Annu. Freq. Control Symp.*, pp. 259–263.
Itoh, M., Gokan, H., and Esho, S. (1982). Fabrication process for surface acoustic wave filters having 0.5 μm finger period electrodes. *J. Vac. Sci. Technol.* **20**, 21–25.
James, S., and Wilson, I. H. (1979). Fine tuning of s.a.w. resonators using argon ion bombardment. *Electron. Lett.* **15**, 683–684.
Jamiolkowski, J., and Sobocinski, J. (1974). Technology of cold-welding of high stability quartz crystal units. *Pr. Inst. Tele- Radiotech.* **18**, 5–15. In Polish.
Judd, G. W., and Thoss, J. L. (1980). Use of apodized metal gratings in fabricating low cost quartz RAC filters. *Ultrason. Symp. Proc.*, pp. 343–347.
Kagiwada, R. S., Yen, K. H., and Lau, K. F. (1978). High frequency SAW devices on AlN/Al$_2$O$_3$. *Ultrason. Symp. Proc.*, pp. 598–601.
Kalev, Kl. A., Metev, S. M., Pushkarova, R. M., Savchenko, S. K., and Stamenvo, K. V. (1981). Laser tuning of low-frequency quartz resonators. *Elektropromst. Priborostr.* **16**, 347–350. In Bulgarian.
Karrer, H. E., and Leach, J. (1969). A quartz resonator pressure transducer. *IEEE Trans. Ind. Electron. Control Instrum.* **IECI-16**, 44–50.
Kasai, T., Noda, J., and Suzuki, J. (1978). Lapping characteristics of LiTaO$_3$ single crystals: study on precision machining of opto-electronic crystals. I. *J. Jpn. Soc. Precis, Eng.* **44**, 1360–1366. In Japanese.
Knolmayer, E. (1981). X-ray goniometry of the modified doubly rotated cuts. *Proc. 35th Annu. Freq. Control Symp.*, pp. 56–59.
Knowles, J. E. (1975). On the origin of the "second level of drive" effect in quartz oscillators. *Proc. 29th Annu. Freq. Control Symp.*, pp. 230–236.
Kobayashi, Y. (1978). Fully automated piezogoniometer (automatic quartz plate classifier). *Proc. 32nd Annu. Freq. Control Symp.*, pp. 317–320.
Koga, I. (1969). Specifying quartz crystal cuts without regard to handedness. *Proc. IEEE* **57**, 2171–2172.

Kosecki, T. (1970). New calibration method for piezoelectric quartz resonators by gold plating. *Prace Inst. Tele-Radiotech.* **14**, 67–70. In Polish.
Kulischenko, W. (1975). Tuning crystals with AJM. *Solid State Technol.* **18**, 20–21.
Kumar, S. (1981). Technological gaps in the development of piezoelectric crystals. *Electron. Inf. & Plann.* **8**, 547–549.
Kusters, J. A. (1970). The effect of static electric fields on the elastic constants of alpha-quartz. *Proc. 24th Annu. Freq. Control Symp.*, pp. 46–54.
Kusters, J. A. (1976). Transient thermal compensation for quartz resonators. *IEEE Trans. Sonics Ultrason.* **SU-23**, 273–276.
Kusters, J. A., and Adams, C. A. (1980). Production statistics of SC (or TTC) crystals. *Proc. 34th Annu. Freq. Control Symp.*, pp. 167–174.
Kusters, J. A., Adams, C. A., and Yoshida, H. (1977). TTCs—further developmental results. *Proc. 31st Annu. Freq. Control Symp.*, pp. 3–7.
Latham, J. I., Shreve, W. R., Tolar, N. J., and Ghate, P. B. (1979). Improved metallization for surface acoustic wave devices. *Thin Solid Films* **64**, 9–15.
Leach, J. G. (1970). 5 MHz BT cut resonators. *Proc. 24th Annu. Freq. Control Symp.*, pp. 117–125.
Lee, P. C. Y., and Wu, K. M. (1977). The influence of support-configuration on the acceleration sensitivity of quartz resonator plates. *Proc. 31st Annu. Freq. Control Symp.*, pp. 29–34.
Liss, F. T., and Richardson, J. F. (1968). Ruggedized quartz oscillator crystals for gun-launched vehicles. Report No. TM-68-23, Harry Diamond Laboratories, U. S. Army Material Command, Adelphi, Maryland 20783, 26pp.
Lorant, M. (1978). New replacement for quartz. *Radio Electron. Constructor* **31**, 529.
Lowe, A. T. (1981). Metallization of quartz oscillators. *Proc. 35th Annu. Freq. Control Symp.*, pp. 48–55.
Lukaszek, T. J. (1970). Mode control and related studies of VHF quartz filter crystals. *Proc. 24th Annu. Freq. Control Symp.*, pp. 126–140.
Lukau, H. (1968). Improved frequency stability and reliability through new vibrating quartz holder. *Siemens-Bauteile-Informationen* **6**, 37–39. In German.
MacDonald, D. B., Shaffer, C. F., Blocker, T. G. III, and Vail, R. C. (1979). Development of a two-step E-beam lithography process for submicron surface acoustic wave (SAW) device fabrication. *Opt. Eng.* **18**, 53–58.
Matistic, A. S. (1976). Quartz-crystal timing accuracy is hard to beat. *Electron. Des.* **24**, 74–79.
Matsuzawa, H., Kamiryo, K., and Kano, T. (1973). Temperature control of quartz crystals for deposited thin-film thickness monitors. *Rec. Electr. Commun. Eng. Conversazione Tohoku Univ. (Japan)* **42**, 211–214. In Japanese.
Mattox, D. M. (1973). Thin film metallization of oxides. *Proc. 27th Annu. Freq. Control Symp.*, pp. 89–97.
McCullough, R. E. (1976). An evaluation of leak test method for hermetically sealed devices. *Proc. 30th Annu. Freq. Control Symp.*, pp. 237–239.
McDermott, J. (1976). Focus on crystals for frequency control. *Electron. Des.* **24**, 40–45.
Merigoux, H., Darces, J. F., and Lamboley, J. (1980). New method to SAW quartz slides. *Proc. 34th Annu. Freq. Control Symp.*, pp. 112–119.
Metcalf, W. S. (1972). Crystal frequency control. *Electron* (12), 15–17.
Metcalf, W. S. (1978). Quartz crystals. *Elektron. Prax.* **13**, 79–80. In German.
Miller, A. J. (1970). Preparation of quartz crystal plates for monolithic crystal filters. *Proc. 24th Annu. Freq. Control Symp.*, pp. 93–103.
Mindlin, R. D. (1968). Optimum sizes and shapes of electrodes for quartz resonators. *J. Acoust. Soc. Amer.* **43**, 1329–1331.

Mitsuyu, T., Ohji, K., Ono, S., Yamazaki, O., and Wasa, K. (1976). Thin-film surface-acoustic-wave devices. *Natl. Tech. Rep. (Japan)* **22**, 905–923. In Japanese.
Nadratowska, B., and Przezdziak, Z. (1969). Etching techniques used in investigations of quartz structure. *Przeglad Elektron* **10**, 287–291. In Polish.
Nemetz, G. E. (1971). Using a pendulum suspension diffractometer to improve precision of X-raying quartz crystals. *Proc. 25th Annu. Freq. Control Symp.*, pp. 134–138.
Ney, R. J., and Hafner, E. (1979). Continuous vacuum processing system for precision quartz crystal units. *Proc. 33rd Annu. Freq. Control Symp.*, pp. 368–373.
Nickols, S. E., and Fay, R. M. (1978). Bonding of piezoelectric materials. *IBM Tech. Disclosure Bull.* **21**, 2986.
Noda, J., Suzuk, J., and Furusawa, Y. (1974). Fabrication of AT-cut quartz for MCF substrates. *Electr. Commun. Lab. Tech. J.* **23**, 79–93. In Japanese.
Nonaka, S., Yuuki, T., and Hara, K. (1971). The current dependency of crystal unit resistance at low drive level. *Proc. 25 Annu. Freq. Control Symp.*, pp. 139–147.
Oguchi, K., and Momosaki, E. (1978). $+5°X$ micro quartz resonator by lithographic process. *Proc. 32nd Annu. Freq. Control Symp.*, pp. 277–281.
Oguchi, K., Shibata, S., and Ogata, T. (1979). +5 degree X micro quartz crystal resonator by lithographic process. *J. Jpn. Soc. Precis. Eng.* **45**, 356–360. In Japanese.
Okano, S. (1977). Quartz crystals—a continuing search. *JEE J. Electron. Eng.* (121), 39–41.
Onoe, M., Kamada, K., Okazaki, M., Tajika, F., and Manabe, N. (1977). 4 MHz AT-cut strip resonator for wrist watch. *Proc. 31st Annu. Freq. Control Symp.*, pp. 48–54.
Oomura, Y. (1976). Miniaturized circular disk R-cut (AT) crystal vibrator by means of side wire mounting. *Trans. Inst. Electron. Commun. Eng. Jpn. E* **59**, 17–18.
Pantani, L. (1975). Temperature stabilization and design criteria in some SAW devices. *Ultrasonics International* (Guildford, Surrey, England: IPC Sci & Technol. Press), pp. 138–141.
Peschl, H. (1979). The quartz [crystal]: Assembly, operation and properties. II. *Funkschau* **51**, 451–454. In German.
Peters, R. D. (1976). Ceramic flat pack enclosures for precision quartz crystal units. *Proc. 30th Annu. Freq. Control Symp.*, pp. 224–231.
Piwonski, W. (1971). Modern technological processes in quartz resonators manufacturing. *Elektronika* (1), 14–17. In Polish.
Pogson, I. (1975). LF quartz crystals in TO-5 case. *Electron. Aust. (Australia)* **37**, 53.
Rankin, D. H. (1972). Vacuum techniques in the quartz crystal industry. *Vacuum* **22**, 377–380.
Reche, J. J. H. (1978). Frequency tuning of quartz resonators by plasma anodization. *Proc. 32nd Annu. Freq. Control Symp.*, pp. 299–303.
Royer, J. J. (1973). Rectangular AT-cut resonators. *Proc. 27th Annu. Freq. Control Symp.*, pp. 30–34.
Schiavone, L. M. (1978). Electrodeless gold metallization for polyvinylidene fluoride films [piezoelectric device application]. *J. Electrochem. Soc.* **125**, 522–523.
Seed, A. (1967). Theoretical studies of microminiature quartz crystal units. *Conference on Frequency Generation and Control for Radio Systems* (London: IEE), Conference Publication No. 31, pp. 29–32.
Seed, A., and Smith, D. W. (1973). Quartz crystals for timepieces. *Electron. Components* **14**, 122–126.
Shanley, C. W., and Dworsky, L. N. (1982). DC plasma anodization of quartz resonators. *Proc. 36th Annu. Freq. Control Symp.*, pp. 108–114.
Sherman, J. H., Jr. (1976). Dimensioning rectangular electrodes and arrays of electrodes on AT-cut quartz bodies. *Proc. 30th Annu. Freq. Control Symp.*, pp. 54–64.
Sherman, J. H., Jr., (1977). Measurement of the characteristic frequency of an AT-cut plate. *Proc. 31st Annu. Freq. Control Symp.*, pp. 108–116.

Sherman, J. H., Jr. (1978). Derivation of a leak specification for a hermetic envelope. *Proc. 32nd Annu. Freq. Control Symp.*, pp. 326–333.

Sherman, J. H., Jr. (1979). Trim sensitivity—A useful characterization of a resonator. *Proc. 33rd Annu. Freq. Control Symp.*, pp. 181–185.

Simmons, G. W., Hicklin, W. H., and Hart, R. K. (1970). Auger spectroscopy in studies of the aging factors of quartz. *Proc. 24th Annu. Freq. Control Symp.*, pp. 111–116.

Simpson, E. E. (1970). Manufacturing high-reliability quartz crystal units under contamination control. *West. Electr. Eng.* **14**, 44–49.

Smagin, A. G. (1974). Frequency correction to 10^{-8} by ruby laser for precision quartz crystals. *Instrum. Exp. Tech.* **17**, 1397–1398.

Smith, H. I. (1977). Surface wave device fabrication. *In* "Surface Wave Filters" (H. Matthews, ed.), pp. 165–217. Wiley: Chichester, England.

Snell, F. E. (1975a). Method for automatically controlling plating rates of material on crystal resonators. *Tech. Dig.* (38), 41–42.

Snell, F. E. (1975b). Method for controlling gold usage in the plating of crystal resonators. *Tech. Dig.* (38), 43–44.

Sobocinski, J., and Marzec, A. (1974). Ultra-high vacuum system for final pumping down of quartz crystal resonators. *Pr. Inst. Tele- Radiotech.* **18**, 55–59. In Polish.

Staudte, J. H. (1968). Micro resonators in integrated electronics. *Proc. 22nd Annu. Freq. Control Symp.*, pp. 226–231.

Staudte, J. H. (1973). Subminiature quartz tuning fork resonator. *Proc. 27th Annu. Freq. Control Symp.*, pp. 50–54.

Suda, P., Zumsteg, A. E., and Zingg, W. (1979). Anisotropy of etching rate for quartz in ammonium bifluoride. *Proc. 33rd Annu. Freq. Control Symp.*, pp. 359–363.

Takeda, I. (1975). Oscillator crystals keep on shrinking as watch demand heats up. *JEE J. Electron. Eng.* **101**, 15–18.

Takeda, I. (1976). Transceiver crystal techniques: painstaking design is essential. E Jap. Electron. Ind. **23**, 60–70.

Tanji, S., and Wakatsuki, N. (1982). Dry processes for TV-IF SAW filter with deep trap. *Proc. 32nd Electron. Components Conf.*, pp. 512–517.

Tanski, W. J. (1979). SAW resonators at 1.29 GHz with Q values approaching the material limit. *Electron. Lett.* **15**, 339–340.

Tanski, W. J. (1981a) Elements of SAW resonator fabrication and performance. *Proc. 35th Annu. Freq. Control Symp.*, pp. 388–394.

Tanski, W. J. (1981b). Surface acoustic wave frequency trimming of resonant and traveling-wave devices on quartz. *Appl. Phys. Lett.* **39**, 40–42.

Thompson, E. C. (1972). Double mount for each electrode of crystal to prevent failure. *Tech. Dig.* (26), 59–60.

Toki, M., Tsuzuki, Y., and Mikami, T. (1978). Precise design of motional inductance of longitudinal mode X-cut quartz crystal resonator. *Trans. Inst. Electron. Commun. Eng. Jpn. E* **61**, 315–316.

Urabe, S., Onuki, K., and Yoshikawa, S. (1979). Fine frequency tuning of SAW devices with MgF_2 thin film evaporation. *Trans. Inst. Electron. Commun. Eng. Jpn. E.* **62**, 557–558.

Van Empel, F. J., Massen, C. H., Arts, H. J. J. M., and Poulis, J. A. (1971). Independent multiple oscillations of a single quartz wafer. *J. Acoust. Soc. Am.* **50**, 1386–1387.

Vasin, L. N., and Shushkov, A. G. (1974). Mechanical processing of quartz resonator plates. *Sov. J. Opt. Technol.* **41**, 477–478.

Vig, J. R. (1975). A high precision laser assisted X-ray goniometer for circular plates. *Proc. 29th Annu. Freq. Control Symp.*, pp. 240–247.

Vig, J., Wasshausen, H., Cook, C., Katz, M., and Hafner, E. (1973). Surface preparation and characterization techniques for quartzs resonators. *Proc. 27th Annu. Freq. Control Symp.*, pp. 98–112.

Vig, J. R., Cook, C. F. Jr., Schwidtal, K., LeBus, J. W., and Hafner, E. (1974). Surface studies for quartz resonators. *Proc. 28th Annu. Freq. Control Symp.*, pp. 96–108.

Vig, J. R., LeBus, J. W., and Filler, R. L. (1975). Further results on UV cleaning and Ni electrobonding. *Proc. 29th Annu. Freq. Control Symp.*, pp. 220–229.

Vig, J. R., LeBus, J. W., and Filler, R. L. (1977a). Chemically polished quartz. *Proc. 31st Annu. Freq. Control Symp.*, pp. 131–143.

Vig, J. R., LeBus, J. W., and Filler, R. L. (1977b). Chemically polished quartz. Report ECOM-4548, U. S. Army Electronics Command, Ft. Monmouth, New Jersey 07703, U.S.A., 37 pp.

Vig, J. R., Brandmayr, R. J., and Filler, R. L. (1979). Etching studies on singly and doubly rotated quartz plates. *Proc. 33rd Annu. Freq. Control Symp.*, pp. 351–358.

Vig, J. R., Washington, W., and Filler, R. L. (1981). Adjusting the frequency vs. temperature characteristics of SC-cut resonators by contouring. *Proc. 35th Annu. Freq. Control Symp.*, pp. 104–109.

Wang, J. S., and Lakin, K. M. (1981). Sputtered AlN films for bulk-acoustic-wave devices. *Ultrason. Symp. Proc.*, pp. 502–505.

Warner, A. Goldfrank, B., Meirs, M., and Rosenfeld, M. (1979). Low "g" sensitivity crystal units and their testing. *Proc. 33rd Annu. Freq. Control Symp.*, pp. 306–310.

Wasshausen, H. (1971). Processing techniques for shock resistant precision quartz crystal units. Report ECOM-3524, U. S. Army Electronics Command, Ft. Monmouth, New Jersey 07703, Available as No. AD-735685 from NTIS.

Werner, J. F., and Dyer, A. J. (1976). The relationship between plate back, mass loading and electrode dimensions for AT-cut quartz crystals having rectangular electrodes operating at fundamental and overtone modes. *Proc. 30th Annu. Freq. Control Symp.*, pp. 40–53.

White, M. L. (1973). Clean surface technology. *Proc. 27th Annu. Freq. Control Symp.*, pp. 79–88.

Wilcox, P. D., Snow, G. S., Hafner, E., and Vig, J. R. (1975). A new ceramic flat pack for quartz resonators. *Proc. 29th Annu. Freq. Control Symp.*, pp. 202–210.

Wilhelmy, H. J. (1979). A man and his work: Jurgen Staudte and the crystal tuning fork. *Elektronik* **28**, 38–42. In German.

Wolfskill, J. M. (1968). Advancements in production of 5 MHz fifth overtone high precision crystal units. *Proc. 22nd Annu. Freq. Control Symp.*, pp. 89–117.

Yamagata, S., and Fukai, I. (1980). Analysis for the spattering electrodes of AT-cut quartz crystal resonator. *Trans. Inst. Electron. Commun. Eng. Jpn. E.* **63**, 677.

Yamashita, S., Echigo, N., Kawamura, Y., Watanabe, A., and Kubota, K. (1978). A 4.19 MHz beveled miniature rectangular AT-cut quartz resonator. *Proc. 32nd Annu. Freq. Control Symp.*, pp. 267–276.

Yasuhara, Y., Yarmaji, N., Kurokawa, T., and Takahashi, K. (1982). Surface acoustic wave devices for consumer use. *IEEE Trans. Consum. Electron.* **CE-28**, 475–481.

Sections 5.1 and 5.2

Adam, J. D. (1982). Magnetostatic wave multichannel filters. *Proc. 36th Annu. Freq. Control Symp.*, pp. 419–427.

Ainger, F. W., Burgess, J. W., Hales, M. C., and Porter, R. J. (1976). The use of lithium tantalate in monolithic crystal filters. *Ferroelectrics* **10**, 75.

Akcakaya, E. (1976). An equivalent circuit applicable to design monolithic crystal filters. *Proc. IEEE Int. Symp. Circuits Syst.*, pp. 359–361.

Akcakaya, E. (1976). New approach to the analysis of monolithic crystal filters. *J. Acoust. Soc. Am.* **60**, 492–502.

Albsmeier, H. (1971). A comparison of the realizability of electromechanical channel filters in the frequency range 12 kHz to 10 MHz. *Frequenz* **25**, 74–79. In German.

Albsmeier, H. (1976). Important process steps in mechanical channel filter fabrication. *Proc. IEEE Int. Symp. Circuits Syst.*, pp. 758–760.

Albsmeier, M. H. (1979). Mechanical filters in communications systems. *Rev. HF* (Belgium) **11**, 44–49.

Albsmeier, H., Gunther, A. E., and Volejnik, W. (1974). Some special design considerations for a mechanical filter channel bank. *IEEE Trans. Commun.* **COM-22**, 935–940.

Allemandou, P. (1979). Design of mechanical filters having attenuation poles. *Onde Electr.* **59**, 59–61. In French.

Althans, W. (1971). New channel-changer with electromechanical filters. *Fernmelde-Praxis* **48**, 923–926. In German.

Amsler, H. (1969). The quartz miniature receiver for Hasler radio paging systems. *Hasler Rev. (Switzerland)* **2**, 20–22.

Amstutz, P. (1981). Electromechanical filters and resonating piezoelectric filters. *Journees d'Electronique 1981, Presses Polytechniques Romandes, Lausanne,* pp. 13–27. In French.

Amstutz, P., Bon, M., Bosc, R., Carru, H., and Loyez, P. (1978). Design and realization of an electromechanical filter model for voice channel. I. *Onde Electr.* **58**, 307–311. In French.

Anderes, S. W. (1972). A polylithic filter channel bank. *IEEE Trans. Commun.* **COM-20**, 48–52.

Anonymous (1971a). Testing resonant reed filters. *Dawe Dig.* **14**, 5–6.

Anonymous (1971b). Testing resonant reed filters. *Electron. Compon.* **12**, 758–759.

Anonymous (1972). The electronmechanical filters of TESLA Strasnice. *Tesla Electron.* **5**, 57–59.

Anonymous (1978). Cheap crystal filter. *Elektor* **4**, 58–59.

Anonymous (1980a). 4.4 MHz crystal filter. *Elektor* **6**, 6–7.

Anonymous (1980b). Designing crystal filters. *Electron. Ind.* **6**, 25–27, 29.

Ardelean, Gh., Niculescu, T., and Simonescu, M. (1968). Optimum filter with magnetostrictive line. *Telecomunicatii* **12**, 481–488. In Rumanian.

Ariga, M., and Sato, M. (1973). A high electromechanical-coupling resonator and its application to filter-synthesis. *Mem. Fac. Technol. Tokyo Metrop. Univ.* (23), 2033–2042.

Arnoldt, M. (1980). Competition for quartz resonators. *Radio Mentor Electron.* **46**, 137–139. In German.

Arranz, T. (1977). Lithium tantalate channel filter for multiplex telephony. *Proc. 31st Annu. Freq. Control Symp.*, pp. 213–224.

Ashida, T. (1971). Eigenfrequencies of monolithic filters. *Electron. Commun. Jpn.* **54**, 41–49.

Ashida, T. (1974). Design and characteristic analysis of monolithic crystal filters. *Electron. Commun. Jpn.* **57**, 1–9.

Ashida, T. (1975). Design of piezoelectric transducers for temperature stabilized mechanical filters. *Electron Commun. Jpn.* **57**, 10–17.

Ballato, A. (1975). The stacked-crystal filter. *Proc. IEEE Int. Symp. Circuits Syst.*, pp. 301–304.

Ballato, A., and Lukaszek, T. (1973a). A novel frequency selective device: The stacked-crystal filter. *Proc. 27th Annu. Freq. Control Symp.*, pp. 262–269.

Ballato, A., and Lukaszek, T. (1973b). Stacked-crystal filters. *Proc. IEEE* **61**, 1495–1496.

Ballato, A., Bertoni, H. L., and Tamir, T. (1974). Systematic design of stacked-crystal filters by microwave network methods. *IEEE Trans. Microwave Theory Tech.* **MTT-22**, 14–25.

Bastelaer, Ph.an (1968). The design of band-pass filters with piezoelectric resonators. *Rev. HF* (Belgium) **7**, 193–206.

Beaver, W. D. (1967a). Theory and design of the monolithic crystal filter. *Proc. 21st Annu. Freq. Control Symp.*, pp. 179–199.

Beaver, W. D. (1967b). Theory and design principles of the monolithic crystal filter. Ph.D. Thesis, Lehigh University.

Beaver, W. D. (1968). Analysis of elastically coupled piezoelectric resonators. *J. Acoust. Soc. Am.* **43**, 972–981.

Beaver, W. D., and Frymoyer, E. M. (1970). The effect of variations in fabrication on the transmission properties of monolithic crystal filters. *IEEE Trans. Sonics Ultrason.* **SU-17**, 59.

Beck, E. (1975). Monolithic crystal filters. *Hasler Rev. (Switzerland)* **8**, 44–49.

Beck, E., Schultze, E., and Meyr, H. (1976a). Chain matrix description of plylithic crystal filters. *Journees d'Electronique et de Mecanique sur Interactions Electronique-Micromecanique (Switzerland)* (Lausanne, Switzerland: Ecole Polytech. Federale de Lausanne), pp. 273–285.

Beck, E., Schultze, E., and Meyr, H. (1976b). An admittance approach to the dual monolithic crystal filter. *Proc. IEEE Int. Symp. Circuits Syst.*, pp. 316–319.

Beletskiy, A. F., Lebedev, A. T., and Ovchinnikov, A. A. (1972). Synthesis of Γ-shaped quartz matched filters for rectangular pulses. *Telecommun. Radio Eng. Pt. 1*, **26**, 45–47.

Belevitch, V., and Kamp, Y. (1969). Theory of monolithic crystal filters using thickness-twist vibrations. *Philips Res. Rep.* **24**, 331–369.

Bernstein, M. (1967). Increased crystal resistance at oscillator noise levels. *Proc. 21st Annu. Freq. Control Symp.*, pp. 244–258.

Berté, M. (1977). Acoustic bulk wave resonators and filters operating in the fundamental mode at frequencies greater than 100 MHz. *Proc. 31st Annu. Freq. Control Symp.*, pp. 122–125.

Bezemer, J. A. (1972). The monolithic quartz crystal filter. *PTT-Bedrijf* **18**, 29–37. In Dutch.

Bezemer, J. A. (1974). The single quartz crystal in the monolithic crystal filter. *Data*, **75**, 1–25. In Dutch.

Bidart, L. (1971). Crystal filter modern conception. *Onde Elec.* **51**, 311–319. In French.

Bidart, L. (1971). Semi-monolithic quartz crystal filters and monolithic quartz filters. *Proc. 25th Annu. Freq. Control Symp.*, pp. 271–279.

Birn, B. (1976). A modified insertion loss theory for mechanical channel filter synthesis. *Proc. IEEE Int. Symp. Circuits Syst.*, pp. 754–757.

Blinchikoff, H. J. (1975). Low-transient intermediate-band crystal filters. *IEEE Trans. Circuits Syst.* **CAS-22**, 509–515.

Bon, M., Bosc, R., and Loyez, P. (1976). New materials for transducers and resonators tolerances assignment in electromechanical filters. *Proc. IEEE Int. Symp. Circuits Syst.*, pp. 739–742.

Bon, M., Bosc, R., and Loyez, P. (1977). New design of electromechanical filters at 128 kHz. *Proc. IEEE Int. Symp. Circuits Syst.*, pp. 239–242.

Bosc, R., and Loyez, P. (1974). Design of an electromechanical filter at 128 kHz in a two step modulation system. *Proc IEEE Int. Symp. Circuits Syst.*, pp. 111–114.

Bosc, R., Collombat, F., Herreng, C., and Loyez, P. (1975). An electromechanical filter 12 channel project. *Echo Rech.* (79), 30–39. In French.

Braun, A. R. (1972). Crystal filter. U.S. Patent 3,656,180.

Braun, A. R. (1973). Resonator interconnections in monolithic filters. U. S. Patent 3,739,304.

Braun, A. R. (1974). Filters and resonators—a review. *IEEE Trans. Sonics Ultrason.* **SU-21**, 219.

Bremon, C. (1978). Single crystal–metal composite transducer for electromechanical filter devices. *Onde Electr.* **58**, 464–469. In French.

Brier, L. (1979). Use of quartz crystals in wide band filters. *Nachrichtentech. Elektron.* **29**, 460–463. In German.

Bucherl, E. (1973). Bandstop filters incorporating double-tuned monolithic crystal units. *Siemens Forsch. Entwicklungsber.* **2**, 238–247.

Bunger, D. A. (1963). Composite tuning fork filters. U. S. Patent 3,437,850.
Burgess, J. W., and Porter, R. J. (1973). Single mode resonance in lithium niobate/lithium tantalate for monolithic crystal filters. *Proc. 27th Annu. Freq. Control Symp.*, pp. 246–252.
Burrascano, P., and Lojacono, R. (1981). Design, realization and alignment methods for monolithic quartz filters. An empirical correction of the elastical analysis. *Note Recens. Not.* **30**, 97–105. In Italian.
Byrne, R. J. (1970). Monolithic crystal filters. *Proc. 24th Annu. Freq. Control Symp.*, pp. 84–92.
Camurri, F., and Costamagna, E. (1973). On the design of monolithic crystal filters. *Alta Freq.* **42**, 341–345.
Carru, H., Reaud-Goud, J., and Villela, G. (1977). Design and modeling of composite metal-monocrystal transducers. *Proc. IEEE Int. Symp. Circuits Syst.*, pp. 243–246.
Caviglia, F., and Gamerro, R. (1977). Linear band-pass filters synthesis method developed for the design of electromechanical filters. *Alta Freq.* **46**, 463–476. In Italian.
Cawley, H. F., Jennings, J. D., Pelc, J. I., Perri, P. R., Snell, F. E., and Miller, A. J. (1975). Manufacture of monolithic crystal filters for A-6 channel bank. *Proc. 29th Annu. Freq. Control Symp.*, pp. 113–119.
Chang, J. H., and Tuteur, F. B. (1968). Adaptive tapped delay line filters. *Proc. 2nd Annu. Princeton Conf. Inf. Sci. Syst.*, pp. 164–168.
Chang, Z.-P., and Barsch, G. R. (1976). Elastic constants & thermal expansion of berlinite *IEEE Trans. Sonics Ultrason.* **SU-23**, 127–135.
Chelmonski, J., Wrobel, T., and Koczynski, B. (1970). Band-stop quartz filter FZ-256 kHz. *Prace Inst. Tele. Radiotech.* **14**, 55–57.
Chelovechkov, A. I., and Sharov, N. V. (1973). The use of narrow-band tuning fork filters in geophysical apparatus. *Geofiz. Appar.* (51), 186–189. In Russian.
Chen, D.-P., Melngailis, J., and Haus, H. A. (1982). Filters based on conversion of surface acoustic waves to bulk plate modes in gratings. *Ultrason. Symp. Proc.*, pp. 67–71.
Chohan, V. C., and Dillon, C. R. (1975). Specification, design and synthesis of F.S.K. filters using double-resonator monolithic crystal filter (MCF) sections. *Arch. Elektron. Ueber-tragungstech.* **29**, 121–124.
Colin, J. E. (1968). Formulae for the calculation of narrow band pass filters with identical piezoelectric crystals and maximally flat attenuation behaviour. *Cables Transm.* **22**, 132–135. In French.
Colin, J. E. (1975). Example of a balanced parametric band-pass filter using piezoelectric crystals at both filter ends. *Cables Transm.* **29**, 427–432. In French.
Costa, P., and Stacchiotti, G. (1974). Polylithic crystal filters—reasons for the choice. *Alta Freq.* **43**, 1040–1043.
Costa, P., Piacentini, G. F., and Stacchiotti, G. (1975). LC conventional, electro-mechanical quartz and polylithic filters. *Telettra* (27), 3–22.
Court, I. N. (1969). Microwave acoustic devices for pulse compression filters. *IEEE Trans. Microwave Theory Tech.* **MTT-17**, 968–986.
Cucchi, S., and Molo, F. (1976). Bridging elements in mechanical filters: design procedure and an example of negative bridging element realization. *Proc. IEEE Int. Symp. Circuits Syst.*, pp. 746–749.
Dailing, J. L. (1980). The overlapping ground—A new monolithic crystal filter configuration. *Proc. 34th Annu. Freq. Control Symp.*, pp. 445–448.
Deckert, J., and Guls, P. (1974). Mechanical filters for channel converters *Tech. Mitt. AEG-Telefunken* **64**, 74–76. In German.
Deschamps, R. (1977). Longitudinal resonance filters with magnetostrictive beams and two oblong holes. *Cables Transm.* **31**, 87–108. In French.

Deschamps, R. G. (1978). Evidence for the existence of poles on the attenuation frequency curve of a magnetostrictive bar filter and calculation of the corresponding characteristics. *C.R. Hebd. Seances Acad. Sci. Ser. B.* **287**, 231–234. In French.
Détaint, J. (1977). Zero temperature coefficient in overtone lithium tantalate thickness-mode resonators. *Electron. Lett.* **13**, 20–21.
Détaint, J. (1981). New materials and devices for piezoelectric filtering of volume waves. *Journees d'Electronique* (Lausanne, Switzerland: Presses Polytechniques Romandes), pp. 145–160. In French.
Détaint, J., and Lançon, R. (1976). Temperature characteristics of high frequency lithium tantalate plates. *Proc. 30th Annu. Freq. Control Symp.*, pp. 132–140.
Détaint, J., Feldmann, M., Henaff, J., Poignant, H., and Toudic, Y. (1979). Bulk and surface acoustic wave propagation in berlinite. *Proc. 33rd Annu. Freq. Control Symp.*, pp. 70–79.
Détaint, J., Poignant, H., and Toudic, Y. (1980). Experimental thermal behavior of berlinite resonators. *Proc. 34th Annu. Freq. Control Symp.*, pp. 93–101.
Détaint, J., Carru, H., Amstutz, P., and Schwartzel, J. (1983). Acoustically coupled resonators: filters and pressure transducers. *Proc. 37th Annu. Freq. Control Symp.*, pp. 239–247.
Dillon, C. R., and Lind, L. F. (1975). Cascade synthesis of monolithic crystal filters possessing finite transmission zeros. *Int. J. Circuit Theory Appl.* **3**, 101–107.
Dillon, C. R., and Lind, L. F. (1976). Cascade synthesis of polylithic crystal filters containing double-resonator monolithic crystal filter (MCF) elements. *IEEE Trans. Circuits Syst.* **CAS-23**, 146–154.
Dillon, C. R., and Mack, L. (1982). Prototype network for crystal filter design. *Saraga Memorial Colloquium on Electric Filters*, 7/1–7.
Dishal, M. (1958). Modern network theory design of a single sideband crystal filter. *IRE Wescon Rec.* **2**, 33.
Dishal, M. (1965). Modern network theory design of single-sideband crystal ladder filters. *Proc. IEEE* **53**, 1205–1216.
d'Albaret, B. (1982). Recent advances in UHF crystal filters. *Proc. 36th Annu. Freq. Control Symp.*, pp. 405–418.
Duchet, C., and Villela, G. (1979). Length expander transducers for mechanical filters of composite metal-lithium niobate structure exhibiting a low temperature coefficient of frequency. *Proc. IEEE Int. Symp. Circuits Syst.*, pp. 900–903.
Dworsky, L. (1981). An improved circuit model for monolithic crystal filters. *IEEE Trans. Sonics Ultrason.* **SU-28**, 283–285.
Dworsky, L. (1981). Monolithic crystal filter design using a variational coupling approximation. *IEEE Trans. Sonics Ultrason.* **SU-28**, 277–283.
Dydyk, M. (1977). Dielectric resonators add Q to MIC filters. *Microwaves* **16**, 150–151, 154–156.
Ecotiere, B. (1973). Mechanical resonant filters: Problems of construction and design. *Cables Transm.* **27**, 126–136. In French.
Egorova, L. V. (1980). The design of crystal rejector filters based on a monolithic structure. *Telecommun. Radio Eng. Pt. 1* **34**, 52–56.
Ernyei, H. H. (1977). Simplified bridge calculation and realization of a pole type mechanical filter. *Proc. IEEE Int. Symp. Circuits Syst.*, pp. 235–238.
Ernyei, H. H. (1978). Miniaturized mechanical filter LTT. *Onde Electr.* **58**, 128–135.
Ernyei, H. H. (1979). Direct solution of resonant line mechanical filter problems with matrices. *Proc. Int. Symp. Circuits Syst.*, pp. 1060–1063.
Ey, K., Hornung, F., and Volejnik, W. (1972). Channel modem features electromechanical filters. *Siemens Rev.* **39**, 293–298.
Ey, K., Hornung, F., and Volejnik, W. (1973). Electromechanical filter channel modem. *Onde Electr.* **53**, 297–302. In French.

Faktor, Z. (1969). Spurious vibrations in component parts of electromechanical filters and their investigation. *Slaboproudy Obzor* **30**, 444–450. In Czech.

Fleischmann, U. (1979). Calibration of quartz filters with adjustable bandwidth. *Funkschau* **51**, 41–42. In German.

Folk, A. (1968). Simple ladder type quartz crystal filter. *Przeglad Telekomun.* 38–42. In Polish.

Foster, S. J., and Redwood, M. (1980). Enlargement of piezoelectric filter bandwidth by parallel connection of coupled resonators. *IEE Proc. G.* **127**, 209–214.

Freris, L. L., and de Carvalho, J. N. (1973). An electromechanical harmonic filter. *International Conference on High Voltage DC and/or AC Power Transmission* (*IEE: London*), pp. 11–15.

Freris, L. L., and de Carvalho, J. N. (1975). Electromechanical filter and its performance in DC transmission systems. *Proc. Inst. Electr. Eng.* **122**, 55–60.

Frymoyer, E. M. (1979). Mechanical filters. *Proc. 33rd Annu. Freq. Control Symp.*, pp. 223–227.

Galvan-Ruiz, J. (1974). Electromechanical filters. *Rev. Telecomun.* **29**, 3–13. In Spanish.

Garrison, J. L., Georgiades, A. N., and Simpson, H. A. (1970). The application of monolithic crystal filters to frequency selective networks. *IEEE Int. Symp. Circuit Theory*, pp. 177–178.

Gehrels, J. (1972). Monolithic crystal filters reduce space requirements by factor of ten. *Can. Electron. Eng.* **16**, 28–30.

Gel'mont, Z. Ya, and Metel'kova, T. F. (1971). Ladder rejection filters with quartz resonators in the series arms. *Elektrosvyaz* **25**, 69–72. In Russian.

Gerber, W. J., Waren, A. D., Pim, K. A., and Curran, D. R. (1965). Hybrid piezoelectric devices. Fourth Quarterly Report on Contract No. DA28-043AMC-00079(E), U. S. Army Electronics Command, Ft. Monmouth, New Jersey.

Glowinski, A. (1970). Integrated quartz filters: synthesis and adjustment problems in monolithic filters. *Proc. International Conference on Advanced Microelectronics* (*Paris: Federation Nationale des Industries Electroniques*), 2 pp.

Glowinski, A., Lançon, R., and Lefevre, R. (1973). Effects of asymmetry in trapped energy piezoelectric resonators. *Proc. 27th Annu. Freq. Control Symp.*, pp. 233–242.

Golenishchev-Kutuzov, V. A. (1981). Narrowband tunable acoustic filter. *Sov. Phys.-Acoust.* (*Engl. Transl.*) **27**, 436–437.

Goodwin, M. W., and Wood, A. F. B. (1974). The use of crystal filters in mobile communications systems. *Communications 74* (London: IPC Business Press), 13-2/1.

Goral, A. (1972). Monolithic filters. *Elektronika* (11), 460–464. In Polish.

Gordon-Smith, D., and Almond, D. P. (1981). Anomalous nonlinearity in quartz crystal filters. *Electron. Lett.* **17**, 207–208.

Gounji, T., Kawatsu, T., Kasai, Y., Takeuchi, T., Tomikawa, Y., and Konno, M. (1983). Timing tank mechanical filter for digital subscriber transmission system. *Proc. 37th Annu. Freq. Control Symp.*, pp. 376–386.

Grant, P. M., Collins, J. H., Darby, B. J., and Morgan, D. P. (1973). Potential applications of acoustic matched filters to air-traffic control systems. *IEEE Trans. Microwave Theory Tech.* **MTT-21**, 288–300.

Grenier, R. P. (1974). Automatic frequency adjustment of monolithic crystal filters. *W. Elec. Eng.* **15**, 15–19.

Grudkowski, T. M., Black, J. F., Reeder, T. M., Cullen, D. E., and Wagner, R. A. (1980). Fundamental mode VHF/UHF bulk acoustic wave resonators & filters on silicon. *Ultrason. Symp. Proc.*, pp. 829–833.

Guenther, A. E. (1972). High-quality wide-band filters of ultrasonic resonators technique and design. *IEEE Trans. Sonics Ultrason.* **SU-19**, 406.

Guenther, A. E. (1972). Remarks on the design of mechanical wide-band filters. *Nachrichtentech. Z.* **25**, 345–351. In German.

Guenther, A. E. (1973a). Electro-mechanical filters—satisfying additional demands. *Int. Symp. on Circuit Theory*, pp. 142–145.
Guenther, A. E. (1973b). High-quality wide-band mechanical filters—theory and design. *IEEE Trans. Sonics Ultrason.* **SU-20**, 294–301.
Guenther, A. E., and Thiele, E. (1980a). Manufacture oriented design of high-performance mechanical filters. *Proc. IEEE Int. Symp. Circuits Syst.*, pp. 631–635.
Guenther, A. E., and Thiele, E. (1980b). Manufacture oriented design of high-performance mechanical filters. *IEEE Trans. Circuits Syst.* **CAS-27**, 1241–1249.
Guenther, A. E., and Traub, K. (1980a). Precise equivalent circuits of mechanical filters. *Proc. IEEE Int. Symp. Circuits Syst.*, pp. 627–630.
Guenther, A. E., and Traub, K. (1980b). Precise equivalent circuits of mechanical filters. *IEEE Trans. Sonics Ultrason.* **SU-27**, 236–244.
Guenther, A. E., Albsmeier, H., and Traub, K. (1979). Mechanical channel filters meeting CCITT specification. *Proc. IEEE* **67**, 102–108.
Haas, W. (1973). Channeling equipment technology using electromechanical filters. *Electr. Commun.* **48**, 16–20.
Haggarty, R. D., Hart, L. A., O'Leary, G. C. (1968). A 10,000 : 1 pulse compression filter using a tapped delay line linear filter synthesis technique. *Electronics and Aerospace Systems Record (IEEE, New York)*, pp. 306–314.
Haine, J. L. (1977). Simple design procedure for single-sideband crystal filters. *Electron. Lett.* **12**, 687–688.
Hales, M. C., and Burgess, J. W. (1976). Design and construction of monolithic-crystal filters using lithium tantalate. *Proc. Inst. Electr. Eng.* **123**, 657–661.
Hales, M. C., and Burgess, J. W. (1976). Wide band monolithic crystal filters using lithium tantalate. *Electrocompon. Sci. Technol.* **3**, 43–49.
Halsig, C. (1976). Mechanical filters of Kombinat VEB Elektronische Bauelemente Teltow and their application. *Proc. IEEE Int. Symp. Circuits Syst.*, pp. 761–766.
Halsig, C. (1977). Mechanical frequency selection-present status of development and perspective. *Nachrichtentech. Elektron.* **27**, 10–13. In German.
Halsig, C. (1979). Mechanical coupling in electromechanical filters. *Proc. IEEE Int. Symp. Circuits Syst.*, pp. 1064–1067.
Halsig, C. (1980). Frequency-selective mechanical components. *Radio Fernsehen Elektron.* **29**, 71–74. In German.
Halsig, C. (1980). Mechanical frequency-selective components. II. *Radio Fernsehen Elektron.* **29**, 160–162. In German.
Hannon, J. J., Lloyd, P., and Smith, R. T. (1970). Lithium tantalate & lithium niobate piezoelectric resonators in the medium-frequency range with low ratios of capacitance & low temperature coefficients of frequency. *IEEE Trans. Sonics Ultrason.* **SU-17**, 239–246.
Hardcastle, J. A. (1976). Some experiments with high-frequency ladder crystal filters. I. Construction. *Radio Commun.* **52**, 896–898, 905.
Hardcastle, J. A. (1977). Some experiments with high-frequency ladder crystal filters. II. Test Equipment. *Radio Commun.* **53**, 29–29.
Hardcastle, J. A. (1977). Some experiments with high-frequency ladder crystal filters. III. *Radio Commun.* **53**, 122–124.
Hardcastle, J. A. (1979). Experience with crystal ladder filters. *Rev. Telegr. Electron.* **67**, 1323–1325. In Spanish.
Hardcastle, J. A. (1979). Ladder crystal filter design. *Radio Commun.* **55**, 116–120.
Hardcastle, J. A. (1979). Third overtone ladder crystal filters. *Radio Commun.* **55**, 1027–1028.
Harris, R. J. (1980). A low cost s.s.b. crystal filter. *New Electron.* **13**, 56.

Hartmann, C. S. (1976). Matched filters. *International Specialist Seminar on the Impact of New Technologies in Signal Processing* (IEE, London), pp. 78–83.

Haruta, K., Lloyd, P., and Hokanson, J. L. (1969). Monolithic crystal filter. II. Normal mode frequencies and displacements. *IEEE Trans. Sonics Ultrason.* **SU-16**, 21.

Herzig, P. A., and Swanson, T. W. (1978a). A polylithic crystal filter for a satellite channel application. *Proc. IEEE Int. Symp. Circuits Syst.*, pp. 64–65.

Herzig, P. A., and Swanson, T. W. (1978b). A polylithic crystal filter employing a Rhodes transfer function. *Proc. 32nd Annu. Freq. Control Symp.*, pp. 233–242.

Herzog, W. (1968). Narrow-band quartz crystal filters with losses. *Proc. Inst. Radio Electron. Eng. Aust.* **29**, 18–24.

Hirst, J., and Vlach, Z. (1969). Frequency characteristics analysis of directly connected filters with piezoelectric elements using digital computers. *Elektrotech. Casopis* **20**, 59–69. In Czech.

Hokanson, J. L. (1969a). Laser machining thin film electrode arrays on quartz crystal substrates. *Proc. 23rd Annu. Freq. Control Symp.*, pp. 163–170.

Hokanson, J. L. (1969b). The monolithic crystal filter: The device, its operation and choice of piezoelectric materials. *Proc. 6th Annu. Integrated Circuits Sem.*, pp. 32–43.

Holt, A. G. J., and Gray, R. L. (1968). Bandpass crystal filters by transformation of a low-pass filter. *IEEE Trans. Circuit Theory*, **CT-15**, 492–494.

Horton, W. H., and Smythe, R. C. (1967). Theory of thickness-shear vibrations, with extensions and applications to VHF acoustically coupled-resonator filters. *Proc. 21st Annu. Freq. Control Symp.*, pp. 160–178.

Horton, W. H., and Smythe, R. C. (1973). Experimental investigation of intermodulation in monolithic crystal filters. *Proc. 27th Annu. Freq. Control Symp.*, pp. 243–245.

Horwood, P. J. (1972). A 5.2 MHz crystal filter for SSB. *Radio Commun.* **48**, 366.

Hribsek, M. F. (1978). High-Q selective filters using mechanical resonance of silicon beams. *IEEE Trans. Circuits Syst.* **CAS-25**, 215–222.

Hribsek, M. F. (1979). The design and application of electromechanical single silicon beam filters. *Proc. 33rd Annu. Freq. Control Symp.*, pp. 173–175.

Izumi, H., Tomikawa, Y., and Konno, M. (1969). Double resonance disk or ring of the contour ((2, 2)) mode and its application to the mechanical filter. *Bull. Yamagata Univ.* **10**, 305–315. In Japanese.

Jain, J. D., and Walther, L. (1970). Applications of piezoelectric, ceramic, torsional resonators in filters. *Hochfrequenztech. Elekroakust.* **79**, 128–132. In German.

Jennings, J. D. (1976). Fine tuning monolithic crystal filters. *Tech. Dig.* (42), 23–24.

Jennings, J. D., and Perri, P. P. (1971). A monolithic channel filter manufacture with a new technology. *Proc. 21st IEEE Electron. Comp. Conf.*, pp. 365–373.

Johnson, R. A. (1966). A twin tee multimode mechanical filter. *Proc. IEEE* **54**, 1961–1962.

Johnson, R. A. (1968). Electrical circuit models of disk-wire mechanical filters. *IEEE Trans. Sonics Ultrason.* **SU-15**, 41–50.

Johnson, R. A. (1970). New single sideband mechanical filters. IEEE WESCON technical papers, (Los Angeles, CA: WESCON), **14**, 10 pp.

Johnson, R. A. (1972). Mechanical filters. *IEEE Trans. Sonics Ultrason.* **SU-19**, 410.

Johnson, R. A. (1973). Mechanical filters. *Proc. IEEE Int. Symp. Circuit Theory*, pp. 402–405.

Johnson, R. A. (1975). The design of mechanical filters with bridged resonators. *Proc. IEEE Int. Symp. Circuits Syst.*, pp. 313–316.

Johnson, R. A. (1976). The design and manufacture of mechanical filters. *Proc. IEEE Int. Symp. Circuits Syst.*, pp. 750–753.

Johnson, R. A. (1977). Mechanical filters take on selective jobs. *Electronics* **50**, 81–85.

Johnson, R. A. (1978). Mechanical filters using disk and bar flexure-mode resonators. *Onde Electr.* **58**, 141–148. In French.
Johnson, R. A. (1983). "Mechanical Filters in Electronics." Wiley, New York.
Johnson, R. A., and Fanthorpe, F. L. (1979). Mechanical filters for single-sideband applications. *Proc. IEEE Int. Symp. Circuits Syst.*, pp. 896–899.
Johnson, R. A. and Guenther, A. E. (1974). Filters and resonators—a review. III. Mechanical filters and resonators. *IEEE Trans. Sonics. Ultrason.* **SU-21**, 244–256.
Johnson, R. A., and Winget, W. A. (1974). FDM equipment using mechanical filters. *Proc. IEEE Int. Symp. Circuits Syst.*, pp. 127–131.
Johnson, R. A., and Yakuwa, K. (1978). Miniaturized mechanical filters. *Proc. IEEE Int. Symp. Circuits Syst.*, pp. 330–335.
Johnson, R. A., Borner, M., and Konno, M. (1971). Mechanical filters—a review of progress. *IEEE Trans. Sonics Ultrason.* **18**, 153–168.
Jumonji, H., Watanabe, N., and Tsukamoto, K. (1975). Design of high performance monolithic crystal filters. *Rev. Electr. Commun. Lab.* **23**, 439–452.
Jungwirt, J. (1974). The design of electro-mechanical filters for telecommunications. *Tesla Electron.* **7**, 3–9.
Jutzi, W. (1970). Active tapped delay line filter. *IBM Tech. Disc. Bull.* **12**, 1359–1360.
Kagawa, Y. (1971). Analysis and design of electromechanical filters by finite element method. *J. Acoust. Soc. Jpn.* **27**, 201–214. In Japanese.
Kagawa, Y., and Yamabuchi, T. (1974). Finite element stimulation of two-dimensional electromechanical resonators. *IEEE Trans. Sonics Ultrason.* **SU-21**, 275–283.
Kallman, H. E. (1940). Transversal filters. *Proc. IRE* **28**, 302–310.
Kaminski, F. (1969). The synthesis of electromechanical chain filters by means of equivalent circuits with distributed constants. *Arch. Elekrotech.* **17**, 449–467. In Polish.
Kaminski, F. (1969). The synthesis of electromechnical chain filters by means of equivalent circuits with distributed parameters. *Rozpr. Elektrotech.* **15**, 717–749. In Polish.
Kaminski, F. (1969). On the synthesis of supernarrow band-pass electromechanical chain filters with regular structure. *Bull. Acad. Polon. Sci. Ser. Sci. Tech.* **17**, 301–104. In Russian.
Kaminski, F. (1973). On synthesis of a certain group of electromechanical filters by means of equivalent networks with distributed or lumped constants. *Arch. Elekrotech.* **22**, 517–521. In Polish.
Kaminski, F. (1973). Synthesis of an electromechanical filter with simple couplers. *Arch. Elektrotech.* **22**, 525–537. In Polish.
Kaminski, F. (1974). Outline of the theory of the regular-structure mechanical and microwave filter synthesis. *Rozpr. Electrotech.* **20**, 295–312. In Polish.
Kaminski, F. (1979). On the possibility of semimonolithic crystal filter synthesis by algebraic method employing equivalent networks with lumped constants. *Pr. Inst. Tele- Radiotech.* (72), 29–38. In Polish.
Kampfhenkel, H. (1973). Iterative synthesis of symmetrical electromechanical filters. *Nachrichtentech. Z.* **26**, 401–107. In German.
Kantor, V. M. (1973). The design of piezoelectric band-elimination filters with given working parameters. *Telecomm. Radio Eng. Pt. 2* **27**, 88–95.
Kantor, V. M., and Lanne, A. A. (1967). The synthesis of low- and high-pass piezoelectric filters from their effective parameters. *Telecomm. Radio Eng. Pt. 1* (4), 6–14.
Kantor, V. M., and Lanne, A. A. (1973). Bandpass limitations of electromechanical filters. *Izv. Vyssh. Vchebn. Zaved. Radioelektron.* **16**, 66–72. In Russian.
Kawana, T., and Kawahata, H. (1979). Preshift mechanical filter for voice frequency telegraph transmission system. *Ultrason. Symp. Proc.*, pp. 119–122.

Kerboull, J. (1978). 2.5 MHz crystal filters for a 12 channel multiplex telphony system. *Onde Electr.* **58**, 458–463. In French.

Kidokoro, M., and Kawana, T. (1977). Mechanical filters take over. *JEE J. Electron. Eng.* (130), 48–52.

Kinsman, B., and D'Alexander, F. (1981). Monolithic crystal bandpass filter. *Motorola Tech. Disc. Bull.* **1**, 6–7.

Kobayashi, M. (1981). Design method of monolithic crystal filters having phase inverting auxiliary electrode. *Trans. Inst. Electron. Commun. Eng. Jpn. Sect. E* **64**, 366.

Kogan, S. S., and Stepanov, A. S. (1971). Electromechanical channel filters. *Telecomm. Radio Eng. Pt. 1* **25**, 44–50.

Koh, Y. (1973). The mechanical filter: evolution to technical maturity. *JEE J. Electron. Eng.* (79), 32–37.

Kohlbacher, G. (1971). Use of electrical low-pass equivalent circuits for the design of multiple-tuned crystal and ceramic filters from monolithic single-filter elements. *Arch. Elek. Ubertrag.* **25**, 492–501. In German.

Kohlbacher, G. (1972). Monolithic quartz and ceramic filters. *Int. Elektron. Rundsch.* **26**, 203–209. In German.

Kohlbacher, G. R. (1972). The design of compact monolithic crystal filters for portable telecommunications equipment. *Proc. 26th Annu. Freq. Control Symp.*, pp. 187–192.

Kohlbacher, G. (1976). Monolithic crystal filters for the VHF range. *Wiss. Ber. AEG-Telefunken* **49**, 248–253. In German.

Kohlhammer, B., and Schuessler, H. (1971). A mechanical torsion filter with piezoelectric transducers as channel and signal filters for a new carrier-frequency system. *Proc. 7th Int. Cong. Acoust.*, pp. 341–344. In German.

Kohlhammer, B., and Schuessler, H. (1971). Remarks on the influence of the mechanical vibration quality on the transmission properties of channel filters. *Frequenz* **25**, 287–288.

Kojima, H., and Sawamoto, K. (1981). A stable quartz crystal pilot filter. *Electr. Commun. Lab. Tech. J.* **30**, 115–123. In Japanese,

Kolb, E. D. (1979). Solubility, crystal growth & perfection of aluminum orthophosphate. *Proc. 33rd Annu. Freq. Control Symp.*, pp. 88–97.

Kollmann, M. (1969). Wide-band crystal filters for higher frequency bands. *Tesla Electron.* **2**, 12–16.

Kollmann, M. (1977). A wide-band crystal filter. *Sdelovaci Tech.* **25**, 99–103. In Czech.

Komori, N. (1977). Crystal filters-designs enhanced to meet expanding requirements. *JEE J. Electron. Eng.* (130), 57–60.

Konno, M., and Tomikawa, Y. (1967). An electro-mechanical filter consisting of a flexural vibrator with double resonances. *Electron. Commun. Jpn.* **50**, 64–73.

Konno, M., and Tomikawa, Y. (1969). Electro-mechanical filters. I. Introduction. *Denshi Tsushin Gakkai Zasshi* **52**, 303–312. In Japanese.

Konno, M., Aoshima, K., and Nakamura, H. (1969). H-shape resonator and its application to electro-mechanical filter. *Bull. Yamagata Univ. (Eng.)* **10**, 261–285. In Japanese.

Konno, M., Tomikawa, Y., Tacano, T., and Izumi, H. (1969). Electromechanical filters using degeneration modes of a disk or a ring. *Trans. Inst. Electron. Commun. Eng. Jpn.*, pp. 19–28. In Japanese.

Konno, M., Yakuwa, K., Yano, T., and Koh, Y. (1978). Electromechanical filters developed in Japan. Part I: General. *Onde Electr.* **58**, 401–408.

Konno, M., Sugawara, S., Tomikawa, Y., and Johnson, R. A. (1979). Mounted free-free flexural bar resonator and mechanical filter. *Proc. Int. Symp. Circuits Syst.*, pp. 1068–1071.

Kopp, H. (1971). A mechanical filter channel bank. *Proc. 7th Int. Conf. Commun.* (New York: IEEE), pp. 6–24.

Kosowsky, D. I. (1955). Synthesis & realization of crystal filters. *Tech. Rep. Res. Lab Electron.*, p. 298.

Kosowsky, D. I. (1958). High-frequency crystal filter design techniques and applications. *Proc. IRE* **46**, 419–429.

Kostarev, V. Y., and Osipov, V. G. (1971). Synthesis of bandpass piezoelectric filters with an attenuation characteristic which is monotonic in the passband and extremal in the stop band. *Telecommun. Radio Eng. Pt. 1* **25**, 61–63.

Krause, G. (1970). A coilless band-pass half-section for realizing quartz poles in the lower cut-off range. *Nachrichtentechnik* **20**, 30–33. In German.

Krause, G. (1971). Design of wide-band pass filters with unsymmetrical damping using resonant crystals, according to the theory of m-chains. II. *Fernmeldetechnik* **11**, 141–145. In German.

Krause, G. (1971). Design of wideband crystal bandpass filters with asymmetrical attenuation on the basis of the theory of recurrent m-sections. *Fernmeldetechnik* **11**, 113–115. In German.

Krause, G. (1971). Matching problems in hand pass crystal filters with sharp cutoffs calculated by the theory of wave parameters. *Nachrichtentechnik* (Germany), **21**, 130–133. In German.

Kunemund, F. (1972). Materials for high-grade electromechanical frequency filters. *Inst. Eng. Aust. Elec. Eng. Trans.* **EE8**, 41–42.

Kunemund, F. (1972). Channel filters with longitudinally coupled flexural mode resonators. *Siemens Forsch.- Entwicklungsben.* **1**, 325–328.

Kunemund, F. L. (1975). Electromechanical filters. *NTG-Fachber.* **54**, 159–170. In German.

Kurth, C. F. (1974). Filter applications in communications and electronics industry. *Proc. 28th Annu. Freq. Control Symp.*, pp. 33–42.

Lajacono, R. (1980). Simplified method for synthesising m.q.f. filters. *IEE Proc. G.* **127**, 57–60.

Lakin, K. M. (1981). Equivalent circuit modeling of stacked crystal filters. *Proc. 35th Annu. Freq. Control Symp.*, pp. 257–262.

Lampe, L. (1977). A new monolithic 10.7 MHz piezo-filter. *Radio Fernsehen Elektron.* **26**, 27–28. In German.

Lançon, R. (1973). Mechanical resonances of asymmetric monolithic structures. *Ann. Telecomm.* **28**, 406–412.

Lane, C. E. (1938). Crystal channel filters for the cable carrier systems. *Bell Syst. Tech. J.* **17**, 125–136.

Lass, M. (1979). The subsequent incorporation of a monolithic quartz filter. *Funkschau* **51**, 1511–1512. In German.

Lechner, D. (1981). Ladder crystal filters for the amateur. *Elektron. Int.* (5), 117–119. In German.

Lee, M. S. (1974). Equivalent network for bridged crystal filters. *Electron. Lett.* **10**, 507–508.

Lee, M. S. (1975). Polylithic crystal filters with loss poles at finite frequencies. *IEEE Int. Symp. Circuits Syst.*, pp. 297–300.

Lefevre, R. (1978). Application of monolithic quartz crystal to the telephone voice channel filter in analog systems. *Onde Electr.* **58**, 475–481. In French.

Lefevre, R. (1979). Monolithic crystal filters with very high Q factor and low spurious level. *Proc. 33rd Annu. Freq. Control Symp.*, pp. 148–158.

Lefevre, R. (1979). State of the art and new developments in piezoelectric monolithic filters. *Rev. HF* **11** (1–2). In French.

Lefevre, R. (1981). Laser processed VHF monolithic crystal filters with one plate integrated matching impedances. *Proc. 35th Annu. Freq. Control Symp.*, pp. 244–249.

Lin, C. C. (1976). Design of symmetrical polylithic crystal filters. *Electron. Lett.* **12**, 202–204.

Lloyd, P. (1971). Monolithic crystal filters for frequency division multiplex. *Proc. 25th Annu. Freq. Control Symp.*, pp. 280, 286.
Lloyd, P. (1971). Monolithic crystal filters. *Proc. 7th Int. Cong. Acoust.*, pp. 309–312.
Lloyd, P., and Haruta, K. (1969). Monolithic crystal filter. I. The theoretical model. *IEEE Trans. Sonics Ultrason.* **SU-16**, 21.
Lojacono, R. (1981). An attempt to improve the analysis and the synthesis of MQF [monolithic quartz filter]. *Journees d'Electronique (Lausanne, Switzerland)*, pp. 135–144.
Lu, S. K. S. (1971). The applications of quartz crystals to discriminators, filters and monolithic filters. *Aust. Electron. Eng.* **4**, 21–24.
Lu, S. K. S. (1977). Comments on 'cascade synthesis of polylithic crystal filters containing double-resonator monolithic crystal filter (MCF) elements.' *IEEE Trans. Circuits Syst.*, **CAS-24**, 274–275.
Lu, S. K. S. (1978). Cascade synthesis of monolithic crystal filters with transmission zeros at finite frequencies. *Electron. Lett.* **14**, 45–46.
Lu, S. K. S. (1979). Cascade synthesis of single-sideband monolithic crystal filters. *IEEE Trans. Circuits Syst.* **CAS-26**, 890–892.
Lueder, E. (1968). Mechanical oscillators in filters for low frequencies. *Arch. Electrotech.* **51**, 351–357. In German.
Lukaszek, T. J. (1971). Mode control and related studies of VHF quartz filter crystals. *IEEE Trans. Sonics Ultrason.* **SU-18**, 238–246.
Lukaszek, T., and Ballato, A. (1980). What SAW can learn from BAW: Implications for future frequency control, selection & signal processing. *Ultrason. Symp. Proc.*, pp. 173–183.
Mailer, H., and Beurle, D. R. (1966). Incorporation of multi-resonator crystals into filters for quantity production. *Proc. 20th Annu. Freq. Control Symp.*, pp. 309–342.
Malinowski, S., and Smith, C. (1972). Intermodulation in crystal filters. *Proc. 26th Annu. Freq. Control Symp.*, pp. 180–186.
Markvoort, J. A. (1974). On the modeling of monolithic quartz crystal filters. *Appl. Sci. Res.* **29**, 361–379.
Martinelli, G., Salerno, M., Masiani, G., and Orlandi, G. (1978). Constrained synthesis for mechanical filters. *Proc. IEEE Int. Symp. Circuits Syst.*, pp. 59–63.
Mason, W. P. (1969a). Equivalent electromechanical representation of trapped energy transducers. *Proc. IEEE* **53**, 1723–1734.
Mason, W. P. (1969b). Constants of a trapped-energy electromechanical transducer made by evaporating a thin layer of a piezoelectric crystal on each side of a quartz plate. *J. Acoust. Soc. Am.* **46**, 687–692.
Mason, W. P., and Thurston, R. N., eds. (1972). "Physical Acoustics," vol. IX. Academic Press, New York.
Masuda, Y., Kawakami, I., and Kobayashi, M. (1973). Monolithic crystal filter with attenuation poles ultilizing 2-dimensional arrangement of electrode. *Proc. 27th Annu. Freq. Control Symp.*, pp. 227–232.
Masuda, Y., Kawakami, I., Kobayashi, M., and Gunj, K. (1974a). Monolithic crystal filters having attenuation poles. *Electron. Commun.* **57**, 8–16. In Japanese.
Masuda, Y., Kawakami, I., and Kobayashi, M. (1974b). Monolithic crystal filters. *Oki Rev.* **38**, 39–44. In Japanese.
Matthes, H. (1976). Crystal band-stop filters with improved spurious resonance behaviour. *J. Circuit Theory Appl.* **4**, 25–42.
Matthews, H., ed. (1977). "Surface Wave Filters, Design, Construction, and Use." Wiley, New York.
Mavrov, N. P. (1978). Problems of optimum design of quartz filter resonators with energy trapping. *Bulg. J. Phys.* **4**, 516–522.

McLean, D. I. (1967a). Physical realization of miniature bandpass filters with single sideband characteristics. *IEEE Trans. Circuit Theory* **CT-14**, 138–159.

McLean, D. I. (1967b). Physical realization of miniature bandpass filters with single frequency or single sideband characteristics. *Proc. 21st Annu. Freq. Control Symp.*, pp. 138–159.

McLean, D. I., Graziani, A. F., and Royer, J. J. (1979). New discrete crystal filters for Bell System analog channel banks. *Proc. 33rd Annu. Freq. Control Symp.*, pp. 166–172.

Means, D. R., and Ghausi, M. S. (1972). Inductorless filter design using active elements and piezoelectric resonators (ceramic resonators, quartz resonators). *IEEE Trans. Circuit Theory* **CT-19**, 247–252.

Metel'kova, T. F., and Stein, M. I. (1976). A quartz-crystal through-connection filter for master groups. *Telecommun. Radio Eng. Pt. 1* **30**, 12–16.

Mifune, H., and Tanaka, T. (1970). Reed filters and their characteristics. *Nat. Tech. Rep.* **16**, 541–550. In Japanese.

Miles, R. H. A., and Cooper, R. G. (1970). Monolithic dual quartz crystal filters. *Electron. Compon.* **11**, 84–85.

Miller, A. J. (1970). Preparation of quartz crystal plates for monolithic crystal filters. *Proc. 24th Annu. Freq. Control Symp.*, pp. 93–103.

Mindlin, R. D. (1966). Studies in the mathematical theory of vibrations of crystal plates. *Proc. 20th Annu. Freq. Control Symp.*, pp. 252–265.

Mindlin, R. D., and Lee, P. Y. (1966). Thickness-shear and flexural vibrations of partially plated crystal plates. *Int. J. Solids Struct.* **2**, 125–139.

Morse, W. C., and Rennick, R. C. (1972). Adjusting frequency of monolithic crystal filters with an automatic vapor plater. *J. Vac. Sci. Technol.* **9**, 28–32.

Mortley, W. S. (1951). FMQ. *Wireless World* **57**, 399–403.

Mortley, W. S. (1957). Frequency-modulated quartz oscillators for broadcast equipment. *Proc. IEE* **104**, Pt. B, 239–249.

Mortley, W. S. (1971). Wave propagation in alpha-quartz dispersive filters. *Marconi Rev.* **34**, 173–206.

Muir, A. J. L. (1973). Monolithic crystal filters. *Syst. Technol.* (17), 16–23.

Muller, O. (1976). Monolithic filters on single chips or analog sampling filters with charge transfer elements (Q-filters). *Elektroniker* **15**, EL6–11. In German.

Nadoliiski, M. M., and Velichkov, B. G. (1980). Equivalent electric scheme of a two-layer piezoelectric low-frequency transformer filter. *Bulg. J. Phys.* **7**, 307–314.

Nakamura, K., and Shimizu, H. (1979). Analysis of two-dimensional energy trapping in monolithic crystal filters. *Trans. Inst. Electron. Commun. Eng. Jpn Sect. E.* **62**, 403–404.

Nakamura, H., Konno, M., and Tanno, K. (1969). Electro-mechanical filter consisting of frame-vibrator with double resonance. *Bull. Yamagata Univ.* **10**, 431–445. In Japanese.

Nakazawa, Y. (1962). High frequency crystal electromechanical filter. *Proc. 16th Annu. Freq. Control Symp.*, pp. 373–390.

Nathanson, H. C., Newell, W. E., Wickstrom, R. A., and Davis, J. R., Jr. (1967). The resonant gate transistor. *IEEE Trans. Electron Devices* **ED-14**, 117–133.

Neubig, B. (1978). Monolithic quartz filters. *Funkschau* **50**, 438–441. In German.

Nonaka, S., Yuuki, T., and Hara, K. (1971). The current dependency of crystal unit resistance at low drive level. *Proc. 25th Annu. Freq. Control Symp.*, pp. 139–147.

O'Clock, G. D., Jr. (1977). Matched filters boost receiver gain by improving the signal-to-noise ratio. *Electron. Des.* **25**, 162–166.

O'Neill, J. F., and Ghausi, M. S. (1967). Design of frequency selective networks using only resonators. *Conference Record Tenth Midwest Symp. of Circuit Theory, Lafayette (New York: IEEE)*, Paper No. X-5.

Ohyama, M., Uehara, K., Yoshida, N., and Yano, T. (1978). N-5000 series FDM channel bank using new mechanical filters. *NEC Res. Dev.* (49), 98–109.
Okuno, K., and Watanabe, T. (1976). A hybrid integrated monolithic crystal filter. *Proc. 30th Annu. Freq. Control Symp.*, pp. 109–118.
Olster, S. H., Oak, I. R., Pearman, G. T., Rennick, R. C., and Meeker, T. R. (1975). A6 monolithic crystal filter design for manufacture and device quality. *Proc. 29th Annu. Freq. Control Symp.*, pp. 105–112.
Ono, M., Tanji, S., and Tominaga, H. (1980). Integrated monolithic crystal filter for citizen band transceivers. *Electrocompon. Sci. Technol.* **8**, 53–59.
Onoe, M. (1973). Low-frequency resonators of lithium tantalate. *Proc. 27th Annu. Freq. Control Symp.*, pp. 42–50.
Onoe, M. (1979. Crystal, ceramic, and mechanical filters in Japan. *Proc. IEEE* **67**, 75–102.
Onoe, M., and Jumonji, H. (1965). Analysis of piezoelectric resonators vibrating in trapped-energy mode. *Electron. Commun.* **48**, 84–93. In Japanese.
Onoe, M. and Spassov, L. (1974). An experiment of two-dimensional monolithic crystal filters. *C.R. Acad. Bulg. Sci.* **27**, 465–468.
Onoe, M., and Yano, T. (1970). Electromechanical wave-separating filters. *Proc. Electron. Compon. Conf. (New York: IEEE)*, pp. 269–276.
Onoe, M., Jumonji, H., and Kobori, N. (1966). High frequency crystal filters employing multiple mode resonators vibrating in trapped energy modes. *Proc. 20th Annu. Freq. Control Symp.*, pp. 266–287.
Orsucci, M., Reggiani, M., and Stacchiotti, G. (1976). Synthesis of bridged polylithic filters. *Alta Freq.* **45**, 747–751.
Oshima, Y. (1971). Quartz crystal filters for export. *Meidensha Rev. Int. Ed.* **37**, 1–9.
Owens, J. M., Smith, C. V., Jr., and Collins, J. H. (1978). Magnetostatic wave bandpass filters and resonators. *Proc. IEEE Int. Symp. Circuits, Syst.*, pp. 563–568.
Ozimek, E. J., and Chai, B. H.-T. (1979). Piezoelectric properties of single-crystal berlinite. *Proc. 33rd Annu. Freq. Control Symp.*, pp. 80–87.
Pang, C. S., Falco, C. M., Kampwirth, R. T., Schuller, I. K., Hudak, J. J., and Anastasio, T. A. (1979). A superconducting RF notch filter. *Proc. Adv. Cryogenic Eng. (New York: Plenum)*, pp. 244–250.
Papadakis, E. P. (1975). Improvements in a broadband electromechanical bandpass filter in the voice band. *IEEE Trans. Sonics Ultrason.* **SU-22**, 406–415.
Peach, R. C., Dyer, A. J., Byrne, A. J., Read, E., and Stevenson, J. K. (1982). Intermediate bandwidth quartz crystal filters—a simple approach. *Proc. 36th Annu. Freq. Control Symp.*, pp. 389–395.
Pearman, G. T., and Rennick, R. C. (1974). Filters and resonators—a review. II. Monolithic crystal filters. *IEEE Trans. Sonics Ultrason.* **SU-21**, 238–243.
Pearman, G. T., and Rennick, R. C. (1977). Unwanted modes in monolithic crystal filters. *Proc. 31st Annu. Freq. Control Symp.*, pp. 191–196.
Pfleiderer, R., and Wollmershauser, P. (1976). Electromechanical pilot filter with improved temperature characteristic. *Proc. IEEE Int. Symp. Circuits Syst.*, pp. 743–745.
Planat, M., Theobald, G., Gagnepain, J.-J., and Siffert, P. (1980). Intermodulation in X-cut lithium tantalate resonators. *Electron. Lett.* **16**, 174–175.
Pochet, J. (1977). Crystal ladder filters. *Wireless World* **83**, 62–63.
Pond, C. W. (1970). Phased matched crystal filters. IEEE WESCON technical papers, (Los Angeles: Wescon), **14**, 6 pp.
Prache, P. M. (1974). Introduction to linear electromechanical networks theory. *Cables Transm.* **28**, 304–327. In French.

Przesmyski, O. (1971). Application of narrow band-stop filters with quartz crystal resonators. *Elektronika* (6), 252-253. In Polish.
Przesmycki, O. (1975). A novel type of a stop-band filter with piezoelectric resonators. *Pr. Inst. Tele Radiotech.* **19**, 51-62. In Polish.
Psenicka, B., and Trnka, J. (1974). Electromechanical filters with poles of attenuation at finity frequencies. *Proc. 5th Colloq. Microwave Commun.* **2**, CT-23/203-12.
Pshenichka, V., and Trnka, I. (1975). Design of electromechanical filters with polar coupling. *Izv. Vyssh. Vchebn. Zaved. Radioelektron.* **18**, 21-26. In Russian.
Reilly, N. H. C., and Redwood, M. (1969). Wave-propagation analysis of the monolithic-crystal filter. *Proc. Inst. Elect. Eng.* **116**, 653-660.
Rennick, R. C. (1973). An equivalent circuit approach to the design and analysis of monolithic crystal filters. *IEEE Trans. Sonics Ultrason.* **SU-20**, 347-354.
Rennick, R. C. (1975). Modelling and tuning methods for monolithic crystal filters. *IEEE Int. Symp. Circuits Syst.*, pp. 309-312.
Reushkin, N. A. (1976). Method of tuning a multiresonator filter. *Telecomm. Radio Eng. Pt. 1* **30**, 25-28.
Revankar, G. N., and Bapat, V. B. (1976). High-Q active crystal bandpass filter. *Stud. J. Inst. Electron. Telecommun. Eng.* **17**, 197-198.
Rhodes, J. D. (1970). A low-pass prototype network for microwave linear-phase filters. *IEEE Trans. Microwave Theory Tech.* **MTT-18**, 290-301.
Rider, L. S. (1970). Microelectronics techniques study. Final Development Rep. On Contract N00039-68-C-2575, General Electric Co., Syracuse.
Rienecker, W. (1980). Mechanical frequency selection in telecommunications. *Funkschau* **52**, 55-58. In German.
Rienecker, W. (1980). Mechanical frequency selection in communication engineering. II. *Funkschau* **52**, 63-65. In German.
Rienecker, W. (1980). Mechanical frequency selection in communications technology. *Antenna* **52**, 356-363. In Italian.
Roberts, D. A. (1971). CdS-quartz monolithic filters for the 100-500 MHz frequency range. *Proc. 25th Annu. Freq. Control Symp.*, pp. 251-261.
Rogozin, Y. I. (1979). Derivation of the acoustic coupling factor in double-resonator monolithic quartz filter structures. *Telecommun. Radio Eng. Pt. 2* **34**, 115-117.
Rupe, D. E., Syler, R. L., and Weber, R. M. (1981). Acoustic filters for impact printers. *IBM Tech. Disc. Bull.* **23**, 3524-3525.
Sandtner, J. (1981). A two-resonator gas-coupled electromechanical filter. *Journees d'Electronique 1981*, pp. 161-167. In French.
Sasaki, E., and Shinozaki, K. (1974). Torsional mode transducer for mechanical filter. *Electr. Commun. Lab. Tech. J.* **23**, 111-124. In Japanese.
Sasaki, E., and Tsukamoto, K. (1974). Resonant Q of quartz crystal resonator for monolithic crystal filter. *Electr. Commun. Lab. Tech. J.* **23**, 67-78. In Japanese.
Sasaki, R., Nagata, T., and Matsushita, S. (1971). Piezoelectric resonators as a solution to frequency selectivity problems in color TV receivers. *IEEE Trans. Broadcast Telev. Receivers* **BTR-17**, 195-201.
Sato, N., Yakuwa, K., Kazama, K., and Fujisaki, M. (1978). A pole-type mechanical filter channel bank. *NTC, Conference Record of the IEEE National Telecommunications Conference*, 30.4/1-5.
Sauerland, F. L. (1969). Design of piezoelectric ladder filters. *Int. Symp. Circuit Theory Dig.* (*New York: IEEE*), p. 9.
Sawamoto, K. (1971). Energy trapping in a lithium tantalate x-cut resonator. *Proc. 25th Annu. Freq. Control Symp.*, pp. 246-250.

Sawamoto, K., and Kondo, S. (1974). Torsional mode mechanical filters. *Electr. Commun. Lab. Tech. J.* **23**, 95–110. In Japanese.

Sawamoto, K., and Niizeki, N. (1970). Zero temperature coefficient of resonant frequency in $LiTaO_3$ length expander bars. *Proc. IEEE* **58**, 1289–1290.

Sawamoto, K., Kondo. S., and Sasaki, E. (1975). A torsional mode mechanical channel filter. *Rev. Electr. Commun. Lab.* **23**, 429–438.

Sawamoto, K., Sasaki, E., Kondo, S., Ashida, T., and Shinozaki, K. (1975). Torsional-mode mechanical filters. *Electron. Commun. Jpn.* **57**, 17–24.

Sawamoto, K., Watanabe, N., and Tsukamoto, K. (1978). A torsional mode even elements pole-type mechanical channel filter. *Electr. Commun. Lab. Tech. J.* **27**, 1689–1702. In Japanese.

Sawamoto, K., Kondo, S., Watanabe, N., Taukamoto, K., Kiyomoto, M., and Ibaraki, O. (1976). A torsional-mode pole-type mechanical channel filter. *IEEE Trans. Sonics Ultrason.* **SU-23**, 148–153.

Sawamoto, K., Watanabe, N., and Tsukamoto, K. (1978). A torsional mode 10-element pole-type mechanical channel filter. *Trans. Inst. Electron. Commun. Eng. Jpn. Sect. E.* **61**, 642–643.

Sawamoto, K.-I., Yano, T., Yakuwa, K., Koh, Y., and Konno, M. (1978). Electromechanical filters developed in Japan. II. Channel EM filters. *Onde Electr.* **58**, 482–487. In French.

Sawamoto, K., Kojima, H., Hirama, K., and Kameyama, K. (1981). A stable narrow band-pass quartz crystal filter at 150 MHz region. *Trans. Inst. Electron. Commun. Jpn. Sect. E.* **64**, 618.

Schaeffler, K. (1967). The piezo-electric tuning fork as a new filter element for the audio-frequency. *BBC Nachr.* **49**, 395–401. In German.

Schneibner, J. (1969). Quartz resonators and piezoceramic vibrators in bandpass and bandstop filters for data technology. II. *Wiss Z. Elektrotech.* **13**, 181–192. In German.

Schrenckenbach, W. (1979). Design and materials requirements for monolithic multiple-electrode crystal filters, using ceramics, with special reference to filters for 10.7 MHz. *Hermsdorfer Tech. Mitt.* **19**, 1701–1704. In German.

Schuessler, H. (1969). Filters with mechanical resonators. *Bull. Assoc. Suisse Elect.* **60**, 216–221. In German.

Schuessler, H. (1971). Consideration about channel filters for a new carrier frequency system with mechanical filters. *Proc. 25th Annu. Freq. Control Symp.*, pp. 262–270.

Schuessler, H. (1974). Filters for channel band filtering with mechanical resonators, quartz crystals and gyrators. *Proc. IEEE Int. Symp. Circuits Syst.*, pp. 106–110.

Schultz, J. J. (1978). Economical diode-switched crystal filters. *CQ Radio Amat. J.* **34**, 33–35, 91.

Seed, A., Wood, A. F. B., and Goodwin, M. W. (1970). Monolithic filters for mobile radio applications. *Proc. Int. Conf. Adv. Microelectron.*, 2 pp.

Sekine, T., and Konno, M. (1976). Narrow band mechanical filter consisting of differential-couplers. *Trans. Inst. Electron. Commun. Eng. Jpn. Sect. E.* **59**, 28.

Sekine, T., Konno, M., and Sugawara, S. (1978). A mechanical filter with two attenuation poles consisting of three resonators. *Trans. Inst. Electron. Commun. Eng. Jpn. Sect. E.* **61**, 549.

Shchebetun, P. D., Sivkov, B. V., and Pashkovskii, N. A. (1975). Measuring the secondary resonances of quartz filter resonators. *Meas. Tech.* **18**, 122–124.

Sheahan, D. F. (1971). Single-sideband filters for short haul systems. *Proc. Mexico Int'l. IEEE Conf. on Systems, Networks and Computers*, pp. 744–748.

Sheahan, D. F. (1973). Crystal filters. *Proc. Int. Symp. Circuit Theory* (*New York: IEEE*), pp. 394–397.

Sheahan, D. F. (1975). Polylithic crystal filters. *Proc. 29th Annu. Freq. Control Symp.*, pp. 120–127.

Sheahan, D. F., and Johnson, R. A. (1975). Crystal and mechanical filters. *IEEE Trans. Circuits Syst.* **CAS-22**, 69–89.

Sheahan, D. F., and Johnson, R. A., eds. (1977). "Modern Crystal and Mechanical Filters." IEEE Press, New York.

Sheahan, D. F., and Schmidt, C. E. (1971). Coupled resonator quartz crystal filters. *Papers presented at the Western Electronic show and convention.* Western Periodicals, N. Hollywood. 8/3 6 pp.

Sherman, J. H., Jr. (1976). Dimensioning rectangular electrodes and arrays of electrodes on AT-cut quartz bodies. *Proc. 30th Annu. Freq. Control Symp.*, pp. 54–64.

Shibayama, K., and Sato, H. (1971). Bar-shaped mechanical filters under the consideration of higher mode effects. *Proc. 7th Int. Cong. Acoust.*, pp. 349–352.

Shockley, W., Curran, D. R., and Koneval, D. J. (1963). Energy trapping and related studies of multiple electrode filter crystals. *Proc. 17th Annu. Freq. Control Symp.*, pp. 88–126.

Siffert, P. (1981). Private Communication.

Siffert, P., and Kerboull, J. (1978). A selective linear phase crystal filter. *Proc. 32nd Annu. Freq. Control Symp.*, pp. 244–249.

Simmonds, T. H., Jr. (1979). The evolution of the discrete crystal single-sideband selection filter in the Bell System. *Proc. IEEE* **67**, 109–115.

Simpson, H. A., Finch, E. D., Jr., and Weeman, R. K. (1971). Composite filter structures incorporating monolithic crystal filters and L–C networks. *Proc. 25th Annu. Freq. Control Symp.*, pp. 287–296.

Singhi, B. M., and Datta, M. R. (1975). Modified crystal selective network. *J. Inst. Electron. Telecommun. Eng. (India)* **21**, 495–498.

Siwa, M. (1978). Crystalline piezoelectric filters. *Elekronika* **19**, 456–462. In Polish.

Smith, W. L. (1968). The application of piezoelectric coupled-resonator devices to communication systems. *Proc. 22nd Annu. Freq. Control Symp.*, pp. 206–225.

Smythe, R. C. (1969). HF and VHF inductorless filters for microelectronic systems. *Proc. IEEE/EIA Electronic Components Conf.*, pp. 115–119.

Smythe, R. C. (1972). Communications systems benefit from monolithic crystal filters. *Electronics* **45**, 48–51.

Smythe, R. C. (1976). VHF monolithic filters. Rpt. No. ECOM-72-0025-F to U. S. Army Electronics Command, Fort Monmouth, New Jersey.

Smythe, R. C. (1978). Some recent advances in integrated crystal filters. *Proc. 32nd Annu. Freq. Control Symp.*, pp. 220–232.

Smythe, R. C. (1979a). Modern crystal filters. *Proc. 33rd Annu. Freq. Control Symp.*, pp. 209–213.

Smythe, R. C. (1979b). Some recent advances in integrated crystal filters. *Proc. IEEE* **67**, 119–129.

Smythe, R. C., and Howard, M. D. (1983). Current trends in crystal filters. *Proc. 37th Annu. Freq. Control Symp.*, pp. 349–353.

Spasov, L., and Borisov, M. (1978). Studies on monolithic quartz filters. *Bulg. J. Phys.* **4**, 523–532.

Spencer, W. J. (1968). A new functional device—The monolithic crystal filter. *IEEE Trans. Mag.* **MAG-4**, 221.

Stearns, C. M., Wanuga, S., and Tehon, S. W. (1977). Multi-mode stacked crystal filter. *Proc. 31st Annu. Freq. Control Symp.*, pp. 197–206.

Stevenson, J. K. (1974). Synthesis of narrowband cascaded crystal-capacitor lattice filters. *Radio Electron. Eng.* **44**, 326–330.

Stevenson, J. K. (1975). Simple design formulae for third and fourth-order crystal lattice filters with finite attenuation poles. *Int. J. Electron.* **38**, 697–710.

Stevenson, J. K. (1976). Design formulae for multi-lattice crystal filters. *Int. J. Electron.* **41**, 105–123.

Stevenson, J. K. (1976). Transformation for modifying the lumped-element equivalent circuit for metal-encapsulated crystals in unbalanced semilattice filters. *Radio Electron. Eng.* **46**, 614–616.

Stewart, J. A. (1971). System considerations for light-route multiplex using crystal filters. *Proc. 7th Int. Conf. Commun.* (*New York: IEEE*), 6-24–29.

Sugawara, S., and Konno, M. (1978). Spurious responses and their suppression in torsional mode mechanical filter. *Trans. Inst. Electron. Commun. Eng. Jpn. Sec. E.* **61**, 737–738.

Sugawara, S., and Konno, M. (1979). Spurious responses and their suppression in the mechanical filter with flexure mode resonators. *Trans. Inst. Electron. Commun. Eng. Jpn. Sect. E.* **61**, 911–912.

Swanson, T. W. (1978). Crystal filter AM–PM conversion measurements. *Proc. 32nd Annu. Freq. Control Symp.*, pp. 250–254.

Sykes, R. A., Smith, W. L., and Spencer, W. J. (1967). Monolithic crystal filters. *IEEE Int. Conv. Rec. Pt. 11*, pp. 78–93.

Sykes, R. A., and Beaver, W. D. (1966). High frequency monolithic crystal filters with possible application to single frequency and single sideband use. *Proc. 20th Annu. Freq. Control Symp.*, pp. 288–308.

Szentirmai, G. (1968). The synthesis of narrowband crystal band-elimination filters, *IEEE Int. Symp. Circuit Theory Dig.* p. 65.

Szentirmai, G. (1970). Problems of crystal filter design. *IEEE Int. Symp. Circuit Theory Dig.*, p. 108.

Szentirmai, G. (1971). Crystal and ceramic filters. *4th Annual Contemporary Filter Design Seminar* (*Univ. Miami*), 59 pp.

Takahashi, I., Yoshida, N., and Ishizaki, Y. (1976). An analysis of a torsional mode transducer for electromechanical filters. *Ultrason. Symp. Proc.*, pp. 602–605.

Tanno, K., Hacchome, T., and Konno, M. (1971). Band eliminators consisting of electromechanical vibrators. *Bull. Yamagata Univ.* (*Eng.*) **11**, 101–122. In Japanese.

Temes, G. C., Mitra, S. K., Editors (1973). "Modern Filter Theory and Design," chaps. 4 and 5. John Wiley & Sons, New York.

Tenen, O. (1971). Crystal filters—their design and engineering. *Proc. Inst. Radio Electron. Eng. Aust.* **32**, A9.

Tenen, O. (1977). Wide-band crystal filter design. *Telecommun. J. Aust.* **27**, 71–80.

Tiersten, H. F. (1969). Electric field effects in monolithic crystal filters. *Proc. 23rd Annu. Freq. Control Symp.*, pp. 56–64.

Tiersten, H. F. (1974a). Analysis of trapped energy resonators operating in overtones of thickness-shear. *Proc. 28th Annu. Freq. Control Symp.*, pp. 44–48.

Tiersten, H. F. (1974b). Analysis of intermodulation in rotated y-cut quartz thickness-shear resonators. *Proc. 28th Annu. Freq. Control Symp.*, pp. 1–4.

Tiersten, H. F. (1975a). Analysis of trapped-energy resonators operating in overtones of coupled thickness-shear and thickness-twist. *Proc. 29th Annu. Freq. Control Symp.*, pp. 71–75.

Tiersten, H. F. (1975b). Analysis of intermodulation in thickness-shear and trapped energy resonators. *J. Acoust. Soc. Am.* **57**, 667–681.

Tiersten, H. F. (1975c). Analysis of nonlinear resonance in rotated y-cut quartz thickness-shear resonators. *Proc. 29th Annu. Freq. Control Symp.*, pp. 49–53.

Tiersten, H. F. (1976a). Analysis of trapped-energy resonators operating in overtones of coupled thickness-shear and thickness-twist. *J. Acoust. Soc. Am.* **59**, 879–888.

Tiersten, H. F. (1976b). An analysis of overtone modes in monolithic crystal filters. *Proc. 30th Annu. Freq. Control Symp.*, pp. 103–108.
Tiersten, H. F. (1976c). Analysis of nonlinear resonance in thickness-shear and trapped-energy resonators. *J. Acoust. Soc. Am.* **59**, 866–867.
Tiersten, H. F. (1977). Analysis of overtone modes in monolithic crystal filters. *J. Acoust. Soc. Am.* **62**, 1424–1430.
Tiersten, H. F., and Mindlin, R. D. (1962). Forced vibrations of piezoelectric crystal plates. *Q. Appl. Math.* **20**, 107–119.
Todorov, M. I., and Dimitriev, A. P. (1978). Device for the automatic selection of piezoceramic resonators for multi-unit filters. *Elektro. Promst. Priborostr.* **13**, 62–64. In Bulgarian.
Todorov, P., and Tsvetanski, R. (1975). Passive integral filters [quartz resonator]. *Tekh. Misul* **12**, 33–39. In Russian.
Tomikawa, Y. (1971). Electromechanical filter with attenuation poles consisting of multi-mode vibrators. *Electron. Commun. Jpn.* **54**, 19–25. In Japanese.
Tomikawa, Y., and Konno, M. (1971). Electro-mechanical filter with attenuation-poles consisting of multi-mode resonators. *Proc. 7th Int. Cong. Acoust.*, pp. 637–640.
Tomikawa, Y., Konno, M., Ogasawara, T. (1969). Multiple modes of a rectangular or triangular plate and its application to the electromechanical filter. *Bull. Yamagata Univ. (Eng.)* **10**, 287–303. In Japanese.
Tomikawa, Y., Sugawara, S., and Havens, D. P. (1976). Resonances in flexure-mode mechanical filters. *Ultrason. Symp. Proc.*, pp. 597–601.
Topolevsky, R., and Burgess, J. W. (1974). Interresonator coupling for overtone monolithic quartz filters. *Electron. Lett.* **10**, 476–477.
Topolevsky, R. B., and Redwood, M. (1975). The electrical characteristics of symmetric and asymmetric monolithic crystal filters: a new analytical approach. *IEEE Trans. Sonics Ultrason.* **SU-22**, 162–168.
Toyama, K., Matsumoto, A., and Nishide, T. (1969). Synthesis of image band-pass filters with two-crystal unsymmetrical lattice structures. *Electron. Commun. Jpn.* **52**, (2), 7–13.
Trnka, J., and Sobatka, V. (1979). Respecting certain manufacturing and technological problems in the design of mechanical filters. *Proc. IEEE Int. Symp. Circuits Syst.*, pp. 1080–1081.
Tsuchida, J., Takao, M., Shirai, K., and Kasai, Y. (1979). Automated mass production of mechanical channel filters [for FDM modern equipment]. *Proc. IEEE Int. Symp. Circuits Syst.*, pp. 1072–1075.
Tsuzuki, Y., Hirose, Y., Takada, S., and Iyima, K. (1971). Holographic investigation of spurious modes of mechanical filters. *Electron. Commun. Jpn.* **54**, 31–38. In Japanese.
Tuladhar, K. K., and Cox, B. D. (1981). Design of 4th-order monolithic crystal filters. *IEEE Proc. G.* **128**, 234–236.
Uno, T. (1975). 200 MHz thickness extensional mode $LiTaO_3$ monolithic crystal filter. *IEEE Trans. Sonics Ultrason.* **SU-22**, 168–174.
Uno, T. (1975). Thickness extensional mode $LiTaO_3$ monolithic crystal filter. *Electr. Commun. Lab. Tech. J.* **24**, 1337–1346. In Japanese.
Uno, T. (1978). A $LitaO_3$ monolithic crystal filter by parallel field excitation. *Trans. Inst. Electron. Commun. Eng. Jpn. Sect. E.* **61**, 915–916.
Uno, T. (1979). VHF $LiTaO_3$ filters. *Proc. IEEE Int. Symp. Circuits Syst.*, pp. 892–895.
Velikin, Y. I., Gel'mont, Z. Y., Zelyakh, E. V., and Ivanova, A. I. (1969). Magnetostrictive ladder filters. *Telecomm. Radio Eng. Pt. 1* **23** (11).
Vig, J. R., LeBus, J. W., and Filler, R. L. (1977). Chemically polished quartz. *Proc. 31st Annu. Freq. Control Symp.*, pp. 131–143.
Vostrova, I. N., and Kantor, V. M. (1975). Monolithic quartz filters. *Telecommun. Radio Eng. Pt. 1*, **29–30**, 61–66.

Vostrova, I. N., Spivak, S. K., and Cherne, K. I. (1973). Equivalent-circuit transformation of a crystal-controlled filter. *Izv. Vyssh. Uchebn. Zaved. Radioelektron.* **16**, 102–104. In Russian.
Waddington, D. E. O'N. (1975). Narrow band crystal filter design. *Marconi Instrum.* **15**, 10–13.
Waki, K., and Miyazaki, S. (1980). Monolithic crystal filter, a complete product. *JEE J. Electron. Eng.* **17**, 62–68.
Wallrabe, A. (1980). A simple quartz filter for amateur construction. *Funkschau* **52**, 60–62. In German.
Wallrabe, A. (1980). A simple home made quartz filter. *Funkschau* **52**, 95–96. In German.
Warner, A. W., and Ballman, A. A. (1967). Low temperature coefficient of frequency in a lithium tantalate resonator. *Proc. IEEE* **55**, 450.
Wasiak, M. (1971). Electro-mechanical 200 kHz channel filter. *Prace Inst. Tele- Radiotech.* **15**, 5–25. In Polish.
Watanabe, H., and Shimizu, H. (1975a). Image parameter design of monolithic filters with two electrode pairs using a high-electromechanical-coupling piezoelectric plate. I. *Rec. Electr. Commun. Eng. Conversazione Tohoku Univ.* **44**, 10–19. In Japanese.
Watanabe, H., and Shimizu, H. (1975b). Image parameter design of monolithic filters with two electrode pairs using a high-electromechanical-coupling piezoelectric plate. III. *Rec. Electr. Commun. Eng. Conversazione Tohoku Univ.* **44**, 42–51. In Japanese.
Watanabe, H., and Shimizu, H. (1976). Image-parameter design of multi-electrode-pair monolithic filters. I. A new method of designing multiple-mode filters. *Rec. Electr. Commun. Eng. Conversazione Tohoku Univ.* **45**, 181–185. In Japanese.
Watanabe, H., and Shimizu, H. (1977). Image-parameter design of multi-electrode-pair monolithic filters. II. Four-electrode-pair monolithic filter. *Rec. Electr. Commun. Eng. Conversazione Tohoku Univ.* **46**, 15–22. In Japanese.
Watanabe, H., and Shimizu, H. (1977). Image-parameter design of multi-electrode-pair monolithic filters. III. Four-electrode-pair bilithic filter. *Rec. Electr. Commun. Eng. Convesazione Tohoku Univ.* **46**, 23–29. In Japanese.
Watanabe, H., and Shimizu, H. (1977). Image-parameter design of two-electrode-pair monolithic filters on high-electromechanical-coupling piezoelectric plates. *Trans. Inst. Electron. Commun. Eng. Jpn. Sect. E.* **E60**, 146–147.
Watanabe, H., and Shimizu, H. (1982). A method of designing multielectrode-pair monolithic filters on high-coupling piezoelectric plates. *Trans. Inst. Electron. Commun. Eng. Jpn. Sect. E.* **65**, 79.
Watanabe, N., and Tsukamoto, K. (1975). High-performance monolithic crystal filters with stripe electrodes. *Electron. Commun. Jpn.* **57**, 53–60.
Watanabe, N., Tsukamoto, K., and Jumonji, H. (1974). Design of high performance monolithic crystal filters. *Electr. Commun. Lab. Tech. J.* **23**, 53–65. In Japanese.
Watanabe, H., Nakamura, K., and Shimizu, H. (1976a). Image-parameter design of monolithic filters with two electrode pairs using a high-electromechanical-coupling piezoelectric plate. IV. *Rec. Electr. Commun. Eng. Conversazione Tohoku Univ.* **45**, 11–16. In Japanese.
Watanabe, M., Kidokoro, M., Kouge, T., Kobessho, T., and Yano, T. (1979). Large scale production of a pole electromechanical channel filter [for channel translating equipment]. *Proc. IEEE Int. Symp. Circuits Syst.*, pp. 1076–1079.
Werner, J. F. (1971). The bilithic quartz-crystal filter. *J. Sci. Technol.* **38**, 74–82.
Werner, J. F., Dyer, A. J., and Birch, J. (1969). The development of high performance filters using acoustically coupled resonators on AT-cut quartz crystals. *Proc. 23rd Annu. Freq. Control Symp.*, pp. 65–75.

White, R. M., and Voltmer, F. W. (1965). Direct piezoelectric coupling to surface elastic waves. *Appl. Phys. Lett.* **7**, 314.
Yakuwa, K. (1969). Electromechanical filters. II. High frequency electromechanical filters. *J. Inst. Electron. Commun. Eng. Jpn.* **52**, 568–577. In Japanese.
Yakuwa, K., Kojima, T., Okuda, S., Shirai, K., and Kasai, Y. (1979). A 128-kHz mechanical channel filter with finite-frequency attenuation poles. *Proc. IEEE* **67**, 115–119.
Yakuwa, K., Okuda, S., Shirai, K., and Kasai, Y. (1977). 128 kHz pole-type mechanical channel filter. *Proc. 31st Annu. Freq. Control Symp.*, pp. 207–212.
Yakuwa, K., and Okuda, S. (1976). Design of mechanical filters using resonators with minimized volume. *Proc. IEEE Int. Symp. Circuits System.*, pp. 790–793.
Yano, T., Futami, T., and Kanazawa, S. (1974). New torsional model electromechanical channel filter. *Proc. European Conf. Circuit Theory Design (IEE, London)* no. (116), 121–126.
Yano, T., Futami, T., and Kanazawa, S. (1975). New torsional mode electromechanical channel filter. *NEC Res. Dev.* (39), 30–36.
Yano, T., Kanazawa, S., Futami, T., and Hayashi, T. (1981). Temperature-stable electromechanical channel filter with temperature-compensation torsional mode transducer. *Jpn. J. Appl. Phys.* **20**, 109–112.
Yee, H. K. H. (1971). Finite-pole frequencies in monolithic crystal filters. *Proc. IEEE* **59**, 88–89.
Yee, H. K. H. (1971). Amplitude distortion compensation in two-pole monolithic crystal filters. *Proc. IEEE* **59**, 1286–1287.
Yen, K. H., Lau, K. F., and Kagiwada, R. S. (1979). Narrowband and wideband shallow bulk acoustic wave filters. *Proc. IEEE Int. Symp. Circuits Syst.*, pp. 629–632.
Yoda, H., Nakazawa, Y., Okano, S., and Kobori, N. (1968). High frequency crystal mechanical filters. *Proc. 22nd Annu. Freq. Control Symp.*, pp. 188–205.
Yoda, H., Nakazawa, Y., and Kobori, N. (1969). High frequency crystal monolithic (HCM) filters. *Proc. 23rd Annu. Freq. Control Symp.*, pp. 76–92.
Yoda, H., Endo, A., and Hasegawa, K. (1975). Crystal filters: with performance up and cost down, they are ready to uncrowd the spectrum. *JEE J. Electron. Eng.* (108), 24–27.
Yuki, T., and Yano, T. (1969). Electromechanical filters. III. Low frequency electromechanical filters. *J. Inst. Electron. Commun. Eng. Jpn.* **52**, 727–732. In Japanese.
Zelenka, J. (1974). A twin-resonator monolithic piezoelectric filter with a common electrode. *Slaboproudy Obz.* **35**, 311–314. In Czech.
Zelenka, J. (1978). A monolithic filter employing thickness dilational modes of vibration *Tesla Electron.* **11**, 105–110.
Zelyakh, E. V., and Krukhmaleva, V. D. (1969). A crystal band-elimination filter. *Radio Eng. Pt. 1* **23**, (5).
Zelyakh, E. V., and Novikov, A. A. (1979). Active piezoelectric rejection filters. *Telecommun. Radio Eng. Pt. 1* **32**, 40–44.
Zima, V., and Dokoupil, S. (1981). A chain of uniform π-networks with piezoelectric resonators in series arms. *Acta Tech. CSAV* **26**, 443–463.
Zmudzki, B. (1979). Piezoelectric tuning fork filters. *Elektronika* **19**, 525–526. In Polish.
Zverev, A. I. (1967). "Handbook of Filter Synthesis." John Wiley & Sons, New York.

Sections 5.3 and 5.4

Akitt, D. P. (1976). 70 MHz surface-acoustic-wave resonator notch filter. *Electron. Lett.* **12**, 217–218.

Akitt, D. P. (1976). High Q acoustic surface wave filters. *Canadian Communications and Power Conference*, pp. 200–201.
Allen, D. E., and Hickernell, F. S. (1981). SAW bandpass filter components for microwave systems. *IEEE MTT-S Int. Microwave Symp. Dig.*, pp. 389–391.
Allen, D. E., and Shepard, J. W. (1977). Surface acoustic wave filters for deep space applications. *Ultrason. Symp. Proc.*, pp. 529–531.
Andrasi, A., Beleznay, F., Puspoki, S., and Serenyi, M. (1982). Realization of acoustic surface wave TV IF filter. *Hiradastechnika* **33**, 299–304. In Hungarian.
Anonymous (1975). IF surface wave filters. *Radio Fernsehen Elektron.* **24**, 607–608. In German.
Anonymous (1975). SAW bandpass filter. *Microwave J.* **18**, 32.
Anonymous (1976). Improved SAW filters minimize sidelobes in chirp radar. *Microwave Syst. News* **6**, 87, 90.
Anonymous (1982). Programmable SAW correlator-PSC-300-120-16. *Mikrowellen Mag.* **8**, 68–70.
Arai, S. (1979). Murata's SAW filters enhance FM tuner selectivity. *JEE J. Electron. Eng.* **16**, 33–35, 151.
Arai, S. (1981). SAW filters finding increased applications in consumer products. *JEE J. Electron. Eng.* **18**, 40–42.
Arai, S., Ieki, H., and Kadota, M. (1977). SAW filters are finding extensive applications. *JEE J. Electron. Eng.* (130), 52–56.
Armstrong, G. A. (1976). The design of SAW dispersive filters using interdigital transducers. *Wave Electron.* **2**, 155–176.
Arranz, T. (1977). Lithium tantalate channel filters for multiplex telephony. *Proc. 31st Annu. Freq. Control Symp.*, pp. 213–224.
Asakawa, K., Itoh, M., Gokan, H., Esho, S., Nishikawa, K., Nishihara, U. (1980). Low loss 0.9–1.9 GHz SAW filters with submicron finger period electrodes. *Proc. IEEE Ultrason. Symp.*, pp. 371–376.
Asakawa, K., Itoh, M., Gokan, H., Esho, S., and Nishikawa, K. (1981). Low loss 1–2 GHz SAW filters with submicron finger period electrodes. *Jpn. J. Appl. Phys. (Japan)*, **20**, 93–97.
Atzeni, C. (1971). Sensor number minimization in acoustic surface-wave matched filters. *IEEE Trans. Sonics Ultrason.* **SU-18**, 193–201.
Atzeni, C., and Masotti, L. (1971). Acoustic surface-wave transversal filters. In "Acoustic Surface Wave and Acousto-optic devices" (T. Kallord, ed.), pp. 69–80, Optosonic Press, New York.
Atzeni, C., and Masotti, L. (1972). Interdigital electrode minimization in acoustic surface wave filters. *IEEE Trans. Sonics Ultrason.* **SU-19**, 401.
Atzeni, C., and Masotti, L. (1972). Weighted intedigital transducers for smoothing of ripples in acoustic-surface-wave bandpass filters. *Electron. Lett.* **8**, 485–486.
Atzeni, C., and Masotti, L. (1973). Band-pass filtering by acoustic planar processing techniques. *Alta Freq.* **42**, 84–92.
Atzeni, C., Masotti, L., and Teodori, E. (1971). Acoustic surface-wave matched filters. *Alta Freq.* **40**, 506–512.
Atzeni, C., Manes, G., and Masotti, L. (1974). Design of surface acoustic wave filters. *Alta Freq.* **43**, 865–875.
Atzeni, C., Manes, G., and Masotti, L. (1975). New insights into SAW filter design. *Proc. Ultrason. Int.*, pp. 133–137.
Atzeni, C., Manes, G., and Masotti, L. (1975). Surface-acoustic-wave phase-interference filters. *Wave Electron.* **1**, 97–104.

Auld, B., (1973). "Acoustic Fields and Waves in Solids," vol. II. Wiley (Interscience), New York.
Auld, B., and Kino, G. (1971). Normal mode theory for acoustic waves and its application to the interdigital transducer. *IEEE Trans. Electron.Devices* **ED-18**, 898–908.
Autran, J.-M., and Maloney, E. D. (1977). Application of CAD techniques to SAW filters. *Electron Eng.* **49**, 51–54.
Bahr, A., and Lee, R. (1973). Equivalent-circuit model for interdigital transducers with varying electrode widths. *Electron. Lett.* **9**, 281–282.
Balek, R., and Bune, V. (1975). Surface elastic wave filters. *Sdelovaci Tech.* **23**, 419–420. In Czech.
Ben Zaken, M., Zilbermann, N., Kogan, E., Romik, P., and Sofer, E. (1980). A UHF fast frequency synthesizer using SAW filters. *Ultrason. Symp. Proc.*, pp. 230–234.
Bergmann, H. (1979). Novelties in acoustical electronics. *Radio Fernsehen Elektron.* **28**, 464–646. In German.
Berzelius. (1979). Filters for surface waves. *Radioind. Elettron. -Telev.* **3**, 197–201. In Italian.
Bidenko, V. A., Grankin, I. M., Nelin, E. A., and Pogrebnyak, V. F. (1979). An approximate parameter analysis of SAW bandpass filters. *Izv. Vyssh. Uchebn. Zaved. Radioelektron.* **22**, 56–60. In Russian.
Bidenko, V. A., Grankin, I. M., Nelin, E. A., and Pogrebnyak, V. P. (1979). Synthesis of surface acoustic wave filters with asymmetric amplitude–frequency response. *Izv. Vyssh. Uchebn. Zaved. Radioelektron.* **22**, 87–88. In Russian.
Biran, A., and Gafni, H. (1981). Nonuniform finger withdrawal in SAW chirp filters. *Proc. MELECON (New York: IEEE)* 2.2.3/1–5.
Biran, A., Chalzel, A., Gafni, H., and Gilboa, H. (1979). Low-sidelobe SAW cascaded filter. *Ultrason. Symp. Proc.*, pp. 570–573.
Biran, A., Chalzel, A., and Gafni, H. (1979). Low-sidelobe SAW cascaded filter. *Proc. Electr. Electron. Eng. Israel* B4-5/1–4.
Boege, T. J., Chao, G., and Drummond, W. S. (1976). Design of arbitrary phase and amplitude characteristics in SAW filters. *Ultrason. Symp. Proc.*, pp. 313–316.
Borisov, M. I., Stojanov, D. V., Velkov, V. V., and Burov, J. I. (1974). Electrically controlled acoustic-surface-wave filters. *Electron. Lett.* **10**, 519–520.
Borner, M., and Kohlbacher, G. (1975). Surface acoustic wave (SAW) devices with coupled resonators. *Wiss. Ber. AEG-Telefunken* **48**, 79–82. In German.
Bourov, J. I., Velkov, V. J., and Stojanov, D. V. (1976). Surface acoustic waves and some applications. *Elektro Promst. Priborostr.* **11**, 252–253. In Bulgarian.
Bray, R., Grubb, J. P., and Ishak, W. (1982). SAW filters using group-type unidirectional transducers: Sources of problems. *Ultrason. Symp. Proc.*, pp. 227–232.
Bristol, T. W. (1974). Acoustic surface-wave-device applications. *Microwave J.* **17**, 25–27, 63.
Bristol, T. W., and Meyer, W. E. (1972). Surface acoustic wave UHF frequency filters. *IEEE Trans. Sonics Ultrason.* **SU-19**, 400–401.
Broux, G., Claes, R., and Deelers, J. J. (1978). The use of filters for surface acoustic waves for the design of circuits for IF stages of television sets. *Antenna* **50**, 369–376. In Italian.
Broux, G., Claes, R., and Deelers, J. J. (1978). The use of surface acoustic wave filters for the design of IF stages of television sets. *Antenna* **50**, 413–417. In Italian.
Broux, G., Claes, R., and Deelers, J. J. (1978). Surface acoustic wave filters for IF-stages in TV receivers. II. Measurement techniques. *Radio Mentor Electron.* **44**, 223–224. In German.
Broux, G., Claes, R., and Deelers, J. J. (1978). Surface-wave filters for TV receiver IF stages. I. Acoustic surface-wave technology. *Radio Mentor Electron.* **44**, 189–192. In German.

Broux, G., Claes, R., and Deelers, J. J. (1978). Surface-wave filters for TV IF stages. III. *Radio Mentor Electron.* **44**, 267–270. In German.
Browning, T. I., Gunton, D. J., Lewis, M. F., and Newton, C. O. (1977). Bandpass filters employing surface skimming bulk waves. *Ultrasonic. Symp. Proc.*, pp. 753–756.
Brummond, B. (1979). Surface-acoustic wave filters for broadcast demodulators. *Electron. Ind.* **5**, 59, 61.
Budreau, A. J., Carr, P. H., and Laker, K. R. (1974). Frequency synthesizer using acoustic surface-wave filters. *Microwave J.* **17**, 65–66, 68–69.
Budreau, A. J., Laker, K. R., and Carr, P. H. (1974). Compact microwave acoustic surface wave filter bank for frequency synthesis. *Proc. Symp. Opt. Acoust. Micro-Electron.*, Polytech. Inst. New York, Brooklyn, pp. 471–483.
Burgess, A. S., and Cole, P. H. (1973). Design of acoustic surface-wave devices using an admittance formalism. *Proc. IEEE Trans. Microwave Theory Tech.* **MTT-21**, 611–618.
Burgess, J. W., Hales, M. C., Airger, F. W., and Porter, R. J. (1973). Single-mode resonance in $LiNbO_3$ for filter design. *Electron. Lett.* **9**, 251–252.
Burgos, J. (1980). An IF picture section using a surface wave filter. *Rev. Esp. Electron.* **27**, 98–102. In Spanish.
Burnsweig, J. (1971). Surface-wave signal processing filters. *Papers presented at the Western Electronics Show and Convention*, 8/4 10 pp.
Campbell, C. K. (1982). Application of the inverse discrete fourier transform to the design of SAW filters with nonlinear phase response. *Ultrason. Symp. Proc.*, pp. 46–49.
Campbell, C. K. (1982). Wide-band linear phase SAW filter design using slanted transducer fingers. *IEEE Trans. Sonics Ultrason.* **SU-29**, 224–228.
Carr, P. H. (1977). Systems applications of SAW filters and delay lines. *AGARD Conf. Proc. No. 230* 3.4/1–12.
Chao, G., Davies, B., and Drummond, W. (1975). Design considerations for nonsymmetrical SAW filters. *Ultrason. Symp. Proc.*, pp. 331–333.
Chauvin, D., Coussot, G., and Dieulesaint, E. (1971). Acoustic-surface-wave television filters. *Electron. Lett.* **7**, 491–492.
Chiba, T. (1979). SAW filters as applied to group-delay equalizer in television rebroadcast transmitters. *Ultrason. Symp. Proc.*, pp. 545–549.
Chiba, T. (1981). Improvement of ripple in pass band of SAW filters. *Trans. Inst. Electron. Commun. Eng. Jpn. Sect. E.* **64**, 517.
Claiborne, L. T. (1974). Surface wave bandpass filters: components and subsystems. *Proc. Symp. Opt. Acoust. Micro-Electron.*, Polytech. Inst. New York, Brooklyn, pp. 461–469.
Claiborne, L. T. (1975). A survey of current SAW device capabilities. *Proc. 29th Annu. Freq. Control Symp.*, pp. 135–138.
Claiborne, L. T., and Staples, E. J. (1972). Surface wave filters in integrated circuits. *Proc. IEEE Trans. Sonics Ultrason.* **SU-19**, 413.
Claiborne, L. T., Hartmann, C. S., Hays, R. M., and Rosenfeld, R. C. (1974). VHF/UHF bandpass filters using SAW device technology. *Microwave J.* **17**, 35–38, 40.
Claiborne, L. T., Hartmann, C. S., Potter, B. R., and Stigall, R. E. (1979). Low loss SAW filters using unidirectional transducer technology. *Proc. Int. Symp. Circuits Syst.*, pp. 609–612.
Claiborne, L. (1971). New surface wave components. *Int. Elektron. Rundsch.* **25**, 206–207. In German.
Coldren, L. A. (1979). Improved temperature stability in SAW resonator filters using multiple coupling paths. *Appl. Phys. Lett.* **35**, 678–680.
Coldren, L. A. (1979). The temperature dependence of SAW resonator filters using folded acoustic coupling. *Ultrason. Symp. Proc.*, pp. 830–835.

Coldren, L. A., and Rosenberg, R. L. (1978a). Multipole SAW resonator filters. *Proc. IEEE Int. Symp. Circuits Syst.*, pp. 548–552.

Coldren, L. A., and Rosenberg, R. L. (1978b). SAW resonator filter overview: Design and performance tradeoffs. *Ultrason. Symp. Proc.*, pp. 422–432.

Coldren, L. A., and Rosenberg, R. L. (1979). Acoustically coupled SAW resonator filters with enhanced out-of-band rejection. *Proc. IEEE Trans. Sonics Ultrason.* **SU-26**, 394–403.

Coldren, L. A., and Rosenberg, R. L. (1979). Surface-acoustic-wave resonator filters. *Proc. IEEE* **67**, 147–158.

Coldren, L. A., Rosenberg, R. L., and Rentschler, J. A. (1977). Monolithic transversely coupled SAW resonator filters. *Ultrason. Symp. Proc.*, pp. 888–893.

Collins, J. (1977). SAW devices emerging into commercial and military hardware. *Microwave Syst. News* **7**, 15–21.

Collins, J., and Owens, J. (1977). SAW devices meet high-rel systems needs. *Microwave Syst. News* **7**, 58, 60, 64, 66, 70.

Colvin, R. D., Carr, P. H., Roberts, G. A., and Charlson, E. J. (1982). Stagger tuning of SAW filters for broad bandwidth and good shape factor. *IEEE Trans. Sonics Ultrason.* **SU-29**, 50–52.

Costanza, S. T., Hagon, P. J., MacNevin, L. A. (1969). Analog matched filter using tapped acoustic surface wave delay line. *IEEE Trans. Microwave Theory Tech.* **MTT-17**, 1042–1043.

Coussot, G. (1975). Filters for surface elastic waves. *Electtrificazione* (11), 521–524. In Italian.

Coussot, G. (1975). Investigation of intermodulation in acoustic-surface-wave filter. *Electron. Lett.* **11**, 116–117.

Coussot, G. (1975). Surface acoustic wave filters. *Rev. Polytech.* (6), 787, 789, 791. In French.

Coussot, G., and Ménager, O. (1975). Experimental investigation of mass-producible acoustic surface wave filter. *Proc. 29th Annu. Freq. Control Symp.*, pp. 181–186.

Coussot, G., and Van D. Driessche, M. (1976). Design of a TV receiver's intermediate frequency stage employing a surface wave filter. *Rev. Tech. Thomson-CSF* **8**, 585–605. In French.

Coussot, G., and Van D. Driessche, M. (1977). Surface [acoustic] wave filters. *Toute Electron.* (418), 37–43. In French.

Coussot, G., and Van D. Driesseche, M. (1977). Filters using elastic surface waves. *Antenna* **49**, 314–320. In Italian.

Craven, G., and Lush, D. M. (1974). Surface-acoustic-wave rejection filters using mode conversion/reconversion. *Electron. Lett.* **10**, 218–219.

Cross, P. S. (1977). Surface acoustic wave resonator-filters using tapered gratings. *Ultrason. Symp. Proc.*, pp. 894–899.

Cross, P. S. (1978). Surface acoustic wave resonator filters using tapered gratings. *Proc. IEEE Trans. Sonics Ultrason.* **SU-25**, 313–319.

Cross, P. S., Schmidt, R. V., and Haus, H. A. (1976). Acoustically cascaded ASW resonator-filters. *Ultrason. Symp. Proc.* pp. 277–280.

Cuozzo, F. C. (1977). Numerical simulation of SAW interdigital filters using an equivalent electrical model. *Ultrason. Symp. Proc.*, pp. 642–647.

Czechowska, Z., and Weiss, K. (1974). The application of surface waves in piezoelectric materials. *Elektronika* **15**, 425–430. In Polish.

Danicki, E. (1972). A 30 MHz filter based on acoustic surface waves. *Bull. Acad. Pol. Sci. Ser. Sci. Tech.* **20**, 289–295.

Danicki, E. (1980). Influence of bulk-wave generation on SAW filter performance. *J. Tech. Phys.* **21**, 405–420.

Darby, B. J., Grant, P. M., and Collins, J. H. (1973). Performance of surface acoustic wave filter modems in noise and interference limited environments. *Proc. Ultrason. Int. Conf.*, pp. 314–321.

Deacon, J. M., Heighway, J., and Jenkins, J. A. (1973). Multistrip coupler in acoustic-surface-wave filters. *Electron. Lett.* **19**, 235–236.

Deacon, J. M., and Heighway, J. (1975). SAW filters for TV receivers. *IEEE Trans. Consum. Electron* **CE-21**, 390–395.

Defranould, Ph., and Desbois, J. (1981). Low loss SAW bandpass filters. *Ultrason. Symp. Proc.*, pp. 17–22.

Desbois, J. (1979). The problem of the accuracy in the fabrication of multipole SAW resonator filters. *Ultrason. Symp. Proc.*, pp. 841–844.

De Vries, A. J. (1977). Surface wave bandpass filters. *In* "Surface Wave Filters" (H. Matthews, ed.), pp. 263–305. Wiley, Chichester, England.

De Vries, A. J., and Adler, R. (1976). Case history of a surface wave TV IF filter for color television receivers. *Proc. IEEE* **64**, 671–676.

De Vries, A. J., Dias, J. F., Rypkema, J. N., and Wojcik, T. J. (1970). Characteristics of surface wave integratable filters (SWIFS). *Proc. Nat. Electron. Conf.*, pp. 537–540.

De Vries, A. J., Dias, J. F., Rypkema, J. N., and Wojcik, T. J. (1971). Characteristics of surface-wave integratable filters (SWIFS). *Proc. IEEE Broadcast. Telev. Receivers* **BTR-17**, 16–23.

Dieulesaint, E., and Hartemann, P. (1973). Acoustic surface wave filters. *Ultrasonics* **11**, 24–30.

Dolbna, E. V., Pis'menetskii, V. A., Khorunzhii, V. A., and Yashkov, O. V. (1977). An orthogonal system of film filters on surface acoustic waves. *Radiotekhnika Kharkov* (43), 42–46. In Russian.

Drummond, B. (1978). A 24 MHz Nyquist SAW filter for the 1450 demodulator. *Tekscope* **10**, 10–12.

Drummond, B. (1979). Surface-acoustic wave filters for broadcast demodulators. *Electron. Ind.* **5**, 59, 61.

Drummond, W. S., and Roth, S. A. (1978). Application of high performance SAW transversal filters in a precision measurement instrument. *Ultrason. Symp. Proc.*, pp. 494–499.

El-Diwany, M. H., and Campbell, C. K. (1977). Modification of optimum impulse response techniques for application to SAW filter design. *Proc. IEEE Trans. Sonics Ultrason.* **SU-24**, 277–279.

Elek, K., and Pfliegel, P. (1980). Synthesis of SAW filters with asymmetrical amplitude and group delay characteristics. *Hiradastechnika* **31**, 321–326. In Hungarian.

Emtage, P. (1972). Self-consistent theory of interdigital transducers. *J. Acoust. Soc. Am.* **51**, 1142–1155.

Engan, H. (1969). Excitation of elastic surface waves by spatial harmonics of interdigital transducers. *IEEE Trans. Electron. Devices* **ED-16**, 1014–1017.

Engan, H. (1976). Interdigital transducer techniques for specialised frequency filters; SAW transversal filters. *Wave Electron.* **2**, 133–154.

Everett, P. (1981). Surface acoustic wave filters. *Int. J. Hybrid Microelectron.* **4**, 240–245.

Evseev, G. S., Nikitin, V. I., Sirotin, G. F., and Ul'yanov, G. K. (1982). Optimising electrode structures of band-pass surface acoustic wave filters. *Izv. Vych. Uchebn. Zaved. Priborostr.* **25**, 69–72. In Russian.

Farnell, G. W. (1977). Elastic surface waves. *In* "Surface Wave Filters" (H. Matthews, ed.), 521 pp. Wiley, Chichester, England.

Feldmann, M., and Henaff, J. (1976). An ASW-filter using two fan-shaped multistrip reflective arrays. *Ultrason. Symp. Proc.*, pp. 397–400.

Feldmann, M., and Henaff, J. (1978). Design of SAW filter with minimum phase response. *Ultrason. Symp. Proc.*, pp. 720–723.

Feldman, M., Henaff, J., and Carel, M. (1976). A.S.W. filter bank using a multistrip reflective array. *Electron. Lett.* **12**, 118–119.

Fujishima, S., Ishiyama, H., Inoue, A., and Ieki, H. (1976). Surface acoustic wave VIF filters for TV using ZnO sputtered film. *Proc. 30th Annu. Freq. Control Symp.*, pp. 119–122.

Fujishima, S., Arai, S., and Ieki, H. (1979). ZnO-glass SAW filters [TV IF filters]. *Proc. Int. Symp. Circuits Syst.*, pp. 625–628.

Furukawa, S., Nakamura, K., Moriizumi, T., and Yasuda, T. (1981). Low loss and temperature-stable SAW filters using $SiO_2/(ZY)LiNbO_3$ structure. *Trans. Inst. Electron. Commun. Eng. Jpn. Sect. E.* **64**, 557.

Furuya, N., Miyama, H., Nakayama, Y., and Kino, Y. (1980). A SAW ring filter with a phase matching electrode. *Ultrason. Symp. Proc.*, **1**, 169–172.

Ganguly, A. K., and Vassell, M. O. (1973). Frequency response of acoustic surface wave filters. I. *J. Appl. Phys.* **44**, 1072–1085.

Ganguly, A. K., Vassell, M. O., and Zucker, J. (1972). A new, first principles approach to acoustic surface wave filter analysis. *IEEE Trans. Sonics Ultrason.* **SU-19**, 400.

Gerard, H. (1969). Acoustic scattering parameters of the electrically loaded interdigital surface wave transducer. *IEEE Trans. Microwave Theory Tech.* **MTT-17**, 1045–1046.

Gerard, H. M. (1977). Surface wave interdigital electrode chirp filters. *In* "Surface Wave Filters" (H. Matthews, ed.), 521 pp. Wiley, Chichester, England.

Gerard, H. M. (1978). Principles of surface wave filter design. *In* "Acoustic Surface Waves" (A. A. Oliner, ed.), 350 pp., Springer-Verlag, Berlin.

Gerard, H. M., and Judd, G. W. (1978). 500 MHz bandwidth RAC filter with constant groove depth. *Ultrason. Symp. Proc.*, pp. 734–737.

Gerard, H. M., Judd, G. W., and Pedinoff, M. E. (1972). Phase corrections for weighted acoustic surface-wave dispersive filters. *IEEE Trans. Microwave Theory Tech.* **MTT-20**, 188–192.

Gerard, H. M., Otto, O. W., and Weglein, R. D. (1974). Wideband dispersive surface wave filters. Report ECOM-73-0110-F, Contract DAAB07-73-C-0110, 51 pp. Available from NTIS.

Gerard, H. M., Yao, P. S., and Otto, O. W. (1977). Performance of a programmable radar pulse compression filter based on a chirp transformation with RAC filters. *Ultrason. Symp. Proc.*, pp. 947–957.

Gerli, D., Tortoli, P., Manes, G. F., Atzeni, C., and Raffini, C. (1982). Analogue SAW transversal filter for baseband processing. *Int. Symp. Circuits Syst.* **2**, 358–361.

Green, J. B., Kino, G. S., Walker, T., and Shott, J. D. (1982). The SAW/FET: A new programmable SAW transversal filter. *Ultrason. Symp. Proc.*, pp. 436–441.

Grice, D. C., Pi. S. C., and Wilk, J. M. (1976). Acoustic surface wave filter for TV tuning circuits. *IBM Tech. Disc. Bull.* **19**, 971–974.

Guenther, A. E. (1976). Rapid approximate analysis of SAW filters. *Proc. IEEE Int. Symp. Circuits Syst.*, pp. 363–366.

Guerard, A., Glowinski, A., and Feldmann, M. (1977). Synthesis by optimisation of acoustic surface wave frequency filters. *Ann. Telecommun.* **32**, 37–48. In French.

Guntersdorfer, M., Veith, R., Eberharter, G., and Bulst, W. E. (1979). VHF wide band SAW filters. *Proc. IEEE Int. Symp. Circuits Syst.*, pp. 613–616.

Gupta, O. S., and Agrawal, N. K. (1980). Review of surface acoustic wave filters. *J. Inst. Eng. (India) Electron. Telecommun. Eng. Div.* **60**, 21–23.

Gupta, O. S., and Sharma, R. S. (1981). Design of SAW filter. *J. Inst. Eng. (India) Electron. Telecommun. Eng. Div.* **61**, 97–98.

Hagon, P. J., and Wheatley, C. E. III (1974). Programmable analog matched filters. *Microwave J.* **17**, 42–45.
Hagon, P. J., Micheletti, F. B., Seymour, R. N., and Wrigley, C. Y. (1973). A programmable surface acoustic wave matched filter for phase-coded spread spectrum waveforms. *IEEE Trans. Microwave Theory Tech.* **MTT-21**, 303–306.
Halgas, F. A., Godfrey, J. T., Sundelin, R., Moore, R. A., Weinert, R. W., and Isaacs, T. J. (1977). Comparative performance of SAW filters on sulfosalt versus quartz substrates. *Ultrason. Symp. Proc.*, pp. 798–802.
Hartemann, P. (1971). Narrow-bandwidth Rayleigh-wave filters. *Electron. Lett.* **7**, 674–675.
Hartemann, P. (1972). Narrow bandwidth Rayleigh wave filters. *IEEE Trans. Sonics Ultrason.* **SU-19**, 401.
Hartemann, P., and Dieulesaint, E. (1969). Acoustic-surface-wave filters. *Electron. Lett.* **5**, 657–658.
Hartmann, C. S. (1973). Weighting interdigital transducers by selective withdrawal of electrodes. *Ultrason. Symp. Proc.*, pp.. 423–426.
Hartmann, C. S. (1974). Acoustic surface wave devices [signal processing functions]. *Commun. Syst. Tech. Conf.*, pp. 122–128.
Hartmann, C. S. (1977). Recent advances in surface acoustic wave devices. *NTC Conf. Rec.*, 16:2/1–5.
Hartmann, C. S. and Wilcus, S. (1983). SAW bandpass filters. *Proc. 37th Annu. Freq. Control Symp.*, pp. 354–360.
Hartmann, C. S., Cheek, T. F., and Vollers, H. G. (1972). VHF/UHF bandpass filters using piezoelectric surface wave devices. *Proc. 26th Annu. Freq. Control Symp.*, pp. 164–170.
Hartmann, C. S., Vollers, H. G., Cheek, T. F. (1972). VHF/UHF bandpass filters using interdigital surface wave transducers. *IEEE Trans. Sonics Ultrason.* **SU-19**, 400.
Hartmann, C. S., Bell, D. T., Jr., and Rosenfeld, R. C. (1973a). Impulse model design of acoustic surface-wave filters. *IEEE Trans. Microwave Theory Tech.* **MTT-21**, 162–175.
Hartmann, C. S., Claiborne, L. T., Buss, D. D., and Staples, E. J. (1973b). Programmable transversal filters using surface waves, charge transfer devices, and conventional digital approaches. *Proc. Intern. Specialist Seminar (London, IEE)*, pp. 102–114.
Haspel, M. (1978). Surface-acoustic-wave dispersive filter implementation of quasi-linear FM matched filter pairs. *Electr. Electron. Eng. Israel 10th Conv.*, pp. 231–236.
Hawkins, D. G., and Thurber, L. J. (1982). The effect of the quartz seed plane on SAW bandpass filter performance. *Ultrason. Symp. Proc.*, pp. 376–378.
Haydl, W. H., and Cross, P. S. (1975). The tuning of surface-acoustic-wave resonator filters with metallisation thickness. *Electron. Lett.* **11**, 252–253.
Haydl, W. H., Sander, W. and Wirth, W. D. (1981). Precision SAW filters for a large phased-array radar system. *IEEE Trans. Microwave Theory Tech.* **MTT-29**, 414–419.
Hays, R. M., and Hartmann, C. S. (1976). Surface-acoustic-wave devices for communications. *Proc. IEEE* **64**, 652–671.
Hays, R. M., Rosenfeld, R. C., and Hartmann, C. S. (1974). One hundred channel selectable surface wave bandpass filter. *Microwave Symp. Dig. Tech. Papers*, p. 236.
Haywood, P. J. (1979). SAW filters. *New Electron.* **12**, 100, 105.
Hazama, K. (1980). Recent SAW devices. *J. Acoust. Soc. Jpn.* **36**, 590–592. In Japanese.
Hazama, K., Yamada, J., Ishigaki, M., and Toyama, T. (1978). Design of mass productive fabrication techniques of high performance SAW TV IF filters. *Ultrason. Symp. Proc.*, pp. 504–508.
Hazama, K., Kishimoto, K., Yuhara, A., Ishigaki, M., and Matsuura, S. (1979). SAW comb filter for TV channel indicating system. *Ultrason. Symp. Proc.*, pp. 550–554.
Heighway, J. (1976). Surface acoustic wave devices. *Wireless World* **82**, 39–43.

Henaff, J. (1981). Filters for charge transfer (CCD) and surface waves (SAW). *Journees d'Electronique. Modern Filtering Techniques*, pp. 49–68. In French.
Henaff, J., Feldmann, M., and Carel, M. (1981). UHF voltage-controlled narrow-bandwidth saw filters. *Proc. 35th Annu. Freq. Control Symp.*, pp. 349–351.
Henaff, J., and Feldmann, M. (1979). Design and capabilities of SAW filters; synthesis and technologies. *Proc. IEEE Int. Symp. Circuits Syst.*, pp. 617–620.
Henaff, J., and Feldmann, M. (1980). Electrically tunable narrow-bandwidth SAW filter. *Ultrason. Symp. Proc.*, pp. 332–335.
Henaff, J., and Feldmann, M. (1980). S.A.W. implemenation of whitening filters for satellite digital communications. *Electron. Lett.* **16**, 124–125.
Henaff, J., Feldmann, M., and Carel, M. (1981). Voltage-controlled SAW filter on berlinite. *Electron. Lett.* **17**, 86–87.
Hewes, C. R., Claiborne, L. T., Hartmann, C. S., and Buss, D. D. (1976). Filtering with analog CCDs and SWDs. *Proc. 30th Annu. Freq. Control Symp.*, pp. 123–128.
Hibino, M., Morimoto, M., and Kobayashi, Y. (1981). SAW filter design by means of factorization of transfer function. *Proc. European Conf. on Circuit Theory and Design, The Hague, Netherlands*, pp. 913–918.
Hickernell, F. S., and Bush, H. J. (1977). The monolithic integration of surface acoustic wave and semiconductor circuit elements on silicon for matched filter device development. *AGARD Conference Proc. No. 230* 3.5/1–13.
Hickernell, F. S., and Bush, H. J. (1978). Monolithic programmable SAW transversal filters. *Proc. IEEE Int. Symp. Circuits Syst.*, pp. 386–389.
Hickernell, F. S., Kline, A. J., Allen, D. E., and Brown, W. C. (1973). Surface elastic wave bandpass filters for frequency synthesis. *IEEE Trans. Microwave Theory Tech.* **MTT-21**, 300–302.
Hickernell, F. S., Olson, D. E., and Adamo, M. D. (1977). Monolithic surface wave transversal filter. *Ultrason. Symp. Proc.*, pp. 615–618.
Hickernell, F. S., Adamo, M. D., DeLong, R. V., Hinsdale, J. G., and Bush, H. J. (1980). SAW programmable matched filter signal processor. *Ultrason. Symp. Proc.*, **1**, 104–108.
Hikita, M., Kinoshita, Y., Kojima, H., and Tabuchi, T. (1980). Sidelobe suppression of SAW resonant filter using twin-turn reflectors. *Electron. Lett.* **16**, 784–785.
Hikita, M., Kinoshita, Y., Kojima, H., and Tabuchi, T. (1980). Phase weighting for low loss SAW filters. *Ultrason. Symp. Proc.*, **1**, 308–312.
Hikita, M., Kinoshita, Y., Kojima, H., and Taubuchi, T. (1980). Resonant SAW filter using surface shear mode on $LiTaO_3$ substrate. *Electron. Lett.* **16**, 446–447.
Hirabayashi, M., Komatsu, Y., and Kanamaru, M. (1981). Surface-acoustic-wave filters with phase-weighted transducers for the output circuit of a consumer VTR. *Jpn. J. Appl. Phys.* **20**, 89–92.
Hirosaki, B. (1977). Systematic jitter reduction effect by inherent delay of acoustic surface wave filter. *Trans. 1st. Electron. Commun. Eng. Jpn. Sect. E.* **59**, 20–21.
Hodur, E. P. (1980). A designer's file: notes on SAW bandpass filters. *Microwave* **19**, 72–76.
Hoffman, G., and Tromp, H. (1979). On the design of VHF and UHF wide-band band pass filters using SAW resonators and LC components. *Proc. IEEE Int. Symp. Circuits Syst.*, pp. 128–129.
Hohkawa, K., and Yoshikawa, S. (1978). Design of surface acoustic wave filters using linear programming technique. *Electr. Commun. Lab. Tech. J.* **26**, 2875–2888. In Japanese.
Hohkawa, K., and Yoshikawa, S. (1978). SAW filter design using linear programming technique. *Rev. Electr. Commun. Lab.* **26**, 755–766.
Hohkawa, K., and Yoshikawa, S. (1981). Multichannel SAW filters employing two dimensionally periodic RDA. *Trans. 1st. Electron. Commun. Eng. Jpn. Sect. E.* **64**, 809–810.

Hohkawa, K., Ishihara, F., and Yoshikawa, S. (1975). Acoustic surface-wave filters without apodisation loss with electrodes with inclined apodisation. *Electron. Lett.* **11**, 259–261.

Hohkawa, K., Yoshikawa, S., and Ishihara, F. (1976). Design of SAW filters using linear programming technique. *J. Acoust. Soc. Jpn.* **32**, 531–539. In Japanese.

Hohkawa, K., Yoshikawa, S., and Ishihara, F. (1979). Surface acoustic wave filters without apodization loss. *Trans. Inst. Electron. Commun. Eng. Jpn. Sect. E.* **62**, 91–92.

Holland, M. G. (1970). Generalized filters using surface ultrasonic waves. *Proc. 24th Annu. Freq. Control Symp.*, pp. 83.

Holland, M. G., and Claiborne, L. T. (1974). Practical surface acoustic wave devices. *Proc. IEEE* **62**, 582–611.

Huber, C., Lane, J., Newman, B. A., Godfrey, J. T., Grauling, C. H., and Moore, R. A. (1977). A slow sidelobe SAW contiguous filterbank using MDC $LiTaO_3$. *Proc. IEEE Ultrasonics Symp.*, pp. 568–572.

Hunsinger, B. J. (1975). Oversampled SAW filter transducers. *Proc. 29th Annu. Freq. Control Symp.*, pp. 177–180.

Hunsinger, B. J. (1979). SAW filter applications in consumer electronics. *Ultrason. Symp. Proc.*, pp. 541–544.

Hunsinger, B. J., and Kansy, R. J. (1975). SAW filter sampling techniques. *IEEE Trans. Sonics Ultrason.* **SU-22**, 270–273.

Ishak, W. S. (1981). Low loss, ultra flat SAW filters using group-type undirectional transducers. *Ultrason. Symp. Proc.*, pp. 7–12.

Ishak, W. S., Karrer, H. E., and Shreve, W. R. (1981). Surface-acoustic-wave delay lines and transversal filters. *Hewlett-Packard J.* **32**, 3–8.

Ishihara, F., Koyamada, Y., and Yoshikawa, S. (1975). Narrow band filters using surface acoustic wave resonators. *Ultrason. Symp. Proc.*, pp. 381–384.

Jack, M. A. (1978). Fast-Fourier-transform processor using SAW chirp filters. *Electron. Lett.* **14**, 634–635.

Jack, M. A., and Paige, E. G. S. (1978). Fourier transformation processors based on surface acoustic wave chirp filters. *Wave Electron.* **3**, 229–247.

Jack, M., Grant, P. M., and Collins, J. H. (1975). New network analyzer employs SAW chirp filters. *Microwave Syst. News* **5**, 62.

Janus, A. R. (1976). Progress report on surface acoustic wave device MMT [filters and delay lines]. *Proc. 30th Annu. Freq. Control Symp.*, pp. 157–166.

Janus, A. R., and Dyal, L. III. (1977). Progress report on surface acoustic wave device MMT-II. *Proc. 31st Annu. Freq. Control Symp.*, pp. 281–284.

Jiru, V., and Smid, V. (1978). A crystal bandpass [filter] with variable bandwidth for telecommunications measuring instruments. *Slaboproudy Obz.* **39**, 173–177. In Czech.

Johansson, J., and Persson, M. (1981). SAW filters already used in millions of TV sets. *Eltek. Aktuell Elekton.* **24**, 30–33. In Swedish.

Jones, W. S., Kempf, R. A., and Hartmann, C. S. (1972). Practical surface wave chirp filter for modern radar systems. *Microwave J.* **15**, 43–46, 48, 50, 76.

Jones, R. R., Schellenberg, J., Tanski, W. J., and Moore, R. A. (1974). Transplexing SAW filters for ECM. I. *Microwaves* **13**, 43–44, 46, 50, 52.

Jones, R. R., Schellenberg, J., Tanski, W. J., and Moore, R. A. (1974). Transplexing SAW filters for ECM. II. *Microwaves* **14**, 68–70, 72–73.

Jordan, P. M., and Lewis, B. (1978). A tolerance-related optimised synthesis scheme for the design of SAW bandpass filters with arbitrary amplitude and phase characteristics. *Ultrason. Symp. Proc.*, pp. 715–719.

Judd, G. (1974). Acoustic pulse compression filters. Report FR-74-14-665, Hughes Aircraft Co., Contract DAAB07-73-C-0121, 51 pp. Available from NTIS.

Karnopp, D., Reed, J., Margolis, D., and Dwyer, H. (1975). Computer-aided design of acoustic filters using bond graphs. *Noise Control Eng.* **4**, 114–118.

Kaverina, G. M., Kleshnev, Y. A., Nikitin, V. I., Sirotin, G. F., and Ul'yanov, G. K. (1974). Ultrasonic surface-wave piezoelectric filters for television receivers. *Telecommun. Radio Eng. Pt. 2* **29**, 121–122.

Kearns, W. J., and Slobodnik, A. J., Jr. (1975). Fabrication of micrometer line width SAW filters using direct optical projection. Report AFCRL-TR-75-0055, RADC, Bedford, Massachusetts, 11 pp.

Kerber, G. L., Wright, R. M., and Wright, J. R. (1976). Surface-wave inverse filter for non-destructive testing. *Ultrason. Symp. Proc.*, pp. 577–581.

Kinoshita, Y., Hikita, M., Tabuchi, T., and Kojima, H. (1979). Broadband resonant filter using surface-shear-wave mode and twin-turn reflector. *Electron. Lett.* **15**, 130–131.

Kishimoto, K., Ishigaki, M., Hazama, K., and Matsuura, S. (1980). SAW comb filter for TV frequency synthesizing tuning system. *Ultrason. Symp. Proc.*, **1**, 377–381.

Kodama, T. (1979). Optimization techniques for SAW filter design. *Ultrason. Symp. Proc.*, pp. 522–526.

Kodama, T. (1979). Optimization techniques for SAW filter design. *Trans. Inst. Electron. Commun. Eng. Jpn. Sect. E* **62**, 630.

Kodama, T., Sato, K., and Uemura, Y. (1981). SAW vestigial sideband filter for TV broadcasting transmitter. *Proc. IEEE Trans. Microwave Theory Tech.* **MTT-29**, 429–433.

Koefoed, N. (1980). Surface wave filters-theory and practice. *Elektronik* (5), 14–16, 18–20. In Danish.

Kogan, E., and Romik, P. (1980). SAW bandpass filters with withdrawal weighted transducers. *Ultrason. Symp. Proc.* **1**, 302–307.

Komatsu, Y., and Yanagisawa, Y. (1977). A surface acoustic wave filter for colour TV receiver VIF. *Proc. IEEE Trans. Electron Devices* **ED-24**, 230–233.

Kosek, M. (1978). Influence of manufacturing inaccuracies on transmission properties of acoustic wave filters. *Tesla Electron.* **11**, 27–28.

Kosek, M. (1980). A band-pass filter with surface acoustic wave. *Sdelovaci Tech.* **28**, 352–257. In Czech.

Kotaka, I., Temmyo, J., Inamura, T., and Yoshikawa, S. (1980). Fine adjustment of wide-bandwidth SAW filter responses. *Trans. Inst. Electron. Commun. Eng. Jpn. Sect. E* **63**, 112–113.

Koyamada, Y., Ishihara, F., and Yoshikawa, S. (1975). Band-elimination filter employing surface-acoustic-wave resonators. *Electron. Lett.* **11**, 108–109.

Koyamada, Y., Ishihara, F., and Yoshikawa, S. (1976). Narrow-band filters employing surface-acoustic-wave resonators. *Proc. IEEE* **64**, 685–687.

Krimholtz, R. (1972). Equivalent circuits for transducers having arbitrary asymmetrical piezoelectric excitation. *IEEE Trans. Sonics Ultrason.* **SU-19**, 427–436.

Kuwano, Y. (1978). Applications of surface acoustic wave filter to TV receiver. *J. Inst. Telev. Eng. Jpn.* **31**, 845–852. In Japanese.

LaGrange, J. B., Daniels, W. D., and Lewandowski, R. L. (1977). SAW devices: The answer for channelized receivers? [Radar]. *Microwave Syst. News* **7**, 87–88, 90, 93–94, 96.

Lagasse, P. E., Vandewege, J., and Naten, T. (1978). Distributed feedback acoustic surface wave resonators for use in large bandwidth filters. *Ultrason. Symp. Proc.*, pp. 438–441.

Laker, K. R., Budreau, A. J., and Carr, P. H. (1976). A circuit approach to SAW filterbanks for frequency synthesis. *Proc. IEEE* **64**, 692–695.

Laker, K. R., Cohen, E., and Slobodnik, A. J., Jr. (1976). Electric field interactions within finite arrays and the design of withdrawal weighted SAW filters at fundamental and higher harmonics. *Ultrason. Symp. Proc.*, pp. 317–321.

Laker, K. R., Cohen, E., Szabo, T. L., and Pustaver, J. A., Jr. (1977). Computer-aided design of withdrawal weighted SAW bandpass, transversal filters. *Proc. IEEE Int. Symp. Circuits Syst.*, pp. 126–130.

Laker, K. R., Cohen, E., Szabo, T. L., and Pustaver, J. A., Jr. (1978). Computer-aided design of withdrawal-weighted SAW bandpass filters. *Proc. IEEE Trans. Circuits Syst.* **CAS-25**, 241–251.

Lakin, K., Joseph, T., and Penunuri, D. (1974). A surface acoustic wave planar resonator employing an interdigital electrode transducer. *Appl. Phys. Lett.* **25**, 363–365.

Langer, E. (1974). Surface wave filters. *6th International Congress on Microelectronics*, 100 pp. In German.

Langer, E. (1976). Surface wave filters for more integration. *Radio Mentor Electron.* **42**, 491–494. In German.

Lardat, C. (1976). Experimental performance of grooved reflective array compressors and resonators. *Ultrason. Symp. Proc.*, pp. 272–276.

Lardat, C. (1978). Surface acoustic wave components. *Rev. HF* **10**, 245–254. In French.

Leedom, D., Krimholtz, R., and Matthaei, G. (1971). Equivalent circuits for transducers having arbitrary even- or odd-symmetry piezoelectric excitation. *IEEE Trans. Sonics Ultrason.* **SU-18**, 128–140.

Lever, K. V. (1976). An appraisal of bandpass filter techniques. *International Specialist Seminar on the Impact of New Technologies in Signal Processing, Aviemore*, pp. 72–77.

Lever, K. V., Patterson, E., Steven, P. C., and Wilson, I. M. (1976). Surface-acoustic-wave matched filters for multifrequency shift keyed communication systems. *Proc. Inst. Electr. Eng.* **123**, 770–774.

Lever, K. V., Patterson, E., Stevens, P. C., and Wilson, I. M. (1976). S.A.W. 350 μs binary phase-shift-keyed matched filter. *Electron. Lett.* **12**, 116–117.

Lever, K. V., Patterson, E., Stevens, P. C., and Wilson, I. M. (1976). Surface acoustic wave matched filters for communications systems. *Radio Electron. Eng.* **46**, 237–246.

Lewis, M. F. (1973). Surface-acoustic-wave devices. *GEC J. Sci. Technol.* **39**, 156–162.

Lewis, M. F. (1973). Surface-acoustic-wave filters employing symmetric phaseweighted transducers. *Electron. Lett.* **9**, 138–140.

Lewis, M. F. (1982). SAW filters employing interdigitated interdigital transducers (I.I.D.T.). *Ultrason. Symp. Proc.*, pp. 12–17.

Lewis, M. F., Lowe, P. J., and Picken, W. G. (1982). MSK SAW filter to complement todays convolver. *Ultrason. Symp. Proc.*, pp. 256–261.

MacDonald, D. B. (1977). Fabrication of L-band pulse compression filters. *Ultrason. Symp. Proc.*, pp. 792–797.

Machida, M., Shibutani, M., Murayama, Y. (1980). ZnO SAW-filter for television PIF circuit. *Trans. Inst. Electron. Commun. Eng. Jpn. Sect. E* **63**, 482–483.

Mader, W. R., and Stocker, H. R. (1982). Extended impulse model for the design of precise SAW-filters on quartz. *Ultrason. Symp. Proc.*, pp. 29–34.

Mader, W., Stocker, H., and Tobolka, G. (1980). Diffraction if TV-IF filters using multistrip couplers. *Ultrason. Symp. Proc.*, pp. 294–297.

Mader, W., Stocker, H., and Veith, R. (1980). Compensation of diffraction effects on group delay time and stopband rejection SAW bandpass filters. *Ultrason. Symp. Proc.*, pp. 391–395.

Maines, J. D. (1972). Advances in surface acoustic wave pulse compression filters. *European Solid State Devices Research Conferences*, p. 181.

Maines, J. D., Moule, G. L., Newton, C. O., and Paige, E. G. S. (1975). A novel SAW variable-frequency filter. *Journees d'Electronique sur Technologie de Points pour le Traitement des Signaux* (*Lausanne, Switzerland: Ecole Polytech*), pp. 215–220.

Malocha, D. C. (1981). Surface wave devices using low loss filter technologies. *Ultrason. Symp. Proc.*, pp. 83–88.

Malocha, D. C., and Wilkus, S. (1978). Low loss capacitively weighted TV IF filter. *Ultrason. Symp. Proc.*, pp. 500–503.

Malocha, D. C., Goll, J. H., and Heard, M.A. (1979). Design of a compensated SAW filter used in a wide spread MSK waveform generator. *Ultrason. Symp. Proc.*, pp. 518–521.

Martin, G. (1978). Surface wave filter with weighting of electrode overlap without beam integration. *Nachrichtentech. Elektron.* **28**, 65–68. In German.

Matsu-ura, S., Hazama, K., and Murata, T. (1981). TV tuning systems with SAW comb filter. *IEEE Trans. Microwave Theory Tech.* **MTT-29**, 434–439.

Matthaei, G. L. (1973). Acoustic surface-wave transversal filters. *IEEE Trans. Circuit Theory* **CT-20**, 459–470.

Matthaei, G., Wong, D., and O'Shaughnessy, B. (1975). Simplification for the analysis of interdigital surface-wave devices. *IEEE Trans. Sonics Ultrason.* **SU-22**, 105–114.

Matthaei, G. L., Wong, D. Y., and O'Shaughnessy, B. P. (1976). Synthesis of two classes of acoustic surface-wave filter tap weights. *IEEE Trans. Microwave Theory Tech.* **MTT-24**, 1–9.

Matthaei, G. L., Savage, E. B., and Barman, F. (1977). Synthesis of acoustic-wave-resonator filters using any of various coupling mechanisms. *Int. Microwave Symp. Dig.*, pp. 328–331.

Matthaei, G. L., Savage, E. B., and Barman, F. (1978). Synthesis of surface-acoustic-wave-resonator band-pass filters using the 'direct-coupled-filter' point of view. *Proc. IEEE Int. Symp. Circuits Syst.*, pp. 541–547.

Matthaei, G. L., Savage, E. B., and Barman, F. (1978). Synthesis of acoustic-surface-wave resonator filters using any of various coupling mechanisms. *IEEE Trans. Sonics Ultrason.* **SU-25**, 72–84.

Matthews, H. (1977). "Surface Wave Filters." Wiley, Chichester, England.

Mayo, R. H. (1981). Surface acoustic waves. *New Electron.* **14**, 33–34.

McCusker, J. H., Perlman, S. S., and Veloric, H. S. (1976). Microsonic pulse filters-replacements for traditional Butterworth designs. *RCA Rev.* **37**, 389–405.

Melngailis, J., and Flynn, G. T. (1974). 16-Channel surface-acoustic-wave grating-filter bank for realtime spectral analyser. *Electron. Lett.* **10**, 107–109.

Melngailis, J., and Williamson, R. C. (1976). Surface-acoustic-wave device for Doppler filtering of radar burst waveforms. *Proc. IEEE MTT-S Int. Microwave Symp.*, pp. 289–291.

Miller, R. L., and De Vries, A. J. (1976). A simple 'building block' method for the design of SAW filters having non-linear phase response. *Ultrason. Symp. Proc.*, pp. 553–557.

Miller, R. L., and De Vries, A. J. (1976). Commercial application of a surface wave television IF filter. *Proc. IEEE MTT-S Int. Microwave Symp.*, pp. 318–320.

Milsom, R. F. (1977). A diffraction theory for SAW filters on nonparabolic high-coupling orientations. *Ultrason. Symp. Proc.*, pp. 827–833.

Milsom, R. F., Murray, R. J., Flinn, I., and Redwood, M. (1981). New orientation of lithium niobate for low bulk-wave degradation of SAW filter stopband. *Proc. IEEE Ultrason. Symposium*, 299–304.

Milsom, R. F., Murray, R. J., Flinn, I., and Redwood, M. (1981). Ultra low bulk orientations of lithium niobate for SAW TV filters. *Electron. Lett. (GB)* **17**, 89–91.

Minowa, J. (1978). A method for accurately adjusting the center frequency of surface acoustic wave filters. *Rev. Electr. Commun. Lab.* **26**, 797–807.

Minowa, J., and Tanaka, K. (1978). Narrow pass band surface acoustic wave filters for transmission system applications. *Rev. Electr. Commun. Lab.* **26**, 1675–1685.

Minowa, J., and Tanaka, K. (1978). Surface acoustic wave filters with narrow pass band. *Electr. Commun. Lab. Tech. J.* **27**, 419–428. In Japanese.

Minowa, J., Morikawa, T., and Abe, H. (1978). Stability for surface acoustic wave filters with narrow passband. *Electr. Commun. Lab. Tech. J.* **27**, 429–440. In Japanese.

Minowa, J., Nakagawa, K., Okuno, K., Kobayashi, Y., and Morimoto, M. (1978). 400 MHz SAW timing filter for optical fiber transmission systems. *Ultrason. Symp. Proc.*, pp. 490–493.

Minowa, J., Nakagawa, K., Okuno, K., Kobayashi, Y., and Morimoto, M. (1980). 800 MHz SAW timing filter for optical fibre transmission system. *Electron. Lett.* **16**, 35–36.

Mitchell, R. F. (1971). Acoustic surface wave filters. *Eurocon Dig.* 2 pp.

Mitchell, R. F. (1971). Acoustic surface-wave filters. *Philips Tech. Rev.* **32**, 179–189.

Mitchell, R. F. (1973). Surface acoustic wave transversal filters: their use and limitations. *International Specialist Seminar on Component Performance and Systems Applications of Surface Acoustic Wave Devices (England: IEE)*, pp. 130–140.

Mitchell, R. F. (1974). Surface acoustic wave devices and applications. IV. Bandpass filters. *Ultrasonics* **12**, 29–35.

Mitchell, R. F. (1976). Basics of SAW frequency filter design: a review. *Wave Electron.* **2**, 111–132.

Mitchell, R. F., and Parker, D. W. (1974). Synthesis of acoustic-surface-wave filters using double electrodes. *Electron. Lett.* **10**, 512.

Mitchell, R. F., and Stevens, R. (1975). Diffraction effects in small-aperture acoustic surface wave filters. *Wave Electron.* **1**, 201–218.

Mitsuyu, T., Ohji, K., Ono, S., Yamazaki, O., and Wasa, K. (1976). Thin-film surface-acoustic-wave devices. *Natl. Tech. Rep. (Japan)* **22**, 905–923. In Japanese.

Mitsuyu, T., Ono, S., and Wasa, K. (1980). 2.2 GHz SAW filters using ZnO/Al_2O_3 structure. *Jpn. J. Appl. Phys.* **20**, 99–102.

Mitsuyu, T., Yamazaki, O., and Wasa, K. (1981). A 4.4 GHz SAW filter using a single-crystal ZnO film sapphire. *Ultrason. Symp. Proc.*, pp. 74–77.

Miyashiro, F. (1978). TV SAW filters utilize X-112° $Y.LiTaO_3$ crystals. *JEE J. Electron. Eng.* (137), 46–47, 56–57, 70.

Moore, P. A., and Redwood, M. (1976). Acoustically coupled surface-wave resonators [filters]. *Electron. Lett.* **12**, 449–450.

Moore, P. A., Murray, R. J., White, P. D., and Garters, J. A. (1982). Surface acoustic wave filters for use in mobile radio. *Radio Electron. Eng.* **52**, 139–144.

Morgan, D. P. (1973). Surface acoustic wave devices and applications. I. Introductory review. *Ultrasonics* **11**, 121–131.

Morgan, D. P. (1978). Microwave acoustic devices. *Proc. 8th European Microwave Conf.*, pp. 378–388.

Morimoto, M., Kobayashi, Y., and Hibino, M. (1980). An optimal SAW filter design using FIR design technique. *Ultrason. Symp. Proc.*, pp. 298–230.

Morita, T. (1981). Quartz SAW filters. *JEE J. Electron. Eng.* **18**, 44–47.

Mortley, W. S. (1971). Wave propagation in alpha-quartz dispersive filters. *Marconi Rev.* **34**, 173–206.

Moule, G. L., Newton, C. O., and Paige, E. G. S. (1977). Performance of a surface acoustic wave variable slope chirp filter. *Ultrason. Symp. Proc.*, pp. 611–614.

Moulic, J. R. (1977). A broadband surface-wave filter with -50 dB stopbands and 1-dB passband ripple. *Ultrason. Symp. Proc.*, pp. 673–674.

Murakami, T. (1978). New color TV receiver with SAW IF filter. *IEEE Trans. Consum. Electron.* **CE-24**, 89–95.

Murphy, P. V. (1975). Surface acoustic wave devices. *J.R. Electr. Mech. Eng.* (25), 63–67.

Murray, R. J., and Neylon, S. (1978). The design, fabrication and performance limitations of narrow-band fast cut-off surface wave filters. *Ultrason. Symp., Proc.*, pp. 482–485.

Murray, R. J., and Read, E. (1977). Surface-acoustic-wave transversal bandpass filters. *GEC J. Sci. Technol.* **44**, 3–12.

Murray, R. J., and Schofield, J. (1980). The use of frequency-selective multistrip couplers in surface acoustic wave transversal filters. *Ultrason. Symp. Proc.*, pp. 288–293.

Murray, R. J., and White, P. D. (1981). SAW components answer today's signal-processing needs. *Electronics* **54**, 120–124.

Murray, R. J., and White, P. D. (1981). Surface acoustic wave devices [a practical guide to their use for engineers]. *Wireless World* **87**, 38–41.

Murray, R. J., and White, P. D. (1981). Surface acoustic wave devices. II. More on bandpass filters, delay lines and oscillators. *Wireless World* **87**, 79–82.

Nabeyama, H., Shimano, M., Hazama, K., Yamada, J., and Noro, Y. (1979). New color television receiver with SAW IF filter and one-chip PIF IC. *IEEE Trans. Consum. Electron.* **CE-25**, 50–59.

Nagy, J. (1974). Computational methods for acoustic surface wave filters. *Proc. 5th Colloquium on Microwave Communication. Vol. III. Electromagnetic Theory* ET-127/227–235.

Nevesely, M. (1979). An electric band-pass filter with elastic surface waves. *Elektrotech. Cas.* **30**, 569–577. In Czech.

Newton, C. O., and Paige, E. G. S. (1976). Surface acoustic wave dispersive filter with variable, linear, frequency-time slope *Ultrason. Symp. Proc.*, pp. 424–427.

Nishikawa, K., Higuchi, Y., and Narahara, K. (1980). Surface acoustic wave VSB filters for TV transmitters. *NEC Res. Dev.* (58), 9–14.

Nishikawa, T., Tani, A., Takeuchi, C., and Minowa, J. (1981). 1.6 GHz SH-type surface acoustic wave filter. *Jpn. J. Appl. Phys.* **20**, 29–32.

Nishiyama, S., and Otomo, J. (1978). 400 MHz SAW filters achieved with $LiNbO_3$. *JEE J. Electron. Eng.* (137), 48–51.

Nonaka, H., Arai, S., and Ieki, H. (1978). ZnO thin films provides economical SAW filters. *JEE J. Electron. Eng.* **137**, 36–38.

Noro, Y., Hazama, K. (1978). Surface acoustic wave filters. *Solid State Phys.* **13**, 359–365. In Japanese.

O'Clock, G. D., Jr., Gandolfo, D. A., and Bush, H. J. (1973). Acoustic surface-wave matched filters using a MOSFET array. *Proc. IEEE* **61**, 1165–1167.

O'Shea, T. F., and Rosenfeld, R. C. (1981). SAW resonator filters with optimized transducer rejection. *Ultrason. Symp Proc.*, pp. 105–110.

Ohyama, S. (1980). The principle and applications of SAW filters. *JEE J. Electron. Eng.* **17**, 48–51, 96.

Ono, M., Wakatsuki, N., and Tominaga, H. (1981). Statistical analysis of the effects of finger dispersion on the electrical characteristics of SAW filters. *Jpn. J. Appl. Phys.* **20**, 123–126.

Otomo, J., Nishiyama, S., Konno, Y., and Shibayama, K. (1977). UHF range SAW filters using group-type uni-directional interdigital transducers. *Proc. 31st Annu. Freq. Control Symp.*, pp. 275–280.

Otto, O. W., and Gerard, H. M. (1978). Nonsynchronous scattering loss in surface-acoustic-wave reflective-array-compression filters. *J. Appl. Phys.* **49**, 3337–3340.

Owens, J. M., and Smith, C. V., Jr. (1979). Surface acoustic wave and magnetostatic wave devices: a status report. *Proc. IEEE Int. Symp. Circuits Syst.*, pp. 568–571.

Paige, E. G. S. (1972). Surface acoustic wave devices: against the digital tide. *European Solia State Devices Research Conf.*, pp. 39–54.
Panasik, C. M. (1981). 250 MHz programmable transversal filter. *Ultrason. Symp. Proc.*, pp. 48–52.
Panasik, C. M. (1982). SAW programmable transversal filter for adaptive interference suppression. *Ultrason. Symp. Proc.*, pp. 100–103.
Panasik, C. M., and Hunsinger, B. J. (1976). Precise impulse response measurement of SAW filters. *IEEE Trans. Sonics Ultrason.* **SU-23**, 239–249.
Parker, D. W. (1976). Acoustic surface wave bandpass filters. *Proc. IEEE Int. Symp. Circuits Syst.*, pp. 689–692.
Parker, D. W. (1977). Acoustic surface wave bandpass filters. *Electron. Power* **23**, 389–392.
Parker, D. W., Pratt, R. G., Smith, F. W., and Stevens, R. (1976). Acoustic surface-wave bandpass filters. *Philips Tech. Rev.* **36**, 29–43.
Parker, D. W., Pratt, R. G., Smith, F. W., and Stevens, R. (1976). Acoustic surface wave filter. *Funk-Tech.* **31**, 811–814. In German.
Parker, D. W., Pratt, R. G., and Stevens, R. (1976). A television IF acoustic-surface-wave filter on bismuth silicon oxide. *Proce. IEEE* **64**, 677–681.
Parker, D. W., Pratt, R. G., Smith, F. W., and Stevens, R. (1977). Acoustic surface-wave bandpass filters. *Mullard Tech. Commun.* **14**, 110–124.
Parker, D. W., Pratt, R. G., Smith, F. W., and Stevens, R. (1977). Acoustic surface-wave bandpass filters. *Electron. Appl. Bull.* **35**, 24–39.
Parker, D. W., Pratt, R. G., Smith, F. W., and Stevens, R. (1977). Filters based on acoustic surface waves. III. *Funk-Tech.* **32**, 4–7. In German.
Peach, R. C., and Dix, C. (1978). A low loss medium bandwidth filter on lithium niobate. *Ultrason. Symp. Proc.*, pp. 509–512.
Peach, R. C., Doggett, N. H., McClemont, F. S., and Katsellis, A. (1981). The diffraction analysis and correction of narrowband SAW transversal filters. *Ultrason. Symp. Proc.*, pp. 58–62.
Penin, J. A. D. (1979). SAW devices as IF filters in TV receivers. *Mundo Electron.* (84), 97–102. In Spanish.
Penunuri, D., and Havens, D. P. (1979). Surface-acoustic-wave filters prove versatile in VHF applications. *Electronics* **52**, 115–120.
Penunuri, D., and Thoss, J. L. (1981). Low-loss, low spurious reflective dot array filter using three-phase transducers. *Ultrason. Symp. Proc.*, pp. 23–27.
Pieper, H., Zidek, E., and Hofmann, H. (1980). Group delay problems in surface wave frequency filters. *Nachrichtentech. Elektron.* **30**, 147–149. In German.
Piwonski, W. (1975). Surface wave piezoelectric filters. *Wiad. Telekomun.* **15**, 7–11. In Polish.
Plass, K. G., (1973). Acoustic surface wave band-stop filter for UHF frequencies. *European Microwave Conf. Vol. II. Brussels* C8.3/3 pp.
Potter, B. R. (1979). L-Band low loss filters. *Proc. 33rd Annu. Freq. Control Symp.*, pp. 396–401.
Potter, B. R. (1979). L-band low loss SAW filters. *Ultrason. Symp. Proc.*, pp. 533–536.
Potter, B. R., and Hartmann, C. S. (1977). Low loss surface-acoustic-wave filters. *IEEE Trans. Parts Hybrids Packag.* **PHP-13**, 348–353.
Potter, B. R., and MacDonald, D. B. (1981). SAW filter technology and applications. *Proc. 35th Annu. Freq. Control Symp.*, pp. 352–357.
Potter, B. R., and Shoquist, T. L. (1977). Multipassband low loss SAW filters. *Ultrason. Symp. Proc.*, pp. 736–739.
Puspoki, S. (1982). A quasi analytical method for acoustic surface wave TV filter's design. *Hiradastechnika* **33**, 241–247. In Hungarian.

Puspoki, S., Rosner, B., and Andrasi, M. (1976). Surface acoustic wave bandpass filters with asymmetric frequency response. *Alta Freq.* **45**, 766–768.

Qiu, P., Shui, Y. A., Chang, D., Jiang, W. H., and Wu, W. Q. (1982). Wideband low loss surface acoustic wave filters. *Ultrason. Symp. Proc.*, pp. 222–226.

Ragan, L. H. (1976). Surface acoustic wave resonator filters. *IEEE MTT-S Int. Microwave Symp.*, pp. 286–288.

Ralston, R. W., and Smythe, D. L. (1979). A SAW/CCD programmable matched filter. *Appl. Phys. Lett.* **35**, 388–390.

Redwood, M., and Topolevsky, R. B. (1975). Coupled-resonator acoustic-surface-wave filter. *Electron. Lett.* **11**, 253–254.

Reilly, J. P., Campbell, C. K., and Suthers, M. S. (1977). The design of SAW bandpass filters exhibiting arbitrary phase and amplitude response characteristics. *IEEE Trans. Sonics Ultrason.* **SU-24**, 301–305.

Romik, P., and Kogan, E. (1981). Electronically programmable surface acoustic wave matched filter. *Proc. MELECON (New York: IEEE)* 2.2.4/1–4.

Ronnekleiv, A., Skeie, H., and Hanebrekk, H. (1973). Design problems in surface wave filters. *International Specialist Seminar on Component Performance and Systems Applications of Surface Acoustic Wave Devices, Aviemore*, pp. 141–151.

Rosa, R. L., and Kerbel, S. J. (1976). Synthesis of transfer functions by parallel-channel SAW filter banks. *Ultrason. Symp. Proc.*, pp. 322–327.

Rosenberg, R. L., and Coldren, L. A. (1977). Fast synthesis of finite loss SAW resonator filters. *Ultrason. Symp. Proc.*, pp. 882–887.

Rosenberg, R. L., and Coldren, L. A. (1979). Crossed-resonator SAW filter: A temperature-stable wider-band filter on quartz. *Ultrason. Symp. Proc.*, pp. 836–840.

Rosenberg, R. L., and Coldren, L. A. (1979). Scattering analysis and design of SAW resonator filters. *Proc. IEEE Trans. Sonics Ultrason.* **SU-26**, 205–230.

Rosenberg, R. L., and Coldren, L. A. (1980). Broader-band transducer-coupled SAW resonator filters with a single critical masking step. *Ultrason. Symp. Proc.*, pp. 164–168.

Rosenfeld, R. C. (1974). Low-loss unidirectional acoustic surface wave filters. *Proc. 28th Annu. Freq. Control Symp.*, pp. 299–303.

Rosenfeld, R. C. (1979). Surface acoustic wave (SAW) bandpass filter review. *Proc. 33rd Annu. Freq. Control Symp.*, pp. 220–222.

Rosenfeld, R. C., and Hartmann, C. S. (1974). Subminiature broadband filters. Report TI-08-74-59 ECOM-72-0326-F, 62 pp. Contract DAAB07-72-C-0326. Available from NTIS.

Rosner, B., Puspoki, S., and Andrasi, A. (1974). Examination of structures of acoustic surface wave filters. *Hiradastechnika* **25**, 333–337. In Hungarian.

Roth, S. A., Zook, J. K., Nicholas, D., Ericson, K., and Drummond, W. S. (1978). Systems considerations in the design of a SAW-filter based television demodulator. *Conference Record of the Twelfth Asilomar Conference on Circuits, Systems and Computers*, pp. 568–572.

Rypkema, J., De Vries, A., and Banach, F. (1975). Engineering aspects of the application of surface wave filters in television IF's. *IEEE Trans. Consum. Electron.* **CE-21**, 105–114.

Sander, W., and Haydl, W. H. (1977). SAW filter application for phased array radar. *AGARD Conference Proceedings No. 230* 3.10/1–11.

Sandy, F., and Parker, T. E. (1976). Surface acoustic wave ring filter. *Proc. 30th Annu. Freq. Control Symp.*, pp. 334–339.

Sandy, F., and Parker, T. E. (1976). Surface acoustic wave ring filter. *Ultrason. Symp. Proc.*, pp. 391–396.

Sardaryan, V. S., and Tatikyan, L. M. (1976). The synthesis of narrow-band acoustic surface wave filter. *Izv. Akad. Nauk Arm. SSR Fiz.* **10**, 397–402. In Russian.
Sato, H., and Otomo, J. (1979). SAW filters cope with new communication systems. *JEE J. Electron. Eng.* **16**, 29–32.
Sato, H., Meguro, T., Yamanouchi, K., and Shibayama, K. (1974). Optimum cut for rotated Y-cut $LiNbO_3$ crystals used as the substrate of elastic surface wave filters. *J. Acoust. Soc. Jpn.* **30**, 549–556. In Japanese.
Sato, H., Yamanouchi, K., Shibayama, K., and Nishiyama, S. (1974). On the design of elastic surface wave filters with no tuning coil. *Proc. 28th Annu. Freq. Control Symp.*, pp. 286–298.
Sato, H., Meguro, T., Yamanouchi, K., and Shibayama, K. (1977). Small ripple acoustic surface wave filter using piezoelectric thin film unidirectional transducer. *Ultrason. Symp. Proc.*, pp. 740–743.
Sato, H., Meguro, T., Yamanouchi, K., and Shibayama, K. (1978). Piezoelectric thin film unidirectional SAW transducer and filter. *Proc. IEEE* **66**, 102–104.
Satoh, H. (1981). Expanding applications expected for lithium niobite SAW filters. *JEE J. Electron. Eng.* **18**, 48–51, 77.
Savage, E. B. (1980). Compensation for nonideal effects in surface acoustic wave interdigital filters, Ph.D. dissertation. University of California, Santa Barbara, California.
Savage, E. B. (1979). Fast computation of S.A.W. filter responses including diffraction. *Electron. Lett.* **15**, 538–539.
Savage, E. B., and Matthaei, G. L. (1979). Compensation for diffraction in SAW filters. *Ultrason. Symp. Proc.*, pp. 527–532.
Savage, E. B., and Matthaei, G. L. (1981). A study of some methods for compensation for diffraction in SAW IDT filters. *IEEE Trans. Sonics Ultrason.* **SU-28**, 439–448.
Schmidt, R. V. (1977). Cascaded SAW gratings as passband filters. *Electron. Lett.* **13**, 445–446.
Schmidt, R. V., and Cross, P. S. (1979). Externally coupled resonator-filter (ECRF). *IEEE Trans. Sonics Ultrason.* **SU-26**, 88–93.
Schmitt, P. (1980). Signal processing by acoustic surface waves. I. *Nachr. Elektron.* **34**, 263–266. In German.
Schmitt, P. (1981). Frequency-dispersive SAW filters for radar pulse compression. *Wiss. Ber. AEG-Telefunken* **54**, 64–78. In German.
Schoenwald, J. S. (1977). Diffraction loss compensation in very low shape factor wideband SAW filters. *Ultrason. Symp. Proc.*, pp. 706–709.
Schoenwald, J. S. (1978). Ultra low shape factor SAW filters using asymmetrically truncated transducers. *Ultrason. Symp. Proc.*, pp. 478–481.
Seguin, M., Knapp-ziller, M., and Foure, J.-L. (1979). Surface acoustic wave filters in telephony carrier systems. *Onde Electr.* **59**, 82–88. In French.
Seifert, F., and Reichard, H. (1978). Reduction of bulk waves in surface acoustic wave (SAW) filters. *Arch. Elektron. Uebertragungstech.*, **32**, 206–208. In German.
Seiler, D. G., Campbell, C. K., and Suthers, M. S. (1977). A technique for measuring and matching the input impedance of a narrow-band surface acoustic wave filter. *IEEE Trans. Instrum. Meas.* **IM-26**, 188–189.
Serra, A. (1979). The technology of surface wave filters. *Rev. Esp. Electron.* **26**, 40–43. In Spanish.
Setsune, K., Isobe, M., and Fujiwara, Y. (1979). Applications of SAW device to TV receiver. *Natl. Tech. Rep. (Japan)* **25**, 588–601. In Japanese.
Sharma, R. S., and Gupta, O. S. (1980). On the acoustic surface wave bandpass filters. *Electro-Technol.* **24**, 17–21.
Shibayama, K., Yamanouchi, K., Sato, H., and Meguro, T. (1976). Optimum cut for rotated Y-cut $LiNbO_3$ crystal used as the substrate of acoustic surface-wave filters. *Proc. IEEE* **64**, 595–597.

Shibayama, K., Yamanouchi, K., and Sato, H. (1978). UHF range surface acoustic wave filters using uni-directional interdigital transducers. *Proc. Jpn. Acad. Ser. B* **54**, 294–299.

Shibayama, K., Yamanouchi, K., and Sato, H. (1979). Recent trends in SAW filters. *Proc. IEEE Int. Symp. Circuits Syst.*, pp. 600–603.

Shibayama, K., Yamanouchi, K., Sato, H., and Meguro, T. (1981). GHz band SAW filter using group-type unidirectional transducer. *IEEE Trans. Sonics Ultrason.* **SU-28**, 91–95.

Shimizu, Y., and Sakaue, T. (1977). A simple technique of fine tuning of surface-acoustic wave filters. *Trans. Inst. Electron. Commun. Eng. Jpn. Sect. E.* **60**, 92.

Shiosaki, T., and Kawabata, A. (1977). 58 MHz surface-acoustic-wave TV-intermediate-frequency filter using ZnO-sputtered film. *Jpn. J. Appl. Phys.* **16**, 483–486.

Shiosaki, T., Ieki, E., and Kawabata, A. (1976). 58-MHz surface-acoustic wave video inter-mediate-frequency filter using ZnO-sputtered film. *Appl. Phys. Lett.* **28**, 475–476.

Shklarsky, D., Das, P. K., and Milstein, L. B. (1980). A SAW filter for SSB waveform generation. *IEEE Trans. Circuits Syst.* **CAS-27**, 464–468.

Shoquist, T. L., Stigall, R. E., and Hays, R. M. (1977). Surface acoustic wave components: products and applications. *EASCON (New York: IEEE)* 18-5A/1 pp.

Shreve, W. R. (1976). Surface wave resonators and their use in narrow-band filters. *Ultrason. Symp. Proc.*, pp. 706–713.

Siebeneicher, K. (1980). Tomorrow's technology for today's filters. *Siemens Comp. (Engl. ed.)* **15**, 167–169.

Sillioc, G. (1981). Piezoelectric filters from hundreds to tens of microns. *Toute Electron.* (468), 48–51. In French.

Singh, A. (1981). Frequency doubling of SAW filter. *Indian J. Pure Appl. Phys.* **19**, 370.

Singh, A. (1982). Surface acoustic wave reflective array compressor filter. *J. Inst. Electron. Telecommun. Eng.* **28**, 119–120.

Sinitsa, V. N., Basov, V. G., Spirin, V. A., Chirkin, N. M., and Kotova, I. F. (1975). An acoustic surface-wave non-dispersive filter. *Izv. Vyssh. Uchebn. Zaved. Radioelektron.* **18**, 98–99. In Russian.

Skeie, H., and Engan, H. (1976). Second-order effects in acoustic surface-wave filters: design methods. *Radio Electron. Eng.* **46**, 207–220.

Skudera, W. J., Jr. (1977). The versatility of the 'in-line' SAW chirp filter. *Proc. 31st Annu. Freq. Control Symp.*, pp. 285–290.

Skudera, W. J., Jr., and Gerard, H. M. (1973). Some practical design considerations of dispersive surface wave filters. *Proc. 27th Annu. Freq. Control Symp.*, pp. 253–261.

Slobodnik, A. J., Jr. (1975). Surface acoustic wave filters at UHF: design and analysis. Report AFCRL-TR-75-0311 (AD-A017 106/6GA), 672 pp. Available from NTIS.

Slobodnik, A. J., Jr., and Laker, K. R. (1975). Improved frequency and time domain response-surface acoustic wave filters for pulse applications. *Proc. IEEE Int. Symp. Circuits Syst.*, pp. 143–146.

Slobodnik, A. J., Jr., and Laker, K. R. (1976). Periodic frequency response SAW filters for a tree approach to many-tone frequency synthesis. *Proc. IEEE MTT-S Int. Microwave Symp.*, pp. 300–302.

Slobodnik, A. J., and Szabo, T. L. (1974). Synthesis of periodic apodized SAW filters in the presence of diffraction. *IEEE Microwave Symp. Dig. Tech. Pap.*, pp. 247–249.

Slobodnik, A. J., Budreau, A. J. Jr., Kearns, W. J., Szabo, T. L., and Roberts, G. A. (1976). SAW filters for frequency synthesis applications. *Ultrason. Symp. Proc.*, pp. 432–435.

Slobodnik, A. J., Jr., Laker, K. R., Szabo, T. L., Kearns, W. J., and Roberts, G. A. (1977). Low sidelobe SAW filters using overlap and withdrawal weighted transducers. *Ultrason. Symp. Proc.*, pp. 757–762.

Slobodnik, A. J., Jr., Roberts, G. A., Silva, J. H., Kearns, W. J., Sethares, J. C., and Szabo, T. L. (1978). UHF switchable SAW filterbanks. *Ultrason. Symp. Proc.*, pp. 486-489.
Slobodnik, A. J., Jr., Fenstermacher, T. E., Kearns, W. J., Roberts, G. A., Silva, J. H., and Noonan, J. P. (1979). SAW Butterworth contiguous filters at UHF. *IEEE Trans. Sonics Ultrason.* **SU-26**, 246-253.
Slobodnik, A. J., Jr., Roberts, G. A., Silva, J. H., Kearns, W. J., Sethares, J. C., and Szabo, T. L. (1979). Switchable SAW filter banks at UHF. *IEEE Trans. Sonics Ultrason.* **SU-26**, 120-126.
Slobodnik, A. J., Jr., Szabo, T. L., and Laker, K. R. (1979). Miniature surface-acoustic-wave filters. *Proc. IEEE* **67**, 129-146.
Slobodnik, A. J., Jr., Silva, J. H., and Roberts, G. A. (1981). SAW filters at 1 GHz fabricated by direct step on the wafer. *IEEE Trans. Sonics Ultrason.* **SU-28**, 105-106.
Smirnov, N. I., and Karavaev, Y. A. (1980). Effect of manufacturing accuracy of electrode on losses in a matched SAW composite-signal filter. *Izv. Vyssh. Uchebn. Zaved. Radioelektron.* **23**, 31-36. In Russian.
Smith, F. W. (1975). Acoustic surface wave bandpass filters. *SERT J.* **9**, 13-15.
Smith, W., Jr. (1973). Minimizing multiple transit echoes in surface wave devices. *Ultrason. Symp. Proc.*, pp. 410-413.
Smith, W., Jr., and Pedler, W. (1975). Fundamental- and harmonic-frequency circuit-model analysis of interdigital transducers with arbitrary metallization ratios and polarity sequences. *IEEE Trans. Microwave Theory Tech.*, **MTT-23**, 853-864.
Smith, W., Jr., and Pedler, W. (1976). Corrections to fundamental- and harmonic-frequency circuit-model analysis of interdigital transducers with arbitrary metallization ratios and polarity sequences. *IEEE Trans. Microwave Theory Tech.* **MTT-24**, 487.
Smith, W., Jr., Gerard, H., Collins, J., Reeder, T., and Shaw, H. (1969). Analysis of interdigital surface wave transducers by use of an equivalent circuit model. *IEEE Trans. Microwave Theory Tech.* **MTT-17**, 856-864.
Smith, W., Jr., Gerard, H., and Jones, W. (1972). Analysis and design of dispersive interdigital surface-wave transducers. *IEEE Trans. Microwave Theory Tech.* **MTT-20**, 458-481.
Smith, W. R. (1977). SAW filters for CPSM spread spectrum communication. *Ultrason. Symp. Proc.*, pp. 524-528.
Smith, W. R. (1980). Performance and applications of surface acoustic wave (SAW) bandpass and dispersive filters. *Proc. Soc. Photo-Opt. Instrum. Eng.* **239**, 170-182.
Smythe, R. C., and Howard, M. D. (1983). Current trends in crystal filters. *Proc. 37th Annu. Freq. Control Symp.*, pp. 349-37.
Snow, P. B. (1977). Matching networks and packaging structures [surface wave filters]. *In* "Surface Wave Filters" (H. Matthews, ed.), pp. 219-261. Wiley, England.
Solie, L. P. (1976). A SAW filter using a reflective dot array (RDA). *Ultrason. Symp. Proc.*, pp. 309-312.
Solie, L. P. (1977). A SAW bandpass filter technique using a fanned multistrip complex. *Appl. Phys. Lett.* **30**, 374-376.
Speiser, J. M., and Whitehouse, H. J. (1971). Surface wave transducer array design using transversal filter concepts. *In* "Acoustic Surface Wave and Acousto-Optic Devices" (T. Kallord, ed.), pp. 81-90. Optosonic Press, New York, VII, 221 pp.
Stahl, A., and Michel, J. P. (1982). Surface acoustic wave devices and optical fibres. *Toute Electron.* (472), 31-33. In French.
Staples, E. J., and Claiborne, L. T. (1973). A review of device technology for programmable surface-wave filters. *IEEE Trans. Microwave Theory Tech.* **MTT-21**, 279-287.
Staples, E. J., Schoenwald, J. S., Wise, J., and Lim, T. C. (1980). Low loss multipole SAW resonator filters. *IEEE MTT-S Int. Microwave Symp. Dig.*, pp. 34-36.

Staples, E. J., Wise, J. Schoenwald, J. S., and Lim, T. C. (1980). SAW resonator 2-pole filters. *Proc. 34th Annu. Freq. Control Symp.*, pp. 273–277.

Stevens, M. R., Parker, D. W., Penna, D. E., and Pratt, R. G. (1975). Acoustic surface wave filters. *International Symposium on Materials for Electronic Components (France)*, p. 291.

Stevens, R., White, P. D., Mitchell, R. F., Moore, P., and Redwood, M. (1977). Stopband level of 2-port SAW resonator filters. *Ultrason. Symp. Proc.*, pp. 905–908.

Stigall, R. E., and Hartmann, C. S. (1977). Phase and magnitude distortion compensation in a low-loss TV IF SAW filter. *Ultrason. Symp. Proc.*, pp. 729–732.

Stocker, H., Kowatsch, M., and Seifert, F. (1978). Technique, construction and application of SAW filters. *Elektronikschau* **54**, 26–30. In German.

Stocker, H. R., Eberharter, G., and Sprengel, H. P. (1980). Nonrecursive transverse filters in surface wave technique of microwave transmission. *NTG-Fachber.* **70**, 216–222. In German.

Stocker, H. R., Veith, R., Willibald, E., and Riha, G. (1981). Surface wave pulse compression filters with long chirp time. *Ultrason. Symp. Proc.*, pp. 78–82.

Stokes, R. B., Lau, K. F., Yen, K. H., Kagiwada, R. S., and Kong, A. M. (1981). Wideband third harmonic chirp filters. *Ultrason. Symp. Proc.*, pp. 28–32.

Stracca, G. B., and Macchiarell, G. (1982). A building block approach to the design of large bandwidth SAW filters. *Alta Freq.* **51**, 52–57.

Sudhakar, P., Bhattacharyya, A. B., and Mathur, B. (1978). SAW bandpass filter with -50 dB sidelobes using unweighted IDTs. *Electron. Lett.* **14**, 437–439.

Sugiyama, K., and Yoskikawa, S. (1977). Design of acoustic surface wave filters using phase-coded transducers. *Trans. Inst. Electron. Commun. Eng. Jpn. Sect. E.* **60**, 142–143.

Sugiyama, K., and Yoshikawa, S. (1977). Design of acoustic surface wave filter using multielectrode pair. *Electr. Commun. Lab. Tech. J.* **26**, 1755–1780. In Japanese.

Suthers, M. S., and Campbell, C. K. (1979). Use of a charge distribution model in SAW bandpass filter design. *Ultrason. Symp. Proc.*, pp. 565–569.

Suthers, M. S., Campbell, C. K., and Reilly, J. P. (1980). SAW bandpass filter design using Hermitian function techniques. *Proc. IEEE Trans. Sonics Ultrason.* **SU-27**, 90–93.

Suzuki, Y., Shimizu, H., Takeuchi, M., Nakamura, K., and Yamada, A. (1976). Some studies on SAW resonators and multiple-mode filters. *Ultrason. Symp. Proc.*, pp. 297–302.

Szabo, T. L., and Laker, K. R. (1979). Interdigital transducer models: their impact on filter synthesis. *IEEE Trans. Sonics Ultrason.* **SU-26**, 321–333.

Szabo, T. L., and Slobodnik, A. J., Jr. (1974). Diffraction compensation in periodic apodized acoustic surface wave filters. *IEEE Trans. Sonics Ultrason.* **SU-21**, 114–119.

Szabo, T. L., Slobodnik, A. J., Jr., and Laker, K. R. (1978). Computer-aided design of low sidelobe transversal filters. *Proc. IEEE International Symposium on Circuits and Systems*, 376–379.

Szabo, T. L., Cohen, E., and Laker, K. R. (1979). Interdigital transducer models for the accurate design of SAW transversal filters. *Proc. IEEE Int. Symp. Circuits Syst.*, pp. 604–608.

Takahashi, S. (1979). $LiTaO_3$ SAW filters [TV IF filters]. *Proc. IEEE Int. Symp. Circuits Syst.*, pp. 621–624.

Takahashi, S., and Kodama, T. (1977). SAW filter TTE source and load impedance dependence estimation. *Trans. Inst. Electron. Commun. Eng. Jpn. Sect. E.* **60**, 141–142.

Takahashi, S., Kodama, T., Myashiro, F., and Ebata, Y. (1977). Television IF surface acoustic wave filter. *Trans. Inst. Electron. Commun. Eng. Jpn. Sect. E.* **60**, 1–2.

Takahashi, S., Hirano, H., Kodama, T., Miyashiro, F., Suzuki, B., Onoe, A., Adachi, T., and Fujinuma, K. (1978). SAW IF filter on $LiTaO_3$ for color TV receivers. *IEEE Trans. Consum. Electron.* **EC-24**, 337–348.

Takshashi, S., Hirano, H., Miyashiro, F., and Kamiyama, H. (1979). $LiTaO_3$ surface acoustic wave filters [colour TV receiver IF filter]. *Toshiba Rev.* (Int. Ed.) (Japan), no. 123, 38–41.

Tancrell, R. H. (1973). Improvement of an acoustic-surface-wave filter with a multistrip coupler. *Electron. Lett.* **9**, 316–317.

Tancrell, R. H. (1974). Analytic design of surface wave bandpass filters. *IEEE Trans. Sonics Ultrason.* **SU-21**, 12–22.

Tancrell, R. H. (1977). Principles of surface wave filter design. *In* "Surface Wave Filters" (H. Matthews, ed.), pp. 109–64. Wiley, Chichester, England.

Tancrell, R. H., and Holland, M. G. (1970). Acoustic surface wave filters. *Ultrason. Symp. Proc.*, pp. 48–64.

Tancrell, R. H., and Holland, M. G. (1971). Acoustic surface wave filters. *Proc. IEEE* **59**, 393–409.

Tani, K. (1978). ZnO thin film SAW filters improve TV images. *JEE J. Electron. Eng.* (137), 39–45.

Tani, K. (1979). SAW filters rationalize TV production. *JEE J. Electron. Eng.* **16**, 24–29.

Tani, K., Senda, K., Niikawa, T., and Tazuke, K. (1978). ZnO surface acoustic wave filter for TV set. *Natl. Tech. Rep. (Japan)* **24**, 134–143. in Japanese.

Tanski, W. J. (1981). Multipole SAW resonator filters: elements of design, fabrication, and frequency trimming. *Ultrason. Symp. Proc.*, pp. 100–104.

Tanski, W. J. (1982). Multipole surface wave resonator filters. *Proc. 36th Annu. Freq. Control Symp.*, pp. 400–404.

Tanski, W. J., Block, M., and Vulcan, A. (1980). High performance SAW resonator filters for satellite use. *Ultrason. Symp. Proc.*, pp. 148–152.

Tanski, W. J., Meyer, P. C., and Solie, L. P. (1982). SAW filters for military and spacecraft applications. *Microwave J.* **25**, 53–62, 66.

Temmyo, J., and Yoshikawa, S. (1978). On the fabrication and performance of SAW delay line filters for GHz SAW oscillators. *IEEE Trans. Sonics Ultrason.* **SU-25**, 367–371.

Temmyo, J., and Yoshikawa, S. (1980). SAW bandpass filter design for 1.6 GHz PCM timing tank applications. *IEEE Trans. Microwave Theory Tech.* **MTT-28**, 846–851.

Terstegge, H. (1970). Acoustical surface waves used in a novel filter unit. *Int. Elektron. Rundsch.* **24**, 279–282. In German.

Theodossiou, L. (1976). Surface acoustic wave filters. *Television* **26**, 645–649.

Thoss, J. L., and Penunuri, D., and Thostenson, M. (1981). Implementation of reflective array matched filters for radar applications. *Ultrason. Symp. Proc.*, pp. 63–68.

Tiemann, J. J., and Young, J. D. (1976). Acoustic surface wave filter using chirp transform for NDT. *Ultrason. Symp. Proc.*, pp. 382–385.

Tiersten, H. F. (1976). Guided acoustic-surface-wave filters. *Appl. Phys. Lett.* **28**, 111–113.

Trankle, E. (1970). The surface-wave effect and its application in filters and ultrasonic delay lines. *Nachrichtentech. Z. (NTZ)* **23**, 436–439. In German.

Tsantes, J. (1979). Surface-acoustic-wave devices replace conventional filter components. *EDN* **24**, 69–78.

Tsukamoto, M. (1978). TTE suppressed surface acoustic wave filter. *Appl. Phy. Lett.* **33**, 559–560.

Tsukamoto, M. (1979). Characteristics of a SAW filter with three inclined and tapered transducers. *J. Appl. Phys.* **50**, 4136–3152.

Tsukamoto, M. (1979). Nonsymmetrical SAW filters with quarter wave reflectors. *IEEE Trans. Sonics Ultrason.* **SU-26**, 423–426.

Tsukamoto, M. (1979). Some types of SAW filters effective for TTE suppression. *Jpn. J. Appl. Phys.* **18**, 1471–1478.

Tsukamoto, M. (1980). Design of SAW filters having inclined and tapered three-transducer configuration. *Jpn. J. Appl. Phys.* **19**, 1291–1296.

Tsukamoto, M. (1980). Experimental results on SAW filters with one bidirectional and two unidirectional transducers. *Jpn. J. Appl. Phys.* **19**, 737–744.

Ugrinovic, K. (1979). Surface acoustic wave in the radar signal processing. *Elektrotechnika Zagreb* (3–6), 199–201. In Croatian.

Unkauf, M. G., Schulz, M. B., Gadoury, J. B. II. (1974). A programmable surface wave ultrasonic matched filter. Report RADC-TR-74-260, 120 pp. Available from NTIS.

Urabe, S., Asakawa, K., Itoh, M., Gokan, H., Esho, S., Nishikawa, K., and Nishihara, T. (1980). Low loss 0.9–1.9 GHz SAW filters with submicron finger period electrodes. *Ultrason. Symp. Proc.*, pp. 371–376.

Valov, V. I. (1979). Surface acoustic-wave filters connected in series. *Izv. Vyssh. Uchebn. Zaved. Radioelektron* **22**, 105–107. In Russian.

Van de Vaart, H., and Solie, L. P. (1977). A SAW pulse compression filter using the reflective dot array. *IEEE Int. Microwave Symp. Dig.*, pp 321–323.

Vasile, C. F. (1977). Two-port description of a SAW filter. *Electron. Lett.* **13**, 326–328.

Vasile, C. F., and Larosa, R. (1972). 1000 bit surface-wave matched filter. *Electron. Lett.* **8**, 479–480.

Veith, R., Kriedt, H., and Rehak, M. (1979). A video IF unit employing a surface wave filter I. *Funkschau* **51**, 226–232. In German.

Veith, R., Kriedt, H., and Rehak, M. (1979). IF amplifier for TV with surface wave filter. *Grundig Tech. Inf.* **26**, 110–114. In German.

Veith, R., Kriedt, H., and Rehak, M. (1981). IF-video section with surface wave filter. *Radioind. Electron.-Telev.* **5**, 83–88. In Italian.

Vlcek, M. (1982). A strip coupling element in filters with a surface acoustic wave. *Slaboproudy Obz.* **43**, 66–71. In Czech.

Vollers, H. G. (1973). High performance surface wave bandpass filters for signal processing applications. *IEEE International Convention and Exposition. vol. VI*, New York, 42.3/4 pp.

Wagers, R. (1976). Transverse electrostatic end effects in interdigital transducers. *Ultrason. Symp. Proc.*, pp. 536–539.

Wagers, R. (1978). Evaluation of finger withdrawal transducer admittances by normal mode analysis. *IEEE Trans. Sonics Ultrason.* **SU-25**, 85–92.

Wakatsuki, N., and Namikata, T. (1976). Suppression of bulk wave in SAW filter by metal over-lay design. *Ultrason. Symp. Proc.*, pp. 332–335.

Waldner, M., Pedinoff, M. E., and Gerard, H. M. (1972). Broadband surface wave nonlinear convolution filters. *IEEE Trans. Sonics Ultrason.* **SU-19**, 399.

Walther, F. G., Budreau, A. J., and Carr, P. H. (1973). Multiple UHF frequency generation using acoustic surface-wave filters. *Proc. IEEE* **61**, 1162–1163.

Wang, S-Y., Hwang, D-M., Chen, T. T., Lai, C. S., and Jiang, Y-F. (1979). A surface-acoustical-wave bandpass filter and its optical probing. *Annu. Rep. Inst. Phys. Acad. Sin.* **9**, 141–150.

Warne, D. H. (1980). Surface acoustic wave filters. *Colloquium on Electronic Filters (England: IEE)*, 10/1–4.

Wasa, K., and Hayakawa, S. (1981). ZnO thin film SAW filter. *Oyo Buturi* **50**, 580–591. In Japanese.

Wearden, T. (1981). SAW devices. *Electron. Prod. Des.* **2**, 55–59.

Webb, D. C. (1975). Surface acoustic wave devices for communications. *International Telemetering Conference*, pp. 581–593.

Webb, D. C. (1978). SAW filters simplify signal sorting. *Microwave Syst. News* **8**, 75, 77–78, 81, 83–84.

Webster, R. T. (1982). Programmable SAW transversal filters. *Microwave J.* **25**, 139–140, 142–146, 148, 150.

Weglein, R. D. (1973). The characteristics of acoustic surface wave grating filters. *European Microwave Conference. vol. II* C8.4/4 pp.

Weglein, R. D. (1976). SAW chirp filter performance above 1 GHz. *Proc. IEEE* **64**, 695–698.

Weglein, R. D., and Otto, O. W. (1974). Characteristics of periodic acoustic-surface-wave grating filters. *Electron. Lett.* **10**, 68–69.

Weglein, R. D., and Wolf, E. D. (1973). The microwave realization of a simple surface wave filter function. *IEEE G-MTT Int. Microwave Symp. Dig. Tech. Pap.*, pp. 120–122.

Weinert, R. W., Isaacs, T. J., Halgas, F., Godfrey, J. T., and Moore, R. A. (1976). Performance of SAW filters on sulfosalt-type substrates. *Ultrason. Symp. Proc.*, pp. 532–535.

White, P. D., and Stevens, R. (1978). Surface acoustic wave resonator filters. *Conference on Radio Receivers and Associated Systems (London: IERE)*, pp. 93–100.

Williamson, R. C. (1977). Reflection grating filters. *In* "Surface Wave Filters" (H. Matthews, ed.), pp. 381–442. Wiley, Chichester, England.

Wise, J., Schoenwald, J., and Staples, E. (1980). Impedance characterization and design of 2-pole hybrid SAW resonator filters. *Ultrason. Symp. Proc.*, pp. 200–203.

Worley, J. C. (1973). Bandpass filtering using surface wave delay line. *IEEE Trans. Sonics Ultrason.* **SU-19**, 401.

Worley, J. C. (1973). Bandpass filters using nonlinear FM surface-wave transducers. *IEEE Trans. Microwave Theory Tech.* **MTT-21**, 302–303.

Yamada, J., Ishigaki, M., Hazama, K., and Toyama, T. (1978). Design and mass productive fabrication techniques of high performance SAW TV IF filter. *Ultrason. Symp. Proc.*, pp. 504–508.

Yamaguchi, M., and Kogo, H. (1976). An elastic surface wave filter for a television receiver with suppressed triple transit echo and multiple passband. *J. Inst. Telev. Eng. Jpn.* **30**, 191–197. In Japanese.

Yamaguchi, M., and Kogo, N. (1976). Design of surface-acoustic-wave filters by taking account of electrical terminations and matching circuits. *Electron. Lett.* **12**, 181–182.

Yamaguchi, M., and Kogo, H. (1978). Design of surface-acoustic-wave television i.f. filters by use of constrained damped least squares. *IEE J. Microwave Opt. Acoust.* **2**, 60–64.

Yamaguchi, M., Temma, T., Takato, K., and Kogo, H. (1977). Frequency response of a surface-acoustic-wave filter using a waveguide. *Electron. Lett.* **13**, 204–205.

Yamaguchi, M., Hashimoto, K., and Kogo, H. (1978). Withdrawal of interdigital electrodes for sidelobe reduction of SAW filters by implicit enumeration algorithm for 0–1 type integer optimization. *Trans. Inst. Electron. Commun. Eng. Jpn. Sect. E.* **61**, 631–632.

Yamaguchi, M., Hashimoto, K., and Kogo, H. (1979). Suboptimization of electroded-withdrawal weighted SAW filters. *IEEE Trans. Sonics Ultrason.* **SU-26**, 53–59.

Yamaguchi, M., Hashimoto, K.-Y., and Kogo, H. (1979). A simple method of reducing side-lobes for electrode-withdrawal weighted SAW filters. *IEEE Trans. Sonics Ultrason.* **SU-26**, 334–339.

Yamaguchi, M., Temma, T., and Kogo, H. (1979). Waveguide-type s.a.w. filter using energy focusing interdigital transducers. *IEE J. Microwave Opt. Acoust.* **3**, 161–168.

Yamanouchi, K., and Shibayama, K. (1978). Low insertion loss acoustic surface wave filters. *Oyo Buturi* **47**, 1170–1175. In Japanese.

Yamanouchi, K., Nyffeler, F. M., and Shibayama, K. (1977). Low insertion loss acoustic surface wave filter using group-type unidirectional interdigital transducers. *J. Acoust. Soc. Jpn.* **33**, 532–539. In Japanese.

Yamanouchi, K., Meguro, T., and Shibayama, K. (1980). Acoustic surface wave filters using new distance weighting techniques. *Ultrason. Symp. Proc.*, pp. 313–316.

Yamanouchi, K., Meguro, T., and Gautam, J. K. (1982). Low-loss GHz range SAW filter using group-type unidirectional transducer-new GUDT and new phase shifter. *Ultrason. Symp. Proc.*, pp. 212–217.

Yamazaki, O., Wasa, K., and Hayakawa, S. (1980). Highly reliable ZnO thin film SAW Nyquist filters for TV. *Ultrason. Symp. Proc.*, pp. 382–385.
Yen, K. H., Lau, K. F., and Kagiwada, R. S. (1978). Shallow bulk acoustic wave filters. *Ultrason. Symp. Proc.*, pp. 680–683.
Yen, K. H., Lau, K. F., Stokes, R. B., Kong, A., and Kagiwada, R. S. (1982). Developments in low-loss, low-ripple SAW filters. *Proc. 36th Annu. Freq. Control Symp.*, pp. 396–399.
Yen, K. H., Stokes, R. B., Lau, K. F., Kong, A., and Kagiwada, R. S. (1982). Low-loss low-ripple SAW filters using three bidirectional centrosymmetric transducers. *Electron. Lett.* **18**, 403–404.
Zelyakh, E. V., and Novikov, A. A. (1977). Active rejection filters with piezoelectric resonators. *Telecomm. Radio Eng. Pt. 1* **31**, 45–50.
Zhang, D. (1978). Suppression of direct transmission in SAW resonator filters by inverse phase IDT. *Acta Phys. Sin.* **27**, 349–352. In Chinese.
Znamenskiy, A. E. (1977). Filters using acoustic surface waves. *Telecommun. Radio Eng. Pt. 1* **31**, 10–15.
Znamenskiy, A. E. (1978). Features of the synthesis of surface acoustic wave filters with given amplitude-frequency response. *Telecommun. Radio Eng. Pt. 2* **33**, 68–70.
Znamenskiy, A. E. (1980). Attenuation of acoustic surface wave filters in the pass-band. *Telecommun. Radio Eng. Pt. 2* **35**, 50–53.
Znamenskiy, A. E., and Krylov, L. N. (1980). Realization of acoustic surface wave broadband filters. *Telecommun. Radio Eng. Pt. 1* **34**, 31–35.
Znamenskiy, A. E., and Muratov, E. S. (1979). Experimental investigation of narrowband matched SAW filters. *Izv. Vyssh. Uchebn. Zaved. Radioelektron.* **22**, 71–73. In Russian.
Znamenskiy, A. E., and Muratov, E. S. (1980). A SAW rejection filter. *Izv. Vyssh. Uchebn. Zaved. Radioelektron.* **23**, 69–72. In Russian.
Znamenskiy, A. E., and Muratov, E. S. (1981). A narrowband matched surface acoustic wave filter with reflectors. *Telecommun. Radio Eng. Pt. 1* **35**, 48–52.
Znamenskiy, A. E., Muratov, E. S., and Gulin, V. N. (1980). Features of the construction of narrowband filters for acoustic surface waves. *Telecommun. Radio Eng. Pt. 2* **35**, 62–67.
Zucker, J., and Ganguly, A. K. (1973). Frequency response of acoustic surface wave filters. II. *J. Appl. Phys.* **44**, 1086–1088.
Zuliani, M., Ristic, V., and Stegeman, G. (1975). Field theory of interdigital transducers. *Ultrason. Symp. Proc.*, pp. 453–457.
Zverev, A. I. (1967). "Handbook of Filter Synthesis." John Wiley and Sons, New York.

Chapter 6

Adams, C. A., and Kusters, J. A. (1978). Improved long-term aging in deeply etched SAW resonators. *Proc. 32nd Annu. Freq. Control Symp.*, pp. 74–76.
Armstrong, J. H., Blomster, P. R., and Hokanson, J. L. (1966). Aging characteristics of quartz crystal resonators. *Proc. 20th Annu. Freq. Control Symp.*, pp. 192–207.
Ballato, A. (1977). Doubly rotated thickness mode plate vibrators. *In* "Physical Acoustics: Principles and Methods" (W. P. Mason and R. N. Thurston, eds.), **13**, 115–181. Academic Press, New York.
Ballato, A., EerNisse, E. P., and Lukaszek, T. J. (1978). Experimental verification of stress compensation in the SC-cut. *Ultrason. Symp. Proc.*, pp. 144–147.
Barton, R. K., and Gratze, S. C. (1975). Surface acoustic wave oscillators: Long term stability. *Electron. Eng. (London)* **47**, 49–51.
Bechmann, R. (1942). Properties of quartz oscillators and resonators in the range from 300 to 5000 kc/s. *Hochfrequenztechnik und Elektroakustik.* **59**, 97–105.

Bell, D. T., Jr. (1977). Aging processes in SAW resonators. *Ultrason. Symp. Proc.*, pp. 851–856.
Bell, D. T., Jr., and Miller, S. P. (1976). Aging effects in plasma etched SAW resonators. *Proc. 30th Annu. Freq. Control Symp.*, pp.358–362.
Belser, R. B., and Hicklin, W. H. (1969). Comparison of aging performance of 5 MHz resonators plated with various electrode metals. *Proc. 23rd Annu. Freq. Control Symp.*, p. 132–142.
Bernstein, M. (1968). Precision measurement of the frequency aging of quartz crystal units. *Proc. 22nd Annu. Freq. Control Symp.*, pp. 232–247.
Bernstein, M. (1970). Precision measurement of the frequency aging of quartz crystal units. Report ECOM-3227 (AD-703841), Army Electronics Command, Fort Monmouth, New Jersey.
Bessol'tsev, V. A., Yefremov, O. N., and Nevolin, V. K. (1976). The effect of the residual gas pressure on the frequency instability of vacuum quartz resonators. *Radio Eng. Electron. Phys.* **21**, 153–154.
Besson, R. (1976). A new piezoelectric resonator design. *Proc. 30th Annu. Freq. Control Symp.*, pp. 78–83.
Bloch, M. B., and Denman, J. L. (1974). Further development on precision quartz resonators. *Proc. 28th Annu. Freq. Control Symp.*, pp. 73–84.
Bruggemann, P., and Muller, F. (1974). A universal aging process for quartz resonators. *Fernmeldetechnik* **14**, 134–136. In German.
Byrne, R. J., and Hokanson, J. L. (1968). Effect of high-temperature processing on the aging behavior of precision 5 MHz quartz crystal units. *IEEE Trans. Instrum. Meas.* **IM-17**, 76–79.
Byrne, R. J., and Reynolds, R. L. (1964). Design and performance of a new series of cold welded crystal unit enclosure. *Proc. 18th Annu. Freq. Control Symp.*, pp. 166–180.
Chaban, A. A. (1967). Instability of elastic oscillations in piezoelectrics in alternating electric fields. *JETP Lett.* (*Engl. Transl.*) **6**, 381–383.
Demchuk, M. I., Dmitriev, S. M., and Chernyavskii, A. F. (1977). Investigation of short-term fluctuations of the frequency in quartz shock-excitation generators. *Meas. Tech.* (*Engl. Transl.*) **20**, 721–725.
Dick, L. A. and Silver, J. F. (1970). Low aging crystal units in temperature compensated oscillator. *Proc. 24th Annu. Freq. Control Symp.*, pp. 141–147.
Dordevic, L., and Samardzija, M. (1968). Aging of quartz crystal units. *Elektroteh. Vestnik* (*Yugoslavia*) **35**, 1–3, 8–11. In Slovenian.
Dybwad, G. L. (1977). Aging analysis of quartz crystal units with Ti Pd Au electrodes. *Proc. 31st Annu. Freq. Control Symp.*, pp. 144–146.
EerNisse, E. P. (1975). Quartz resonator frequency shifts arising from electrode stress. *Proc. 29th Annu. Freq. Control Symp.*, pp. 1–4.
Efremov, O. N., Lyubimov, L. A., Fomicheva, Z. I., Shin, V., and Yaroslavskii, M. I. (1975). Aging of precision quartz resonators. *Meas. Tech.* (*Engl. Transl.*) **18**, 442–443.
Engdahl, J., and Matthey, H. (1975). 32 kHz quartz unit for high precision wrist watch. *Proc. 29th Annu. Freq. Control Symp.*, pp. 187–194.
Forrer, M. P. (1969). A flexure-mode quartz for an electronic wrist watch. *Proc. 23rd Annu. Freq. Control Symp.*, pp. 157–162.
Gagnepain, J.-J. (1976). Fundamental noise studies of quartz crystal resonators. *Proc. 30th Annu. Freq. Control Symp.*, pp. 84–91.
Gerber, E. A., and Sykes, R. A. (1966). State-of-the-art quartz crystal units and oscillators. *Proc. IEEE* **54**, 103–116.
Gerber, E. A., and Sykes, R. A. (1974). State of the art-quartz crystal units and oscillators. *In* "Time and Frequency" (B. E. Blair, ed.). pp. 43–64. National Bueran of Standards, Monograph 140, Washington, D.C.

Gilden, M., Montress, G. K., and Wagner, R. A. (1980). Long-term aging and mechanical stability of 1.4 GHz SAW oscillators. *Ultrason. Symp. Proc.*, pp. 184–187.
Gniewinska, B. (1971). The effect of crystal current level on frequency stability of quartz crystal oscillators. *Pr. Inst. Tele. Radiotech.* **15**, 73–79. In Polish.
Hafner, E., and Blewer, R. S. (1968a). Low aging quartz crystal units. *Proc. IEEE* **56**, 366–368.
Hafner, E., and Blewer, R. S. (1968b). Quartz crystal aging. *Proc. 22nd Annu. Freq. Control Symp.*, pp. 136–154.
Helmick, C. N., Jr., and White, D. J. (1978). Observations of aging and temperature effects on dielectric-coated SAW devices. *Ultrason. Symp. Proc.*, pp. 580–585.
Hoffman, D. M. (1974). The structure and properties of thin metal films. *Proc. 28th Annu. Freq. Control Symp.*, pp. 85–88.
Holland, R. (1974). Nonuniformly heated anisotropic plates: II. frequency transients in AT and BT quartz plates. *Ultrason. Symp. Proc.*, pp. 592–598.
Jamiolkowski, J. (1971). High stability quartz crystal resonator aging equipment. *Pr. Inst. Tele. Radiotech.* **15**, 45–59. In Polish.
Jaroslavsky, M. I., and Lavrentsov, V. D. (1982). Long-term frequency variations of quartz crystal units under different environmental conditions. *Ferroelectrics* **43**, 51–56.
Kanbayashi, S., Okano, S., Hirama, K., Kudama, T., Konno, M., and Tomikawa, Y. (1976). Analysis of tuning fork crystal units and application into electronic wrist watches. *Proc. 30th Annu. Freq. Control Symp.*, pp. 167–174.
Latham, J. I., and Saunders, D. (1978). Aging and mounting developments for SAW resonators. *Ultrason. Symp. Proc.*, pp. 513–517.
Lubentsov, V. F., and Shustrova, L. A. (1969). Effect of high-temperature annealing on the parameter of Y (xy) cut of quartz bars. *Meas. Tech. (Engl. Transl.)* **12**, 1127–1128.
Lukaszek, T. J., and Ballato, A. (1980). What SAW can learn from BAW. *Ultrason. Symp. Proc.*, pp. 173–183.
Lukaszek, T. J., and Ballato, A. (1979). Resonators for severe environments. *Proc. 33rd Annu. Freq. Control Symp.*, pp. 311–321.
Lysakowska, M. (1979). Mid-stability quartz crystal units. *Pr. Inst. Tele. Radiotech.* **72**, 63–66. In Polish.
Malakhov, A. N., and Solin, N. N. (1969). Amplitude and frequency fluctuations of quartz oscillators. *Izv. Vyssh. Ichebn. Zaved. Radiofiz* **12**, 529–537. In Russian.
Parker, T. E. (1977a). Aging characteristics of SAW controlled oscillators. *Ultrason. Symp. Proc.*, pp. 862–866.
Parker, T. E. (1977b). Current developments in SAW oscillator stability. *Proc. 31st Annu. Freq. Control Symp.*, pp. 359–364.
Parker, T. E. (1978). Frequency stability of surface wave controlled oscillators. *Proc. IEEE Circuits and Systems Symp.*, pp. 558–562.
Parker, T. E. (1980). Analysis of aging on SAW oscillators. *Proc. 34th Annu. Freq. Control Symp.*, pp. 292–301.
Parker, T. E. (1982a). Development of precision SAW oscillators for military applications. *Proc. 36th Annu. Freq. Control Symp.*, pp 453–458.
Parker, T. E. (1982b). Precision surface acoustic wave (SAW) oscillators. *Ultrason. Symp. Proc.*, pp. 268–274.
Parker, T. E. (1983a). Very long period random frequency fluctuations in SAW oscillators. *Proc. 37th Annu. Freq. Control Symp.*, pp. 410–414.
Parker, T. E. (1983b). Random and systematic contributions to long-term frequency stability in SAW oscillators. *Ultrason. Symp. Proc.*, pp. 257–262.
Parker, T. E., and Schulz, M. B. (1975). Stability of SAW controlled oscillator. *Ultrason. Symp. Proc.*, pp. 261–263.

Quesada, V. (1977). Study of the instability of quartz resonators. *Mem. Soc. Astron. Ital.* **48**, 659–663. In Italian.

Schoenwald, J. S., Harker, A. B., Ho, W. W., Wise, J., and Staples, E. J. (1980). Surface chemistry related to SAW resonator aging. *Ultrason. Symp. Proc.*, pp.193–199.

Schoenwald, J. S., Wise, J., and Staples, E. J. (1981). Absolute and differential aging of SAW resonator pairs. *Proc. 35th Annu. Freq. Control Symp.*, pp. 383–387.

Shreve, W. R. (1977). Aging in quartz SAW resonators. *Ultrason. Symp. Proc.*, pp. 857–861.

Shreve, W. R. (1980). Active aging of SAW resonators. *Ultrason. Symp. Proc.*, pp. 188–192.

Shreve, W. R. (1982). Private communication.

Shreve, W. R., Kusters, J. A., and Adams, C. A. (1978). Fabrication of SAW resonators for improved long term aging. *Ultrason. Symp. Proc.*, pp. 573–579.

Shreve, W. R., Bray, R. C., Elliot, S., and Chu, Y. C. (1981). Power dependence of aging in SAW resonators. *Ultrason. Symp. Proc.*, pp., 94–99.

Silver, J. F., and Dick, L. A. (1970). Low aging crystal units for use in temperature compensated oscillators. *Proc. 24th Annu. Freq. Control Symp.*, pp. 141–147.

Simpson, P. A., and Morgan, A. H. (1959). Quartz crystals at low temperatures. *Proc. 13th Annu. Freq. Control Symp.*, pp. 207–231.

Smagin, A. G. (1975). A low-temperature oscillator with a frequency stability 4×10^{-14}. *Instrum. Exp. Tech. (Engl. Transl.)* **18**, 1853–1855.

Smagin, A. G. (1977). Very stable crystal oscillators aging less than 10^{-10} per month. *Instrum. Exp. Tech. (Engl. Transl.)* **20**, 188–189.

Smolarki, A., and Wojcicki, M. (1975). Quartz crystal resonator aging meter, *Pr. Inst. Tele. Radiotech.* **19**, 77–84. In Polish.

Stokes, R. B. (1982). Propagation loss effects on SAW oscillator aging. *Ultrason. Symp. Proc.*, pp. 275–278.

Stokes, R. B., and Delaney, M. J. (1983). Aging mechanisms in SAW oscillators. *Ultrason. Symp. Proc.*, pp. 247–256.

Sykes, R. A., Smith, W. L., and Spencer, W. L. (1963). Studies on high precision resonators. *Proc. 17th Annu. Freq. Control Symp.*, pp. 4–27.

Valdois, M., Besson, J., and Gagnepain, J.-J. (1974). Influence of environment conditions on a quartz resonator. *Proc. 28th Annu. Freq. Control Symp.*, pp. 19–32.

Vig, J. R. (1977). Resonator aging. *Ultrason. Symp. Proc.*, pp. 848–849.

Vig, J. R., LeBus, J. W., and Filler, R. L. (1975). Further results on UV cleaning and Ni electrobonding. *Proc. 29th Annu. Freq. Control Symp.*, pp. 220–226.

Wainwright, A. E., Walls, F. L., and McCaa, W. D. (1974). Direct measurements of the inherent frequency stability of quartz crystal resonators. *Proc. 28th Annu. Freq. Control Symp.*, pp. 177–180.

Warner, A. W. (1963). Use of parallel-field excitation in the design of quartz crystal units. *Proc. 17th Annu. Freq. Control Symp.*, pp. 248–266.

Warner, A. W., Fraser, D. B., and Stockbridge, C. D. (1965). Fundamental studies of aging in quartz resonators. *IEEE Trans. Sonics Ultrason.* **SU-12**, 52–59.

Wilcox, P. D., Snow, G. S., Hafner, E., and Vig, J. R. (1975). A new ceramic flat pack for quartz resonators. *Proc. 29th Annu. Freq. Control Symp.*, pp. 202–210.

Yaroslavskii, M., Sorokin, K. V., Lavrova, T. P., Karaul'nik, A. E., Petrozhitskaya, I. N., Motin, P. E., Shin, V., Efremov, O. N., and Lyubimov, L. A. (1977). High stability quartz resonators. *Meas. Tech. (Engl. Transl.)* **20**, 1175–1177.

Yoda, H., Ikeda, H., and Yambe, Y. (1972). Low power oscillation for electric wrist watch. *Proc. 26th Annu. Freq. Control Symp.*, pp. 140–147.

Index to Volumes 1 and 2

Bold faced numerals indicate the volume in which the following pages appear. For subjects that appear on two or more consecutive pages, only the first page in that range is listed.

A

Absorber, nonlinear, **2:** 188
Absorption
　cell, **2:** 188
　dip, **2:** 183
　of gases, **1:** 275
Acceleration, due to gravity, **2:** 269
　sensitivity, **1:** 182
　transducer, **2:** 292
Accuracy, **2:** 126
　atomic standards, **1:** 126
Acoustic loss, **1:** 148, 150
Acoustic relaxation, **1:** 150, 154
Acoustic properties, modification, **1:** 280
Acoustic transmission, line, **1:** 249
Acoustical coupling, **1:** 199
Active hydrogen maser, **2:** 162
Additive noise, **2:** 275, 278
Admittance. *See also* Immittance
　acoustic, **1:** 233
　crystal, **2:** 50, 51
　transducer, **1:** 251
AGC, **2:** 59
Aging, **1:** 271
　causes of, **1:** 273
　influence of mounting and adhesives, **1:** 282
　influence of operating power, **1:** 282
　isolation of causes, **1:** 277
　as one-rate process, **1:** 277
　SAW delay lines, **1:** 283
　of surface-wave (SAW) devices, **1:** 279
　of SAW resonators with etched groove arrays, **1:** 281
　tuning fork, **1:** 273
Aging processes, **1:** 273
Aging rate
　for low-frequency crystal, **1:** 272
　of thickness-shear resonators, **1:** 273
Air-abrasive unit, **1:** 169
Airgap electrodes, **1:** 277
Allan variance, **2:** 201
Alpha-quartz. *See also* Quartz
　Al-hole center, **1:** 43
　aluminum concentration, **1:** 44
　conductivity, **1:** 44
　determination of Q, **1:** 41
　dielectric loss, **1:** 44
　electrical conductivity, **1:** 43
　electron spin resonance, **1:** 43
　infrared absorption, **1:** 41, 44
　interstitial ions, **1:** 42
　radiation hardness, **1:** 44
　resonator resistance, **1:** 44
　specific heat, **1:** 40
　sweeping effectiveness, **1:** 42
　thermal conductivity, **1:** 40
　thermal diffusivity, **1:** 40

Aluminum phosphate (berlinite), **1:** 20
Admittance circle, **2:** 9
Ammonia, **2:** 149
Ammonia maser, **2:** 115, 169
Amplifier
  sustaining, **2:** 55
  phase errors, **2:** 60
Amplitude
  measurements, **2:** 32
  of oscillation, **2:** 289
  transmission factor, **1:** 134
Analysis, statistical, **2:** 197
Anelastic absorption, **1:** 157
Anelastic loss, **1:** 157
Anelastic process, **1:** 150
Anelasticity, **1:** 36
Angle, crystallographic, **1:** 71. *See also* Cut
Anisotropy. *See* Crystal
Annealing, **1:** 153
Annealing process, **1:** 154
Anodization, **1:** 174
Antiresonance, **2:** 15
Antisymmetric modes, **1:** 201
Apodization, **1:** 263
Apodization weighting, **1:** 264
Application in science of atomic standards, **2:** 174
  gravitational potential, **2:** 174
  gravitational red shift, **2:** 174
  relativity, **2:** 174
Application in technology of atomic standards, **2:** 175
  communication, **2:** 175
  navigation, **2:** 175
Approximation. *See also* Simulation; Theory
  Mindlin, **1:** 60
  quasi-static, **1:** 52
  thin plate, **1:** 56, 60
    power series, **1:** 90, 93
    trigonometric, **1:** 93
Array, reflective, **1:** 126
Array grooves, **1:** 122, 127
AT cut, **1:** 182, 213
Atmospheric effects, **2:** 251
Atom detector, **2:** 142
Atomic clock, **2:** 115
Atomic hydrogen maser, **2:** 161
  frequency stability, **2:** 163
Atomic clock, **2:** 116
  accuracy, **2:** 116
  long-term measurement, **2:** 219
  primary standards, **2:** 116
  time-keeping potential, **2:** 125
Atomic resonator, **2:** 122
Atomic standard, passive, **2:** 126
Atomic time, international (TAI), **2:** 268
Atomic transition, **2:** 177
Autoclave, **1:** 26
Axis. *See* Coordinate

B

B mode, **2:** 295
Balanced bridge, **2:** 25, 27, 30, 40
Balanced-bridge measurement, **2:** 36
Bandpass shaping, **1:** 238
Bandstop filters, SAW, **1:** 266
Bandwidth
  hardware, **2:** 205
  software, **2:** 205
  transition, **1:** 269
Bar, thin, narrow, **1:** 78, 80
Barium oxide, **2:** 148
Baseband, **1:** 258
Baseline interferometry, **2:** 263
BAW oscillator, **2:** 47
Beam optics, **2:** 139
Beam trajectories, **2:** 139
Berlinite ($AlPO_4$), **1:** 219
Bidirectional loss, **1:** 243
Bimorph, **1:** 83, 87
Bismuth germanium oxide, **1:** 19
Boundary condition, **1:** 52, 67, 79, 86, 93, 104
Bragg diffraction, **1:** 165
Breit–Rabi equation, **2:** 116
Bridge attenuation, **2:** 29
Bridge unbalance, **2:** 29
Broadening
  homogeneous, **2:** 182
  inhomogeneous, **2:** 182
BT cut, **1:** 182
Buffer gas, **2:** 151
Bulk-wave model, **1:** 232, 241, 245, 257
  advanced, **1:** 249
Bulk-acoustic-wave oscillator. *See* BAW oscillator
BVA_2 technology, **2:** 296

C

C mode, **2**: 295
Cadmium sulfide, **1**: 213
Capacitance
  clamped, high-frequency, **1**: 80
  free, low-frequency, **1**: 80
  interdigital transducer, **1**: 236
  motional, **1**: 106
  ratio, **1**: 71
  shunt, **1**: 106
Capacitive weighting, **1**: 263
Carrier frequency, **1**: 232, 259
Cavity
  design, **1**: 132
  dielectrically loaded, **2**: 281, 284
  Fabry–Perot, **1**: 122, 129
  lead and niobium, **2**: 281
  lead–ceramic, **2**: 284
  niobium, frequency stability, **2**: 282
  phase shift, **2**: 147, 149
  pulling, **2**: 141, 161
  sapphire sphere, **2**: 284
  spacing, **1**: 126
  stabilization, **2**: 162
  superconducting, environmental sensitivity, **2**: 283
  superconductive, **2**: 281
Center frequency
  grating-reflection band, **1**: 133
  stopband, **1**: 128
Centrosymmetric crystal, **1**: 3
Ceramic filters, **1**: 187
Cesium frequency standard, **2**: 186
Cesium maser, **2**: 166
Cesium resonator, **2**: 236
Cesium standards, accuracy, **2**: 236
Cesium-beam standard, **2**: 143
C field, **2**: 117
Characteristic frequencies, **2**: 11
Characteristic parameters, **2**: 11
Charge compensator, **1**: 148
Chebyshev polynominals, **1**: 254
Chemical etching, **1**: 213
Chemical polishing, **1**: 32
Chi-squared distribution, **2**: 211
Christoffel, **1**: 58, 65, 74
CI meter. *See* Crystal impedance meter
Circuit. *See* Network
Cissoidal impedance plot, **2**: 10
Cleaning technology, **1**: 170
Clear-access signal, **2**: 261
Clock flyover mode, **2**: 261
Clock transition, **2**: 117
Clocks, portable, **2**: 251
Coaxial techniques, **2**: 253
Coefficient, temperature, **1**: 110. *See also* Constant
Coefficients as function of angle, **2**: 101
Coherence, **2**: 179
Coherent observation time, **2**: 122
Collimation, **2**: 137
Collimator, **2**: 159
Collisions, **2**: 151
Coloration of quartz, **1**: 155
Common-view approach, **2**: 262
Common-view technique, **2**: 236
Communication satellite, **2**: 260
Comparison of frequency standards with regard to volume, weight, power demand, and selling price, **2**: 174
  cesium beam, **2**: 174
  crystal oscillator, **2**: 174
  hydrogen maser, **2**: 174
  rubidium gas cell, **2**: 174
Compensation, in dual-mode oscillator, **2**: 289
Compliance tensor, **1**: 9
Confinement, of the atom, **2**: 122
Constant
  elastic
    fourth-order, **1**: 57
    stiffened, **1**: 59
    third-order, **1**: 56, 58
  material, higher order, **1**: 55
  quartz, piezoelectric, **1**: 107
Constitutive equations, **1**: 8, 246
Constitutive relations, **1**: 120
Contaminant, **1**: 170
Convention, sign, **1**: 54
Conversion method, multiple, **2**: 227
Cooling, of atoms or ions, **2**: 170
Coordinate
  laboratory, **1**: 64
  material, **1**: 54
  normal. *See* Mode, normal; Eigenvector
  rotation, **1**: 55
  time, **2**: 269
Coriolis acceleration, sensitivity, **2**: 292
Cost-effectiveness, of clock transports, **2**: 252
Coupled-mode analysis, **1**: 133

Coupled-mode formalism, **1:** 128
Coupled-resonator device, **1:** 199
Coupled strip resonator, **1:** 206
Coupling, elastic, **1:** 51, 82, 90
Coupling coefficient, **1:** 120; **2:** 6
    electromechanical (piezoelectric), **1:** 69, 71, 75, 80, 83, 85, 100
    piezoelectric, **1:** 213
Coupling factor, effective, **1:** 253
Coupling-of-modes technique, **1:** 127
Crossed-field bulk-wave model, **1:** 245
Crossed-field model, **1:** 247, 250
Cryogenic-oscillator transmission filter, **2:** 283
Crystal
    doubly rotated, **2:** 102
    drive level, **2:** 55
    motional resistance, **2:** 55
    overtone mode, **2:** 60
    of rhomboid geometry, **1:** 277
    semiconducting, **1:** 52. *See also* Resonator; Cut; Constant
Crystal-controlled oscillator, **2:** 47
Crystal filter, **1:** 187
    acoustically coupled, **1:** 191
    bandpass, **1:** 189
    discrete-resonator, **1:** 192
    monolithic, **1:** 187
    two-pole, narrow-band, **1:** 196
    wide-band, symmetrical lattice, **1:** 196
Crystal header, **1:** 175
Crystal holder, **1:** 175
Crystal impedance meter, **2:** 3, 19
Crystal oscillator, **2:** 51
    frequency stability, **2:** 100
    packaged, **2:** 51, 52
    temperature compensated, **2:** 51
    temperature controlled, **2:** 52
    voltage controlled, **2:** 51
Crystal resonator, **2:** 2
    in electrical circuits, **2:** 4
    electrodeless, **2:** 80
Crystal systems, **1:** 3
Crystal unit, **2:** 7
Crystallographic axis, **1:** 4
Customer–vendor relationship, **2:** 298
Current
    drive, **1:** 56
    resonator, **1:** 75, 79, 86, 89
Cutting crystals, **1:** 163
Cutoff frequency, **1:** 203, 207

Cut, doubly rotated, **1:** 167
Cut, quartz. *See also* Resonator; Mode
    AT, BT, **1:** 111
    CT, DT, **1:** 77
    GT, **1:** 50
    IT, **1:** 111
    LC, **1:** 111
    RT, **1:** 111
    SC, **1:** 69, 111
    X, Y, **1:** 111
Cyclotron orbit, of electron, **2:** 189
Czochralski technique, **1:** 15, 19

D

Deep-space experiments, **2:** 268
Defect center, substitutional Al^{3+}, **1:** 148
Deformation. *See also* Displacement
    elastic, **1:** 79
    finite, **1:** 57
Degree of freedom, **2:** 213
Delay
    differential, **2:** 236
    temperature coefficient, **1:** 120
Delay line, **1:** 267
    frequency response, **1:** 134
    (YZ) $LiNbO_3$, **1:** 253
Delay-line filter, **1:** 121
Delay-line system, **2:** 222
Density
    α-quartz, **1:** 12
    aluminum phosphate, **1:** 21
    bismuth germanium oxide, **1:** 20
    lithium niobate, **1:** 14
    lithium tantalate, **1:** 14
Desorption of gases, **1:** 275
Dewar flask, **2:** 105
Detector, atomic beam device, **2:** 137
Deuterium maser, **2:** 166
Dicke regime, **2:** 121
Dielectric cavity, **2:** 166
Dielectric constants, **1:** 5
Differential propagation, **2:** 263
Diffraction, SAW, **1:** 262
Digital computation methods, **2:** 63
Digital measurement, **2:** 217
Digital signal processing, **2:** 198
Dipole, **2:** 138
    double, **2:** 138
    multiple, **2:** 138

Dipole movement, **2:** 138
  electric, **2:** 138
  magnetic, **2:** 138
Dipole optics, **2:** 139
Diode, step-recovery, **2:** 135
Discriminator, **2:** 222
Disc resonator, **1:** 222
Dislocations, **1:** 29
Dispersion relations, **1:** 51, 60, 82, 94, 247
Displacement. *See also* Deformation
  elastic, **1:** 79, 90, 92
  electric, **1:** 52, 78, 234
Displacement effects, **1:** 152
Dissemination techniques, **2:** 271
Dissociation, of H_2, **2:** 160
Distortion, in FM, **2:** 131
Diurnal variations, **2:** 237
Doppler broadening, **2:** 182
Doppler effect, **2:** 120
  first-order, **2:** 120
  second-order, **2:** 120
Doppler frequency, **2:** 183
Doppler velocity, **2:** 183
Doppler width, **2:** 183
Double-heterodyne system, **2:** 135
Drift rate, **2:** 80
Drive. *See* Current
Drive level, **2:** 35
Dual-beam device, **2:** 144
Dual-mixer technique, **2:** 229
Dynamic frequency–temperature effect, **2:** 37

E

E^1 center. *See* Oxygen vacancy
Effects
  temperature, **1:** 110
  thermal. *See* Effects, temperature
  magnetic, **1:** 61
Eigenvalue, **1:** 64, 67, 74
Eigenvector, **1:** 65, 67, 74
Eight-pole filter, **1:** 209
Elastic compliance, **1:** 7
Elastic constant, **1:** 6
  changes, **1:** 153
  compliance, **1:** 12
    α-quartz, **1:** 12
    lithium niobate, **1:** 14
    lithium tantalate, **1:** 16

  nonlinear, **1:** 218
  stiffness, **1:** 12
    α-quartz, **1:** 12
    aluminum phosphate, **1:** 21
    bismuth germanium oxide, **1:** 20
    lithium niobate, **1:** 14
    lithium tantalate, **1:** 16
Elastic nonlinearity, **1:** 218
Elastic stiffness, **1:** 7
Electric displacement, **1:** 5
Electric field, **1:** 5
Electric polarization, **1:** 5, 8
Electric susceptibility, **1:** 5
Electric-dipole transition, **2:** 169
Electrical analog, of a mechanically vibrating system, **2:** 4
Electrical measurements, on crystal resonators, **2:** 16
Electrical noise, **2:** 51
Electrode, **1:** 87
  interdigital (IDT), **1:** 120
  mass loading, **1:** 100
Electrode positioning, **1:** 260
Electrode withdrawal, **1:** 263
Electromechanical coupling constant, **1:** 241, 255
Electron bombardment, **1:** 174
Electron–hole pair, **1:** 148
Electron multiplier, internal, **2:** 145
Electron spin resonance, **1:** 33, 155
Elliptic integral, **1:** 236
Emissivity factor, **2:** 106
Enantiomorphous form, **1:** 3
Enclosure, **1:** 175
  alumina ceramic, **1:** 274
Energy trapping, **1:** 93, 99. *See also* Resonator
Environmental effects, **1:** 182; **2:** 173
  cesium beam, **2:** 173
  crystal oscillator, **2:** 173
  hydrogen maser, **2:** 173
  rubidium gas cell, **2:** 173
Equation
  constitutive, **1:** 52, 60, 66, 73, 78, 81, 92, 104
  differential, **1:** 53, 92, 104
  frequency, **1:** 51, 68
  graphical solution, **1:** 69
Equivalent circuit, SAW IDTs, **1:** 250
Equivalent electrical circuit, **2:** 2, 4, 6, 43
  elements, **2:** 14

Equivalent motional inductance, **2:** 49
Error-correcting routines, **2:** 31
Etch, **1:** 167
Etched grooves, **1:** 126
Etch tunnels, quartz, **1:** 30, 32
Etching, **1:** 32
Evaporation, **1:** 174
Excitation
  lateral, of plates, **1:** 64, 73
  parallel-field. *See* Excitation, lateral
  parallel to length, in bars, **1:** 78
  perpendicular to length in bars, **1:** 80, 87
  thickness, of plates, **1:** 63, 81, 83
Extension, of frequency measurements, **2:** 186
Extrapolation, of shorter tests, **1:** 281
Extraterrestrial time and frequency comparison, **2:** 253

F

Fabrication, SAW devices, **1:** 178
Fabrication facility, crystals, **1:** 176
Feedback, **2:** 54
  positive, **2:** 47
Field electric, **1:** 52, 78, 92
Figure, of merit, **2:** 14
Filter
  asymmetric-amplitude, **1:** 263
  bulk-acoustic-wave (BAW), **1:** 187, 230
  discrete-resonator, **1:** 191
  disk–wire, **1:** 225
  dispersive, **1:** 262
  electromechanical, **1:** 187, 221
  extensional-mode, **1:** 224
  half-lattice, **1:** 193
  hermitian-baseband, **1:** 263
  lattice, **1:** 193
  mechanical, **1:** 188
  monolithic crystal, **1:** 93, 102
  nondispersive, **1:** 262
  SAW resonator, **1:** 230
  surface-acoustic wave (SAW), **1:** 187
  tapped delay-line. *See* Filter, transversal
  television IF. *See* Television IF filter
  three-phase, **1:** 263
  torsional mode, **1:** 226
  transversal, **1:** 187
  tuning fork, **1:** 224
  two-phase, **1:** 263
  (YZ)LiNbO$_3$, response curve, **1:** 256

Filter measurements, **2:** 19
Filter synthesis, impulse response, **1:** 263
Filter transfer formation, **2:** 194
Flexure-mode bars and plates, **1:** 223
Flicker
  of frequency floor, **2:** 168
  of phase noise, **2:** 124
Flicker floor, **2:** 123
Flop-in, flop-out system, **2:** 143
Flow
  generalized, **1:** 64
  heat, **1:** 61
  power, **1:** 60
Fluorescence, **2:** 150
Forbidden transitions, **2:** 189
Force. *See also* Stress
  generalized, **1:** 64
  static, **1:** 58
Fourier frequency, **2:** 193
Fourier transform, **1:** 121, 239, 250, 256, 258; **2:** 199
Fractional frequency accuracy, **2:** 173
  cesium beam, **2:** 173
  crystal oscillator, **2:** 173
  hydrogen maser, **2:** 173
  rubidium gas cell, **2:** 173
Fractional frequency stability, **2:** 172, 242
  cesium beam, **2:** 172
  crystal oscillator, **2:** 172
  hydrogen maser, **2:** 172
  rubidium gas cell, **2:** 172
Fractional linewidth, **2:** 121, 178
Frequency
  antiresonance, **1:** 66, 69, 102, 106
  cutoff, **1:** 95
  fractional, **2:** 195
  inharmonic, **1:** 98, 100
  instantaneous, **2:** 196
  mean, **2:** 207
  of oscillation, **2:** 48
  resonance, **1:** 66, 69, 98, 102, 106
  torsional, **1:** 76
Frequency accuracy, **2:** 47
Frequency adjustment, varactor diode, **2:** 86
Frequency anomalies, **1:** 217
Frequency change, **1:** 149
  short-term, **1:** 271
Frequency deviation, fractional, **2:** 200
Frequency division, one-step, **2:** 189
Frequency domain, **2:** 203

Frequency drift, **2:** 207
  aging, **2:** 124
Frequency–drive-level effect, **2:** 35
Frequency fluctuations, random, **1:** 284
Frequency instability, sources, **2:** 48
Frequency-lock servo, **2:** 128
  frequency modulation, **2:** 129
  integrator, **2:** 130
  phase modulator, **2:** 130
Frequency measurement, **2:** 219
  interpretation, **2:** 93
  long-term drift, **2:** 93
  short-term measurement, **2:** 93
  optical regime, **2:** 186
Frequency multiplication, **2:** 231
Frequency reproducibility, **2:** 182
Frequency response, hermitian, **1:** 261
Frequency stability, **1:** 180; **2:** 47, 182, 193, 195
  long-term, **2:** 57
  measurement, **2:** 89
    beat-frequency method, **2:** 90
    dual-mixer technique, **2:** 90
  short-term, **2:** 93
    definition, **2:** 93, 94
    power spectral density function, **2:** 95
    time domain representation, **2:** 95
Frequency–temperature
  and aging measurements, **2:** 37
  rate of change, **2:** 103
Frequency–temperature dependence, of a quartz crystal resonator, **2:** 17
Frequency synthesis, **2:** 227
  stabilized-laser, **2:** 187
Frequency syntonization accuracy, **2:** 271
Frequency transfer, **2:** 252
Frequency transients, stress induced, **2:** 294

G

Gain curve, laser, **2:** 181
Gas cell, **2:** 150
Generalized motional arm reactances, **2:** 41
Generic sources, of measurement errors, **2:** 3
Getter, **2:** 144
Global high-precision comparison, of clocks, **2:** 264
Global positioning system (GPS), **2:** 260
Geostationary meteorological satellite, **2:** 266
Geostationary operational environmental satellite (GOES), **2:** 253
Geostationary satellite, **2:** 270
GOES satellite time code, **2:** 257
Goniometer, **1:** 167
Gradients, thermal, **1:** 55, 60
Graphite-coated surface, **2:** 144
Grating positioning, **1:** 132
Grating reflector, **1:** 123, 126
Green's function, **1:** 249
Groove depth, **1:** 131
Ground-wave accuracy, **2:** 248
Ground-wave propagation, **2:** 248
Group delay, **1:** 265
Growth
  berlinite ($AlPO_4$), **1:** 22
  $LiNbO_3$, **1:** 15
  $LiTaO_3$, **1:** 17

H

Heater location, **2:** 104
Heisenberg's uncertainty relationship, **2:** 122
Heterodyne technique, **2:** 220
Hexapoles, **2:** 138
High-polymer coating, **2:** 158
High-Q LC circuits, **2:** 275
Hilbert transform, **1:** 236, 251
Holder
  cold-welded metal, **1:** 275
  glass, **1:** 275
Holder design, **1:** 274
Hole, **1:** 148
Hole burning, **2:** 182
Hole-compensated Al center, **1:** 151
Hole-compensated centers, **1:** 156
Homodyne technique, **2:** 221
Hydrogen anneal, **1:** 278
Hydrogen effects, **1:** 154
Hydrogen maser, **2:** 159
Hydrogen storage beam tube, **2:** 168
Hydrothermal growth, **1:** 25, 29
Hyperfine energy level, **2:** 118
Hysteresis, **2:** 38, 111

I

Iconoscope, **1:** 165

IDT
  acoustic reflection, **1:** 257
  admittance, **1:** 130, 273
  aperture, **1:** 243
  apodization, **1:** 132
  bandwidth, **1:** 242
  double electrode, **1:** 236
  electrical Q, **1:** 242
  frequency response, **1:** 134
  frequency specification, **1:** 259
  impedance element, **1:** 268
  impulse response, **1:** 121
  maximum coupling, **1:** 242
  perturbation, **1:** 263
  self-resonant, **1:** 268
  series resistance, **1:** 131
  simple electrode, **1:** 236
  size, **1:** 131
  spurious resonator modes, **1:** 131
  synchronous frequency, **1:** 133
  three-phase, **1:** 265
  transverse modes, **1:** 132
  unapodized, **1:** 130
  unwanted modes, **1:** 132
IF substitution method, **2:** 32
Imaging technique, **1:** 262
Immittance
  elastic wave, **1:** 65
  matrix, **1:** 64
  normalized, **1:** 70
  resonator, **1:** 62, 75, 79, 82, 102
Immittance diagram, **2:** 2
Immittance plot, **2:** 36
Impedance. *See also* Immittance
  characteristic, **1:** 248
  crystal, **2:** 51
  mechanical, **1:** 246
Impedance analyzer, **2:** 33
Impedance circle, **2:** 9
Impulse model, **1:** 233, 239
  delta-function, **1:** 257
  sine-wave, **1:** 257
Impulse response
  baseband, **1:** 259
  finite, **1:** 258
  real, **1:** 260, 261
Impurity content, **1:** 148
Impurity defects, model, **1:** 148
Impurities, effect of, **1:** 278
Inductance, motional, **1:** 106, 112

Inductorless bandwidth, **1:** 211
Inductorless limit, **1:** 211
Inelasticity, **1:** 217
Inflection point, **2:** 61
Infrared absorption, **1:** 148
Initial aging, **1:** 273
Initial stabilization period, **1:** 273, 278
In-line model, **1:** 247
Insertion loss, **1:** 232, 242; **2,** 69
  SAW delay line, **2:** 69
  SAW resonator, **2:** 71
Insulation, foam, **2:** 105
Interaction region, **2:** 139, 145
Interdigital transducer, **1:** 230. *See also* IDT
Interferometry, very-long-baseline (VLB1), **2:** 268
Intermediate-band design, **1:** 198
Intermodulation, **1:** 216
International atomic time (TAI), **2:** 234
International symbols, **1:** 4
Interstitial impurities, **1:** 154
Inversion transition, **2:** 169
Inverter, admittance, **1:** 251
Ion
  individual, **2:** 189
  in Penning trap, **2:** 189
Ion-etch process, **1:** 138
Ion implantation, **1:** 126
Ion milling, **1:** 179, 213
Ion mobility, **1:** 183
Ion pump, **2:** 144
Ion storage, **2:** 169
Ion trap, radio-frequency, **2:** 170
Ionic current, **1:** 157
Ionization potential, **2:** 148
Ionizing radiation, **1:** 39, 147, 183
Ionosphere, **2:** 237, 242
Ionosphere propagation errors, **2:** 262
IR studies, **1:** 155
Isolation amplifier, **2:** 21, 30
Isotopic overlap, **2:** 157

J

Jaumann network. *See* Filter, half-lattice

K

Kirchhoff, **1:** 90
Krypton light source, **2:** 179

L

Lamb dip, **2:** 183
Lapping saw, **1:** 164
Laser
  CO_2, **2:** 187
  color-center, **2:** 188
  diode, **2:** 149
  dye, **2:** 149, 188
  F-center, **2:** 189
  fine and coarse tunable, **2:** 189
  HCN, **2:** 186
  He–Ne, **2:** 180, 186
    with CO_2 cell, **2:** 185
    with I_2 cell, **2:** 185
    with Methane cell, **2:** 185
    with Ne cell, **2:** 184
  H_2O, **2:** 187
  millimeter-wave, **2:** 188
  potential role, **2:** 179
  saturation-absorption stabilized, **2:** 186
  tunable, **2:** 188
Laser–methane cell combination, **2:** 182
Laser signals, pulsed, **2:** 264
Laser stabilization, **2:** 182
Laser synchronization, **2:** 263
Lateral field resonator, **1:** 276
Lateral field excitation, **1:** 220
LC-cut, **1:** 182
Legendre polynominal, **1:** 236
Length standard, **2:** 178
Length and time standard, combined, **2:** 178
Level control, automatic (ALC), **2:** 54, 56, 84
LF (low frequency) broadcast, **2:** 242
Lifetime, resonator-limited, **2:** 173
  cesium beam, **2:** 173
  crystal oscillator, **2:** 173
  hydrogen maser, **2:** 173
  rubidium gas cell, **2:** 173
Lift-off process, **1:** 138
Light shift, **2:** 158
Line shape, Lorentzian, **2:** 128
Line width, **2:** 121
  natural, **2:** 121
Lithium niobate, **1:** 11, 219, 241
  YZ-cut, **1:** 241, 243, 248
  128° Y-cut, **1:** 266
Lithium tantalate, **1:** 11, 213, 220
  doubly rotated cuts, **1:** 220
  rotated-Y-cut monolithic filters, **1:** 220
  rotated-Y-cut resonators, **1:** 220
  X-cut, **1:** 220
  Z-cut, **1:** 220
Load capacitor, **2:** 12, 24, 35
Long-term aging, **1:** 179. *See also* Aging
Long-term drift, **2:** 48. *See also* Aging
Long-term stability, **2:** 125
  atomic standards, **2:** 125
Loop filter, **2:** 225
Loop gain vector, **2:** 48
Loop-phase conditions, **2:** 57
Loop-phase error, **2:** 20
Loran-C, **2:** 235, 248
Lorentzian line shape, **2:** 194
Loss–bandwidth relation, **1:** 243
Loss–fractional bandwidth, **1:** 244
Lumped-element-equivalent electrical circuit, **2:** 41
Lumped-mass spring system, **2:** 5

M

Magnetic dipole moment, **2:** 119
Magnetic dipole transition, **2:** 117
Magnetic field dependency, **2:** 118
Magnetic field environment, **2:** 252
Magnetic field inhomogeneities, **2:** 146
Magnetic hyperfine splitting, **2:** 119
Magnetic hyperfine transition, **2:** 116
Magentic shielding, **2:** 117
Magnetostrictive ferrites, **1:** 221
Magnetostrictive transducer, **1:** 222
Majorana transition, **2:** 147
Maser, oscillating threshold, **2:** 161
Maser oscillator, **2:** 135
  active, **2:** 126
Mass changes, **2:** 17
Mass loading, **1:** 204, 274
  crystal resonator, **2:** 288
Mass spectrometer, **2:** 145
Material. *See also* Crystal
Material changes, **1:** 274
Material quality, **1:** 157
Matrix, **1:** 9, 54, 61, 65, 78, 81, 83, 85, 88.
  See also Tensor; Immittance
Maximum admittance, **2:** 16
Maxwell distribution, **2:** 121, 138
Measurement hierarchy, **2:** 197
Memory, **2:** 110
Mercury resonances, **2:** 170

Metallization, **1:** 181
Meteorological monitor, **2:** 289
Meteorological satellite system, **2:** 257
Methane cell, **2:** 181
Metrology, **2:** 171
  astronomical time, **2:** 171
  atomic time, **2:** 171
  coordinated universal time, **2:** 171
  international atomic time, **2:** 171
  radio astronomy, **2:** 171
  timekeeping, **2:** 171
  very-long-baseline interferometry, **2:** 171
Microcircuit bridge, **2:** 29
Microcircuit chip resistor, **2:** 24
Microwave cavity, **2:** 141
Microwave interrogation, **2:** 139
Mindlin, **1:** 60, 90
Minimum impedance, **2:** 16
Mirror, equivalent, **1:** 129
Mixer efficiency, **2:** 128
Mobility analogy, **1:** 221
Mode
  A,B,C, thickness, **1:** 68, 71
  contour, **1:** 77
    flexure, **1:** 51, 77
  coupled, **1:** 51, 82, 90
  dilatation, **1:** 52
  extension, **1:** 77, 80, 93
  face-shear, **1:** 81, 90, 93
  flexure, **1:** 83, 87, 90, 93
  normal, **1:** 62, 64, 74
  resonance, **1:** 50
  shear, fast, slow, **1:** 68
  single, **1:** 65
  spectrum, **1:** 56; **2:** 33
  thickness, **1:** 63, 73
  thickness shear, **1:** 51, 72, 93
  thickness twist, **1:** 93
  torsion, **1:** 51, 75
Mode coupling, nonlinear, **1:** 216
Mode spacing, **1:** 207
Modified crystal resonator, **2:** 7
Modulation, square-wave, **2:** 132
Molecular flow, **2:** 137
Molecular hydrogen, **2:** 159
Monolithic crystal filter, **1:** 199
Mössbauer effect, **2:** 121
Motional arm, **2:** 5, 38
Motional-arm resonance, **2:** 14
Motional energy, **1:** 157

Motional resistance, **2:** 49
Mounting, **1:** 175
Mounting structure, **2:** 7
Mounting system, **1:** 272
Mount, type of, **1:** 180
Multistrip coupler, **1:** 265

N

Na defect, **1:** 150
Narrow-band (NB) design, **1:** 198
Natural linewidth, **2:** 122
NAVSTAR. *See* Global Positioning System
Navy navigational satellite system (NNSS), **2:** 258
Network
  distributed, **1:** 104
  equivalent electric, **1:** 87, 103
  lumped element, **1:** 103
  multiport, **1:** 65
Network analyzer, **2:** 23, 30, 36
Noise, **2:** 204, 206
  amplitude, **2:** 195
  flicker frequency, **2:** 204
  flicker phase, **2:** 204
  in oscillators, **2:** 193
  perturbing the phase, **2:** 276
  phase, **2:** 195
  phase-fluctuation, **2:** 84
  pseudo-random (PRN), **2:** 265
  in SAW oscillators, **2:** 72
    flicker noise, **2:** 74
    phase noise, **2:** 72
  random-walk frequency, **2:** 204
  white frequency, **2:** 204
  white phase, **2:** 204
  wideband, **2:** 54
Noise bandwidth, **2:** 194
Noise floor, **1:** 124
Noise processes, **2:** 94
  frequency-scintillation noise, **2:** 94, 96
  frequency white noise, **2:** 94, 96
Noise model, **2:** 196
Noise modulation, **2:** 195
Noise pedestal, **2:** 195
Noise processes, **2:** 94
  phase-scintillation noise, **2:** 94, 96
  phase white noise, **2:** 94, 96
  random-walk-frequency noise, **2:** 94, 96
Nonlinear effects, **1:** 216, 227

Normal-mode analysis, **1:** 234
Normal-mode model, **1:** 233
Normal-mode theory, **1:** 239, 255
Notch filter, **1:** 266
  interferometer, **1:** 267
  (YZ) $LiNbO_3$, **1:** 267
N-sample variance, **2:** 201
Nuclear spin, **2:** 119
Nyquist frequency, **2:** 200

O

Offset methods, **2:** 35
Omega navigation system, **2:** 248
Omega transmitters, **2:** 250
Operational-satellite techniques, **2:** 253
Optical absorption, **1:** 153
Optical detection, **2:** 150
Optical fibers, **2:** 253
Optical frequencies, measurement, **2:** 185
Optical pumping, **2:** 115, 149, 152
Optical transition, **2:** 121
Oriascope, **1:** 165
Oscillating magnetic field, **2:** 121
Oscillator
  all-cryogenic, parametric, **2:** 283
  conditions for SAW oscillation, **2:** 68
  feedback, **2:** 66
  gain control, **2:** 85
  mechanical effects, **2:** 88
    acceleration, **2:** 88
  miniature, **2:** 60
  miniature integrated circuit, **2:** 64
  modified Pierce, **2:** 82
  multifrequency SAW, **2:** 76
  precision quartz crystals, **2:** 79
  temperature-compensated (TCXO), **2:** 108
    analog compensation, **2:** 109
    digital compensation, **2:** 10
    electrical compensation, **2:** 108
    microprocessor compensation, **2:** 111
    thermal loss, **2:** 105
    thermistor-network configurations, **2:** 109
    varactor, **2:** 108
  temperature compensation, of SAW, **2:** 75
  temperature-controlled, **2:** 101
    amplifier as heater, **2:** 105
    crystal oven control circuit, **2:** 102
    double oven, **2:** 102
    dual mode oscillator, **2:** 105
    location, of heater and thermistor, **2:** 104
    rate, of frequency change, **2:** 103
    single oven, **2:** 102
    stabilization time, **2:** 107
    time required to heat crystal, **2:** 107
    uniformity of temperature, **2:** 104
Oscillator circuits, **2:** 53
  bridge, **2:** 54
  Butler, **2:** 60
  common base, **2:** 59
  common collector, **2:** 58
  emitter coupled, **2:** 60
  modified piece, **2:** 56
Oven, **2:** 87
  single-stage, **2:** 87
  temperature control, **2:** 101
  two-chamber, **2:** 137
Oxygen vacancy, **1:** 153

P

Palladium leak, **2:** 160
Parallel field. *See* Lateral field
Parallel resonance, **2:** 15
Paramagnetic center, **1:** 151
Parasitic inductance, **2:** 20
Parseval's theorem, **2:** 194
Passive hydrogen maser, **2:** 162
Path delay, **2:** 254, 257, 266
  variation, **2:** 256
Path-delay correction, **2:** 264
Pattern definition process, **1:** 179
Pattern generator, **1:** 261
Penning trap, **2:** 170
Period measurement, **2:** 219
Performance, of crystal resonators, **2:** 16
Performance model, **2:** 197
Permittivity, **1:** 5
Permittivity constant
  α-quartz, **1:** 12
  aluminum phosphate, **1:** 21
  bismuth germanium oxide, **1:** 20
  lithium niobate, **2:** 14
  lithium tantalate, **1:** 16
Perturbations, acoustic, **1:** 126
Phase
  linearity, **1:** 264
  transfer, **2:** 82
Phase angle, **2:** 18
Phase condition, loop, **2:** 82

Phase detector, **2:** 222
Phase-difference measurement, **2:** 218
Phase error, **1:** 232; **2:** 18, 23
Phase fluctuation, **2:** 51
Phase-lock servo, **2:** 128
Phase-locked loop, **2:** 128, 223
Phase measurement, **2:** 32
Phase multiplier, **2:** 231
Phase noise, **1:** 135
  excess, **1:** 216
  white, **2:** 124
Phase shift
  per collision, **2:** 165
  in crystal network, **2:** 49
  SAW oscillators, **2:** 68
Phase shifter, **2:** 222
Phase slope, **2:** 57, 82
Phase spectrum, **2:** 195
Phase stability, **2:** 48, 60
Photolithography, **1:** 138, 180
Photon recoil, **2:** 121
Photon transformer, **2:** 150
Photoresist, **1:** 138, 178
Piezoelectric ceramics, **1:** 221
Piezoelectric ceramic transducers, **1:** 223
Piezoelectric constants, **1:** 7
Piezoelectric crystal resonator, **2:** 38
Piezoelectric devices, as circuit elements, **2:** 2
Piezoelectric loading, **1:** 204
Piezoelectric matrix. *See also* Tensor; Matrix
  α-quartz, **1:** 9
  bismuth germanium oxide, **1:** 10
  lithium niobate, **1:** 9
Piezoelectric strain coefficients, **1:** 8
Piezoelectric strain constant, **1:** 13
  α-quartz, **1:** 13
  aluminum phosphate, **1:** 21
  lithium niobate, **1:** 15
  lithium tantalate, **1:** 17
Piezoelectric stress coefficients, **1:** 8
Piezoelectric stress constant, **1:** 13
  α-quartz, **1:** 13
  aluminum phosphate, **1:** 21
  bismuth germanium oxide, **1:** 20
  lithium niobate, **1:** 15
  lithium tantalate, **1:** 17
Piezoelectric tensor, **1:** 9. *See also* Tensor; Matrix
Piezoelectrically stiffened resonator, **2:** 41
Piezoelectrically unstiffened resonator, **2:** 41

Pi-network, **2:** 21, 31, 40
Planck's equation, **2:** 122
Planetary lap, **1:** 168
Plasma etching, **1:** 179
Plate, piezoelectric, **1:** 63, 73, 81, 83
Point-contact diodes, **2:** 186
Point groups, **1:** 3
Polariscope, **1:** 165
Polarization, **1:** 2; **2:** 157
Polishing, **1:** 168
  chemical, **1:** 168, 180
Population difference, **2:** 123
Portable clock, **2:** 269
Port
  electrical, **1:** 247
  mechanical, **1:** 247
Position-location system, **2:** 261
Potential
  electric, **1:** 52, 92
  electrostatic, **1:** 237, 239
  Rayleigh wave, **1:** 232
  transducer, **1:** 236
Power flow, acoustic, **1:** 240
Power-law model, **2:** 203
Power spectral density, one-sided, **2:** 231
Power spectrum, **2:** 194
Poynting theorem, **1:** 235
Poynting vector. *See* Flow, power
Pressure changes, **1:** 274
Pressure sensitivity, **1:** 183
Primary frequency standard, **2:** 235
Primary loop, **2:** 135
Processing techniques, SAW, **1:** 138
Propagation delay, **2:** 237, 266
Propagation loss, **1:** 120
Pulling factor, **2:** 168
Pump
  cryogenic, **1:** 172
  diffusion, **1:** 171
  ionization, **1:** 171
  roughing, **1:** 173
  turbo, **1:** 171

Q

Q, **2:** 142. *See also* Quality factor
  cavity, **2:** 142
  electrical, **1:** 237, 242
  line, **2:** 142
  material, **1:** 163

Quadrupoles, **2:** 138
Quality assurance, **2:** 299
Quality factor, **1:** 50, 102; **2:** 14, 38
  atomic frequency standard, **2:** 121
  SAWR, **1:** 124
Quantum transition, **2:** 178
Quartz, **1:** 2, 11. *See also* α-quartz
  acoustic loss, **1:** 28, 33, 36, 38
  Al–Li$^+$ center, **1:** 38
  Al–Na$^+$ center, **1:** 37
  aluminum-related centers, **1:** 33
  anelastic loss, **1:** 36
  Al-hole center, **1:** 33
  charge compensation, **1:** 33
  conductivity, **1:** 156
  coordinate system, **1:** 24
  cultured, **1:** 162, 166
  defects, **1:** 25, 28
  dielectric loss, **1:** 33, 36, 38
  dielectric relaxation, **1:** 156
  dislocations, **1:** 29
  electrical conductivity, **1:** 39
  electrodiffusion, **1:** 35
  electrolytically swept cultured,
    1: 149
  fault surfaces, **1:** 29
  growth, **1:** 25, 27
  high-Q, **1:** 149
  high-Q cultured, **1:** 28
  infrared absorptions, **1:** 27, 33
  internal friction, **1:** 36
  interstitial impurities, **1:** 32
  lithium-doped, **1:** 149
  mobility of interstitials, **1:** 39
  natural, **1:** 149, 165
  neutron-irradiated, **1:** 153
  oxygen vacancy center, **1:** 35
  phase transition, **1:** 25
  point defects, **1:** 32, 35
  quality factor Q, **1:** 36
  radiation response mechanism, **1:** 38
  relaxation time, **1:** 37
  ST-cut, **1:** 241, 248; **2:** 75
  structure, **1:** 23
  thermal properties, **1:** 39
  trimming, **1:** 25
Quartz crystal oscillator, **2:** 126
Quartz crystal units, **2:** 3
Quartz resonator, **2:** 80
  AT-cut, **2:** 60, 80
  double-rotated cuts, **2:** 81
  GT-cut, **2:** 80
  overtone mode, **2:** 61, 81
  SC-cut, **2:** 62, 80
  X-cut, 80
Quartz thermometer, **2:** 294

R

Rabi cavity, **2:** 139
Rabi pedestal, **2:** 140, 146
Radar filters, **1:** 230
Radiation, influence on aging, **1:** 279
Radiation conductance, **1:** 238, 252, 255
Radiation effect, **1:** 147
Radiation loss, **2:** 106
Radio broadcast services, **2:** 242
Radio interference, **2:** 273
Radioactive lifetime, **2:** 151
Ramsey cavity, **2:** 140
Ramsey pattern, **2:** 145
Rayleigh wave, **1:** 119, 232, 234, 246,
    252
  amplitude, **1:** 234
  potential, **1:** 240
Reactance, acoustic, **1:** 268
Recoil shift, **2:** 120
Recovery time, **1:** 278
Reference channel, **2:** 23
Reference oscillator, **2:** 126
Reflection
  acoustic, **1:** 249, 257, 264
  bandwidth, **1:** 128
  magnitude, **1:** 128
  phase, **1:** 129
Reflection coefficient, **2:** 16, 23, 25
  grating, **1:** 129
Rejection, out-of-band, **1:** 264
Relation. *See* Equation; Condition
Relativistic corrections, **2:** 268
Relaxation, exponential, **2:** 163
Relaxation time, **2:** 151
Remote frequency calibration, **2:** 250
Remote synchronization, **2:** 268
Replication technique, **1:** 261
Requirements, for oscillation, **2:** 48
Resistance
  anomalous, **1:** 216
  motional, **1:** 113
Resistance anomalies, **1:** 217

Resolution, of balanced-bridge measurements, **2:** 29
Resonance, **2:** 15
  nonlinear, **1:** 216
  with load capacitor, **1:** 15
Resonance curve, **2:** 38
Resonance line
  homogeneous, **2:** 158
  inhomogeneous, **2:** 157
Resonance range, **2:** 9
Resonance spectrum, atomic, **2:** 140
Resonator. *See also* Cut; Mode
  AT, **1:** 94, 111
  BT, **1:** 94, 111
  bulk acoustic wave (BAW), **2:** 3
  contoured, **1:** 100
  CT, **1:** 77, 94
  doubly rotated, **1:** 71, 111
  DT, **1:** 77, 94
  E, F, **1:** 94, 111
  electrodeless, **1:** 182
  extensional mode, **1:** 272
  face-shear mode, **1:** 272
  flexure-type, **1:** 272
  GT, **1:** 50, 94, 111
  IT, **1:** 111
  LC, **1:** 111; **2:** 294
  mass loaded, **1:** 100
  RT, **1:** 111
  SC, **1:** 69, 111; **2:** 295
  sensitivity to temperature gradients, **2:** 102
  surface acoustic wave (SAW), **2:** 3
  trapped energy, **1:** 93, 99
  tuning fork, **1:** 94
  width-shear, **1:** 272
  X, **1:** 111
  Y, **1:** 111
Resonator environment, **2:** 39
Resonator immittance, **2:** 39
Resonator measurement, **2:** 2
Response time, **2:** 296
Rochelle salt, **1:** 2
Rubidium isotope, **2:** 153
Rubidium maser, **2:** 165
Rubidium standard, **2:** 261
  filter cell, **2:** 156
  integrated gas cell, **2:** 156
  photocell, **2:** 156
  pressure shift, **2:** 154
  temperature coefficient, **2:** 154

S

Sample Allan variance, **2:** 214
Sample variance, **2:** 200
Sawing, **1:** 163
SAW, harmonic responses, **1:** 256
SAW bandpass filters, **1:** 257
SAW bandstop filter, **1:** 257. *See also* Bandstop filters, SAW
SAW delay line
  effective Q, **1:** 142
  maximum unloaded Q, **1:** 140
SAW delay-line filter, **1:** 122
SAW delay line oscillator, **2:** 68
SAW filter
  bandwidth, **1:** 243
  insertion loss, **1:** 242, 244
SAW IDT
  dispersive, **1:** 253
  impulse response model, **1:** 232, 239, 254
  in-line model, **1:** 245
  $LiNbO_3$ substrate, **1:** 269
  ST-cut quartz, **1:** 269
  synthesis, **1:** 252
SAW (Surface acoustic wave) oscillator, **2:** 66
SAW reflection, **1:** 254
SAW regeneration, **1:** 254
SAW resonator (SAWR), **1:** 122
  acceleration sensitivity, **1:** 144
  advantages, **1:** 144
  and BAWR, comparison, **1:** 123
  cavity losses, **1:** 135
  fabrication method, **1:** 137
  loss
    conversion of energy, **1:** 137
    coupled to atmosphere, **1:** 137
    diffraction, **1:** 136
    geometrical nonuniformities, **1:** 137
    ohmic, **1:** 135
    material, **1:** 135
    radiation, **1:** 136
    scattering, from imperfections, **1:** 137
    scattering, into bulk waves, **1:** 136
  maximum unloaded Q, **1:** 140
  minimum series resistance, **1:** 142
  one-port, **1:** 125
  performance specification, **1:** 133
  Q, unloaded, **1:** 144
  stability, long-term, **1:** 144

SAW resonator *(continued)*
   temperature stability, **1:** 144
   two-port, **1:** 125
SAW resonator (SAWR) oscillator, **2:** 70
   aging rate, **1:** 143
   force sensitivity, **1:** 143
   new cuts and materials, **1:** 142
   stability, short-term, **1:** 143
   ST quartz, **1:** 142
   temperature stability, **1:** 142
SAW transduction, **1:** 232, 252
SAW trimming, **1:** 138
SAWR model, **1:** 133
Satellite ephemeris error, **2:** 263
Satellite techniques, **2:** 258, 273
Saturated absorption, **2:** 181
Saturated gain, **2:** 183
Scattering-parameter measurements, **2:** 31
SC-cut, **1:** 159, 182, 276
Schering bridge, **2:** 27
Scaling, **1:** 175
Second definition, **2:** 115, 268
Secondary loop, **2:** 135
Sensitivity, to mass changes, **2:** 289
Sensors, **2:** 289
   acceleration, **2:** 291
   chemical, **2:** 289
   environmental, **2:** 289
   force, **2:** 290
   gas, **2:** 289
   hydrostatic pressure, **2:** 291
   temperature, **2:** 293
Separated-oscillatory-field technique, **2:** 140
Series resonance frequency, **2:** 49
Servo electronics, **2:** 127
Servo loop, **2:** 127
Short-term stability, **2:** 124
   atomic standards, **2:** 124
Shot noise, **2:** 124
Sidelobe, time-domain, **1:** 262
Sidelobe ratio, **1:** 232
Signal-to-noise consideration, **2:** 186
Signal-to-noise ratio, **2:** 125
   atomic standards, **2:** 125
   optimum, **2:** 135
Simulation. *See also* Approximation; Theory
   computer, **1:** 56
   finite element, **1:** 56
   Green's function, **1:** 56
   variational, **1:** 56

Sky-wave accuracy, **2:** 248
Slave oscillator, **2:** 127
Source, for atoms or molecules, **2:** 137
Spacelab experiment, **2:** 264
Spatial averaging, **2:** 158
Specification, **2:** 297
   basic, **2:** 299
   military, **2:** 300
Spectral density, **2:** 193, 199
   one-sided, **2:** 198
Spectrum, two-sided, **2:** 193
Speed, of light, **2:** 180
Spin-exchange cavity tuning, **2:** 162
Spin-exchange collisions, **2:** 154, 162
Spurious modes, **2:** 17, 25, 33
Sputter etching, **1:** 179
Sputtering, **1:** 174
Stability
   averaged, **2:** 186
   long term
      BAWR and SAWR, **1:** 124
      SAW, **1:** 138
   of ocillators, **2:** 72
      influence of temperature, **2:** 75
      long-term, **2:** 76
      short-term, **2:** 72
   short-term, **2:** 80, 186
Standard, for frequency, time, and length, unified, **2:** 180
Standard frequency- and time-signal broadcasts, **2:** 238
Standards, **2:** 299
State selection
   optical, **2:** 152
   spatial, **2:** 138
State selector, **2:** 137, 142
Static capacitance C_0, **2:** 38
Static charge, **1:** 183
Static frequency–temperature effect, **2:** 37
Stern and Gerlach experiment, **2:** 119
Stiffness tensor, **1:** 9
Storage bulb, **2:** 159
Storage vessel, **2:** 150
Stored-ion technique, **2:** 170
Strain, elastic, **1:** 52, 92
Strain changes, **1:** 274
Strain tensor, **1:** 7, 9
Stress
   compressional, **1:** 6
   elastic, **1:** 52, 78, 92. *See also* Force

Stress *(continued)*
  mechanical, **1:** 2
  shear, **1:** 6
  tensile, **1:** 6
Stress relaxation, **1:** 150; **2:** 17
Stress tensor, **1:** 7, 9
Stripline oscillator, **2:** 279
Structural defects, **1:** 18
Sub-Doppler observation, **2:** 181
Substitution elements, **2:** 20
Substitution measurement, **2:** 27
Substitution technique, **2:** 39
Substitutional Al^{3+} defect, **1:** 150
Substrate, **1:** 180
Surface-acoustic-wave oscillator. *See* SAW oscillator
Surface-acoustic-wave resonator. *See* SAWR
Surface contouring, **1:** 169
Surface ionization, **2:** 148
Surface layer, **2:** 151
Surface wave, backward-traveling, **1:** 126
Sweep rate, **2:** 33
Sweeping process, **1:** 148, 156
Sweeping technique, **1:** 35
Symmetric modes, **1:** 201
Symmetry, **1:** 54
Synchronization error, **2:** 207
Synchronous detectors, **2:** 32
Synthesis, SAW bandpass filter, **1:** 257
Synthetic quartz, Z-growth, **1:** 151
Syntonization, **2:** 273

T

Tandem lattice configuration, **1:** 196
Tandem two-pole configuration, **1:** 209
TDRS (tracking and data relay satellite), **2:** 265
Television IF filter, **1:** 265
  zinc oxide on glass, **1:** 265
  $LiNbO_3$ cuts, **1:** 265
Television filters, **1:** 230, 265
Television signals, precise T/F comparison methods, **2:** 250
Temperature coefficient
  elastic compliance, **1:** 12
    α-quartz, **1:** 12
    lithium niobate, **1:** 14
    lithium tantalate, **1:** 16

elastic stiffness, **1:** 12
  α-quartz, **1:** 12
  aluminum phosphate, **1:** 21
  lithium niobate, **1:** 14
  lithium tantalate, **1:** 16
permittivity, **1:** 12
  α-quartz, **1:** 12
  lithium niobate, **1:** 14
  lithium tantalate, **1:** 16
piezoelectric strain constant, **1:** 13
  α-quartz, **1:** 13
  lithium niobate, **1:** 15
  lithium tantalate, **1:** 17
piezoelectric stress constant, **1:** 13
  α-quartz, **1:** 13
  aluminum phosphate, **1:** 21
  lithium niobate, **1:** 15
  lithium tantalate, **1:** 17
Temperature-compensated dielectric-resonator material, **2:** 284
Temperature control
  and compensation. *See* Oscillator
  of oscillators, **2:** 86
Temperature-control circuit, **2:** 87
Temperature control techniques, **2:** 61
Temperature gradients, **1:** 159
Temperature influence, on aging, **1:** 279
Temperature measurements, **2:** 293
Temperature resolution, **2:** 295
Temperature sensitivity, SAW filters, **1:** 232, 267
Temperature stability requirements, **2:** 61
Temperature stabilization rate, **2:** 53
Temperature transients, **2:** 107, 111
Tensor, **1:** 55. *See also* Matrix
  fourth-rank, **1:** 7
  second-rank, **1:** 7
Tensor notation, **1:** 9
Terrestrial-based T/F services, **2:** 259
Terrestrial methods, **2:** 237
Terrestrial time comparison, **2:** 266
Test channel, **2:** 23
Test network, **2:** 31
Test oscillator, **2:** 19
T/F dissemination experiments, **2:** 266
Thallium, **2:** 148
Theory. *See also* Approximation; Simulation
  linear, **1:** 53
  nonlinear, **1:** 56
Thermal conductivity, **1:** 19; **2:** 106

Thermal effects, **1:** 158
Thermal expansion
　α-quartz, **1:** 12
　aluminum phosphate, **1:** 21
　lithium niobate, **1:** 14
　lithium tantalate, **1:** 16
Thermal gradient effect, **1:** 276
Thermal isolation, **2:** 86
Thermal noise, **2:** 275
Thermal resistance, **2:** 107
Thermistor location, **2:** 104
Thickness-shear vibration, **1:** 202
Thickness-shear (TS) wave, **1:** 204
Thickness-twist (TT) wave, **1:** 204
Thin-film measurement, **2:** 288
Time
　and frequency comparison, **2:** 272
　and frequency standard, **2:** 178
　between resynchronization, **2:** 172
　　cesium beam, **2:** 172
　　crystal oscillator, **2:** 172
　　hydrogen maser, **2:** 172
　　rubidium gas cell, **2:** 172
Time code, **2:** 254
Time comparison, point-to-point, **2:** 260
Time-comparison experiments, **2:** 264
Time delay, **1:** 258
Time-difference measurement, **2:** 218
Time domain, **2:** 203
Time domain characterization, **2:** 123
Time domain envelope, **1:** 259
Time-interval counter, **2:** 230
Time-interval measurement, **2:** 217
Time stability, **2:** 193, 272
Time synchronization, **2:** 237
Time transfer, **2:** 251
Time-transfer accuracy, **2:** 265, 271
Tolerance, **2:** 297
Tomography, light scattering, **1:** 32
Tool-made sample, **2:** 298
Transducer, **1:** 221, 234
　apodized, **1:** 253, 258
Transducer admittance, **1:** 235
Transducer impedance, **1:** 233
Transducer metallization, source of relaxation, **1:** 284
Transducer potential, **1:** 238, 240
Transfer oscillator, **2:** 229
Transfer phase, **2:** 60
Transfer standard, **2:** 26, 236

Transformation, similarity, **1:** 64
Transient effects, **1:** 154
Transient frequency change, **1:** 154
Transient thermal effects, **1:** 154
Transit or Nova satellite system, **2:** 258, 264
Transit time, **2:** 140
Transition temperature, **2:** 282
Translation technique, **2:** 203
Transmission, **2:** 237
　HF- and medium-frequency, **2:** 237
　low- and very-low-frequency, **2:** 237
Transmission bridge, **2:** 25, 30
Transmission line, **2:** 25, 30
Transmission-line form, **1:** 246
Transmission-line model, **1:** 126
Transmission network, **2:** 4, 40
Transmission test set, **2:** 22
Transponder channel, **2:** 260
Trap characterization, **1:** 156
Trapped-energy analysis, **1:** 202
Trapped ion, **2:** 169
Trimming, final frequency, **1:** 174
Trim-to-frequency, **1:** 169, 179, 181
Triple transit, **1:** 249, 264
TTC-cut, **1:** 159
Tuning forks, **2:** 275
　flexural vibrations, **2:** 277
　length-extensional vibrations, **2:** 277
　quartz, **2:** 276
　temperature sensitivity, **2:** 278
Tunnel-diode oscillator, **2:** 279
Turning point, **2:** 101
Turnover temperature, **2:** 61, 81, 86
TV frame synchronization, **2:** 250
TV methods, **2:** 250
TV time sychronization, **2:** 266
Twinning, **1:** 165
Two-sample (Allan) variance, **2:** 123
Types, of attributes, **2:** 298

U

Ultraviolet laser **2:** 167
Unit variations, **2:** 298
Unity gain point, **2:** 136
Universal coordinate time (UTC), **2:** 234
Unperturbed radiation state, **2:** 122
UV-ozone cleaning, **1:** 180
Unwanted modes, suppression, **1:** 209

V

Vacuum deposition, **1:** 170
Vacuum system, **1:** 173
Varactor, **2:** 27
Vector-ratio meter, **2:** 21, 31, 40
Velocity
  elastic particle, **1:** 52
  wave, **1:** 79
Velocity distribution, **2:** 141
Velocity perturbation, **1:** 242
Very-low-frequency (VLF) broadcast, **2:** 242
Very-narrow-band (VNB) design, **1:** 198
Vibration, modes of. *See* Modes
Vibrational transition, **2:** 169
Voltage, mean-square, **2:** 193
Voltage spectrum, **2:** 194
Voltage-controlled (crystal) oscillator (VCO or VCXO), **2:** 127

W

Wafer processing, **1:** 137
Wall coating, **2:** 158
Wall collisions, **2:** 165
Wall shift, **2:** 165
Wave
  elastic, nonlinear, **1:** 56
  longitudinal, **1:** 120
  plane, **1:** 66
  transverse (shear), **1:** 120
  velocity, **1:** 79, 128
Wavelength standard, **2:** 179
Wavenumber, lateral, **1:** 93. *See also* Dispersion relations
Weak coupling approximation, **1:** 233, 237, 240
Wide-band design, **1:** 198
Window function, **2:** 199

X

X-ray diffraction, **1:** 166
X-ray topography, **1:** 18, 30, 32

Z

Zeeman level, **2:** 163
Zeeman transition, **2:** 118
Zero crossing, **2:** 252
Zero-field transition, **2:** 119
Zero-phase condition, **2:** 22
Zero-phase π network, **2:** 22, 35
Zero-phase technique, **2:** 34
Zero reactance, **2:** 39
ZnO film, **2:** 214